Solar Cell Nanotechnology

Scrivener Publishing
100 Cummings Center, Suite 541J
Beverly, MA 01915-6106

Publishers at Scrivener
Martin Scrivener (martin@scrivenerpublishing.com)
Phillip Carmical (pcarmical@scrivenerpublishing.com)

Solar Cell Nanotechnology

Edited by

Atul Tiwari, Rabah Boukherroub, and Maheshwar Sharon

Scrivener Publishing

WILEY

Co-published by John Wiley & Sons, Inc. Hoboken, New Jersey, and Scrivener Publishing LLC, Salem, Massachusetts.
Published simultaneously in Canada.

For general information on our other products and services or for technical support, please contact our Customer Care Department within the United States at (800) 762-2974, outside the United States at (317) 572-3993 or fax (317) 572-4002.

Wiley also publishes its books in a variety of electronic formats. Some content that appears in print may not be available in electronic formats. For more information about Wiley products, visit our web site at www.wiley.com.

For more information about Scrivener products please visit www.scrivenerpublishing.com.

Cover design by Russell Richardson

Library of Congress Cataloging-in-Publication Data:

ISBN 978-1-118-68625-6

Printed in the United States of America

10 9 8 7 6 5 4 3 2 1

Contents

Preface

Developments in human civilization have revolved around the consumption of energy. Historically, control over energy has been directly related to the centralization of power. The natural abundance of fossil fuels has helped sovereign nations in gaining better control over sociological developments. The absence of such natural resources has forced the remaining societies to greatly depend on the lucky ones that have them. In some cases, the developments have slowed to a crawl, and even basic needs have not been fulfilled. Therefore, materials scientists and engineers have taken up the challenge to reduce our dependence on fossil fuel resources. The invention of methods to harness solar energy has given new hope to the nations with limited or no fossil fuel reserves. In the last few decades the idea of technologies based on solar cells has been well established. New techniques of materials synthesis and their integration in novel engineering designs have helped the industry produce solar cells with high energy efficiency. The use of high purity silicon has dominated the solar cell industry for several decades. It appears that the efficiency of silicon-based solar cells has achieved maximum potential and no further improvements are expected. The high cost of extra pure silicon and saturation in energy efficiency has motivated scientists to look for new materials and technologies that are efficient and relatively cost effective.

Journals are flooded with research articles with claims about materials and technologies that might be able to produce highly efficient solar cells. It is therefore important to summarize the literature in order to draw meaningful conclusions based on the latest research findings. Since silicon-based solar cells are immensely popular, various books are available on the subject. It was therefore decided to look at other types of solar cells, which are either in the development phase or are likely competitors of silicon solar cells.

The purpose of our efforts has been to collect articles from various experts in the field with this view in mind. This book is composed of seventeen chapters, each of which was written by an expert in their field. Topics are broadly designed to cover dye-sensitized types of solar cells and their related problems, layered types of solar cells, application of lithography in solar cells, and luminescent solar and plasmonic light trapping. Graphene being the most recent discovery, its application in organic types of solar cells is covered in this book. Finally, analytical modeling and electrical circuit design, which are another important aspect of solar cell development, are also included. The final section of the book appends to the series of articles written on putative future trends in this area. This section is specifically written for advanced practitioners to help them channel their expertise in the desired direction.

We expect that readers will gain an in-depth knowledge in new areas of solar cells, which are not commonly known and for which literature is scarcely available. This title has been edited for a broad readership that includes scholars and researchers from diverse backgrounds, and for commercial sectors looking for innovative solar cell materials and related technologies. This could be a solitary reference book for researchers in materials science, engineering and nanotechnology. We are confident that readers will gain invaluable information that can be utilized not only in research and development, but also in commercial manufacturing of solar cell panels.

Atul Tiwari, PhD
Rabah Boukherroub, PhD
Maheshwar Sharon, PhD

Part 1
CURRENT DEVELOPMENTS

1

Design Considerations for Efficient and Stable Polymer Solar Cells

Prajwal Adhikary[1], Jing Li[2], and Qiquan Qiao[1,*]

[1]*Center for Advanced Photovoltaics, Department of Electrical Engineering and Computer Sciences, South Dakota State University, Brookings, South Dakota, USA*
[2]*Chongqing Institute of Green and Intelligent Technology, Chinese Academy of Sciences, Chongqing, China*

Abstract
Organic photovoltaics (OPVs) are gaining more interest due to their potential for low fabrication cost, mechanical flexibility and light weight. Not long ago, lower power conversion efficiency and inferior stability compared to their inorganic counterparts were considered major issues associated with OPVs. Today, OPVs with an efficiency as high as 12% have been achieved and stability under ambient conditions has also been significantly improved, especially with the devices using inverted architecture. A major advantage of using OPVs over traditional thin film inorganic solar cells with comparable efficiency (~14–15%), is their excellent performance in real-life environments like high temperature and low light conditions. In this chapter, we discuss recent advances in realizing high performance and stable OPV devices. The chapter consists of three major parts: the role of interfacial layer for efficient Bulk heterojunction (BHJ) solar cells, the selection of interfacial layer for stable and longer life time OPVs and major interfacial materials used for conventional and inverted device architectures.

Keywords: Organic solar cells, inverted structure, normal structure, interfacial layer, cell stability

**Corresponding author*: qiquan.qiao@sdstate.edu

Atul Tiwari, Rabah Boukherroub, and Maheshwar Sharon (eds.) Solar Cell Nanotechnology, (3–40) 2014 © Scrivener Publishing LLC

1.1 Introduction

1.1.1 Background

Organic and organic/inorganic hybrid solar cells have attracted a lot of interest due to their solution-based processing and low cost [1–15]. A completely new field of conducting polymers emerged in the 1970s when Shirakawa, MacDiarmid and Heeger found that simple doping of polyacetylene with a controlled amount of halogen (Cl, Br, I) could lead to a dramatic increase in conductivity (eleven orders of magnitude) of the film in which electrical properties could be tuned from insulator to semiconductor to metal [16]. In early days, it was fundamentally challenging to achieve high performance organic photovoltaic (OPV) cells by sandwiching a single organic layer between two metal electrodes, in which it is difficult for excitons generated in the active layer to reach the metal interface with the diffusion length of only 10–20 nm [17]. A major breakthrough in OPVs was achieved in 1986 when Tang introduced the concept of bilayer in which copper phthalocyanine as p-type and a perylene derivative as n-type, two organic materials, were sandwiched between metal electrodes [18]. This concept increased the OPVs cell efficiency to 1%, basically due to improved interface for exciton dissociation.

Later, it was found that better charge transport materials with high electron affinity are also required to have their band levels (HOMO and LUMO) aligned with respect to most donor-type polymers. In 1993, Sariciftci *et al.* reported C_{60} which possesses higher electron affinity and mobility and soon established itself as a widely used acceptor [19]. Halls *et al.* reported an increase in photocurrent as high as 20 times with the use of C_{60} as acceptors with MEH-PPV [20]. A similar increase in photocurrent was reported by Morita *et al.* when they used C_{60} with P3AT [21] .

It was known that in a bilayer configuration, exciton dissociates at the donor-acceptor interface but the diffusion length of exciton is about 10 nm which limits their dissociation from the photoexcitation sites outside their diffusion length. Therefore, only a small fraction of polymer actually contributes to photocurrent generation. This limitation of bilayer structure was overcome in 1994 when Yu *et al.* made the first bulk heterojunction organic solar cell by dissolving MEH-PPV and C_{60} which showed photosensitivity of an order magnitude higher than pure polymer-based solar cells [22].

Major impedance in further improvement of PSC efficiency was due to large bandgap (~ 2 eV) of most polymers which had narrow light absorption range and poor hole mobility [23]. Synthesis of soluble polythiophenes, especially poly(3-hexylthiophene) (P3HT), was one step further in the field of OPVs. Currently, morphology optimization of P3HT/PCBM-based PSCs has increased the PCEs by higher than 4% [24]. An efficiency of over 6% was reported when PCDTBT, a low bandgap polymer that utilized cyclopentadithiophene unit as the donor block in the polymer chain, was developed by Leclerc *et al.* which incorporated the use of TiO_x layer as optical spacer [25]. Efficiency above 7–8% was achieved when low bandgap polymers designed by Yu *et al.* based on thieno-thiophene (TT) and benzodithiophene (BDT) alternating units were synthesized. Higher V_{oc} of 0.74 V was achieved owing to lowered HOMO level of polymer, and low bandgap helped to harvest more light which led to J_{sc} of 14.5 mA/cm^2 [26].

Major requirements for large area roll to roll (R2R) processing in order to make organic photovoltaic commercially successful are high efficiency and stability. Various strategies can enhance performance parameters of the cell (V_{oc}, J_{sc}, FF) including choice of materials, efficient light harvesting mechanisms and morphological optimizations. On the other hand, improvement in device stability requires understanding degradation mechanisms of individual components of the OPV devices. Also, utilization of inverted devices instead of conventional architecture could help us achieve more stable OPV devices as it allows us to use materials that are more stable. For example, the acidic nature of PEDOT:PSS used as hole transport layer in conventional devices hampers stability as it etches the bottom ITO. On the other hand, low work function metals on the cathode side oxidize when exposed to ambient condition. The generally used vacuum-deposition-based technique for top electrode increases the cost.

Therefore, a technique that could overcome the above mentioned hindrances is to use inverted device structure, in which charge collecting electrodes are reversed, i.e., ITO could be used as cathode for collecting electrons and top electrodes could be used for collecting holes which are just opposite to that of regular BHJ structure. First, inverted structure does not require acidic PEDOT:PSS. Top electrodes could be any high work function metal (e.g. Ag, Au) which increase device stability, as these metals are air stable. Also, oxidation of silver into silver oxide has been reported to improve hole extraction efficiency as AgO has Fermi level closer to HOMO of polymer. Second, use of n-type metal oxides such as ZnO and TiO_2 as electron transport layer saves the active layer from UV-induced degradation as most of the UV light is absorbed by this layer, and also the Fermi level of these metal oxides that lies close to PCBM LUMO provides additional interfacial area for exciton dissociation. Third, the use of solution-based metal could be employed in the form of top electrode which can avoid vacuum-based deposition technique, thus reducing cost. Fourth, inverted structure could take advantage of naturally favored vertical phase separation of active layer where PCBM rich layer lies towards ITO and polymer rich layer lies on the top, which improves overall device performance. Finally, reports suggest that in the modeling of optical field distribution in the different device structures, higher current density from inverted structures resulted due to enhanced absorption of incident light in the active layer as there is no PEDOT:PSS and Ca layer as in normal structure which is responsible for parasitic absorption of light [27]. Currently, the highest efficiency reported (9.2%) so far in single junction organic BHJ solar cells is based on inverted structure.

1.1.2 Theory

Semiconductor properties shown by polymers are due to the pi conjugation, the alternating single and double (or triple) bonds between the carbon atoms. Ground state of carbon has $1s^22s^22p^2$ configuration in which, single bonds are associated with σ-bonds consisting of localized electrons and double bonds are associated with both σ-bonds and π-bonds where π-bonds consist of π-electrons which can delocalize along the conjugation length responsible for charge transport in carbon-based semiconductors.

Figure 1.1 shows the bonds in ethene molecule. Two σ-bonds connect to each of the hydrogen atoms and one σ-bond is formed between the carbon atoms. The fourth valence electron of each carbon atom situated in the respective p_z orbital, which is oriented perpendicular to the sp_2 orbitals, form a π bond, which is weaker than the σ-bond due to lesser overlap of the p_z orbitals. Along with the formation of σ and π bonds, orbitals are split into bonding and antibonding (denoted with *) orbitals. In accordance with Pierls instability, formation of two delocalized band states takes place in a polymer known as highest occupied molecular orbitals (HOMOs) and lowest unoccupied molecular orbitals (LUMOs). Difference in energy level

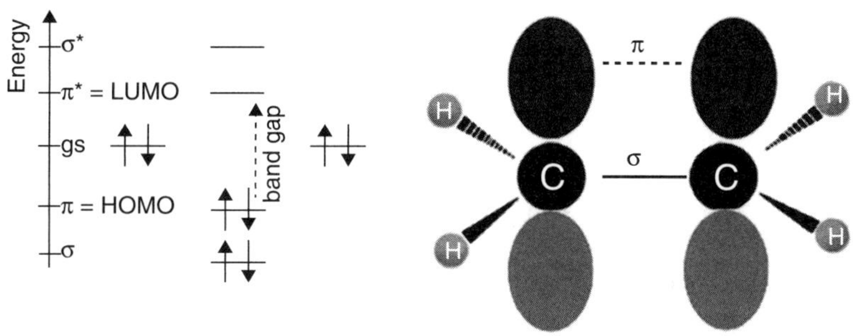

Figure 1.1 Schematic showing σ and π bond formation in an ethene molecule (left) and energy band diagram showing HOMO and LUMO energy levels (right).

between HOMO and LUMO gives rise to bandgap of the polymer. Generally, energy level differences are in the order of 1–3 eV [28].

In a typical inorganic solar cell, incident photon generates electron-hole pair which easily dissociates into electron and hole in the presence of ambient thermal energy. However, in organic semiconductors incident photon generates electron-hole pairs known as excitons which are rather tightly bound with energies of 0.3 to 1eV which are impossible to break in the presence of ambient thermal energy. Notably, ambient thermal energy is approximately 0.025 eV which is far lower than the energy required to break excitons into electrons and holes owing to lower dielectric permittivity (ε= 3–4) in an organic material. This energy requirement for breaking excitons makes it imperative to use a second material which could provide a built-in potential sufficient enough to dissociate excitons. Most often in an OPV cell, primary light absorber or donor material are the sites where excitons are formed. Difference in the molecular orbital energies between donor and acceptor materials helps to overcome the coulombic attraction and hence dissociation of electron and hole occurs at the donor-acceptor interface. In order to be dissociate, these excitons must find interface which is the region where donor and the acceptor meet.

1.1.2.1 Photovoltaic Processes in Donor-Acceptor (D-A) System

Buckminsterfullerene (e.g. C_{60}) is widely used acceptor material due to its higher electron affinity. One important factor is the diffusion length that an exciton can travel to reach the D-A interface before it can recombine. Generally the D-A ratio and phase separation, carrier mobility, and bulk-morphology are the factors responsible for efficient exciton dissociation. Figure 1.2 depicts formation of excitons upon exposure to light. These excitons will not readily convert into electrons and holes after they reach the interface between donor and the acceptor within ~10 nm. As shown in Figure 1.2, excitons formed in the donor side get dissociated into electrons and holes when excitons find the interface. There are two substeps before charge carriers can move to their respective electrodes (electrons towards cathode, hole towards anode). First, there is the charge transfer (CT) state in which even the electron reaches the acceptor but are still bound together and might recombine as geminate recombination because they both still belong to the same exciton. A non-geminate recombination would have occurred if these charge carriers had been separated at the interface with electrons on the acceptor LUMO and holes on the donor HOMO got recombined with opposite charge originating from a different exciton present at the interface. Second, the true dissociation state, known as charge separation (CS) state, in which electrons stay at the LUMO of the acceptor and holes stay in the HOMO of the donor [29].

Charge carriers once separated can now make their way to the respective electrodes through a combination of drift and diffusion mechanisms. They may still have to go through a competing recombination process [30]. Figure 1.3 shows the ultrafast charge transfer time during which carriers are dissociated with electrons transferring into acceptor and holes staying at the donor. All the absorbed photons do not necessarily lead to mobile carriers; some photogenerated carriers

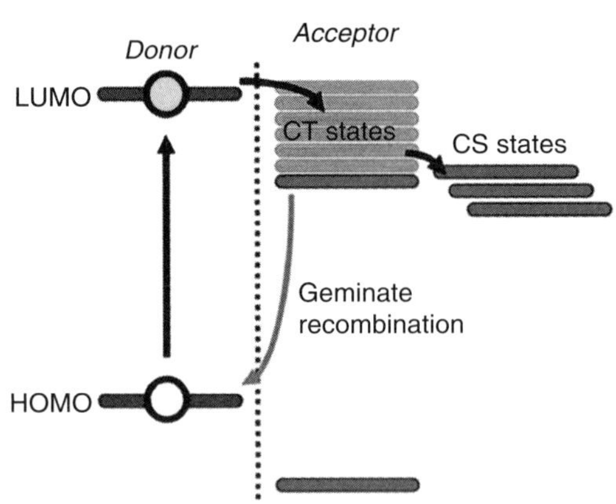

Figure 1.2 Energy diagram illustrating excited electrons while transferring from LUMO of donor to the LUMO of acceptor creating a charge separated (CS) state. Reproduced with permission from ref. [29].

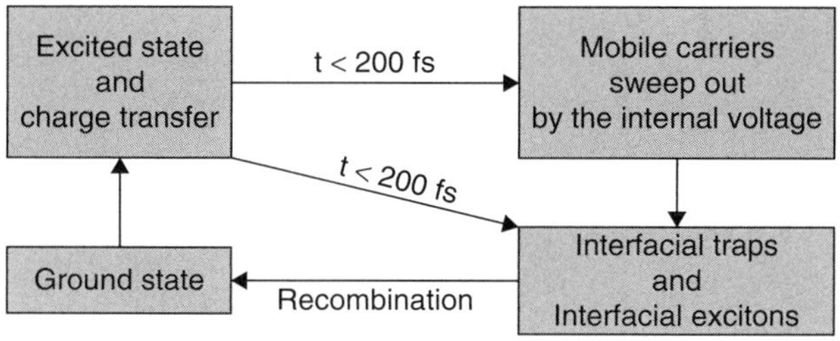

Figure 1.3 Ultrafast time dynamics during the operation of the BHJ solar cell. The excited charge carriers could either be swept out by the internal voltage or get trapped and recombine back into the ground state.

fall into interfacial traps or form bound interfacial charge transfer excitons which can lead to carrier recombination [31]. For efficient charge extraction, recombination time must be longer compared to the charge transport time (total time taken by the carriers from the point of dissociation to the point when they are collected at the electrode).

1.1.2.2 *Equivalent Circuit Diagram of a PV Cell under Illumination*

Equivalent circuit of a simplified solar cell is shown in Figure 1.4. A series resistance (R_s) represents charge carrier transport resistance and shunt resistance (R_{sh}) represents leakage current. The J-V characteristics of a solar cell can be described as Eq. 1.1,

$$J = J_0\left\{\left[exp\frac{q(V - JR_sA)}{nK_BT}\right] - 1)\right\} + \frac{V - JR_sA}{R_{sh}A} - J_{ph} \tag{1.1}$$

where k_B is Boltzmann's constant, T is temperature, q is elementary charge, A is device area, n is ideality factor of the diode, J_0 is reverse saturation current density, and J_{ph} is photocurrent. The J-V curves and photovoltaic parameters including V_{oc} and FF strongly depend on the n, J_0, R_s, and R_{sh} [32].

1.1.2.3 *Parameters Governing Performance of Solar Cells*

Power conversion efficiency (PCE) of an organic solar cell depends upon how well the light is absorbed, exciton generated and diffused to the D-A interface, separated into electrons and holes, and

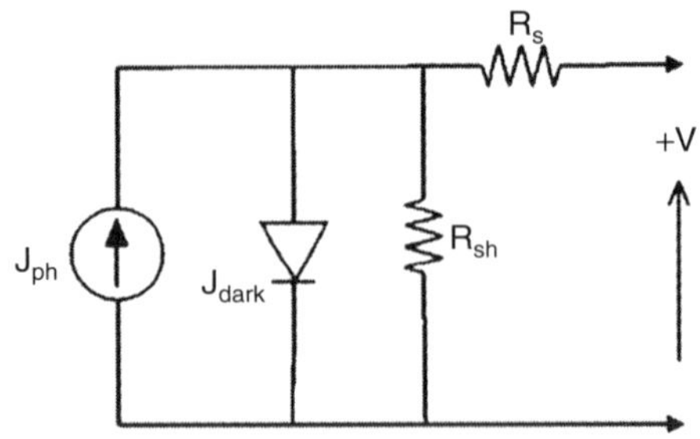

Figure 1.4 Equivalent circuit diagram of a PV cell under illumination. Reproduced with permission from ref. [8].

finally collected at the respective electrode. The external quantum efficiency (EQE) is defined as the ratio between charge carriers at the electrode upon light illumination to the number of incident photons at a particular wavelength [33]. EQE can be expressed as Eq. 1.2,

$$\text{EQE}(\lambda) = \eta_A(\lambda) \times \eta_{ED}(\lambda) \times \eta_{CS}(\lambda) \times \eta_{CC}(\lambda) \tag{1.2}$$

where, λ = wavelength of the incident photon;
$\eta_A(\lambda)$ = absorption efficiency of the organic material, a ratio of excitons generations from HOMO to LUMO upon light excitation to the total incident photons;
$\eta_{ED}(\lambda)$ = exciton diffusion efficiency, which is the ratio of number of excitons that diffuse and reach the D-A interface to number of total excitons generated;
$\eta_{CS}(\lambda)$ = charge separation efficiency and denotes the ratio of the number of excitons that successfully undergo the CT process to the number of excitons that have reached the D–A interface;
$\eta_{CC}(\lambda)$ = charge collection efficiency and is defined by the ratio of number of charge carriers that have been collected at the electrodes to the number of charge carriers that has undergone CT process.

The PCE (η) of a photovoltaic device can be expressed as Eq. 1.3,

$$\eta = (J_{sc} \times V_{oc} \times FF)/P_{in} \tag{1.3}$$

where, J_{sc}, V_{oc}, FF and P_{in} are short-circuit current density, open-circuit voltage, fill factor and power of incident light.

Current density is dependent upon the bandgap of the organic material, thickness and morphology of the active layer. Theoretically, J_{sc} is defined by Eq. 1.4,

$$J_{sc} = \frac{hc}{q} \int \frac{Psun(\lambda) \cdot EQE(\lambda)}{\lambda} d\lambda \tag{1.4}$$

where, h is the Planck's constant, c the speed of light, q the elementary charge, and P_{sun} the solar irradiance.

Open circuit voltage (V_{oc}) is the maximum potential available from the cell which occurs at zero current. In an organic system, theoretically V_{oc} is determined by the energy level difference between HOMO of the donor and LUMO of the acceptor.

Fill factor (FF) denotes the squareness of the IV curve is governed by several factors including carrier mobility, balanced charge transport, and planar polymer structure for molecular packing in the organic photovoltaic devices [34–36].

1.2 Role of Interfacial Layer for Efficient BHJ Solar Cells

An ideal interfacial layer is the one that provides Ohmic contact with minimum contact resistance between the active layer and respective electrode. Once the charge carriers are dissociated in the active layer, efficient charge sweep-out from active layer to respective electrode is dependent on choice and electrical property of the interfacial layer. Nature of electrical contact provided by the interfacial layer has significant impact on the key performance parameters (e.g. J_{sc}, V_{oc}, FF) of the organic solar cell, hence the overall power conversion efficiency.

Choice of materials has become an important criterion to optimize the electronic/electrical properties of the interface for efficient performance of organic solar cells. Several classes of interfacial materials including metal oxides, conjugated semiconductor electrolytes, self-assembled structures, crosslinkable materials, and graphene-based materials which require efficient integration and compatibility with other layers in multilayer devices have been previousy tried [37]. The role of an interfacial layer is described below.

1.2.1 Role of Interfacial Layer on V_{oc}

As mentioned above, maximum V_{oc} achievable in a BHJ polymer solar cell is the difference in energy level between HOMO of the donor and LUMO of the acceptor (V_{oc1} in Figure 1.5). However, for this to happen there must be an Ohmic contact formed between the active layer and the electrodes (both anode and cathode). Various interfacial effects such as dipole formation, charge transfer, and formation of interfacial states can occur depending upon their interaction between the contacts. The static work function difference between the two electrodes in which active layer is sandwiched might create an internal electric field which can reduce the V_{oc} of the cell due to the formation of Schottky contact (V_{oc2} in Figure 1.5). Formation of Schottky contact at the interface and its role in device performance is discussed in the last part of this section.

An integer charge transfer (ICT)-based model is widely used for understanding interfacial effects in OPVs. Negative ICT state (E_{ICT-}) represents Fermi-level pinning on acceptor ($E_{F,e}$) defined as the energy gained by adding one electron to an organic molecule, and positive ICT state (E_{ICT+}) represents Fermi-level pinning on donor

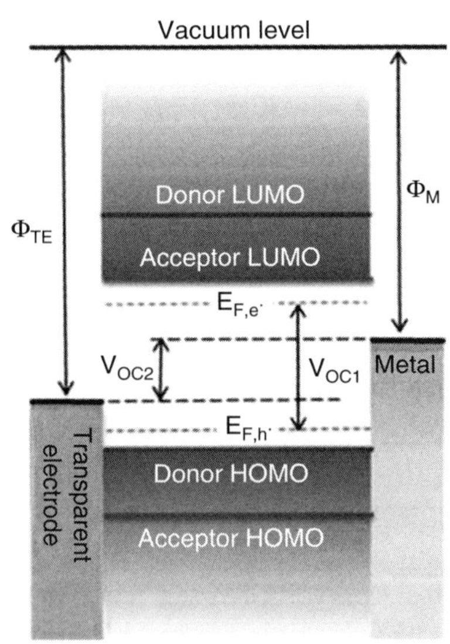

Figure 1.5 Energy level diagram of a BHJ OPV device where organic layer is sandwiched between transparent electrode and metal electrode. Reproduced with permission from ref. [38].

($E_{F,h}$) defined as energy required to take away one electron from organic molecule. As shown in Figure1.5, an electron can transfer from the substrate to the organic semiconductor when the Fermi level of the substrate is lower than $E_{F,e}$ of organic semiconductor. Also, when the Fermi level of the substrate is higher than $E_{F,h}$ holes will transfer from substrate to organic semiconductor [39, 40].

According to a study conducted by Mihailetchi *et al.* on a normal structure device, various metal cathodes such as LiF/Al, Ag, Au, and Pd were used as a top electrode in a device configuration of ITO/PEDOT:PSS/OC_1C_{10}-PPV:PCBM/metal cathode, and they found that only LiF/Al formed Ohmic contact with the active layer while other metals formed non-Ohmic contact. In the case of Ohmic contacts, V_{oc} was dependent on HOMO of the donor and LUMO of the acceptor, while for non-Ohmic contact V_{oc} was in agreement with work function difference between the electrodes based on MIM (metal-insulator-metal) model [41].

A similar study on inverted device structure were carried out by Hau *et al.* in which various metal cathodes including Ca/Al, Al, Ag, Cu, Au and Pd were used in a device configuration of ITO/ZnO-NPs/C_{60}-SAM/P3HT:PCBM/metal anodes as shown in Figure 1.6 [42]. They found that V_{oc} of the cells were reduced with a decrease in work function of the anode metals as shown in Figure 1.7. However with the insertion of a PEDOT:PSS layer between the active layer and metal anode, maximum V_{oc} was achieved despite the choice

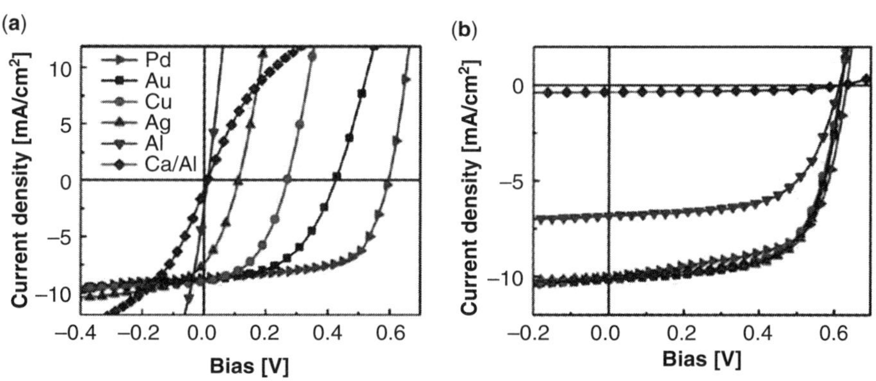

Figure 1.6 J-V characteristics of inverted device with ZnO-Nps/C_{60}-SAM as electron transport layer fabricated (a) without a PEDOT:PSS layer (top), (b) with a PEDOT:PSS layer using various top metal electrodes (bottom). Reproduced with permission from ref. [42].

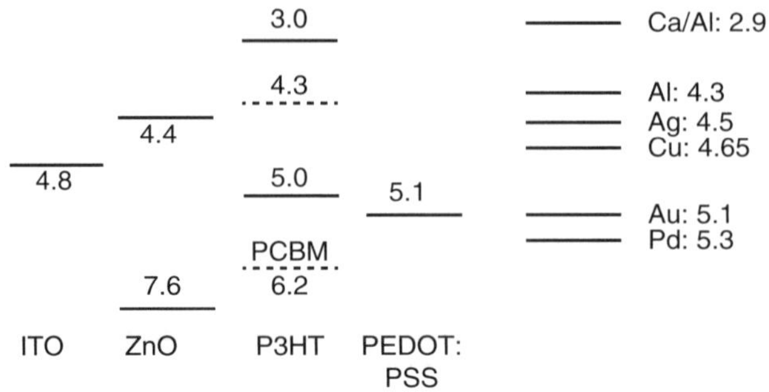

Figure 1.7 Band diagram of inverted device structure ITO/ZnO–NPs/C60–SAM/P3HT: PCBM/metal electrodes (Ca/Al, Al, Ag, Cu, Au, Pd) solar cells.

of metals. This result indicated that PEDOT:PSS provided Fermi-level pinning to the E_{ICT+} of the active layer for better Ohmic contact formation. The absence of Ohmic contact provided by PEDOT:PSS devices showed V_{oc} dependent on work function difference of the electrodes based on MIM model.

Hau *et al.* also studied the role of SAMs in an ITO/PEDOT:PSS/P3HT:PCBM/ZnO/metal-based device and found that devices without SAMs could lead to poor performance mainly due to the formation of Schottky contact at the interface between ZnO and metal [43]. Schottky contact acts as an electron injection barrier as shown in Figure 1.8, with inclusion of SAM injection barrier reduced to zero forming an Ohmic contact. Formation of Ohmic contact and higher device efficiency with inclusion of SAM is due to the appropriate dipoles and better chemical bonding between SAM/metal interfaces.

1.2.2 Influence on Active Layer Vertical Morphology Based on underneath Interfacial Layer

Optimum phase separation between donor and acceptor domains in a BHJ system is required for efficient exciton dissociation, charge transport and charge collection. Ideally donor-acceptor should not exceed 10 nm which is the exciton diffusion length to avoid exciton recombination. Recently, vertical phase separation has been found to play a crucial role in obtaining efficient charge transport pathways by forming interpenetrating network between the active layer and respective electrodes. Studies show that when active layer is spin coated on top of a substrate, donor and acceptor are not equally distributed

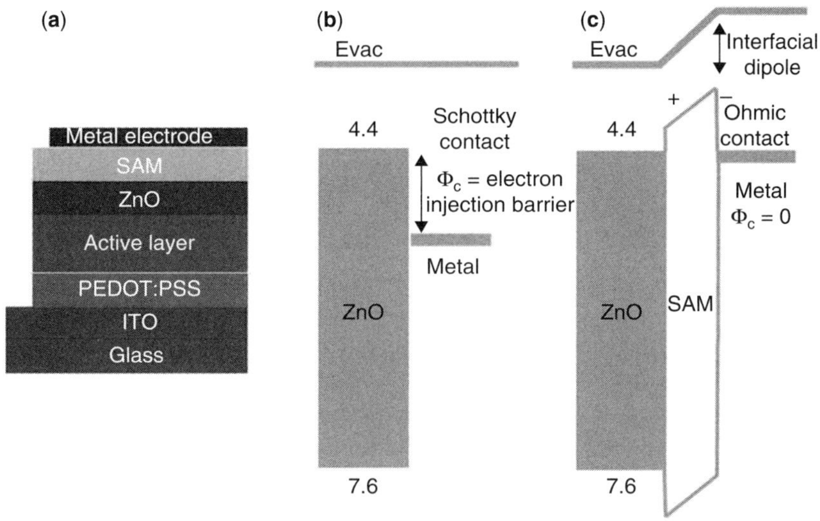

Figure 1.8 (a) Device architecture, (b) formation of Schottky contact between ZnO and metal in absence of SAM, and (c) formation of Ohmic contact between ZnO and metal.

throughout the film. Depending upon the choice of solvent, rate of drying, and thermal and vapor annealing conditions, morphology of an active layer could be altered. Germack *et al.* used surface sensitive near-edge X-ray absorption fine structure (NEXAFS) spectroscopy and found when active layer consisting of P3HT with surface energy of 26.9 mN/m^2 and PCBM with surface energy of 37.8 mN/m^2 were coated on substrates with higher surface energy value (γ), the buried interface was selectively PCBM-rich, whereas when active layer was coated on substrates with lower γ value, they found the buried interface was selectively P3HT rich. However, the top side of the active layer was always P3HT rich owing to lower γ value of air [44].

Subiah *et al.* carried out XPS- and AES-based study to find out the vertical profile of a PDTS-BTD:$PC_{71}BM$-based system [45]. As shown in Figure 1.9, C composition was found to be higher in the bottom than in the top. And, S composition was higher in the top and lower in the bottom.

Also, in their work a comparative study was performed on normal and inverted structure BHJ cells; inverted structure yielded higher PCE owing to higher photocurrent due to better charge collection. EQE spectra in Figure 1.10 shows higher current yield in inverted device structure than in conventional geometry.

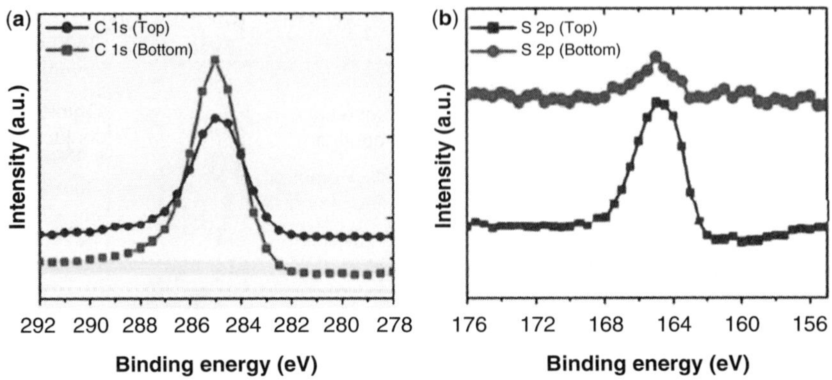

Figure 1.9 XPS spectra of (a) C 1s region and (b) S 2p region obtained from top and bottom surfaces of the PDTS-BTD:$PC_{71}BM$ films. Reproduced with permission from ref. [45].

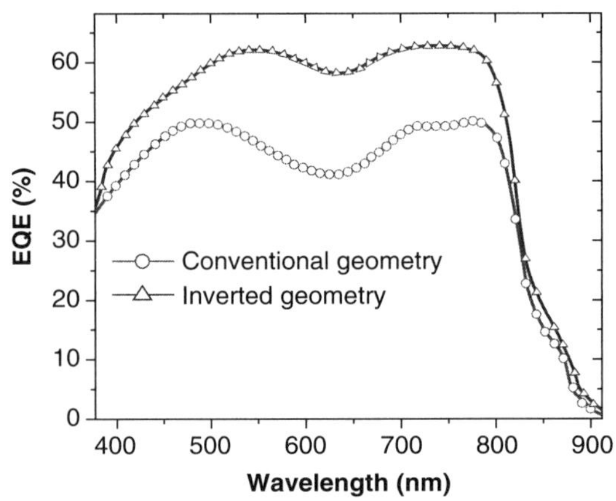

Figure 1.10 EQE spectra of conventional and inverted geometry OPV devices based on PDTS-BTD:$PC_{71}BM$. Reproduced with permission from ref. [45].

1.2.3 Light Trapping Strategies and Plasmonic Effects for Efficient Light Harvesting

Typically a thicker active layer can help to absorb more sunlight. However, the low charge carrier mobility in organic materials limits the thickness of active layer to be around ~100 nm so as to minimize recombination. Therefore we simply need a technique where light absorption is maximized keeping thickness of the active layer constant. Various strategies have been put forward which can enhance light harvesting. Conventionally, light trapping was based on total internal reflection effects achieved by roughening of the entrance interface. Light could stay in longer interaction distance with the active material thereby increasing absorption. This technique could improve light enhancement by $4n^2$, where n is the refractive index of the active layer. To overcome this limitation, a highly ordered and periodic nanostructure-based method are being explored which can theoretically enhance the light absorption up to $12 \times 4n^2$ [46].

Reflection and absorption of the interfacial layer play a crucial role in the photovoltaic performance of an organic cell. Hadipour *et al.* have conducted a systematic study using various buffer layers on a P3HT:PCBM-based system [47]. Figure 1.11 shows simulated electric field intensity within P3HT:PCBM for incident light wavelength $\lambda = 550$ nm, where higher optical field intensity is observed in devices without use of Ca because Ca has less back reflectivity compared to the others. This indicated that one of the most widely used cathode buffer layers in conventional BHJ possesses parasitic absorption which can reduce photocurrent by 25%. A study regarding absolute optical reflection of 60 nm thin Ag, Al, Ca electrodes and its effect in J_{sc} by Pandey *et al.* showed that Ag reflects 96%, Al reflects 90% and Ca reflects 68% of light and is directly proportional to the J_{sc} [48].

Hadipour *et al.* also showed that higher refractive index material such as MoO_3 can act as an optical spacer and requirement of its optimum thickness for higher device performance. An optical spacer in an organic solar cell inserted between the active layer and the reflective electrode which can redistribute the optical electric field.

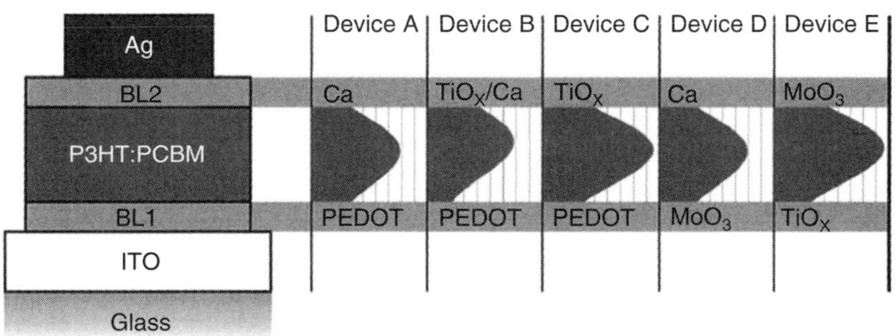

Figure 1.11 Conventional structure (A, B, C, and D) and inverted structure (E). Various buffer layers employed in the study are shown in the figure. The active layer consists of 80 nm thick P3HT:PCBM blend for all cell structures simulated electric field intensity within the P3HT:PCBM layers for incident light of wavelength $\lambda = 550$ nm. Reproduced with permission from ref. [47].

Another interesting work regarding optical spacer carried out by Gilot *et al.* using ZnO (39 nm) between active layer and top electrode revealed the necessity of inclusion of ZnO as an optical spacer for maximum optical field to be placed in the active area [49]. Calculated optical electric field for light of different wavelengths with a 40 nm thick active layer with and without ZnO is shown in Figure 1.12. Absorption peak of P3HT:PCBM film is at about 550 nm, and we should focus on wavelengths between 500–600 nm. In the figure (bottom), the device with ZnO let the optical electrical field be maximum in the active layer while without ZnO, the maximum optical field is not placed exactly in the active area. However, for thicker active layer this effect was shown to be detrimental as the active layer itself is in the most effective position of the optical electric field.

Similarly, metallic nanoparticles have been reported that can enhance performance of an OPV cell through surface plasmonic

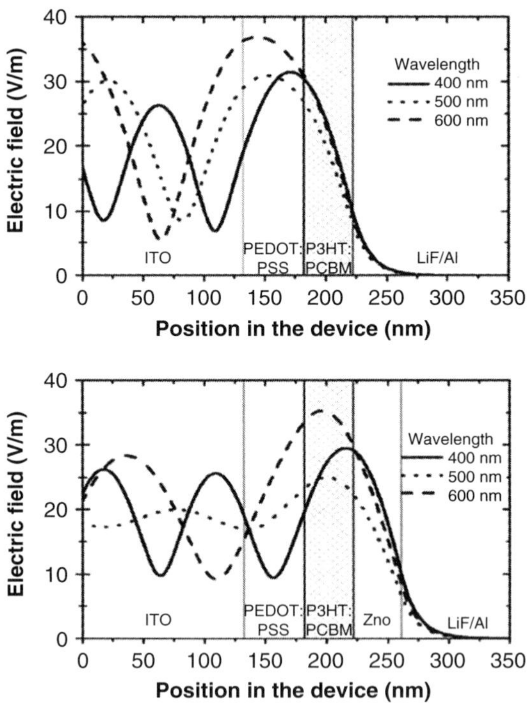

Figure 1.12 Calculated optical electric field intensity for 400,500 and 600 nm wavelength incident light on device without (top) and with 39 nm ZnO (bottom). Active layer consisted of 40 nm P3HT:PCBM. Reproduced with permission from ref. [49].

effects. Wu *et al.* blended Au nanoparticles onto PEDOT:PSS based on device configuration of ITO/PEDOT:PSS-Au NPs/ P3HT:PCBM /cathode metal [50]. They found incorporation of Au NPs on poly(3,4-ethylenedioxythiophene) doped by poly(styrenesulfonate) (PEDOT: PSS) enhanced light absorption induced by localized surface plasmon resonance (LSPR). LSPR occurs when frequency of the incident light matches with the resonance peak of the noble metallic materials which can then enhance electromagnetic field near the surface of the metallic NPs, thus improving light trapping inside the active material. Also in their work, dynamic PL measurements showed that the LSPR effect reduced the lifetime of photogenerated excitons in the active blend indicating reduction of geminate recombination due to interplay between metallic NPs and excitons leading to enhanced rate of exciton dissociation and hence improvement in charge transfer process. Figure 1.13(a) shows light trapping through forward scattering due to LSPR induced by Au NPs, while Figure 1.13(b) indicates enhancement of the electromagnetic field due to LSPR.

1.2.4 Morphology Control of Active Layer and ETL by Processing

Ngo *et al.* reported a comparative study on active layer deposition technique that can play a significant role in morphology of film [51]. Their study was based on spin coating vs spray coating as shown in Figure 1.14. Figure 1.14(a) shows dotted polymer

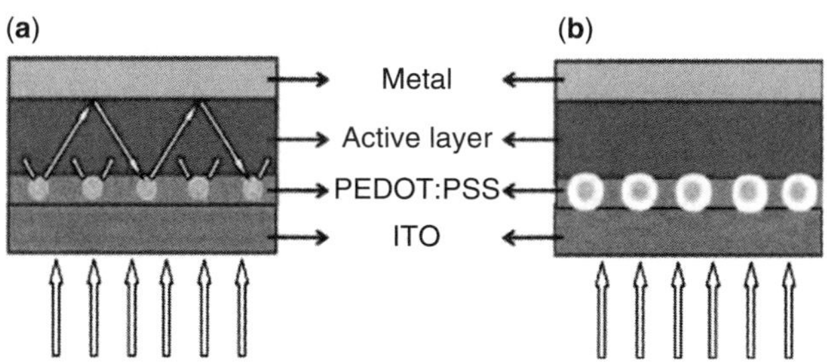

Figure 1.13 Schematic illustrating induction of LSPR due to forward scattering of light in presence of Au NP. (a) Increase in optical path length due to light trapping. (b) Schematic representation of the local enhancement of the electromagnetic field. Reprinted with permission from ref. [50]. Copyright © 2011 American Chemical Society.

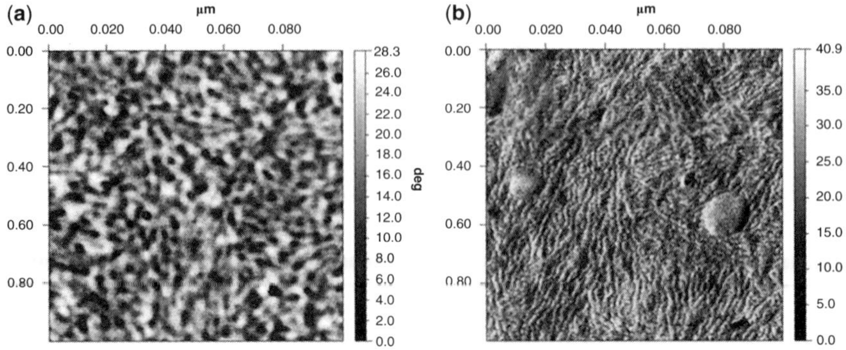

Figure 1.14 Phase diagram indicating (a) active layer deposited via spin coating and (b) spray coating techniques. Reproduced with permission from ref. [51].

morphology in active layer while Fig. 1.14(b) shows self-assembled fibrillar structure arising mainly due to longer evaporation time taken by the solvents. These fibrillar structures contribute in enhancing charge carrier mobility by providing lesser resistant charge transport pathway leading to higher J_{sc} and FF. In the same work, they also mentioned that the morphology of widely used ZnO as ETL could be controlled into more compact and smoother film. However, deposition of both ETL and active layer via spray coating technique led to lower efficiency due to poor interface formed between these two layers. Nevertheless, utilization of simple airbrush technique to spray coat the active layer and ETL can be modified for large-scale manufacturing of polymer solar cells at a much lower cost.

1.3 Selection of Interfacial Layer for Stable and Longer Lifetime

Stability is a major issue for organic solar cells to compete with existing inorganic solar technology. Longer lifetime is the prerequisite for commercial application. Polymer solar cells degrade both under illumination and in the dark. Therefore it is very important to understand the origins of degradation. An organic cell is susceptible to both chemical and physical degradations. Chemical degradation includes diffusion of oxygen and moisture into the active layer, oxidation of metal electrode, chemical degradation of the ITO electrode, degradation of PEDOT:PSS layer, and photo-oxidation of polymers. Major physical degradation includes change in morphology of the donor and acceptor molecules which may not remain the same as they were during the device fabrication process. Change in ambient temperature could easily change the micro/nano phase separation between donor and acceptor due to slow diffusion or recrystallization of the components inside active layer thereby reducing PCE of the cell.

1.3.1 Stability of Active Layer Materials

Degradation of active layer obviously leads to degradation of power conversion efficiency. It is important to understand the mechanisms involved in degradation of active layer and methods to improve it. Light-induced degradation is one of the major degradation processes of active layer besides oxidation of polymers in the presence of ambient oxygen and moisture. In poly-phenylenevinylene (PPV)-based polymers that were extensively used for OPV research, photochemical decomposition is from the side chain and vinylene moiety degradation [52]. Initially, highly reactive singlet oxygen was thought to be the source of degradation. However, Chambon *et al.*, through their transient absorption spectroscopy, showed that formation of superoxide oxygen anion during electronic transfer of photoinduced carriers acts like electron acceptors and oxidize the polymer [53]. Similarly with the most popular polymer P3HT, it was previously thought that the singlet oxygen was responsible for photodegradation, but it is now known that it is the hydroperoxide formation at the benzylic position that initiates side chain oxidation as shown in Figure 1.15.

Manceu *et al.* studied photodegradation of polymers in which they exposed samples to light (under 1 sun and ambient air). A normalized number of absorbed photons versus ageing time was used for comparing stability for various polymers [55]. They then ranked monomers, from which polymers are made, according to their

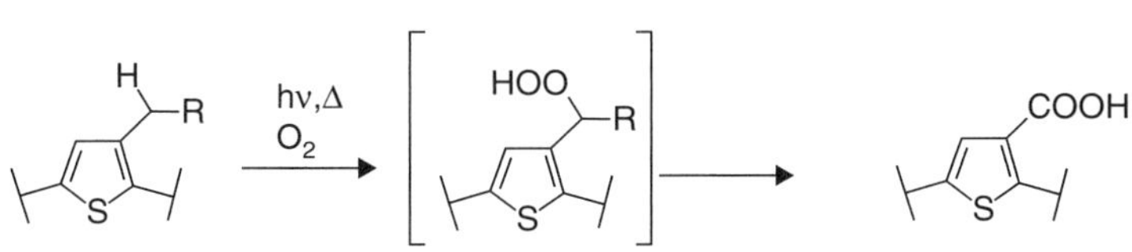

Figure 1.15 Photooxidation mechanism in P3HT. Reproduced with permission from ref. [54].

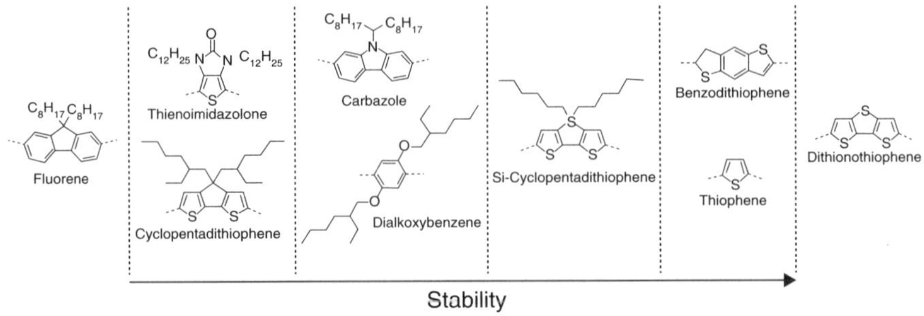

Figure 1.16 Stability order of various donor groups. Reproduced with permission from ref. [55].

stability to light as shown in Figure 1.16. They found that donor groups with side chains are more susceptible to degradation and substitution of a carbon with silicon improves stability.

Another study carried out by Xia *et al.* based on UV absorptions, PL spectra, and FT-IR spectra, showed that photovoltaic performance of aryl side chains is more stable than alkyl side chains [56].

Rate of degradation as studied by Kumar *et al.* depends upon different solvents used for desired morphology control [57]. Their work shows the choice of solvents could broaden the effective density of states which then controls the rate of degradation of active layer. Rate of degradation also depends upon regioregularity and molecular packing [58]. Ebadian *et al.* used 94% and 98% regioregular (RR) P3HT and found that although lower RR P3HT initially has lower PCE compared to higher RR P3HT, after a long period of time lower RR P3HT-based devices showed higher PCE, indicating electron charge transport as the major factor in determining the degradation rate [59]. In the same study, better charge transport due to higher carrier mobility was shown in higher RR P3HT as electron transport plays a critical role in polymer degradation hence leading to poor device stability.

Since side chains are responsible for degradation of the polymer, Liu *et al.* introduced thermo-cleavage technique in which a carboxylic ester was attached at the 3-position to every second thiophene moiety [60]. This allowed researchers to get rid of the ester group after heating to 200°C. This method paved the way for obtaining thermally stable bulk heterojunction [61]. A similar study carried out by Krebs *et al.* using P3MHOCT:PCBM after thermal cleavage yielded P3CT:PCBM and later showed better stability even after 4000 hours of full sun exposure [62].

1.3.2 Stability of Metal Electrodes

Aluminum and calcium are commonly used cathode metals in conventional BHJ solar cells. Although they show stable performance when stored in an inert environment, their lower work function leads to oxidation when exposed to ambient conditions converting them into insulating material, thereby lowering PCE to almost no photovoltaic response. Another mode of degradation is diffusion of oxygen and water through pinholes and aluminum grains respectively present in the thermally evaporated aluminum electrode into the organic layer. As shown in Figure 1.17, water can eventually result in the homogenous photo-oxidation of organic material, while molecular oxygen can lead to photo-oxidation of polymers at places near microscopic pinholes [54, 63]. AFM-based studies show photo-oxidation of the organic material leads to protrusion towards the outside [64].

This problem could be solved using different metal electrodes, most possibly Ag, as in inverted device structure which has higher work function and is less susceptible to oxygen and moisture.

1.3.3 Stability of Transparent Electrode

ITO is generally used as a transparent electrode in BHJ OPV devices. It is known that PEDOT:PSS layer deposited on top of ITO leads to degradation of the interface between the two due to acidic nature of PEDOT:PSS. Also, in one study led by Norrman *et al.* ITO/PEDOT:PSS could promote the degradation of the active layer in which PSS could diffuse to other parts of the device with a reaction

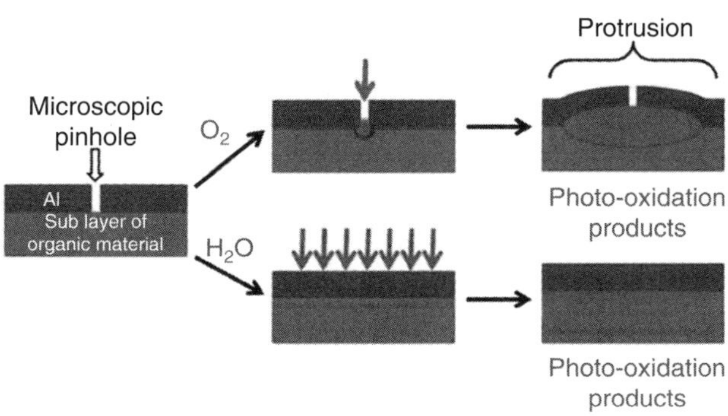

Figure 1.17 Schematic representing modes of water and molecular oxygen diffusion. Reproduced with permission from ref. [54].

that can lead to degradation of the overall PCE [65]. Another study done by Schäfer *et al.* showed change in work function of the ITO upon UV irradiation leading to lowering of V_{oc}, which they conclude is a reversion of the ITO work function [66]. *Kanai et al.* later showed that such an effect can be overcome by introducing MoO_3 between the ITO and organic layer [67].

1.3.4 Stability by Electron Transport Layers (ETLs)

Besides energy level alignment for efficient electron transport from acceptor LUMO to the cathode electrode, electron transport layer (ETL) is also reported for improving device lifetime. LiF has also been widely used as electron transport layer which can improve device stability due to moisture blocking ability [68]. TiO_x is another popular material used as ETL, which could be applied both for energy alignment and better stability despite its lower conductivity in amorphous form. Li *et al.* showed inclusion of 20 nm TiO_x in a device structure consisting of ITO/PEDOT:PSS/P3HT:PCBM/TiO_x/Al improved PCE compared to those without TiO_x. Also, they showed that devices without TiO_x degraded significantly after continuous exposure to UV light; however the cells with TiO_x layer exhibited negligible degradation. TiO_x layer contains organic moieties coming from sol gel-based deposition. They conducted an IR- and ESR-based study on UV irradiated TiO_x layer. Results showed that alkoxide functionalities Ti-OR (OR= alkoxide) present in the sol gel derived TiO_x layer get photo-oxidized into CO_2 and H_2O by utilizing O_2 causing oxygen gas scavenging and saving the organic layer lying underneath [69]. A study conducted by Jin *et al.* showed the role of thin CdSe (8 nm) layer inbetween organic layer and metal electrode could enhance the lifetime of an OPV device by blocking the moisture and oxygen when exposed to air.

ZnO is one of the most widely used ETL with high electron mobility in inverted device structure. Ferreira *et al.* used ZnO as ETL in conventional structure and compared stability with devices without ETL [70]. He found that devices with ZnO as ETL exhibited longer lifetime retaining almost 60% of the initial efficiency after 78 days of exposure to air. TEM-based analysis suggested that the reaction between metal electrode (Al) and active layer was minimized with the inclusion of the ZnO sol gel layer in between. Puetz *et al.* doped ZnO with indium (IZO) and found that it not only improved PCE but also stability [71]. They used IZO between active layer and

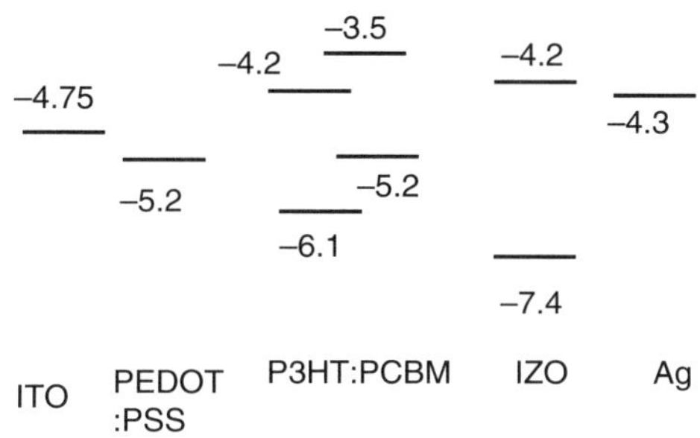

Figure 1.18 Energy level diagram showing IZO acting as an electron transport layer barrier for holes and preventing further recombination losses.

metal electrode in a device with configuration (ITO/P3HT:PCBM/IZO/Ag) and found that PCE is enhanced compared to a device without the IZO. An energy level diagram is shown in Figure 1.18. However, although improvement in V_{oc} was just 50mV, it demonstrated better Ohmic contact provided by inclusion of IZO layer which led to higher PCE. Also, cells with IZO layer showed longer stability even after stored under ambient conditions due to improved barrier effect exhibited by IZO layer.

1.3.5 Stability by Hole Transport Layers (HTLs)

PEDOT:PSS is the most commonly used hole transport layer in regular BHJ devices. One of the key issues associated with PEDOT:PSS is its acidic nature that can etch the underneath ITO in the long run leading to degradation in PCE. Kim *et al.* showed that pH of PEDOT:PSS could be adjusted by addition of NaOH [72]. They showed the optimum amount of NaOH (0.2 molar ratios) could remove sulfonic acid group (up to 23%) with device performance still remaining the same and lifetime increased by about 25%. Another disadvantage of using PEDOT:PSS is its hygroscopic nature which can ultimately oxidize the active layer. Girroto *et al.* compared PCE of devices made of three different hole transport layers, namely: PEDOT:PSS, thermally evaporated MoO_3 and sol-gel derived MoO_3. In their experiments, devices with different hole transport layer exhibited similar performance. However, stability of MoO_3-based devices showed a longer lifetime compared to PEDOT:PSS [73]. Use of sol-gel derived MoO_3 is beneficial in achieving a low cost technique suitable for R2R manufacture of solar devices. Another approach to replace PEDOT:PSS is to use mixed oxides as reported by Huang *et al.* in which they

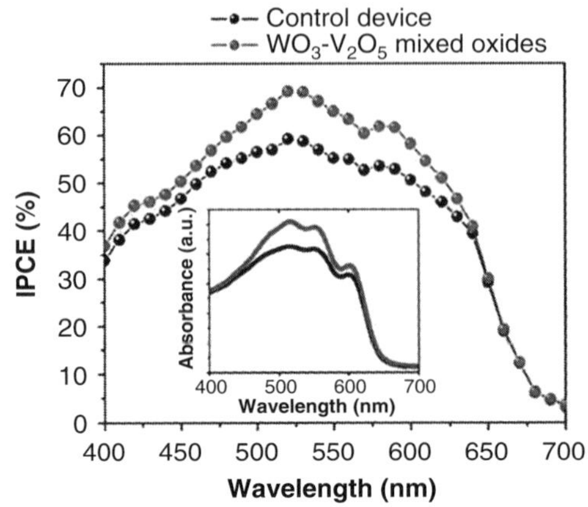

Figure 1.19 Incident photon current efficiency (IPCE) spectra illustrating with and without mixed oxides. Inset shows absorption spectra of both devices. Reproduced with permission from ref. [74].

used WO_3-V_2O_5 with remarkable stability in inverted device structure [74]. Their cells retained more than 90% of initial efficiency even after 1000 h of air exposure. Mixed oxide can suppress the leakage current as higher shunt resistance could be achieved with its inclusion. Also from incident photon current efficiency (IPCE) as shown in Figure 1.19, higher IPCE is achieved for mixed oxides resulting mainly due to improvement in the absorption of incident light due to better light trapping for same thickness of active layer.

1.4 Materials Used as Interfacial Layer

1.4.1 Conventional Solar Cell Devices

1.4.1.1 Cathode and Electron Transport Layers

Al is widely used metal cathode in conventional BHJ PSCs due to its abundance and because its energy level (work function) matches with PCBM LUMO. Lower work function of Al leads to easy oxidation into Al_2O_3 when it comes in contact with atmospheric oxygen leading to deterioration in performance [75]. Also, if Al is directly deposited on top of active layer it can react with the organic materials, causing instability of the Al electrode [76]. Therefore, other interfacial material has to be inserted between Al and active layer for better performance and stability.

Generally, materials used between Al and active layer are Ca, Ba, and Mg for efficient electron extraction from PCBM LUMO to the metal electrode [22, 77]. These low work function materials lower the energy barrier by Fermi level pinning mechanism, hence forming better Ohmic contact with PCBM LUMO and top electrode despite much lower work function than PCBM LUMO. This contact is thought to be formed via charged interfacial states, which improve the FF and V_{oc} of the BHJ [78]. An energy level diagram of various low work function metals are shown in Figure 1.20. Due to higher sensitivity of these materials to moisture, device performance degrades when exposed under atmospheric conditions.

LiF is another widely used electron transport layer between Al and active layer which offers good Ohmic contact with the organic material because the location of Fermi level of LiF is close to LUMO of PCBM [79]. Gao *et al.* used C_{60}:LiF composite in between active layer comprising P3HT:PCBM and Al and found improvement in device performance due to efficient hole-blocking arising from highest occupied molecular orbital (HOMO) of C_{60}; these devices also showed better stability [80].

Caesium carbonate (Cs_2CO_3) offers efficient electron extraction when incorporated between active layer and Al electrode [81]. Chen *et al.* compared PCE of devices with and without Cs_2CO_3 and found significant improvement in device performance. He carried out an XPS-based study and concluded that the n-doping effect on active layer due to electron transfer from the cathode interlayer ultimately enhances electron injection and collection and thereby overall PCE [82].

Sol-gel derived n-type metal oxides such as titanium suboxide (TiO_x) and zinc oxide (ZnO) are widely used interlayers for electron

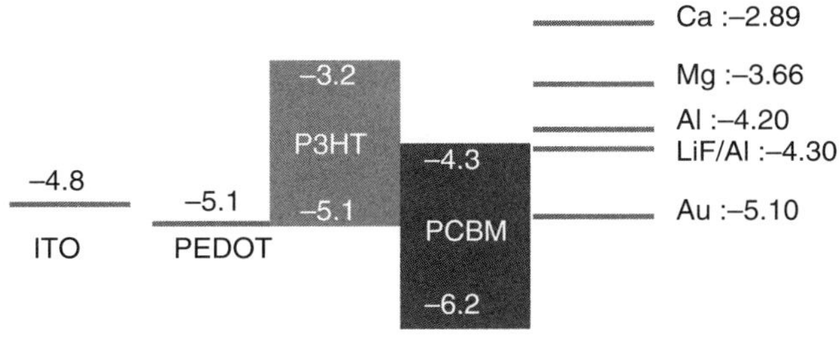

Figure 1.20 Energy band diagram showing work function of various metal cathodes in conventional device architecture.

extraction. Conduction band and valence band of TiO_x is at about -4.4 eV and -8.1 eV [83]. Conduction band edge makes it suitable for electron extraction from PCBM LUMO and valence band edge provides good hole-blocking property from P3HT HOMO. Also, TiO_x layer plays a role of optical spacer which enhances better light harvesting in the active layer [25]. Despite such advantages, TiO_x layer has poor electron mobililty of about 1.7 x 10^{-4} cm^2 V^{-1} s^{-1} which is lower than PCBM and can limit the performance.

Compared to TiO_x, ZnO possesses high electron mobility (~ 0.066 cm^2 V^{-1} s^{-1}) and Fermi level ~ 4.4 eV [84]. Higher electron mobility makes it a better choice as an electron transport layer. Similar to TiO_x, valence band at about -7.5 eV allows efficient hole-blocking from P3HT HOMO. Introduction of ZnO as an optical spacer has also been studied by Gilot *et al.*, which showed an enhanced absorption at the active layer when thin (<60 nm) of ZnO is used in ITO/PEDOT:PSS/P3HT:PCBM/ZnO/LiF/Al-based device configuration [85].

Yip *et al.* showed a simple solution method consisting of self-assembled monolayer (SAM) on top of ZnO serving multiple purposes including enhancement of exciton-dissociation efficiency, passivating inorganic-surface trap states, and optimizing organic layer lying on top of it. This layer can also overcome unfavorable energetic and poor chemical interface between the metal oxide and organic layer leading to improvement in the Ohmic contact thereby improving J_{sc}, V_{oc}, FF and PCE [86]. By appropriately choosing SAM modifier layer, other high work function metals could also be used which can improve device stability.

1.4.1.2 *Anode and Hole Transport Layers*

ITO has dominated the OPV market as transparent conducting oxide for use in front contact of solar cells [87]. High transparency and high conductivity of ITO have made it very popular. It suffers from two major drawbacks: (1) scarcity of indium and (2) brittleness [88]. Also, ITO is generally deposited via slow vacuum-based technique which is not just time consuming but also expensive compared to solution-based methods [88]. Although solution-based ITO nanowires and nanoparticles could be used for cost effectiveness, they offer inferior opto-electronic properties as compared to sputtered ITO [89]. Therefore, other alternatives such as conducting polymers [90], metal nanostructures [91], carbon nanotubes [92], and graphene [93] are actively being investigated. Barnes *et al.*

conducted a comparative analysis on various TCOs and concluded the best metrics for comparison is J_{loss} as defined in Eq. 1.5 for evaluation of electro-optical performance of the TCO. Their analysis demonstrates silver nanowire network could be a better alternative than carbon-based networks in terms of performance [88].

$$J_{loss} = \frac{q}{c}\int_{\lambda 1}^{\lambda 2} G(\lambda)\left(1-T(\lambda)\right)QE(\lambda)d\lambda \tag{1.5}$$

where QE(λ) is the quantum efficiency spectrum of the PV device, q is the fundamental charge, h is Plank's constant, c is the speed of light, $G(\lambda)$ is the wavelength-dependent AM 1.5G solar flux and $T(\lambda)$ is the wavelength-dependent electrode transmittance, and λ_1 and λ_2 are chosen to correspond with spectral response of the PV device.

PEDOT:PSS is the most commonly used hole transport layer in PSCs as it increases the work function of ITO for effective hole transport and improves the contact property between active layer and ITO. The acidic nature of PEDOT:PSS is detrimental to device lifetime as it easily etches the ITO [94]. Kim *et al.* employed self-assembled PEDOT between spin coated PEDOT and ITO, which not just avoided corrosion of ITO but also inhibited the diffusion of indium into the PEDOT layer [95]. Since PEDOT consists of acidic polystyrene sulfonate polymer which corrodes ITO, the corroded indium ions can then migrate towards PEDOT leading to deterioration of its hole transporting properties. With self-assembled PEDOT underneath, prevention of ITO corrosion is achieved. Similarly, Xiao *et al.* used ethylene glycol on top of PEDOT:PSS which caused conformational changes in PEDOT chains and lowered the energy barrier for charge hopping among PEDOT chains, which led to increase in conductivity of PEDOT:PSS leading to higher J_{sc}. This treatment also changed the morphology of the above active layer and notably increased surface roughness, thereby increasing contact area with the top electrode that lowered contact resistance and hence improved FF in ITO/PEDOT:PSS/EG/P3HT:PCBM devices compared to untreated PEDOT:PSS devices [96].

Besides PEDOT:PSS, p-type semiconducting transition metal oxides such as molybdenum oxide(MoO_3), vanadium oxide (V_2O_5), nickel oxide (NiO), and tungsten oxide (WO_3) are widely used in PSCs as hole transport/electron blocking layer due to their Fermi level being closed to HOMO of most donors, and their conduction band of around -2.5 eV easily blocks electrons that might arrive at the ITO from acceptor LUMO.

Thermally evaporated MoO_3 was compared with PEDOT:PSS by Shrotriya *et al.* and it was found that optimum thickness (5 nm) of MoO_3 is required for higher PCE than that which could be obtained using PEDOT:PSS [97]. High efficiency conventional BHJ PSC utilizing MoO_x as hole transport layer was reported by Sun *et al.* which performed better than devices fabricated using PEDOT:PSS. Improvement in PCE using MoOx was achieved due to close value of index of refraction between the active layer (n=2.1) and MoO_x (n=2.0), while that of PEDOT:PSS was 1.4. This matching of refractive index in the case of MoOx led to electric field intensity being redistributed inside the active layer and than compared to the device based on PEDOT:PSS [98]. Another important advantage of using MoO_3 is improvement in device lifetime as compared to one using PEDOT:PSS. In the same work, when devices made with MoO_3 were kept in air for 720 h, 50% of the initial performance was retained, while in the case of a device made with PEDOT:PSS, degradation to 10% of initial performance was seen after 480 h of air exposure.

PCE as high as 5.2% based on P3HT:PCBM system was achieved by Irwin *et al.* when NiO was employed as hole transport layer using pulsed laser deposition technique [99]. They incorporated crystalline cubic NiO as confirmed by glancing-angle X-ray diffraction (GA-XRD). AFM- and SEM-based studies showed NiO could significantly planarize the surface due to decrease in RMS roughness from 4–5 nm for bare glass/ITO to 1–1.5 nm for glass/ITO/NiO surface. As compared to devices fabricated with PEDOT:PSS, devices using NiO showed higher performance due to improved electron blocking provided by the later device. Most importantly, NiO-based devices showed better stability under ambient conditions than PEDOT:PSS-based devices. Cost effective solution-based NiO was deposited with similar performance achieved using PEDOT:PSS by Steirer *et al.*[100].

Tan *et al.* achieved higher PCE using solution processed WO_3 as compared to acidic PEDOT:PSS [101]. Thermal annealing at 150°C allowed tungsten (VI) isopropoxide to decompose into WO_3. Their work shows although WO_3 has lower transmittance than PEDOT:PSS in the 350–500 nm range, better light distribution in the active layer resulted in higher J_{sc} and hence overall PCE. The refractive index value of tungsten oxide (~1.84) is closer to the active layer (2.1) making light distribution in the active layer better than in devices with PEDOT:PSS (1.4) as HTL [102].

Huang *et al.* showed self-assembled multilayers consisting of $–CF_3$ groups of treated ITO surface could improve device performance when compared to PEDOT:PSS. This was achieved due to an increase in ITO work function from –4.7 to –5.12 eV after modification, which is then close to HOMO of P3HT; this improves hole injection efficiency, as increase in work function is more energetically favorable, and hence the performance of the OPV device [103]. Their soft-imprinting method which is different than immersion technique as used conventionally for coating CF_3- is beneficial for mass production.

1.4.2 Inverted Device Structure

1.4.2.1 Cathode and Electron Transport Layers

It is the polarity of ITO that can be modified to make it function either as anode to collect holes or as cathode to collect electrons. In inverted structure BHJ PSCs, various interfacial layers could be employed on top of ITO to match the work function with PCBM LUMO so that electron could be efficiently extracted. A general ETL not only transports the electrons from PCBM LUMO to ITO but also modifies the work function of ITO so that it serves as a cathode. Various inorganic and organic materials used for tuning ITO work function include ZnO, TiO_x, Cs_2CO_3, Ca, Al_2O_3, and poly [(9,9-bis(3'-(N,N-dimethylamino)propyl)-2,7-fluorene)-alt-2,7-(9,9–dioctylfluorene)] (PFN) [104].

ZnO is one of the most common materials utilized as ETL in inverted device structure. White *et al.* were the first to use solution processed ZnO using zinc acetate dissolved in 96% 2-methoxy ethanol and 4% ethanolamine as precursor. ZnO prepared from this method yielded similar photoconductivity as compared to sputtered ZnO, which was expensive and time consuming [105]. Small *et al.* reported ZnO-PVP [poly (vinyl pyrrolidone)]-based composite which was shown to have uniform distribution of ZnO nanoclusters that improved the cell performance compared to those where no PVP is used. They also conducted the UV ozone treatment on top of ZnO-PVP that improved the efficiency, indicating optimum surface treatment time could lead to the highest performance [106].

A study conducted by Adhikary *et al.* showed optimum UV ozone treatment on sol-gel prepared ZnO as ETL could improve device performance of an inverted BHJ polymer solar cell. They compared three different conditions of UV ozone exposure time on ZnO: unexposed, optimally exposed (5 minutes) and overexposed (20 minutes). The unexposed ZnO film was thought to be contaminated by organic residues originating from the sol-gel method acting as recombination centers. Optimally exposed ZnO film yielded highest JV characteristics, while as UV ozone treatment time was increased beyond optimal time it led to the formation of p-type defects (oxygen interstitials) which lowered the electron extraction efficiency of the layer and hence resulted in poor deivce efficiency. JV characteristics for three different UV ozone treatment conditions are shown in Figure 1.21.

Lower conductivity of the intrinsic ZnO layer requires low thickness to be deposited on ITO. Such lower thickness imposes problems during large-scale production. Also, it is difficult to achieve lower density of pin holes and uniform coverage. Therefore, doping ZnO with Al, which can increase the conductivity by order of three, provides researchers an opportunity to increase the film thickness while still not losing performance [108].

Ameri *et al.* carried out a comparative study on normal and inverted device structure, with inverted device structure having TiOx as ETL [109]. Their optical modeling and experimental results showed up to 15% higher EQE and J_{sc} in the case of inverted device

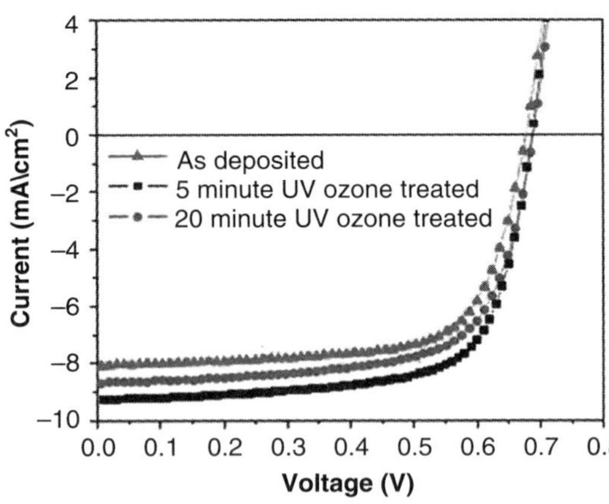

Figure 1.21 J-V curves for PDPP3T:PC_{60}BM inverted solar cell with different UV ozone treatment time on ZnO films. Reproduced with permission from ref. [107].

structure, most likely due to better contact selectivity provided by TiO_x layer leading to reduced surface recombination. Mor *et al.* showed that vertically aligned TiO_2 nanotubes could provide highly ordered architectures by additional interfaces that could be achieved between polymer (donor)-TiO_2 along with donor-acceptor. Improved charge separation yielded higher efficiency [110]. Ko *et al.* studied UV-sensitivity of TiO_x layer and found that bimolecular recombination could be suppressed by lowering resistivity of the layer which can drastically improve the performance of the inverted PSCs [111]. Solution processed doping of TiO_x layer by silver nanoparticles was reported by Nie *et al.* which can improve the conductivity and charge carrier mobility similar to those achieved by UV-illumination [112].

Solution-processed CsF and Cs_2CO_3 could also be used as work function modifier of ITO. Use of Cs_2CO_3 can significantly reduce the work function of ITO from 4.7 to 3.06 eV [113]. Also, Cs_2CO_3-based inverted P3HT:PCBM device has achieved efficiency as high as 4.2% [114] Water soluble conjugated polymer electrolytes such as WPF-6-oxy-F could also be applied as electron transport layer which can reduce the work function of ITO [115].

1.4.2.2 Anode and Hole Transport Layers

PEDOT:PSS has been used as HTL in inverted solar cells. A major challenge in using PEDOT:PSS as HTL in inverted devices is coating on top of active layer due to its hydrophobic nature. Weickert *et al.* showed spray coating could improve homogeneity of the PEDOT:PSS layer [116]. Their method yielded stable devices (>80 days) with PEDOT:PSS providing good Ohmic contact between organic layer and top metal electrode. Huang *et al.* used modified PEDOT:PSS as shown in Figure 1.22 by doping it with D-sorbital which acted like conductive glue in which two devices (top and bottom) were attached together to form transparent solar cell [113]. This method could also help to achieve self-encapsulation technique for highly stable solar cells.

Thermally evaporated n-type metal oxides such as MoO_3 and WO_3 could also be incorporated in inverted device structures to function as efficient hole transport/electron blocking buffer layers [74]. Tao *et al.* used WO_3 and reported that optimum thickness of WO_3 hole selective layer (HSL) is required for better performance of the inverted device [117].

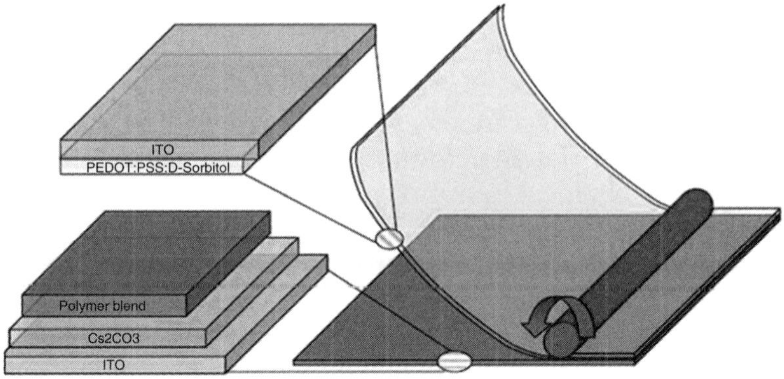

Figure 1.22 Schematic showing modified PEDOT:PSS with D-sorbitol for transparent solar cell. Reproduced with permission from ref. [113].

Gao *et al.* used graphene oxide (GO) as HTL in inverted device structure [118]. Deposition of a thin layer of GO on top of polymer leads to the doping of conjugated polymer that improved the conductivity as high as 3.7 S m^{-1}. Also, 2D structure of GO was shown to prohibit it from penetrating the bulk of conjugated polymers. This surface doping could significantly improve charge transport across the metal/polymer interface.

1.5 Conclusion and Outlook

Commercial viability of OPV devices requires high efficiency and stability in ambient conditions. Performance parameters of the device depend upon choice of materials, efficient light harvesting mechanisms and morphological optimizations. Understanding degradation mechanisms of individual components in OPV devices can help us choose materials that are robust without sacrificing efficiency. Inverted BHJ device architecture seems to be more promising than conventional BHJ devices when it comes to realizing efficient, low cost, flexible alternatives suitable for R2R fabrication of organic polymer solar cells.

Acknowledgement

This work was supported in part by NSF CAREER (ECCS-0950731), NSF EPSCoR (Grant No. 0903804) and the State of South Dakota,

NASA EPSCoR (No. NNX13AD31A), 3M Nontenured Faculty Award, and SDBoR CRGP grant.

References

1. M. Siddiki, J. Li, D. Galipeau, and Q. Qiao, *Energy and Environmental Sciences*, Vol. 3, 867–883, 2010.
2. M.K. Siddiki, S. Venkatesan, D. Galipeau, and Q. Qiao, *ACS Appl. Mater. Interfaces*, Vol. 5, 1279–1286, 2013.
3. M.K. Siddiki, S. Venkatesan, M. Wang, and Q. Qiao, *Sol. Energy Mater. Sol. Cells*, Vol. 108, 225–229, 2013.
4. P. Taranekar, Q. Qiao, J. Jiang, K.S. Schanze, and J.R. Reynolds, *Journal of the American Chemical Society*, Vol. 129, 8958 - 8959, 2007.
5. Y. Xie, Y. Bao, J. Du, C. Jiang, and Q. Qiao, *Phys. Chem. Chem. Phys.*, Vol. 14, 10168–10177, 2012.
6. Y. Xie, Y. Li, L. Xiao, Q. Qiao, R. Dhakal, Z. Zhang, Q. Gong, D. Galipeau, and X. Yan, *The Journal of Physical Chemistry C*, Vol. 114, 14590–14600, 2010.
7. T. Xu, Q. Chen, D.-H. Lin, H.-Y. Wu, C.-F. Lin, and Q. Qiao, *Journal of Photonics for Energy*, Vol. 1, 011107, 2011.
8. T. Xu, and Q. Qiao, *Energy Environ. Sci.*, Vol. 4, 2700–2720, 2011.
9. T. Xu, S. Venkatesan, D. Galipeau, and Q. Qiao, *Solar Energy Materials and Solar Cells*, Vol. submitted and being reviewed, 2011.
10. T. Xu, M. Yan, J. Hoefelmeyer, and Q. Qiao, *RSC Advance*, Vol. 2, 854 - 862, 2011.
11. W. Zhang, H. Wang, B. Chen, X. Bi, S. Venkatesan, Q. Qiao, and S. Yang, *J. Mater. Chem.*, Vol. 22, 24067–24074, 2012.
12. J. Li, M. Yan, Y. Xie, and Q. Qiao, *Energy Environ. Sci.*, Vol. 4, 4276–4283, 2011.
13. S. Shao, K. Zheng, T. Pullerits, and F. Zhang, *ACS Applied Materials & Interfaces*, Vol. 5, 380–385, 2012.
14. F.S.R. Yunfei Zhou, Ying Yuan, Hans-Frieder Schleiermacher, Michael Niggemann, Gerald A. Urban, and Michael Krüger, *Appl. Phys. Lett.*, Vol. 96, 013304, 2010.
15. M.E. Yunfei Zhou, Clemens Veit, Birger Zimmermann, Frank Rauscher, Phenwisa Niyamakom, Seyfullah Yilmaz, Ines Dumsch, Sybille Allard, Ullrich Scherf, Michael Krüger, *Solar Energy Materials & Solar Cells*, Vol. 95, 1232–1237, 2011.
16. C. Chiang, C. Fincher Jr, Y. Park, A. Heeger, H. Shirakawa, E. Louis, S. Gau, and A.G. MacDiarmid, *Physical Review Letters*, Vol. 39, 1098–1101, 1977.
17. L.J. Rothberg, M. Yan, F. Papadimitrakopoulos, M.E. Galvin, E.W. Kwock, and T.M. Miller, *Synthetic Metals*, Vol. 80, 41–58, 1996.
18. C.W. Tang, *Applied Physics Letters*, Vol. 48, 183–185, 1986.
19. N. Sariciftci, L. Smilowitz, A. Heeger, and F. Wudl, *Synthetic Metals*, Vol. 59, 333–352, 1993.
20. J.J.M. Halls, K. Pichler, R.H. Friend, S.C. Moratti, and A.B. Holmes, *Synthetic Metals*, Vol. 77, 277–280, 1996.

21. S. Morita, A.A. Zakhidov, and K. Yoshino, *Solid State Communications*, Vol. 82, 249–252, 1992.
22. G. Yu, K. Pakbaz, and A. Heeger, *Applied Physics Letters*, Vol. 64, 3422–3424, 1994.
23. Z. Bao, A. Dodabalapur, and A.J. Lovinger, *Applied Physics Letters*, Vol. 69, 4108–4110, 1996.
24. G. Li, V. Shrotriya, J. Huang, Y. Yao, T. Moriarty, K. Emery, and Y. Yang, *Nature materials*, Vol. 4, 864–868, 2005.
25. S.H. Park, A. Roy, S. Beaupré, S. Cho, N. Coates, J.S. Moon, D. Moses, M. Leclerc, K. Lee, and A.J. Heeger, *Nature Photonics*, Vol. 3, 297–302, 2009.
26. Y. Liang, Y. Wu, D. Feng, S.-T. Tsai, H.-J. Son, G. Li, and L. Yu, *Journal of the American Chemical Society*, Vol. 131, 56–57, 2008.
27. S. Albrecht, S. Schäfer, I. Lange, S. Yilmaz, I. Dumsch, S. Allard, U. Scherf, A. Hertwig, and D. Neher, *Org. Electron.*, Vol. 13, 615–622, 2012.
28. C. Deibel, and V. Dyakonov, *Reports on Progress in Physics*, Vol. 73, 096401, 2010.
29. W. Chen, M. Nikiforov, and S. Darling, *Energy & Environmental Science*, Vol., 2012.
30. S.R. Cowan, R. Street, S. Cho, and A. Heeger, *Physical Review B*, Vol. 83, 035205, 2011.
31. C.J. Brabec, S. Gowrisanker, J.J.M. Halls, D. Laird, S. Jia, and S.P. Williams, *Advanced Materials*, Vol. 22, 3839–3856, 2010.
32. A. Jain, and A. Kapoor, *Solar Energy Materials and Solar Cells*, Vol. 86, 197–205, 2005.
33. Y.-W. Su, S.-C. Lan, and K.-H. Wei, *Materials Today*, Vol. 15, 554–562, 2012.
34. D. Chen, F. Liu, C. Wang, A. Nakahara, and T.P. Russell, *Nano Letters*, Vol. 11, 2071–2078, 2011.
35. M. Campoy-Quiles, T. Ferenczi, T. Agostinelli, P.G. Etchegoin, Y. Kim, T.D. Anthopoulos, P.N. Stavrinou, D.D.C. Bradley, and J. Nelson, *Nature materials*, Vol. 7, 158–164, 2008.
36. D. Chen, A. Nakahara, D. Wei, D. Nordlund, and T.P. Russell, *Nano Lett.*, Vol. 11, 561–567, 2010.
37. L.-M. Chen, Z. Xu, Z. Hong, and Y. Yang, *Journal of Materials Chemistry*, Vol. 20, 2575–2598, 2010.
38. H.-L. Yip, and A.K.-Y. Jen, *Energy & Environmental Science*, Vol. 5, 5994–6011, 2012.
39. C. Tengstedt, W. Osikowicz, W.R. Salaneck, I.D. Parker, C.-H. Hsu, and M. Fahlman, *Applied Physics Letters*, Vol. 88, 053502–053502-053503, 2006.
40. P. Sehati, S. Braun, L. Lindell, X. Liu, L.M. Andersson, and M. Fahlman, *Selected Topics in Quantum Electronics, IEEE Journal of*, Vol. 16, 1718–1724, 2010.
41. V.D. Mihailetchi, P.W.M. Blom, J.C. Hummelen, and M.T. Rispens, *Journal of Applied Physics*, Vol. 94, 6849–6854, 2003.
42. S.K. Hau, K.M. O"Malley, C. You-Jung, H.-L. Yip, H. Ma, and A.K.Y. Jen, *Selected Topics in Quantum Electronics, IEEE Journal of*, Vol. 16, 1665–1675, 2010.
43. H.-L. Yip, S.K. Hau, and A.K.-Y. Jen, *Solar & Alternative Energy*, Vol., 2009.
44. D.S. Germack, C.K. Chan, B.H. Hamadani, L.J. Richter, D.A. Fischer, D.J. Gundlach, and D.M. DeLongchamp, *Applied Physics Letters*, Vol. 94, 233303–233303, 2009.

45. J. Subbiah, C.M. Amb, J.R. Reynolds, and F. So, *Solar Energy Materials and Solar Cells*, Vol. 97, 97–101, 2012.
46. D.-H. Ko, J.R. Tumbleston, A. Gadisa, M. Aryal, Y. Liu, R. Lopez, and E.T. Samulski, *Journal of Materials Chemistry*, Vol. 21, 16293–16303, 2011.
47. A. Hadipour, D. Cheyns, P. Heremans, and B.P. Rand, *Advanced Energy Materials*, Vol. 1, 930–935, 2011.
48. A.K. Pandey, P.E. Shaw, I.D.W. Samuel, and J.-M. Nunzi, *Applied Physics Letters*, Vol. 94, 103303–103303, 2009.
49. J. Gilot, I. Barbu, M.M. Wienk, and R.A. Janssen, *Applied Physics Letters*, Vol. 91, 113520–113520-113523, 2007.
50. J.-L. Wu, F.-C. Chen, Y.-S. Hsiao, F.-C. Chien, P. Chen, C.-H. Kuo, M.H. Huang, and C.-S. Hsu, *ACS Nano*, Vol. 5, 959–967, 2011.
51. S.V. Evan Ngo, David Galipeau and Qiquan Qiao, *IEEE Transactions on Electron Devices*, Vol., 2013.
52. S. Chambon, A. Rivaton, J.-L. Gardette, M. Firon, and L. Lutsen, *Journal of Polymer Science Part A: Polymer Chemistry*, Vol. 45, 317–331, 2007.
53. S. Chambon, A. Rivaton, J.-L. Gardette, and M. Firon, *Journal of Polymer Science Part A: Polymer Chemistry*, Vol. 47, 6044–6052, 2009.
54. M. Jørgensen, K. Norrman, S.A. Gevorgyan, T. Tromholt, B. Andreasen, and F.C. Krebs, *Advanced Materials*, Vol. 24, 580–612, 2012.
55. M. Manceau, E. Bundgaard, J.E. Carlé, O. Hagemann, M. Helgesen, R. Søndergaard, M. Jørgensen, and F.C. Krebs, *Journal of Materials Chemistry*, Vol. 21, 4132–4141, 2011.
56. Y. Xia, J. Tong, B. Li, C. Wang, and H. Liu, *Journal of Polymer Science Part A: Polymer Chemistry*, Vol. 49, 2969–2979, 2011.
57. A. Kumar, Z. Hong, S. Sista, and Y. Yang, *Advanced Energy Materials*, Vol. 1, 124–131, 2011.
58. M. Manceau, S. Chambon, A. Rivaton, J.-L. Gardette, S. Guillerez, and N. Lemaître, *Solar Energy Materials and Solar Cells*, Vol. 94, 1572–1577, 2010.
59. S. Ebadian, B. Gholamkhass, S. Shambayati, S. Holdcroft, and P. Servati, *Solar Energy Materials and Solar Cells*, Vol. 94, 2258–2264, 2010.
60. J. Liu, E.N. Kadnikova, Y. Liu, M.D. McGehee, and J.M. Fréchet, *Journal of the American Chemical Society*, Vol. 126, 9486–9487, 2004.
61. S.A. Gevorgyan, and F.C. Krebs, *Chemistry of materials*, Vol. 20, 4386–4390, 2008.
62. F.C. Krebs, and H. Spanggaard, *Chemistry of materials*, Vol. 17, 5235–5237, 2005.
63. M. Jørgensen, K. Norrman, and F.C. Krebs, *Solar Energy Materials and Solar Cells*, Vol. 92, 686–714, 2008.
64. K. Norrman, S.A. Gevorgyan, and F.C. Krebs, *ACS Applied Materials & Interfaces*, Vol. 1, 102–112, 2008.
65. K. Norrman, N.B. Larsen, and F.C. Krebs, *Solar Energy Materials and Solar Cells*, Vol. 90, 2793–2814, 2006.
66. S. Schäfer, A. Petersen, T.A. Wagner, R. Kniprath, D. Lingenfelser, A. Zen, T. Kirchartz, B. Zimmermann, U. Würfel, X. Feng, and T. Mayer, *Physical Review B*, Vol. 83, 165311, 2011.
67. Y. Kanai, T. Matsushima, and H. Murata, *Thin Solid Films*, Vol. 518, 537–540, 2009.

68. D. Gao, M.G. Helander, Z.B. Wang, D.P. Puzzo, M.T. Greiner, and Z.H. Lu, *Advanced Materials*, Vol. 22, 5404–5408, 2010.
69. J. Li, S. Kim, S. Edington, J. Nedy, S. Cho, K. Lee, A.J. Heeger, M.C. Gupta, and J.T. Yates Jr, *Solar Energy Materials and Solar Cells*, Vol. 95, 1123–1130, 2011.
70. S.R. Ferreira, P. Lu, Y.-J. Lee, R.J. Davis, and J.W. Hsu, *The Journal of Physical Chemistry C*, Vol. 115, 13471–13475, 2011.
71. A. Puetz, T. Stubhan, M. Reinhard, O. Loesch, E. Hammarberg, S. Wolf, C. Feldmann, H. Kalt, A. Colsmann, and U. Lemmer, *Solar Energy Materials and Solar Cells*, Vol. 95, 579–585, 2011.
72. H. Kim, S. Nam, H. Lee, S. Woo, C.-S. Ha, M. Ree, and Y. Kim, *The Journal of Physical Chemistry C*, Vol. 115, 13502–13510, 2011.
73. C. Girotto, E. Voroshazi, D. Cheyns, P. Heremans, and B.P. Rand, *ACS Applied Materials & Interfaces*, Vol. 3, 3244–3247, 2011.
74. J.-S. Huang, C.-Y. Chou, and C.-F. Lin, *IEEE electron device letters*, Vol. 31, 332–334, 2010.
75. S.K. Hau, H.-L. Yip, O. Acton, N.S. Baek, H. Ma, and A.K.-Y. Jen, *Journal of Materials Chemistry*, Vol. 18, 5113–5119, 2008.
76. M. Logdlund, and J.L. Bredas, *The Journal of Chemical Physics*, Vol. 101, 4357–4364, 1994.
77. Y.S. Eo, H.W. Rhee, B.D. Chin, and J.-W. Yu, *Synthetic Metals*, Vol. 159, 1910–1913, 2009.
78. C.J. Brabec, A. Cravino, D. Meissner, N.S. Sariciftci, T. Fromherz, M.T. Rispens, L. Sanchez, and J.C. Hummelen, *Advanced Functional Materials*, Vol. 11, 374–380, 2001.
79. S.E. Shaheen, C.J. Brabec, N.S. Sariciftci, F. Padinger, T. Fromherz, and J.C. Hummelen, *Applied Physics Letters*, Vol. 78, 841–843, 2001.
80. D. Gao, M.G. Helander, Z.-B. Wang, D.P. Puzzo, M.T. Greiner, and Z.-H. Lu, *Advanced Materials*, Vol. 22, 5404–5408, 2010.
81. J. Huang, W.-J. Hou, J.-H. Li, G. Li, and Y. Yang, *Applied Physics Letters*, Vol. 89, 133509–133509-133503, 2006.
82. F.-C. Chen, J.-L. Wu, S.S. Yang, K.-H. Hsieh, and W.-C. Chen, *Journal of Applied Physics*, Vol. 103, 103721–103721-103725, 2008.
83. J.Y. Kim, S.H. Kim, H.H. Lee, K. Lee, W. Ma, X. Gong, and A.J. Heeger, *Advanced Materials*, Vol. 18, 572–576, 2006.
84. H.L. Yip, S.K. Hau, N.S. Baek, H. Ma, and A.K.Y. Jen, *Advanced Materials*, Vol. 20, 2376–2382, 2008.
85. J. Gilot, I. Barbu, M.M. Wienk, and R.A.J. Janssen, *Appl. Phys. Lett.*, Vol. 91, 113520, 2007.
86. H.-L. Yip, S.K. Hau, N.S. Baek, H. Ma, and A.K.Y. Jen, *Advanced Materials*, Vol. 20, 2376–2382, 2008.
87. S.-I. Na, Y.-J. Noh, S.-Y. Son, T.-W. Kim, S.-S. Kim, S. Lee, and H.-I. Joh, *Applied Physics Letters*, Vol. 102, 043304–043304-043305, 2013.
88. T.M. Barnes, M.O. Reese, J.D. Bergeson, B.A. Larsen, J.L. Blackburn, M.C. Beard, J. Bult, and J. van de Lagemaat, *Advanced Energy Materials*, Vol. 2, 353–360, 2012.
89. S. Heusing, P. de Oliveira, E. Kraker, A. Haase, C. Palfinger, and M. Veith, *Thin Solid Films*, Vol. 518, 1164–1169, 2009.

90. S.I. Na, S.S. Kim, J. Jo, and D.Y. Kim, *Advanced Materials*, Vol. 20, 4061–4067, 2008.
91. C.H. Chung, T.B. Song, B. Bob, R. Zhu, H.S. Duan, and Y. Yang, *Advanced Materials*, Vol. 24, 5499–5504, 2012.
92. Z. Wu, Z. Chen, X. Du, J.M. Logan, J. Sippel, M. Nikolou, K. Kamaras, J.R. Reynolds, D.B. Tanner, and A.F. Hebard, *Science*, Vol. 305, 1273–1276, 2004.
93. K.S. Kim, Y. Zhao, H. Jang, S.Y. Lee, J.M. Kim, K.S. Kim, J.-H. Ahn, P. Kim, J.-Y. Choi, and B.H. Hong, *Nature*, Vol. 457, 706–710, 2009.
94. Y.-H. Kim, S.-H. Lee, J. Noh, and S.-H. Han, *Thin Solid Films*, Vol. 510, 305–310, 2006.
95. Y.-H. Kim, S.-H. Lee, J. Noh, and S.-H. Han, *Thin Solid Films*, Vol. 510, 305–310, 2006.
96. T. Xiao, W. Cui, J. Anderegg, J. Shinar, and R. Shinar, *Organic Electronics*, Vol. 12, 257–262, 2011.
97. V. Shrotriya, G. Li, Y. Yao, C.-W. Chu, and Y. Yang, *Applied Physics Letters*, Vol. 88, 073508–073508-073503, 2006.
98. Y. Sun, C.J. Takacs, S.R. Cowan, J.H. Seo, X. Gong, A. Roy, and A.J. Heeger, *Advanced Materials*, Vol. 23, 2226–2230, 2011.
99. M.D. Irwin, J.D. Servaites, D.B. Buchholz, B.J. Leever, J. Liu, J.D. Emery, M. Zhang, J.-H. Song, M.F. Durstock, and A.J. Freeman, *Chemistry of materials*, Vol. 23, 2218–2226, 2011.
100. K.X. Steirer, J.P. Chesin, N.E. Widjonarko, J.J. Berry, A. Miedaner, D.S. Ginley, and D.C. Olson, *Organic Electronics*, Vol. 11, 1414–1418, 2010.
101. Z.a. Tan, L. Li, C. Cui, Y. Ding, Q. Xu, S. Li, D. Qian, and Y. Li, *The Journal of Physical Chemistry C*, Vol. 116, 18626–18632, 2012.
102. E. Özkan, and F.Z. Tepehan, *Solar Energy Materials and Solar Cells*, Vol. 68, 265–277, 2001.
103. L.-C. Huang, H.-W. Liu, C.-W. Liang, T.-R. Chou, L. Wang, and C.-Y. Chao, *Soft Matter*, Vol. 8, 1467–1472, 2012.
104. Z. He, C. Zhong, S. Su, M. Xu, H. Wu, and Y. Cao, *Nature Photonics*, Vol. 6, 593–597, 2012.
105. M. White, D. Olson, S. Shaheen, N. Kopidakis, and D.S. Ginley, *Applied Physics Letters*, Vol. 89, 143517–143517-143513, 2006.
106. C.E. Small, S. Chen, J. Subbiah, C.M. Amb, S.-W. Tsang, T.-H. Lai, J.R. Reynolds, and F. So, *Nature Photonics*, Vol. 6, 115–120, 2011.
107. Prajwal Adhikary, Purna P Maharjan, David Galipeau, and Qiquan Qiao, *IEEE Transactions on Electron Devices*, Vol., 2013.
108. T. Stubhan, H. Oh, L. Pinna, J. Krantz, I. Litzov, and C.J. Brabec, *Organic Electronics*, Vol. 12, 1539–1543, 2011.
109. T. Ameri, G. Dennler, C. Waldauf, P. Denk, K. Forberich, M.C. Scharber, C.J. Brabec, and K. Hingerl, *Journal of Applied Physics*, Vol. 103, 084506–084506-084506, 2008.
110. G.K. Mor, K. Shankar, M. Paulose, O.K. Varghese, and C.A. Grimes, *Applied Physics Letters*, Vol. 91, 152111–152113, 2007.
111. D.-H. Ko, J.R. Tumbleston, M.-R. Ok, H. Chun, R. Lopez, and E. Samulski, *Journal of Applied Physics*, Vol. 108, 083101–083101-083106, 2010.

112. W. Nie, R.C. Coffin, and D.L. Carroll, *International Journal of Photoenergy*, Vol. 2013, 2013.
113. J. Huang, G. Li, and Y. Yang, *Advanced Materials*, Vol. 20, 415–419, 2008.
114. H.-H. Liao, L.-M. Chen, Z. Xu, G. Li, and Y. Yang, *Applied Physics Letters*, Vol. 92, 173303, 2008.
115. G. Jo, S.-I. Na, S.-H. Oh, S. Lee, T.-S. Kim, G. Wang, M. Choe, W. Park, J. Yoon, and D.-Y. Kim, *Applied Physics Letters*, Vol. 97, 213301, 2010.
116. J. Weickert, H. Sun, C. Palumbiny, H.C. Hesse, and L. Schmidt-Mende, *Solar Energy Materials and Solar Cells*, Vol. 94, 2371–2374, 2010.
117. C. Tao, S. Ruan, G. Xie, X. Kong, L. Shen, F. Meng, C. Liu, X. Zhang, W. Dong, and W. Chen, *Applied Physics Letters*, Vol. 94, 043311–043311-043313, 2009.
118. Y. Gao, H.L. Yip, K.S. Chen, K.M. O'Malley, O. Acton, Y. Sun, G. Ting, H. Chen, and A.K.Y. Jen, *Advanced Materials*, Vol. 23, 1903–1908, 2011.

2

Carbazole-Based Organic Dyes for Dye-Sensitized Solar Cells: Role of Carbazole as Donor, Auxiliary Donor and π-linker

A. Venkateswararao and K. R. Justin Thomas*

Organic Materials Laboratory, Department of Chemistry, Indian Institute of Technology, Roorkee, India

Abstract

Carbazole has been widely used as a building block in the construction of organic sensitizers suitable for use in dye-sensitized solar cells (DSSC). Due to its electron-rich nature, it is often used as donor, auxiliary donor or a bridging segment between the donor and acceptor. Carbazole can be functionalized at various positions (C1, C2, C3, C6, C7, C8, and N9) with ease and offers a wide variety of modes for linking it with other fragments in a dye. In this review, we survey the synthesis, optical properties and DSSC performance of the carbazole-containing organic dyes. We illustrate with examples that the nature and connecting position of carbazole play a significant role in determining the functional efficacy of the dyes. Careful incorporation of carbazole in a sensitizer tunes the HOMO and LUMO energy levels to produce high conversion efficiency. Generally, the organic dyes containing carbazole as donor (8.3%) or auxiliary donor (8.5%) exhibited better efficiency than the dyes featuring carbazole as π-linker (7.2%). Also, it has been observed that panchromatic absorbing dyes can be developed by incorporation of low bandgap chromophores to enhance the power conversion efficiency in these classes of dyes.

Keywords: Energy conversion, dye-sensitized solar cells, organic sensitizers, carbazole, absorption, electrochemistry, TDDFT computations

**Corresponding author*: krjt8fcy@iitr.ernet.in, krjt8fcy@yahoo.co.in

Atul Tiwari, Rabah Boukherroub, and Maheshwar Sharon (eds.) Solar Cell Nanotechnology, (41–96) 2014 © Scrivener Publishing LLC

2.1 Introduction

Growing demand for clean energy production from renewable energy sources in the last few decades has led to newer technological breakthroughs such as photovoltaics. Among the photovoltaic methods, those that have been developed to harvest the solar energy, heterojunction solar cells, bulk heterojunction solar cells [1] and dye-sensitized solar cells (DSSC) [2] are notable due to their promising characteristics and unlimited practical applications. Though the silicon-based solar cells [3] are commercially available, dye-sensitized solar cells have received intense interest due to their potential advantages like low production cost, transparency, multicolor options, flexibility, light weight and short energy payback time [4]. On the contrary, the bulk heterojunction solar cells have not been demonstrated with reasonable efficiency suitable for practical device manufacturing. They suffer largely due to the limited number of *n*-type organic semiconductors available.

DSSC is typically composed of five major components (Figure 2.1), namely, (i) transparent conducting oxide (TCO), (ii) a mesoporous semiconductor metal oxide (such as TiO_2 or ZnO) film, (iii) a sensitizer, (iv) an electrolyte layer (I_2/I^- couple, Co(II)/Co(III) couple or Fc/Fc^+ couple), (v) counter electrode (platinum or carbon coated glass substrate). Five basic processes which occur during the operation of

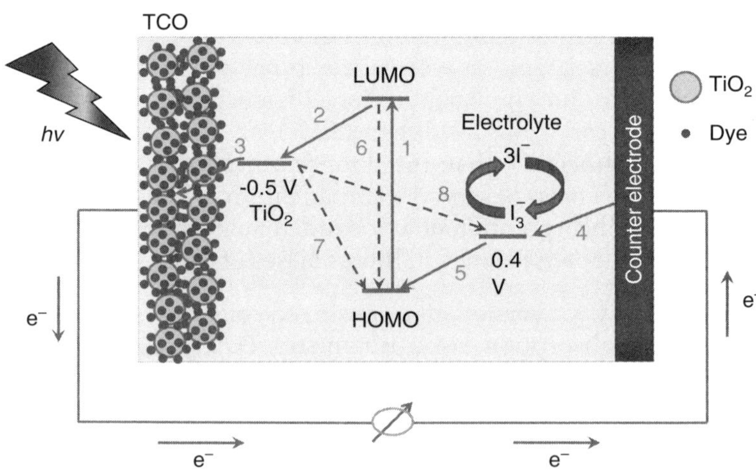

Figure 2.1 Schematic representation of DSSC showing different components and the processes involved during the operation.

the DSSC are instrumental in governing the efficiency of the device. They are: (1) absorption of solar light by the dye anchored on TiO_2, (2) injection of electrons into the conduction band (CB) of TiO_2, (3) charge collection at the anode, (4) regeneration of redox couple by electron injection from cathode, (5) reduction/regeneration of the oxidized dye by redox couple, (6) relaxation of the excited dye to ground state by non-radiative processes such as internal conversion, aggregation-induced quenching, etc., (7) recombination of electrons from the CB of TiO_2 with the oxidized dye, (8) recombination of electrons in CB of TiO_2 with the electrolyte. The first five processes complete the cyclic flow of electrons and lead to the generation of electricity. The last three processes are detrimental for the efficiency of the DSSC. In order to avoid such undesirable processes the properties of dyes must be fine tuned to possess favorable HOMO and LUMO levels [5–7] and structural features to impede unwanted electron passage between the components in DSSC. So the structure-property relationship evaluation for the organic dyes is of utmost importance to further the research outcome in this frontier area to the application level.

Though several chromophores such as triarylamine [8], coumarin [9–11], fluorene [12–14], phenothiazine [15, 16], benzothiadiazole [17, 18], benzotriazole [18, 19], indoline [20, 21], etc., have been used in the assembly of organic dyes, carbazole-based organic materials are exceptionally interesting due to their promising properties such as relatively low oxidation potentials, thermal and photostability and hole-conducting capabilities [22, 23]. Moreover, a variety of nuclear positions available for functionalization expand the scope and utility of carbazole-derived materials. Carbazole containing molecular materials have been successfully demonstrated for application as hole-transporting emitters in organic light-emitting diodes (OLED) [24–28], organic semiconductor in thin-film transistors (TFT) [29–31] and electron donors in photovoltaics [32, 33]. The first report on the use of carbazole-based organic dyes came in 2006 and was authored by Hara *et al.* Several dyes featuring carbazole as either donor or π-linker have been synthesized and demonstrated as sensitizers in DSSC [34]. Though several recent reviews devoted to the development of organic dyes are available in literature, a complete structure-property correlation on dyes featuring carbazole is not yet made. In this review, we collect the information available on this emerging class of dyes featuring carbazole either as donor or bridging unit and try to evolve design principles one may adopt to arrive at promising materials.

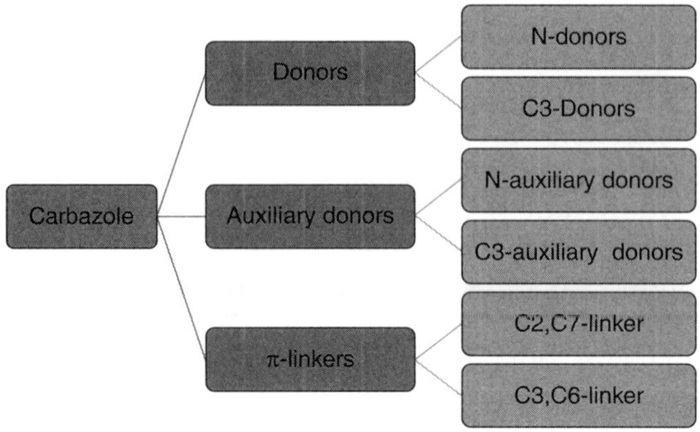

Figure 2.2 Categorization based on the functional role of carbazole in organic dyes suitable for DSSC.

Due to the electron richness of the system it has been used as a donor in dyes. Monofunctionaliztions both at C3 and N9 have been used to tether the carbazole as donor while difunctionalizations at C2, C7 and C3, C6 for exploiting it as bridging segment. Due to its electron donating character carbazole normally strengthens the donor-acceptor interaction in dipolar compounds and benefits the optical properties such as absorption wavelength and molar extinction coefficient. It is interesting to examine the effect of structural differences arising due to functionalization at various nuclear positions of carbazole on the vital properties such as absorption and oxidation propensity which play a major role in determining the performance of DSSCs. Depending on the role of carbazole in an organic dye, we classify such organic dyes into three categories: donors, auxiliary donors, π-linkers. Further within a group, they are classified, for convenience of treatment, according to their nuclear substitution pattern (Figure 2.2).

2.2 Carbazole as a Donor for Dye-Sensitized Solar Cells

Electron-donating ability of carbazole is well established though it is less basic than most of the analogous triarylmines [35]. Due to this, carbazole-based materials are used as hole-transporting emitters in organic light-emitting diodes [24–28]. Additionally, most of the carbazole derivatives inherit high-lying triplet state (T_1) and wide bandgap. They are beneficial for them to serve as hosts for deep-blue-emitting singlet as well as triplet emitters [36]. Though there is a tradeoff in the absorption properties of carbazole-based dyes, when compared to the analogous triarylamine-based dyes, the former dyes due to rigidity inhibit non-radiative relaxation pathways in the excited state and planarity ensures facile charge migration toward acceptor due to favorable orbital overlap across the conjugation. Though carbazole can be functionalized at several nuclear positions, substitutions at C3 and N9 have been found to be successful in extracting the electrons from carbazole.

2.2.1 Carbazole as Donor via C3-Position

The general molecular configuration of the dyes belonging to this category is displayed in Figure 2.3. The substitution of carbazole at 3-position was often accomplished via the bromination of carbazole by using *N*-bromosuccinimide in suitable solvents such as dimethylformamide (DMF) [37], tetrahydrofuran (THF) [38] or dichloromethane (DCM) [39]. *N*-alkylation or arylation is useful to promote the solubility or enhance the donor strength. *N*-alkylation on carbazole is achieved by simple nucleophilic substitution with alkyl halide in the presence of a base aided by a phase transfer catalyst [40]. Though the *N*-arylation can be achieved by simple nucleophilic substitution with aryl fluorides [41], less activated aryl substrates required harsh conditions involving Ullmann protocol [42] or palladium-catalyzed C-N coupling reaction strategies [43].

Simple carbazole-based dyes (**D1-D3** in Figure 2.4) containing carbazole as donor and cyanoacrylic acid as acceptor have been reported by Kim *et al.* [44] They have systematically investigated the effect of a number of chromophores on the performance of the DSSCs. The short circuit current density (J_{sc}) of the dyes increased from **D1** (1.29 mA cm^{-2}) to **D2** (2.31 mA cm^{-2}) due to the increase of molar extinction

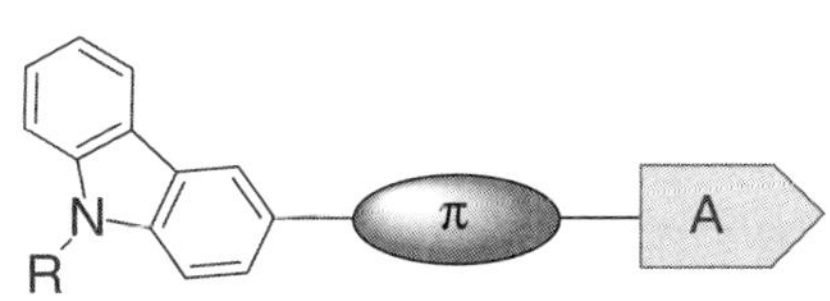

Figure 2.3 Molecular composition of organic dyes with carbazole as donor via C3-position.

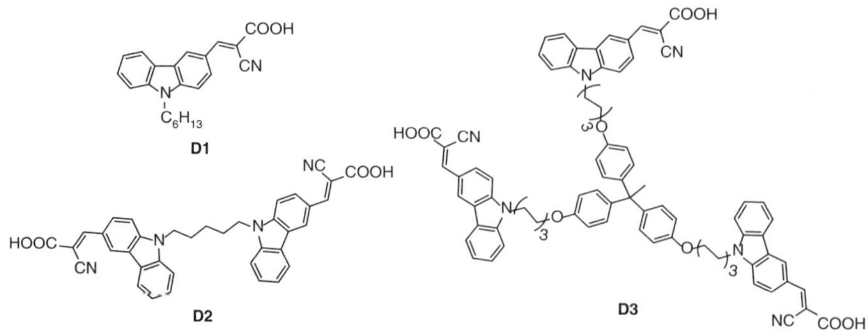

Figure 2.4 Structures of the simple carbazole dyes having multiple light-harvesting chromophores.

coefficient by increasing the number of carbazole units, thereby increasing the light harvesting efficiency. The first excited state potential increased by increasing the number of chromophore units which facilitated the facile electron injection into conduction band (CB) of TiO_2. The open circuit voltage of the devices increased monotonously from 540 mV (**D1**) to 715 mV (**D3**) due to the suppression of charge recombination on introduction of a bulky structure and negative shift in Fermi level of TiO_2 caused by the accumulation of electron density in the conduction band of TiO_2 by efficient electron injection.

Oligothiophene units have been found to act as an effective conjugation pathway in the organic dyes possessing donor-acceptor architecture. By combining carbazole donor with oligothiophene bridges, Ko *et al.* synthesized several dyes (**D4–D9** in Figure 2.5) showing interesting optical and electrochemical trends [45]. Their dyes showed variations in *N*-substituent, conjugation pathway and acceptor unit. Firstly, *N*-substituents such as 9,9-dimethylfluoren-2-yl or 4-(2,2-diphenylvinyl)phenyl group do not affect the nature of the frontier orbitals significantly. The HOMO of the dyes is mainly located on the carbazole and thiophene segments. However, electron-rich *N*-substituents lowered the HOMO energy level and facilitated the regeneration of oxidized dyes. In addition, the absorption of the dyes are red-shifted when compared to the simple dyes without thiophene in the conjugation (**D1** and **D2**). All these favorable properties contributed to the enhancement of device parameters. It is also interesting to note that although the substitution of cyanoacrylic acid unit (**D4** and **D5**) with rhodanine group (**D6** and **D7**) improved the absorption characteristics, it did not produce good photocurrent because of less favored electron injection from dye to CB of TiO_2. The dyes **D8** and **D9** generated

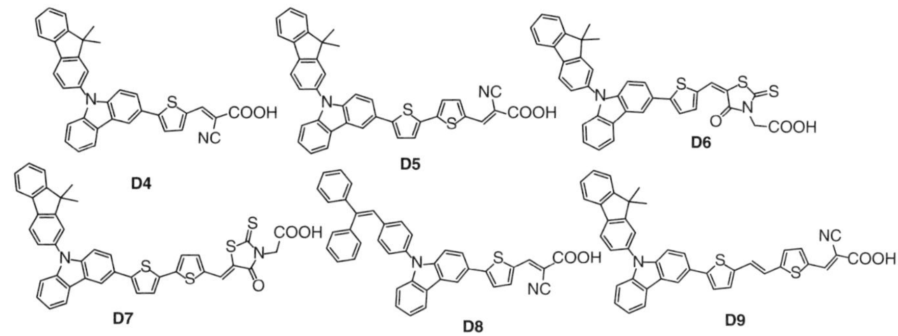

Figure 2.5 Dyes (**D4–D9**) containing carbazole as C3-donor, oligothiophene conjugation and various *N*-substituents.

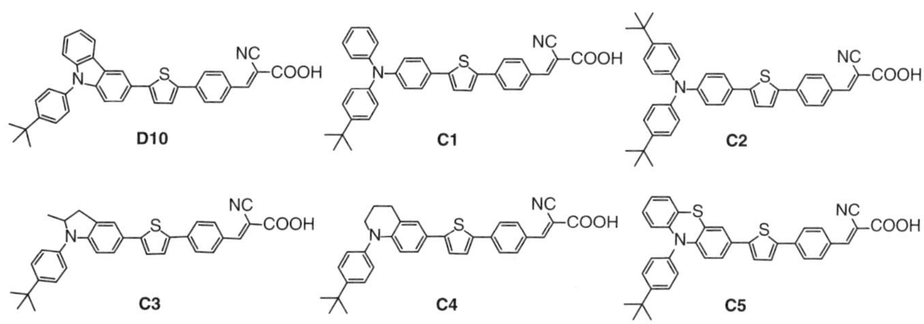

Figure 2.6 Organic dyes containing carbazole, phenothiazine, triphenylamine, indoline or tetrahydroquinoline donor.

lower photocurrent than **D4** and **D5**, probably arising due to the lowered LUMO of former compounds which reduce the electron injection efficiency. Slightly better device efficiency realized for **D8** when compared to **D9** may be attributed to the competing *cis-trans* isomerization pathways present for the dyes containing vinylene linkers.

Chow *et al.* reported a series of dyes featuring different donors such as triphenylamine, indoline, tetrahydroquinoline, carbazole and phenothiazine [46]. All the dyes possess 4-*tert*-butylphenyl unit as a common group. The dye **D10** with a carbazole unit as a donor displayed better photon-to-electron conversion efficiency (6.72%) among all dyes. The success of the dye **D10** in DSSC is attributed to the red-shifted absorption displayed by the dye on adsorption on TiO_2. It is worth noting here that **D10** displayed the blue-shifted absorption when compared to other dyes. Addition of CDCA as a co-adsorbent did not improve the device performance. All these observations indicate the absence of aggregation at the surface of TiO_2. On comparison with **D5**, irrespective of the variation in *N*-substitution, the replacement of thiophene in **D5** with phenyl unit in **D10** produced higher device efficiency. This is probably due to the higher LUMO level of **D10** which benefits electron injection and leads to accumulation of electrons at CB of TiO_2.

Chen *et al.* designed and synthesized two dyes **D11** and **D12** (Figure 2.7) based on *N-tert*-butyl phenyl functionalized 3-*tert*-butyl-carbazole [47]. The molecules were designed to unravel the structural elements required to suppress the intermolecular interactions at the surface of TiO_2 which would retard the aggregation propensity. The efficiency of **D12** (3.75%) was slightly higher than **D11** (3.62%) due to the presence of hexyl chains on thiophene in **D12**. This suppresses the electron transfer from TiO_2 to redox electrolyte and results in the increase of V_{oc} from 570 mV to 660 mV. However, the marginal increase in efficiency is due to the poor absorption characteristics of **D12** triggered by the non-planar conjugation pathway which led to small photocurrent generation. So it appears that structural tweaking for the suppression of electron recombination as demonstrated for **D12** may not be ideal. Also, the efficiency of **D11** is less than that observed for **D5** and **D10** due to the presence of low-lying LUMO in **D11** which shrinks the gap between the CB of TiO_2 and LUMO of the dye and results in low electron injection efficiency.

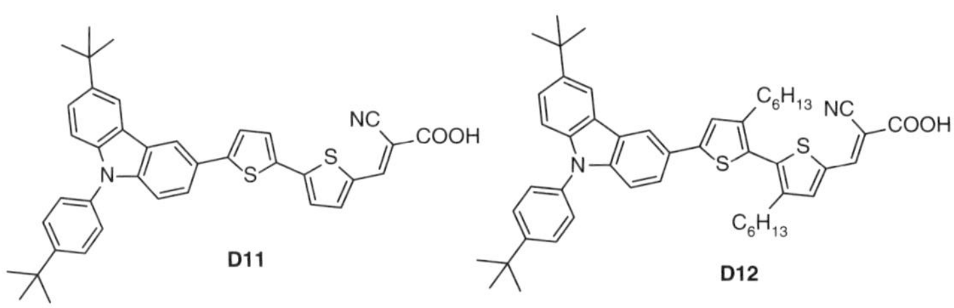

Figure 2.7 Carbazole dyes with bithiophene bridge.

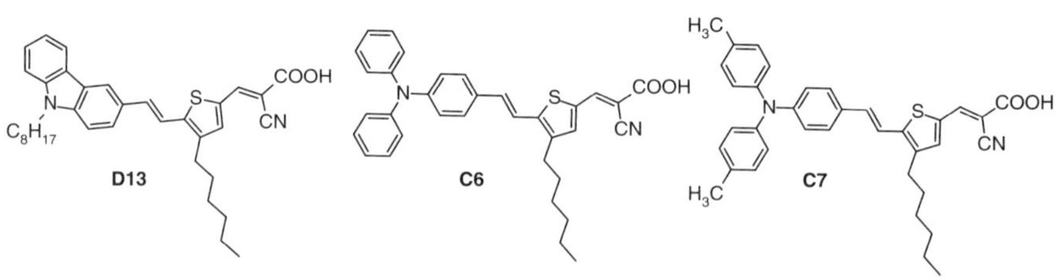

Figure 2.8 Dyes containing 3-vinylcarbazole unit and other comparable systems.

Bin *et al.* synthesized three dyes featuring 9-octylcarbazole, triphenylamine and dimethyl substituted triphenylamine as donor, vinylene 3-hexylthiophene as spacer and cyanoacrylic acid as acceptor [48]. The dye **D13** possessing carbazole framework as donor showed exceptional efficiency in DSSC (6.33%) due to high V_{oc} (700 mV). The hexyl group on thiophene helps to suppress the molecular interactions and keeps the electrolyte away from the TiO_2 layer. The **D13** showed higher efficiency (η = 6.33%) than the similar compound with triphenylamine donor (**C6**; η = 5.48%). However, **D13** showed inferior efficiency compared to (*E*)-2-cyano-3-(5-(4-(di-*p*-tolylamino) styryl)-4-hexylthiophen-2-yl)acrylic acid (**C7**; η = 6.72%). The higher efficiency observed for **C7** is attributed to the red-shifted absorption which leads to higher IPCE at longer wavelength region. The methyl groups in **C7** increase the electron richness of the donor and facilitate the donor-acceptor interaction in the dye. The benefit of vinyl group in these dyes is understandable on comparing the optical properties of these dyes with **D4**. Introduction of vinyl bridge elongated the conjugation and produced much intense and red-shifted absorption profile for **D13** (Table 2.1).

Yamamoto *et al.* reported a set of dyes (Figure 2.9) with carbazole, indoline derivative as donor, acceptor of cyanoacrylic acid connected by dithienothiophene linker [49]. Contradictory to the previously mentioned observations, the carbazole-based dyes showed relatively less efficiency in DSSC (Table 2.1). The bandgap of the carbazole dyes are high compared to related indoline dyes (Figure 2.9) which led to poor absorption in the longer wavelength region.

Grätzel *et al.* reported three dyes containing carbazole or phenothiazine donors (Figure 2.10) [50]. A dye, **D15**, having 4,4-didodecyl-4*H*-cyclopenta[1,2-*b*:5,4-*b*′]dithiophene as a linker displayed longer wavelength absorption peaking at 532 nm with a very high molar extinction coefficient (57500 M^{-1} cm^{-1}). The carbazole dye **D15** exhibited higher efficiency (7.50%) when compared to a similar dye (7.00%) containing phenothiazine as a donor (**C10**). This discrepancy is attributed to the higher molar absorption coefficient realized for the former (**D15**) which led to larger photocurrent. When compared with **D14**, the performance of **D15** is better due to the presence of low lying HOMO in the latter which facilitates the dye regeneration.

The role of alkyl chains present in the conjugation pathway in retarding the recombination of the injected electrons in the

Table 2.1 Optical, electrochemical and photovoltaic performance parameters of dyes containing carbazole as donor.

Dye	λ_{max}, nm (ε_{max}, M^{-1} cm^{-1})	E_{ox}, V (*vs* NHE)	E_{ox}^{*}, V (*vs* NHE)	J_{SC} (mA cm^{-2})	V_{OC} (mV)	*ff*	η (%)	*Ref*
D1	349 (21406)	1.44	−1.44	1.29	540	0.60	0.42	[44]
D2	352 (38344)	1.40	−1.45	1.97	565	0.62	0.69	[44]
D3	350 (63902)	1.33	−1.53	2.31	715	0.65	1.07	[44]
D4	411 (27000)	1.52[a]	−1.13[a]	9.83	740	0.70	5.02	[45]
D5	435 (30060)	1.29[a]	−1.13[a]	11.50	680	0.66	5.15	[45]
D6	469 (12300)	1.54[a]	−0.79[a]	3.76	620	0.77	1.79	[45]
D7	489 (6012)	1.64[a]	−0.71[a]	1.39	550	0.73	0.55	[45]
D8	406 (21500)	1.58[a]	−1.05[a]	7.85	690	0.71	3.87	[45]
D9	451 (29500)	1.54[a]	−0.90[a]	9.11	610	0.68	3.76	[45]
D10	428 (40000)	0.91	−1.55	14.63	685	0.67	6.70	[46]
C1	433 (32600)	0.73	−1.76	13.32	660	0.61	5.43	[46]
C2	439 (34000)	0.69	−1.77	13.96	670	0.66	6.14	[46]
C3	460 (28900)	0.51	−1.77	12.60	630	0.61	4.87	[46]
C4	454 (26400)	0.54	−1.72	8.48	570	0.65	3.14	[46]

Table 2.1 (*Cont.*)

Dye	$\lambda_{max,}$ nm (ε_{max}, M^{-1} cm^{-1})	E_{ox}, V (*vs* NHE)	E_{ox}^{*}, V (*vs* NHE)	J_{SC} (mA cm^{-2})	V_{OC} (mV)	*ff*	η (%)	*Ref*
C5	446 (33900)	0.56	–1.83	14.12	680	0.64	6.32	[46]
D11	465 (47900)	1.33	–0.74	10.02	570	0.64	3.62	[47]
D12	417 (18500)	1.35	–0.77	9.44	660	0.63	3.75	[47]
D13	474 (32000)	0.89	–1.33	12.73	700	0.71	6.33	[48]
C6	507 (26000)	1.02	–1.07	11.69	680	0.69	5.48	[48]
C7	512 (29000)	0.95	–1.11	13.93	710	0.68	6.72	[48]
D14	428 (35000)	0.59[b]	–1.35[b]	13.45	618	0.68	5.64	[49]
C8	481 (40000)	0.51[b]	–1.37[b]	15.85	589	0.67	6.11	[49]
C9	485 (41000)	0.46[b]	–1.40[b]	16.17	595	0.66	6.38	[49]
D15	532 (57500)	1.15	–0.96	13.10	770	0.73	7.50	[50]
C10	532 (46500)	0.97	–1.23	12.90	774	0.70	7.00	[50]
D16	443 (42100)	–	–	10.20	640	0.72	4.70	[34]
D16	485 (40100)	0.99	–0.78	10.56	670	0.63	4.46	[53], [51]

(*Continued*)

Table 2.1 (*Cont.*)

Dye	λ_{max}, nm (ε_{max}, $M^{-1} cm^{-1}$)	E_{ox}, V (*vs* NHE)	E_{ox}^{*}, V (*vs* NHE)	J_{SC} (mA cm^{-2})	V_{OC} (mV)	*ff*	η (%)	*Ref*
D17	463 (41400)	–	–	10.90	720	0.71	5.60	[34], [52]
D17	480 (38800)	1.07	–0.83	9.90	750	0.67	4.97	[53], [51], [54]
D17	480 (38800)	0.98	–0.97	10.50	748	0.70	5.50	[55]
D18	473 (35800)	–	–	12.40	720	0.71	6.30	[34]
D18	480 (38400)	0.96	–0.89	10.67	690	0.68	5.01	[53], [51]
D18	480 (38400)	–	–	11.00	690	0.64	5.00	[54]
D18	479 (38400)	0.83	–1.07	11.10	790	0.71	6.20	[55]
D19	474 (33100)	0.86	–0.85	5.72	740	0.48	2.03	[53]
D20	473 (38600)	0.83	–0.87	6.40	730	0.47	2.20	[53]
D21	478 (39300)	0.98	–0.74	10.41	740	0.46	3.54	[53]
D22	–	–	–	11.90	707	0.67	5.61	[51]
D23	471 (46900)	1.09	–0.61	11.38	700	0.60	4.78	[53]
D24	483 (36200)	–	–	14.50	760	0.71	7.8	[54]

Table 2.1 (*Cont.*)

Dye	λ_{max}, nm (ε_{max}, M^{-1} cm^{-1})	E_{ox}, V (*vs* NHE)	E_{ox}^{*}, V (*vs* NHE)	J_{SC} (mA cm^{-2})	V_{OC} (mV)	*ff*	η (%)	*Ref*
D24	483 (36200)	0.95	–0.93	11.00	802	0.71	6.30	[55]
D25	480 (38500)	–	–	14.50	730	0.69	7.30	[54]
D26	480 (35200)	0.96	–0.93	10.80	785	0.71	6.00	[55]
D27	481 (35200)	0.95	–1.01	11.70	825	0.69	6.60	[55]
D28	473 (39700)	0.90	–0.99	11.50	826	0.67	6.40	[55]
D29	474 (35900)	0.92	–1.02	8.15	720	0.72	4.20	[55]
D29	474 (35900)	0.92	–1.02	10.20	770	0.65	5.10	[56]
D30	473 (38900)	0.97	–1.12	9.71	760	0.72	5.30	[56]
D31	472 (33900)	1.08	–1.00	9.73	770	0.72	5.40	[56]
C11	472 (28200)	1.12	–0.95	7.63	640	0.72	3.50	[56]
C12	471 (35600)	1.01	–1.06	6.89	690	0.75	3.60	[56]
C13	474 (39100)	0.95	–1.15	8.00	670	0.72	3.90	[56]
C14	513 (31800)	0.66	–1.26	8.74	570	0.74	3.70	[56]

(*Continued*)

Table 2.1 (*Cont.*)

Dye	λ_{max}, nm (ε_{max}, $M^{-1} cm^{-1}$)	E_{ox}, V (*vs* NHE)	E_{ox}^{*}, V (*vs* NHE)	J_{SC} (mA cm^{-2})	V_{OC} (mV)	*ff*	η (%)	*Ref*
C15	498 (34000)	0.75	–1.17	7.82	650	0.66	3.40	[56]
C16	509 (34100)	0.67	–1.21	6.22	690	0.70	3.00	[56]
D32	485 (43900)	0.93	–0.94	8.70	750	0.65	4.30	[57]
D33	480 (41500)	0.94	–0.94	7.30	720	0.69	3.60	[57]
D34	374 (12300)	1.59	–1.09	4.00	1156	0.80	3.68	[43]
D35	381 (7800)	1.46	–1.35	4.71	1091	0.79	4.05	[58]
D36	438 (71000)	1.55	–0.89	8.47	470	0.60	2.39	[59]
D37	468 (73200)	1.55	–0.78	10.90	400	0.57	2.48	[59]
D38	429 (20900)	1.38	–0.96	7.12	939	0.78	5.22	[43]
D39	416 (27400)	1.48	–1.15	5.83	1033	0.77	4.61	[58]
D40	378 (10800)	1.45	–1.38	3.88	1077	0.78	3.28	[58]
D41	409 (41536)	2.00	–0.59	2.19	601	0.63	0.82	[60]
D42	375 (27038)	1.84	–0.98	1.71	634	0.64	0.70	[60]

Table 2.1 (*Cont.*)

Dye	λ_{max}, nm (ε_{max}, $M^{-1} cm^{-1}$)	E_{ox}, V (*vs* NHE)	E_{ox}^*, V (*vs* NHE)	J_{SC} (mA cm^{-2})	V_{OC} (mV)	*ff*	η (%)	*Ref*
D43	420 (16420)	–	–	7.03	523	0.61	2.23	[61]
D44	393 (29000)	1.51	−1.23	6.32	530	0.72	2.41	[62]
D45	393 (28300)	1.11	−1.60	12.20	600	0.77	5.64	[62]
D46	386 (96900)	0.97	−1.80	13.00	630	0.75	6.14	[62]
D47	441 (37500)	1.01	−1.64	13.70	680	0.70	6.52	[62]
D48	603	1.17	−0.75	8.00	473	0.68	2.59	[63]
D49	612 (13400)	0.93	−1.00	8.17	411	0.66	2.22	[63]

[a] oxidation potential measured with dye adsorbed on TiO_2 film. [b] oxidation potential measured with photo emission yield spectrometer

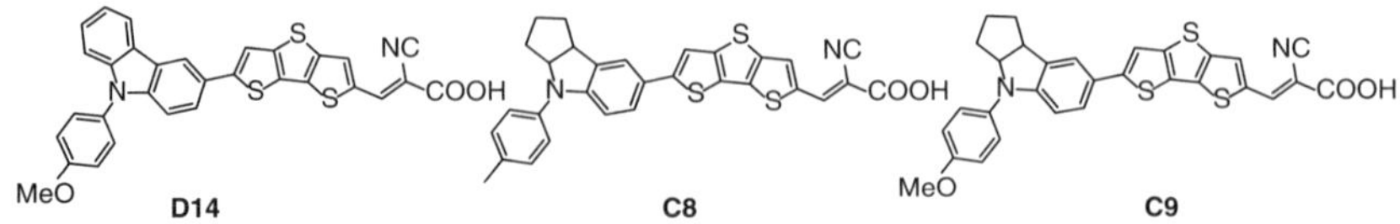

Figure 2.9 Carbazole and indoline dyes containing dithienothiophene linkers.

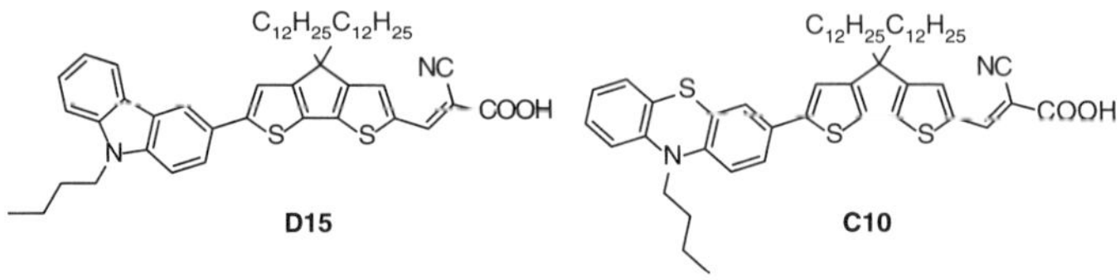

Figure 2.10 Carbazole and phenothiazine dyes containing cyclopentadithiophene linkers.

conduction band of TiO_2 with the electrolyte has been investigated by Hara *et al.* by designing some interesting organic dyes (Figure 2.11) [34]. They have synthesized organic dyes featuring carbazole donor, terthiophene to hexithiophene segments in the conjugation and cyanoacrylic acid acceptor. It was found that incorporation of hexyl chains on the thiophene suppresses the aggregation and stops I_3^- from reaching the TiO_2 layer. This led to an enhancement in the electron lifetime and charge collection efficiency. For instance, the V_{oc} realized for **D17** is significantly larger than that observed for the device based on **D16**. The dye containing four 3-hexylthiophene units (**D18**) exhibited higher efficiency (6.3%) in the series. Studies on interfacial electron transfer kinetics revealed that the larger molecular size of dye and two adsorption sites lowered the reorganizing energy of dye and also led to wider distribution of charges which controls I_3^- concentration at TiO_2 layer [51]. The device based on **D18** while using ionic liquid electrolytes showed good long-term stability under visible light irradiation at 60 °C [52]. The same group later observed that the number of thiophene units in the oligothiophene conjugation pathway and the position of the hexy functionalization influence the dye loading and aggregate thickness [53]. The dyes **D19–D21** possess two hexyl chains at different positions and form thick 3D aggregates on TiO_2 film, while **D17** and **D18** containing three- and four-hexyl chains form 2D dense packing. This reveals that dye loading and thickness of aggregates can be easily optimized by appropriate alkyl group introduction. Interestingly, the dyes **D22** and **D23** which possess

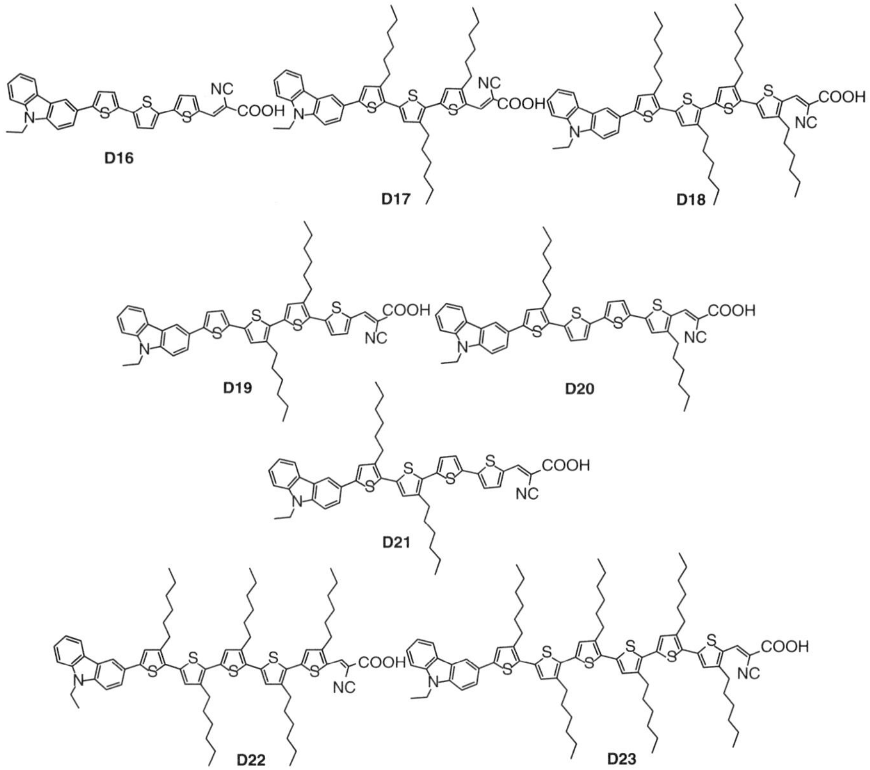

Figure 2.11 Carbazole-based dyes possessing oligothiophene in the conjugation pathway.

five and six hexyl substituted thiophenes produced inferior efficiency in the devices than the dyes **D17** and **D18**.

It has been found by Mori *et al.* that alkyl chains present on the thiophene unit and the nature of alkoxyphenyl fragment attached to the carbazole nucleus (Figure 2.12) significantly altered the V_{oc} of the devices [54]. Incorporation of those groups improved the sphericity of the dyes, facilitated the wider distribution of charged atoms and provided good packing at the surface of TiO_2, and were efficient in blocking I_3^- ions from approaching the dyes. This minimized the dark current and increased the V_{oc} significantly. Thus the dyes analogous to **D17** and **D18** but possessing additional hexyloxyphenyl group on carbazole, **D24** and **D25**, contributed to the increase in electron lifetime and V_{oc} (Table 2.1) due to the negatively charged oxygen atoms repelling the approach of I_3^- ions. Supporting this hypothesis, reducing the alkyl chains to propyl on thiophene

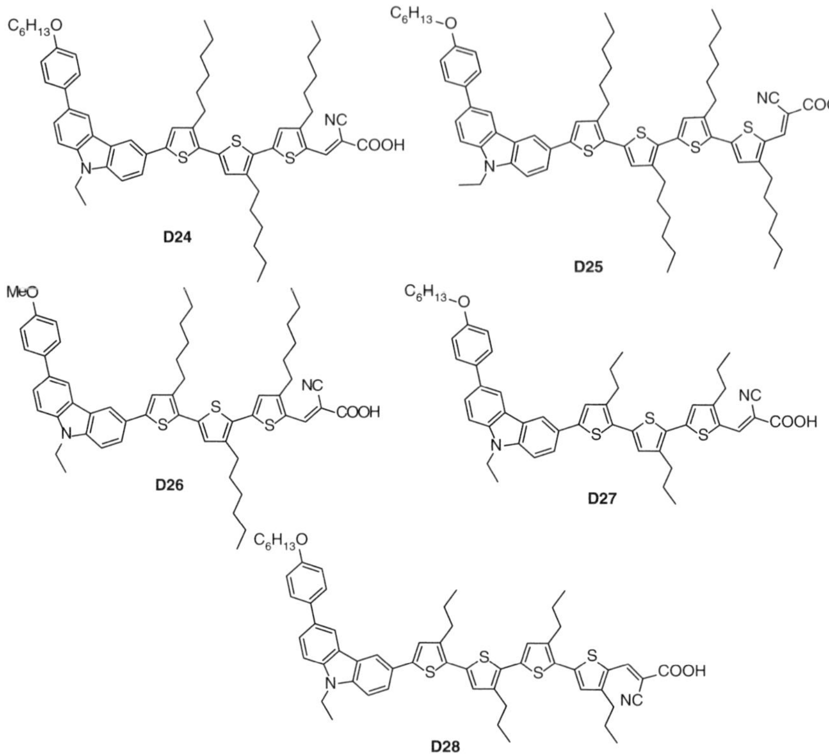

Figure 2.12 Organic dyes designed to unravel the importance of alkyl chain length.

and methyl in alkoxyphenyl units produced a counter effect. Due to the optimized structural components in **D24**, it exhibited very high efficiency (8.10%) in DSSC.

Mori *et al.* further improved the efficiency of the devices based on dyes **D17-D18**, **D24** and **D26–D28** by using cobalt complex as redox couple [55]. Cobalt complexes when used as redox partner in DSSC retard the recombination and contribute to the enhancement in V_{oc}. Since the oxidation potential of cobalt redox couple is lower than that of iodide couple, the gap between the CB of TiO_2 and oxidation potential of cobalt complex increases which results in higher V_{oc}. On using cobalt complex, the V_{oc} of **D17** and **D18** increased (720 mV to 790 mV) significantly compared to those observed while using iodide couple. Among the dyes, **D28** showed promising efficiency (7.90%).

The nature of the *N*-substituent on carbazole have been found to significantly influence the optical and dye-loading efficacy of the dyes. Koumura *et al.* prepared three dyes (Figure 2.13) with different *N*-substituents such as ethyl (**D29**), phenyl (**D30**), *p*-methoxyphenyl (**D31**) groups [56]. The dye **D30** possessed higher molar extinction coefficient among the dyes attributed to the phenyl

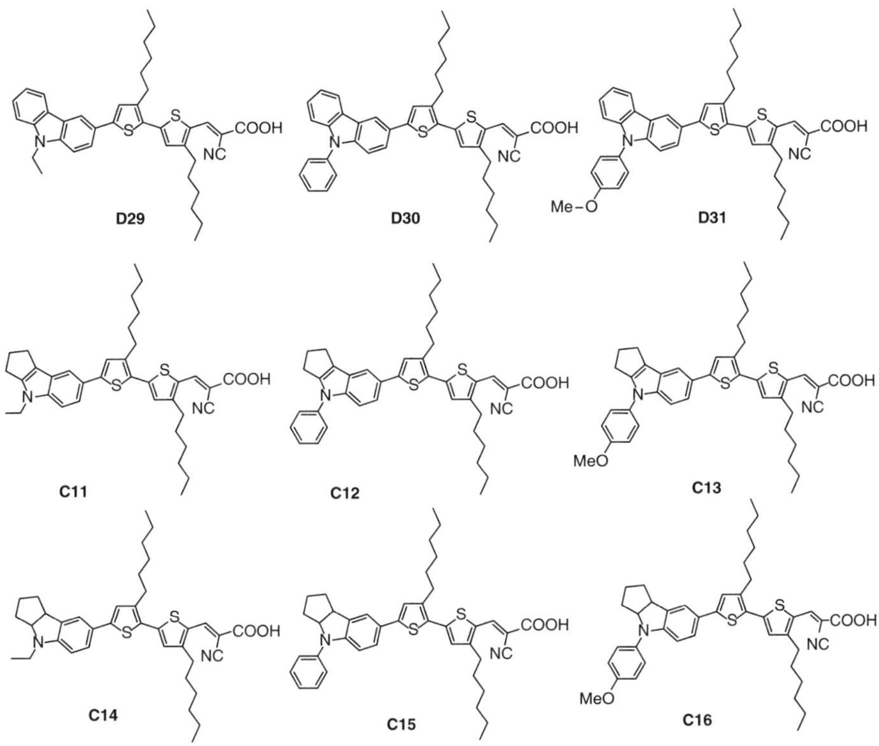

Figure 2.13 Dyes showing difference in *N*-substitution.

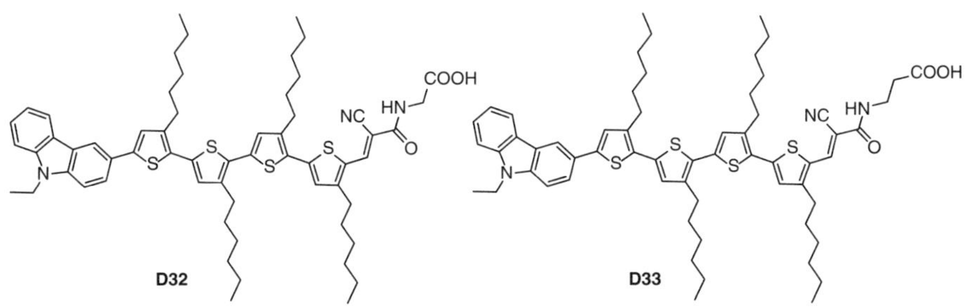

Figure 2.14 Structures of the dyes (**D32** and **D33**) containing carbazole as donor via 3-position.

group present on nitrogen. This combined with the superior dye-loading characteristics led to larger photocurrent in the DSSC of **D30**. Interestingly, the analogous dyes derived from indole and indoline showed inferior performance when compared to the carbazole dyes due to less dye loading and poor blocking effect which results in shorter electron lifetimes. The carbazole-based dyes have a special property of surface coverage on TiO_2 electrode which helps in improving the dye loading.

Mori *et al.* examined the influence of non-conjugated bridge between the donor framework and anchoring group by applying it to a previously successful dye, **D18** [57]. Thus the newly synthesized dyes, **D32** and **D33**, contain glycine and β-alanine as anchoring group. The J_{sc} of **D32** and **D33** is much less than the original dye **D17**. The increase in the length of non-conjuagating segment decreases the short-circuit current due to low electron injection efficiency (J_{sc} of **D32** is 8.7 mA cm^{-2} and of **D33** is 7.30 mA cm^{-2}). Despite low J_{sc} observed for the dyes, the V_{oc} of **D32** (750 mV) is slightly larger due to less dispersion force with similar blocking effect of dye compared with **D18** (700 mV) and **D33** (720 mV) under the same conditions.

2.2.2 Carbazole as Donor and Linked through *N*9-position

A large number of organic dyes (see Figure 2.15 for a schematic representation) have been reported where carbazole is a donor attached to the π-linker through *N*9-position. Carbazole can be easily functionalized at *N*9-position either by *N*-arylation or alkylation. *N*-arylation on carbazole can be achieved by Ullmann coupling reaction involving a copper catalyst and aryl halide, preferably an iodide, in the presence of a base in polar solvent. This method is cheap and convenient when compared to the palladium catalyzed protocols. This class of dyes has the following advantages: (1) the cation radical

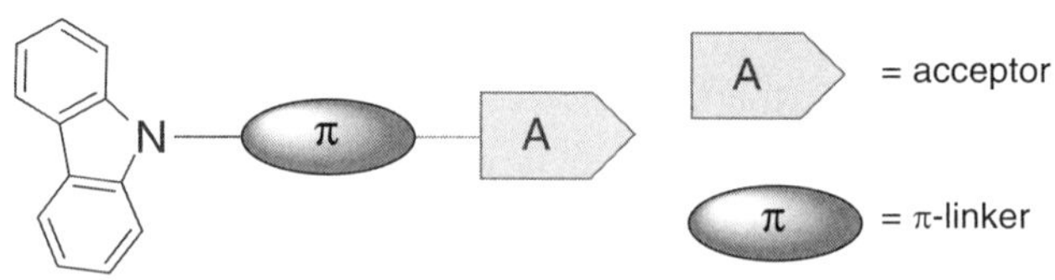

Figure 2.15 Structural composition of organic dyes with *N*9-linked carbazole donor

formed on removal of electron is easily stabilized by chromophores present of C3 and C6 positions, and (2) these dyes normally aggregate at the surface of TiO_2 but increase the dye loading due to the presence of compact planar structural elements. They can be compared to diphenylamine end-capped organic dyes which normally possess a nonplanar arrangement around the amine nitrogen.

Sun *et al.* synthesized **D34** and **D38** dyes and employed them as sensitizer in DSSC utilizing Br^-/Br_3^- redox couple [43]. The DSSC using **D34** exhibited higher conversion efficiency (3.68%) than the one containing (*E*)-2-cyano-3-(4-(diphenylamino)phenyl)acrylic acid (η = 2.74%). The absorption of **D38** (429 nm) was red-shifted when compared to **D34** (374 nm). This is due to extension of conjugation with vinylthiophene. The J_{sc} of **D38** is higher because of better absorption parameters which further improved the efficiency (5.22%) [58]. The Br^-/Br_3^- redox couple has been found to improve the V_{oc} of the devices and the device with **D38** has achieved record V_{oc} of 1.156 V, which is the highest for DSSCs utilizing organic dyes. Introduction of *tert*-butyl group on the C3 and C6 position of carbazole has been used to fine tune the HOMO and LUMO of the dyes. Thus the compound **D35** showed a bathochromic shift in the absorption spectrum when compared to the similar compound **D34** which lacks the *tert*-butyl substituents. However, the molar extinction coefficient took a beating for the *tert*-butylated derivatives. Furthermore, the low-lying LUMO of **D35** facilitates the injection of electrons to the CB of TiO_2 thereby increasing J_{sc} (4.71 mA cm^{-2}) which results in an overall higher efficiency (4.05%) when compared to **D34**. But replacement of thiophene in **D38** with phenyl group gave a dye **D39** which exhibited inferior performance in DSSC due to low-lying HOMO [58]. This makes the regeneration of the dye difficult, hence the V_{oc} of the dye (1.033 V) decreased significantly when compared to the DSSC composed of **D38**. Additionally, substitution of carbazole for *tert*-butyl groups (**D40**) was found to be futile as the dye showed significant blue-shifted absorption [58].

Zafer *et al.* found that insertion of thiophene (**D36**) and bithiophene (**D37**) segments in the conjugation pathway of the dye **D34**, is beneficial to red shift the absorption wavelength [59]. Though the light harvesting property of **D37** increased substantially due to the presence of bithiophene in the conjugation pathway, the device did not yield a favorable efficiency (2.48%) due to low V_{oc} observed for the cell. The diminished V_{oc} may be attributed to the pronounced recombination of electrons with the electrolyte. However, the dye

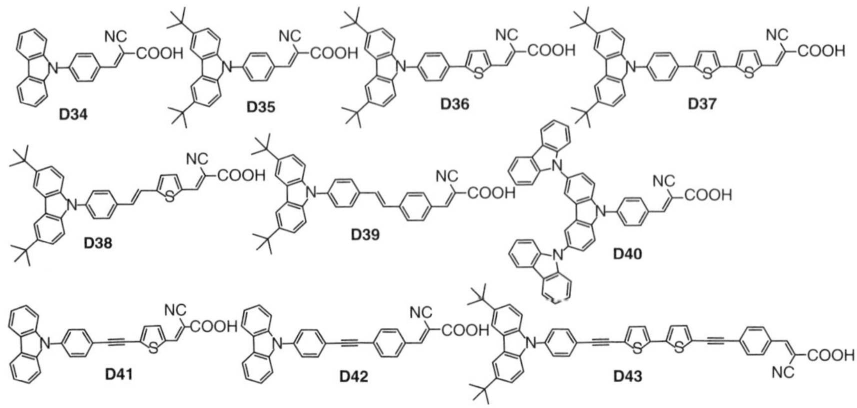

Figure 2.16 Organic dyes with *N*-substituted carbazole as donor and varied conjugation pathways.

D37 showed better device efficiency than **D36** due to the broader spectral response which increased its J_{sc}.

Giribabu *et al.* reported two dyes (**D41** and **D42**) featuring acetylene extended conjugation between the carbazole donor and cyanoacrylic acid acceptor [60]. The thiophene-containing dye (**D41**) displayed longer wavelength absorption and larger photocurrent when compared to the phenylene derivative (**D42**). However, these acetylene-linked derivatives are far inferior to the corresponding vinylene derivatives (**D38** and **D39**). A similar trend has been observed earlier for triarylamine-based organic dyes containing fluorenylacetylene linkers.

Castellano *et al.* reported an interesting dye (**D43**) with elongated conjugation comprising bithiophene and phenylacetylene segments [61]. It exhibited better absorption characteristics when compared to the similar dyes with carbazole in place of bithiophene. It points out that the bithiophene segment is electron richer than the carbazole unit. So the bithiophene segment benefits the donor-acceptor interaction. Despite promising absorption qualities, **D43** exhibited poor PCE when compared to **DL1** and **DL2** (shown in Figure 2.44 ahead). The poor performance of **D43** is assigned to low V_{oc} which may originate from the dye aggregation. Dye aggregation may open electron recombination pathways.

Ko *et al.* reported a series of dyes (**D44–D47**, Figure 2.17) based on 3,6-disubstituted carbazole donor and demonstrated their

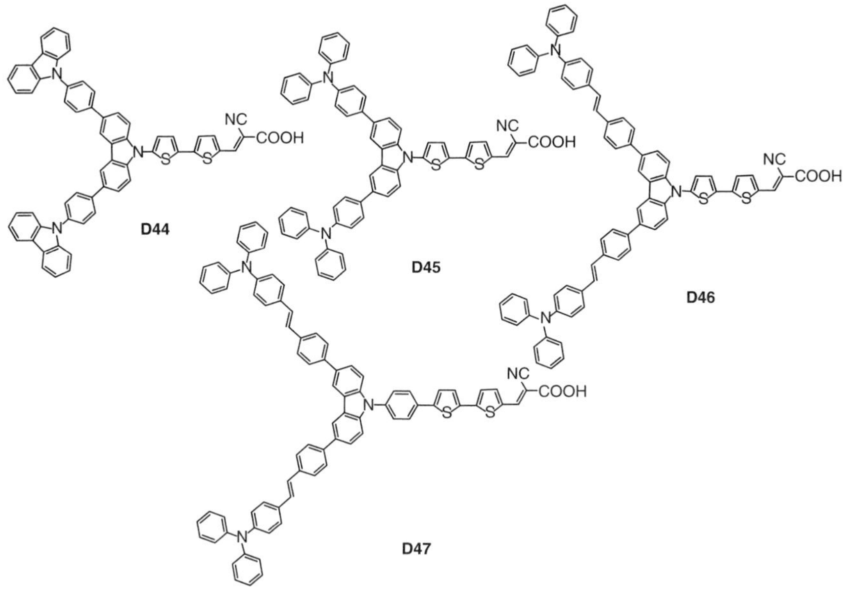

Figure 2.17 Dyes featuring carbazole as donor and auxiliary donor.

function as efficient and stable sensitizers in DSSC [62]. They have shown that by proper functionalization at the C3 and C6 positions, the LUMO of the molecules can be fine tuned. The presence of triphenylamine auxiliary donors in **D45–D47** raises the LUMO when compared to the carbazole analog **D44**. The higher LUMO of **D45–D47** provided more thermodynamic driving force for the injection of electron from the excited state of dyes to CB of TiO_2. This resulted in higher J_{sc} for the devices based on these dyes when compared to a dye, **D44**, containing triphenylamine as auxiliary donor. The dyes containing vinylene linkages (**D46** and **D47**) exhibited high molar extinction coefficients in the absorption spectra. Despite similar absorption properties, the dye **D44** showed very low V_{oc} when compared to **D45**. Probably, the high oxidation potential of **D44** is hampering the regeneration of the dyes and reduces the electron density at CB of TiO_2. The longer conjugation pathway present in **D47** led to a red-shifted absorption, which results in comparatively higher J_{sc} (13.70 mA cm^{-2}) and efficiency (6.52%) in the series. The authors found that the DSSC based on dye **D47** is quite stable up to 1000 h under visible light soaking at 60 °C.

Wu *et al.* first prepared and characterized another interesting class of organic dyes (Figure 2.18) integrating carbazole and squaraine

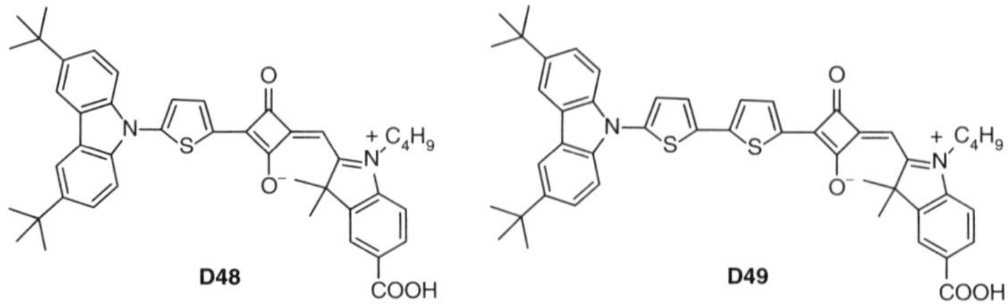

Figure 2.18 Squaraine-based organic dyes with carbazole donor group.

chromophores to realize panchromatic absorption [63]. In their design they have used carbazole as donor and thiophene/bithiophene bridged squaraine as acceptor and anchoring unit. The dye possessing bithiophene linker, **D49**, showed red-shifted absorption with higher molar extinction coefficient when compared to **D48** which contains thiophene linker. The DSSCs of **D48** and **D49** displayed higher J_{sc} and larger V_{oc} when compared to the corresponding dyes without carbazole donor.

2.3 Carbazole as a π-Linker

The efficiency of DSSC is mainly affected by two processes, namely, dye aggregation and charge recombination. Since the organic dye structure plays a major role in the two above processes, molecular engineering is an effective way to find the ways to deter them. It has also been generally accepted that altering the nature of the conjugation pathway that bridges the donor and acceptor fragments in organic dyes is an effective way to fine tune the electronic properties of the dyes. In the construction of organic dyes, scientists have used a wide variety of rigid polyaromatic and heteroaromatic hydrocarbons such as fluorene [12–14], fluorenone [64], fluoranthene [65], pyrene [66], naphthyl [67, 68], anthracene [69–71], oligophenylene [72, 73], oligothiophene [74], fused thiophenes [8], tetrahydrophenanthrene [75], dibenzothiophene [76], dibenzofuran [76], phenothiazine [15, 16] and phenoxazine [77, 78], etc. Generally, conjugation pathways comprising electron-rich chromophores have been found to benefit the absorption characteristics. However, very strong electron donors may destabilize the LUMO. So a chromophore rendering balanced optimization of electronic properties is an ideal bridge. Carbazole, being moderately electron rich, has been widely explored as π-linker in the design of organic dyes. Due to several functionalization possibilities, carbazole has been used as a linker by 2,7-disubstitution or 3,6-disubstitution. Use of 2,7-disubstituted carbazole in the bridge helps to enhance the conjugation and serve as effective charge-delocalizing segment compared to the 3,6-disubsituted carbazole. It has been found that 2,7-disubstitution leads to elongation of conjugation pathway and is beneficial for absorption properties. It was also found that the 2,7-carbazole-bridged dyes exhibited red-shifted absorptions and lower oxidation potentials than the corresponding fluorene- and phenylene-bridged dyes.

2.3.1 Carbazole as a Bridge via C2, C7 Positions

The general molecular design of dyes with 2,7-disubstituted carbazole linker are illustrated schematically in Figure 2.19. In some cases a central carbazole served as donor for dianchoring bis(acceptor) chromophores.

The efficiency of DSSC depends on the nature of the anchoring group present in the dye. A strongly interacting anchoring unit may possess a favorable dipole at the interface between the TiO_2 film and dye, and facilitate the electron injection into the CB of TiO_2. Though several anchoring units such as hydroxyl, carboxylic acid, cyano-acrylic acid, or rhodanine-3-acetic acid have been tested, cyanoacrylic acid has been found to be more successful. Recently, Harima *et al.* synthesized a series of dyes (**L3–L6**) featuring diphenylamine donor, 2,7-disubstituted carbazole linker, and pyridine as anchoring unit [79]. They have also prepared for comparison, dyes with the same structural units, but possessing conventional carboxylic acid anchors. The efficiency of the pyridine-based dyes **L3** and **L4** (1.04–1.15%) are reasonably higher than those of the dyes **L1** and **L2** (0.91-0.97%) with carboxylic acid anchor. However, insertion of

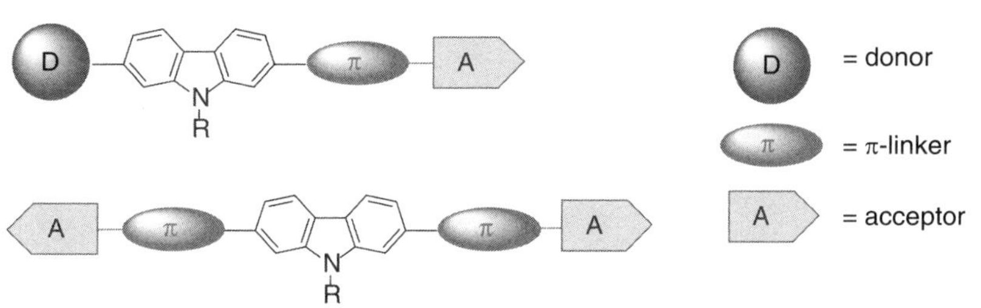

Figure 2.19 Molecular compositions of organic dyes with carbazole as linker via C2,C7-positions.

thiophene in the conjugation pathway (**L5** and **L6**) increased the light harvesting property and led to larger J_{sc} (> 5.63 mA cm^{-2}) and efficiencies (> 1.84%). A slight increase in the power conversion efficiency (η = 1.81-2.35%) was observed on incorporation of additional carboxylic acid anchoring group to the above mentioned architecture (**L7** and **L8** in Figure 2.20) [80]. These dyes exhibited higher J_{sc} (5.16-7.04 mA cm^{-2}) when compared to **L1–L7,** showing that carboxylic acid anchoring units provide stronger interaction with TiO_2.

New organic dyes (**L9** and **L10** shown in Figure 2.21) containing diphenylamine donor, 2,7-carbazolyl π-linker and cyanoacrylic acid acceptor have been synthesized and demonstrated as sensitizers for DSSC [81]. They possess red-shifted absorption with high molar extinction coefficient when compared to the corresponding dyes with fluorene or phenyl units in the conjugation. The dye-sensitized solar cells fabricated using these dyes as sensitizers

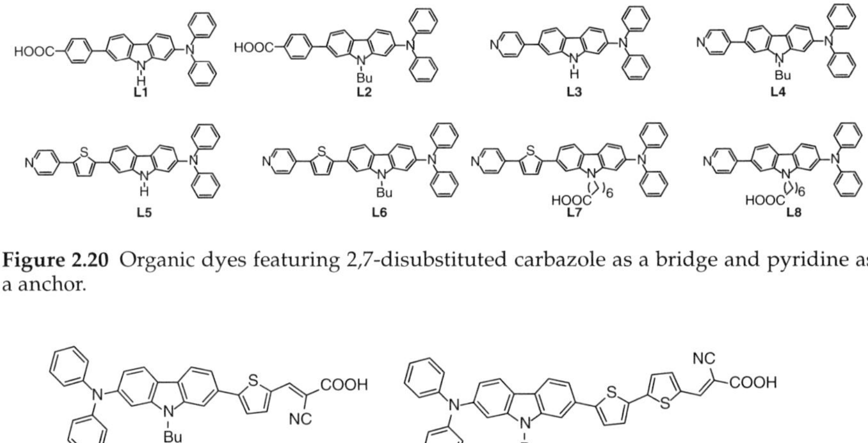

Figure 2.20 Organic dyes featuring 2,7-disubstituted carbazole as a bridge and pyridine as a anchor.

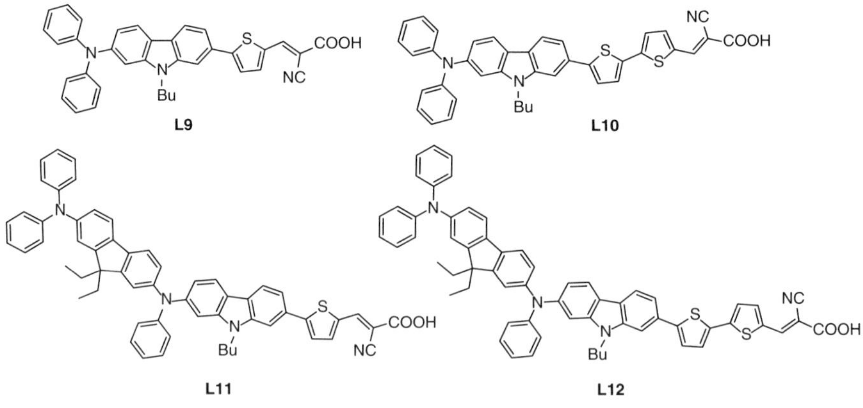

Figure 2.21 Organic dyes featuring 2,7-diaminofluorene donor and 2,7-disubstituted carbazole linker.

exhibited power conversion efficiencies as high as 6.8% which increased to 7.20% when chenodexoycholic acid (CDCA) was used to impede dye aggregation. The DSSCs with ionic liquid electrolytes displayed marked stability over 1000 h.

Thomas *et al.* successfully combined the 2,7-diaminofluorene donor and 2,7-disubstituted carbazole linker to obtain two organic dyes, **L11** and **L12** (Figure 2.21) [82]. They exhibited longer wavelength absorption (Table 2.2) than the corresponding dyes with fluorene and phenylene linkers due to the enhanced electron richness in the conjugation pathway imparted by the carbazole unit. However, as mentioned earlier, it also had a counter effect in lowering the LUMO energy of the dyes which resulted in poor electron injection efficiency. Due to this the less conjugated dye, **L10**, showed higher conversion efficiency (4.62%) when compared to **L11** (4.20%).

Lin *et al.* reported di-anchoring dyes (Figure 2.22, **L13-L15**) with 2,7-disubstituted carbazole as a central donor [83]. These rod-shaped molecules showed promising optical properties which could be tuned by altering the oligothiophene spacer length and inserting alkyl chain on the 3-position of thiophene. The dye with extended conjugation (**L14**) exhibited most red-shifted absorption profile while the dye possessing 3-hexyl thiophene showed shorter wavelength absorption due to the non-planarity induced to accommodate the hexyl chain. This dye inherited a high lying LUMO which increased the electron injection efficiency and photocurrent density.

2.3.2 Carbazole as a Linker via C3, C6 Positions

Organic dyes featuring 3,6-disubstituted carbazole as π-linker are easily achieved by dibromination followed by palladium catalyzed cross-coupling reactions. Most of the organic dyes in this category either used carbazole as linker and donor or as electron-rich conjugating segment (Figure 2.23). Because the direct electronic delocalization between the substituents present on C3 and C6 positions may be limited, 3,6-disubstitution on carbazole cannot be treated as ideal conjugation.

Organic dyes containing pyrrole units in the conjugation are limited. Lin *et al.* synthesized several dyes in which the conjugation bridge was composed of pyrrole, thiophene and carbazole or other aromatic segments [84]. Diphenylamine end-cap served as a donor while cyanoacrylic acid acted as acceptor. It is interesting to note that the carbazole-based dye (**L16**) exhibited the most red-shifted

Table 2.2 Optical, electrochemical and photovoltaic performance parameters of dyes containing carbazole as linker.

Dye	λ_{max}, nm (ε_{max}, M^{-1} cm^{-1})	E_{ox}, V (*vs* NHE)	E_{ox}^{*}, V (*vs* NHE)	J_{SC} (mA cm^{-2})	V_{OC} (mV)	*ff*	η (%)	*Ref*
L1	374 (34900)	0.93	–2.12	2.96	503	0.61	0.91	[79], [80]
L2	376 (34300)	1.00	–2.03	3.07	520	0.61	0.97	[79], [80]
L3	372 (30200)	0.97	–2.15	3.16	524	0.63	1.04	[79], [80]
L4	375 (33000)	1.02	–2.07	3.35	522	0.62	1.15	[79], [80]
L5	394 (48100)	0.93	–1.93	5.80	540	0.60	1.89	[79], [80]
L6	396 (49600)	0.97	–1.87	5.63	548	0.60	1.84	[79], [80]
L7	375 (33000)	1.01	–2.08	5.16	568	0.62	1.81	[80]
L8	396 (48700)	0.97	–1.87	7.04	568	0.59	2.35	[80]
L9	473 (29300)	1.15	–1.18	15.30	660	0.60	6.05	[81]
L10	489 (34300)	1.10	–1.17	18.10	630	0.60	6.83	[81]
L11	492 (31000)	0.91	–1.09	12.60	634	0.58	4.62	[82]
L12	493 (33200)	0.90	–1.05	11.80	636	0.56	4.20	[82]
L13	480 (79600)	0.92	–1.26	10.12	620	0.68	4.26	[83]
L14	484 (97900)	0.74	–1.34	11.74	600	0.63	4.41	[83]
L15	470 (82500)	0.89	–1.28	7.44	570	0.67	2.82	[83]

Table 2.2 (*Cont.*)

Dye	$\lambda_{max,}$ nm (ε_{max}, M^{-1} cm^{-1})	E_{ox} , V (*vs* NHE)	E_{ox}^{*}, V (*vs* NHE)	J_{SC} (mA cm^{-2})	V_{OC} (mV)	*ff*	η (%)	*Ref*
L16	497 (7050)	1.07	–0.99	12.93	580	0.64	4.80	[84]
C17	437 (35600)	1.05	–1.29	13.47	600	0.59	4.77	[84]
C18	472 (38400)	1.08	–0.93	14.20	570	0.60	4.79	[84]
C19	475 (52700)	1.11	–1.09	18.14	610	0.56	6.16	[84]
C20	475 (51000)	1.12	–1.08	16.79	640	0.58	6.18	[84]
C21	462 (53800)	1.08	–1.02	13.54	600	0.64	5.25	[84]
L17	450 (26360)	0.64	–1.92	7.61	589	0.63	2.81	[85]
C22	445 (24040)	0.62	–1.85	9.95	597	0.61	3.59	[85]
C23	489 (25480)	0.55	–1.63	10.60	658	0.70	4.91	[85]
L18	412 (21000)	–	–	5.27	711	0.72	2.69	[86]
C24	466 (14000)	–	–	12.18	826	0.65	6.55	[86]
L19	453 (33900)	1.37	–1.13	9.36	712	0.67	4.47	[87]
L20	470 (54000)	1.17	–1.13	10.52	701	0.68	5.01	[87]
L21	470 (58500)	0.80	–1.35	12.66	610	0.62	4.82	[83]

(*Continued*)

Table 2.2 (*Cont.*)

Dye	λ_{max}, nm (ε_{max}, $M^{-1} cm^{-1}$)	E_{ox}, V (*vs* NHE)	E_{ox}^*, V (*vs* NHE)	J_{SC} ($mA\ cm^{-2}$)	V_{OC} (mV)	*ff*	η (%)	*Ref*
L22	455 (89200)	0.77	−1.50	10.63	640	0.67	4.55	[83]
L23	468 (102400)	0.78	−1.41	9.99	550	0.66	3.64	[83]
L24	438 (35526)	1.76	−0.78	7.64	656	0.68	4.04	[88]
C25	458 (38578)	1.57	−0.79	7.01	612	0.63	3.16	[88]
L25	379 (50800)	1.22	−1.14	14.12	774	0.58	6.37	[89]
C26	425 (27100)	1.22	−1.19	12.00	776	0.56	5.24	[89]
L26	372 (30000)	1.21	−1.57	4.20	674	0.67	1.88	[89]
L27	327 (21900)	1.54	−1.95	1.84	492	0.63	0.57	[90]
L28	378 (46100)	1.22	−1.75	4.72	556	0.61	1.61	[90]
L29	422 (21400)	0.89	−1.49	9.36	786	0.72	5.29	[76]
C27	406 (28600)	1.00	−1.60	6.02	669	0.75	3.03	[76]
C28	414 (28000)	0.91	−1.57	7.50	704	0.72	3.84	[76]
L30	442 (27000)	0.88	−1.48	11.20	757	0.70	5.91	[76]
C29	384 (23000)	0.98	−1.72	6.83	666	0.74	3.36	[76]
C30	407 (20200)	0.89	−1.63	7.59	745	0.75	4.21	[76]

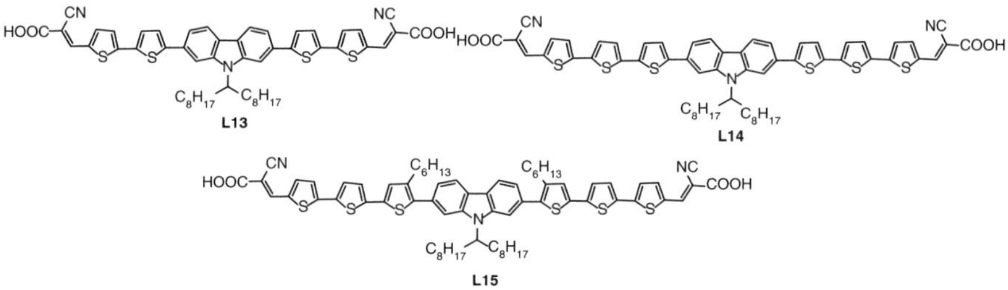

Figure 2.22 Dianchoring organic dyes with 2,7-disubstituted carbazole donor.

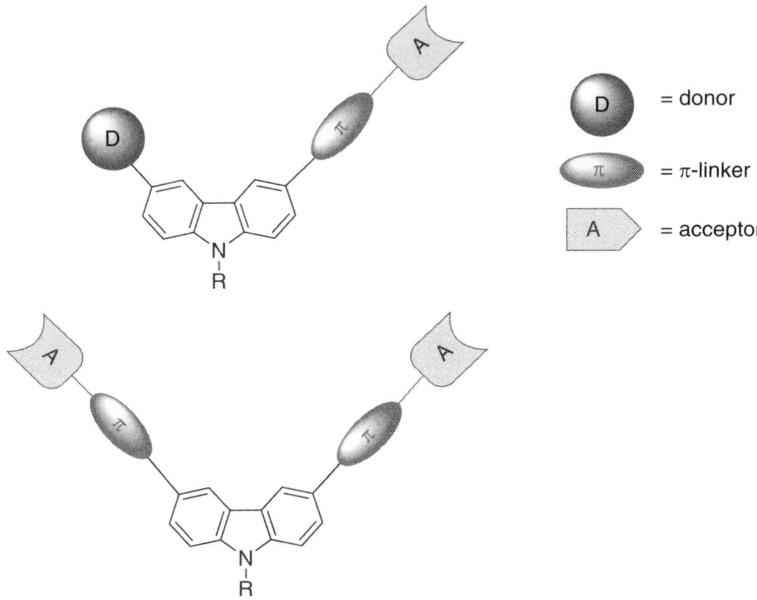

Figure 2.23 Molecular architecture of organic dyes containing 3,6-disubstituted carbazole as π-linker.

absorption in the series. However, the molar extinction coefficient realized for the charge transfer transition is significantly lower. Though the electron richness of carbazole increases the donor-acceptor interaction, the broken conjugation because of 3,6-disubstitution on carbazole diminishes the electronic transition probability. **L16** showed an overall efficiency of 4.8% with J_{sc} = 12.93 mA cm^{-2}, V_{oc} = 580 mV and *ff* = 0.64. The comparatively low J_{sc} observed for **L16** is attributed to less light harvesting capability and low-lying LUMO level.

Sun *et al.* reported a series of dianchoring dyes (Figure 2.25) with rhodanine as an acceptor, and several bridging units such as

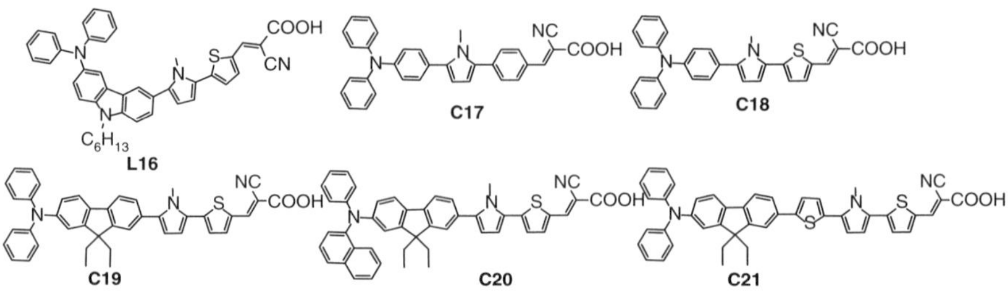

Figure 2.24 Organic dyes containing 3,6-disubstituted carbazole as π-linker and related compounds.

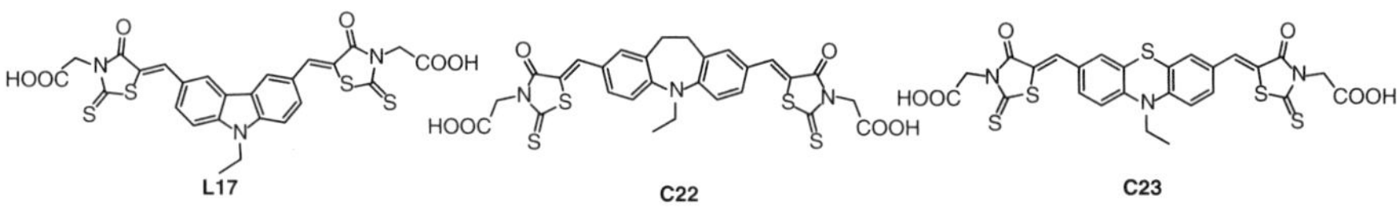

Figure 2.25 Organic dyes containing electron-rich heteroaromatics as π-linker and donor.

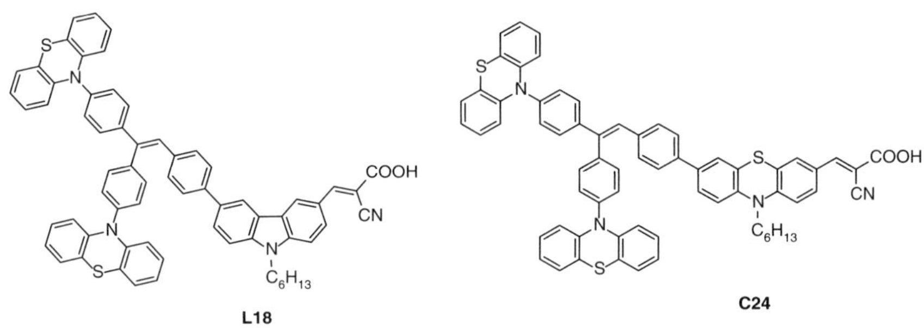

Figure 2.26 Organic dyes containing phenothiazine auxiliary donors and carbazole π-linker.

carbazole, iminodibenzyl and phenothiazine [85]. The efficiency of the devices assumed the order according to the linker as: iminodibenzyl (**C22**) < carbazole (**L17**) < phenothiazine (**C23**). The light-harvesting ability of the dye containing phenothiazine dye is marginally better due to improved absorption characteristics and balanced HOMO/LUMO energy levels.

Kuang *et al.* reported a dye (**L18**) with triphenylethylene chromophore bearing phenothiazine as donor, 3,6-disubstituted carbazole linker and cyanoacetic acid acceptor [86]. The dye achieved an efficiency of 2.69% in the DSSC. This is higher than that reported for the dye, **D1**, which lacks the auxiliary donor and exemplifies the requirement of electron-rich donors for carbazole-based organic dyes.

The stability and adsorption of dyes on TiO_2 were essential for practical applications. In order to improve affinity of the dyes towards TiO_2, Suranna *et al.* designed a new series of di-anchoring dyes (**L19** and **L20**) based on carbazole linker and oligothiophene spacers [87]. The dye **L20** showed red-shifted absorption ($\Delta\lambda$=17 nm) and high molar extinction coefficient when compared to **L19** due to the elongated conjugation. The higher IPCE plateau of **L20** (> 60%) in the 400–550 nm region produced J_{sc} of 10.52 mA cm^{-2} and led to high efficiency (5.01%) in DSSC. Lin *et al.* extended this concept and synthesized a series of dyes with carbazole linker (**L21**–**L23**) [83]. They have studied the effect of planarity on the performance of DSSC by introducing hexyl chains on thiophene units. It is interesting to note that the 3,6-disubstituted dyes (**L21–L23**) showed more red-shifted absorption than the corresponding 2,7-disubstituted analogs (**L13–L15**). This is probably due to direct conjugation with nitrogen of carbazole in the former case, which acts as auxiliary donor. The hexyl chain substitution in **L22** lowered the absorption wavelength due to the non-planarity. But this is helpful in hampering the intramolecular charge recombination and leads to an increase in V_{oc} (640 mV). Anandan *et al.* extended the conjugation of **L19** by incorporation of cyanovinlylene between thiophene and carbazole and produced **L24** dye [88]. The cyanovinylene substitution in **L24** lowered the LUMO level when compared to that in **L19**. This contributed to the reduction in photovoltage (V_{oc} = 660 mV) and lower conversion efficiency (4.04%). However, the dye **L24** displayed higher efficiency in DSSC when compared to the corresponding diphenylamine dye (**C25,** η = 3.16%) owing to higher dye loading for **L24** facilitated by planar carbazole and higher electron lifetime (~ 4 times).

Zhou *et al.* reported novel dyes (**L25** and **L26**, Figure 2.28) with carbazole linker and triarylamine donor [89]. The dyes featured different anchoring units such as cyanoacrylic acid (**L25**) and pyridine (**L26**). Compared to **C26** (η = 5.24%) where biphenyl bridging unit was there in place of carbazole, **L25** showed higher J_{sc} (14.12 mA cm^{-2}) and higher conversion efficiency (6.37%). Pyridine-based dye, **L26** displayed significantly lower conversion efficiency (η = 1.88%) due to smaller dye loading, huge dark current and pronounced charge recombination. This again points out that pyridine unit cannot serve as good anchoring unit. Structural modifications are required to tune

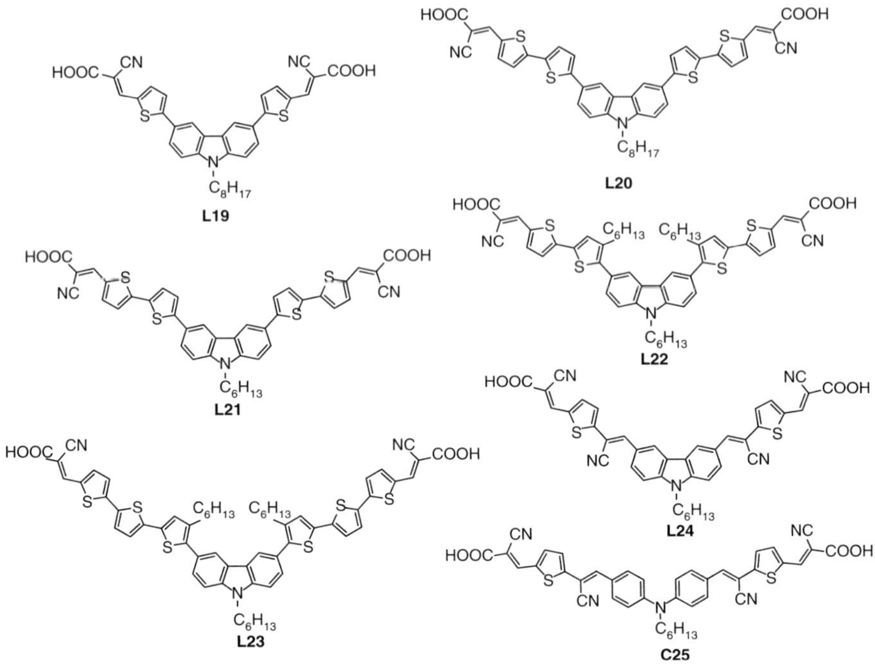

Figure 2.27 Organic dyes containing carbazole as π-linking donor and oligothiophene spacers.

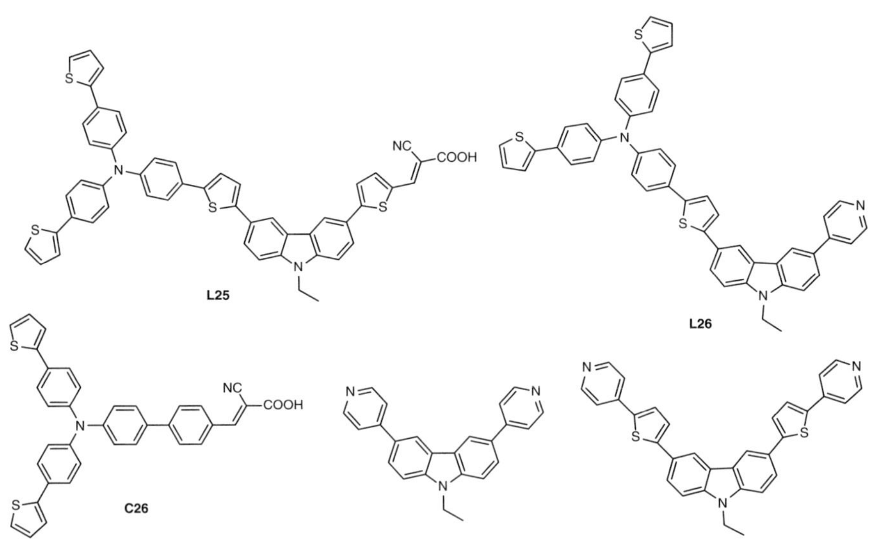

Figure 2.28 Other examples in which carbazole act as linker.

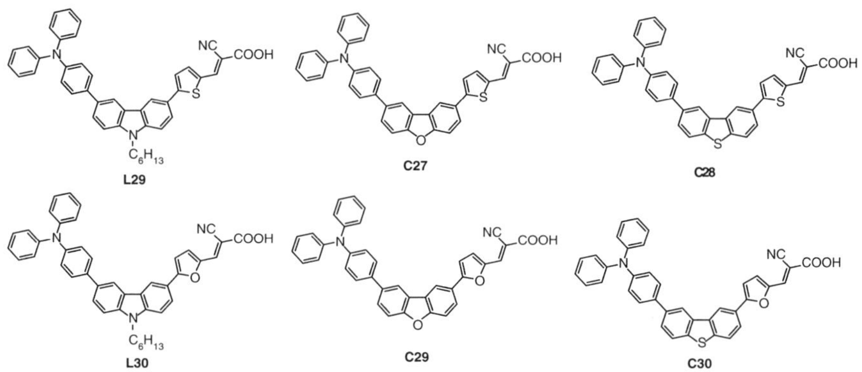

Figure 2.29 Compounds containing carbazole linker and triphenylamine auxiliary donor and related compounds.

its anchoring properties. Harima *et al.* thought that di-anchoring pyridine dye may solve this problem. They have prepared two dyes (**L27** and **L28**) with chelating-like pyridine units [90]. **L28** displayed red-shifted absorption due to elongated conjugation. However, the efficiency of the devices that made use of these dyes is below par.

Recently, Su *et al.* reported a series of D-π-A dyes featuring different linkers such as carbazole, dibenzofuran and dibenzothiphene [76]. Triphenylamine served as donor and cyanoacrylic acid as acceptor. Carbazole dyes (**L29** and **L30**) showed red-shifted absorption when compared to other dyes (**C27–C30**) due to the strong electron-donating nature of carbazole. The broader IPCE of carbazole dyes manifested in higher J_{sc} (9.36-11.20 mA cm^{-2}), further shifted to high J_{sc} of (11.09-13.89 mA cm^{-2}) in optimized dye bath solvent of acetonitrile. The substitution of hexyl chain on carbazole strongly hampers the intermolecular charge recombination and raises the V_{oc} (730-769 mV). **L30** showed best conversion efficiency (7.09%) with J_{sc} = 13.89 mA cm^{-2}, V_{oc} = 769 mV and *ff* = 0.66 in acetonitrile solvent bath.

2.4 Carbazole as Auxiliary Donor for DSSC

The auxiliary donor (AD) concept has been widely used to improve the donor-acceptor interaction and consequently the charge migration efficiency from donor to acceptor. In particular, the auxiliary donor when it is in conjugation with the donor can help to stabilize the cation radical formed on removal of electron by delocalizing the charge. Suitably selected auxiliary donor has been found to enrich the optical and electrochemical properties and led to improvement in the DSSC efficiency. The advantages of using carbazole as AD in DSSC dyes are: (a) elongation in conjugation leads to red-shifted absorption, (b) transition probability of charge transfer transition is increased due to push-push donor architecture, (c) hydrophobic nature of carbazole and bulkiness of the overall dye helps to stop the back electron transfer, and (d) the delocalization/stabilization of cation radical increases. Carbazole can be deployed as auxiliary donor by C2, C3, or *N*9-substitution.

2.4.1 Carbazole as Auxiliary Donor via C2-position

The generally adopted structural composition of the organic dyes with carbazole as auxiliary donor via C2-substitution is illustrated in Figure 2.30. Since the conjugation of the donor is extended to carbazole nitrogen, the donor strength may not experience uplift, but due to extended π-conjugation, delocalization of charge is favorable.

The effect of carbazole as auxiliary donor on simple triphenylamine-based, metal-free organic dyes has been studied by Wang *et al.* They have prepared three new dyes (Figure 2.31, **AD1–AD3**) based on a triphenylamine framework and featuring one or two carbazole or acceptor units in a dye [91]. The dyes (**AD1–AD3**) displayed red-shifted (15–24 nm) absorption when compared to the parent dyes (**AC1–AC3**) [92] which established the conjugating effect of carbazole segment. Moreover, introduction of two carbazole units in a dye (**AD3**) benefits the absorption significantly compared to the dye (**AD2**) containing two acceptor units. Probably in the former dye, due to the presence of two carbazole auxiliary donors, the donor-acceptor interaction is enhanced, while in the later dye, it is bifurcated between the two acceptors. Finally, **AD3** displayed higher power conversion efficiency

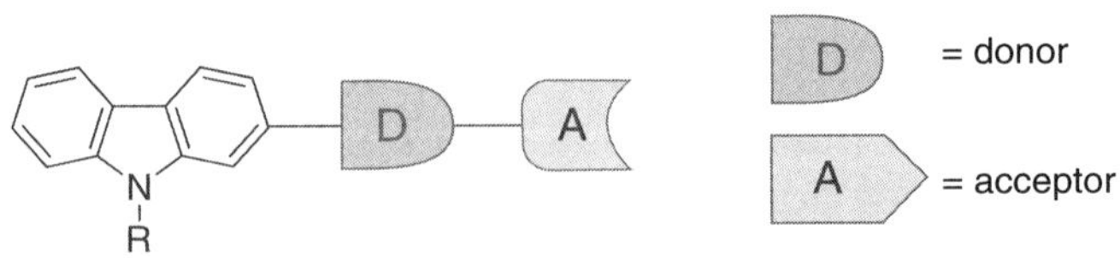

Figure 2.30 Molecular configuration of organic dyes using C2-carbazole as auxiliary donor.

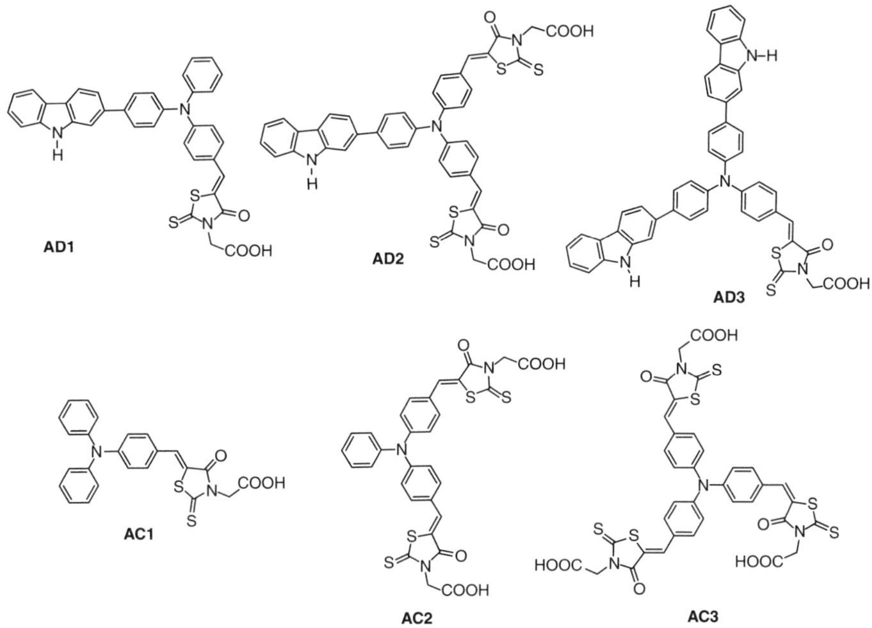

Figure 2.31 Dyes **AD1–AD3** with carbazole as auxiliary donor and related parent dyes.

(4.63%) due to larger photocurrents (J_{sc} = 12.70 mA cm^{-2}) when compared to **AD1** and **AD2**.

2.4.2 Carbazole as Auxiliary Donor via C3-Position

Carbazole has been successfully employed as auxiliary donor via C3-substitution due to the facile chemical functionalization and a wide variety of reactions that are possible at this position. Either the carbazole has been used as a part of triarylamine donor or linked via an extended conjugation with the amine donor (Figure 2.32). In the former configuration the oxidation potential of the dyes are negatively shifted indicating the donor support of carbazole.

Though the organic dyes featuring trifluorenylamine and 2,7-diaminofluorene donors have been demonstrated as successful sensitizers in DSSC, their carbazole counterparts remained unexplored. Freeman *et al.* synthesized organic materials (Figure 2.33) possessing dicarbazolylamine unit and compared their optical and electronic properties with the corresponding diphenylamine analogs [93]. The bis-carbazole dyes exhibited

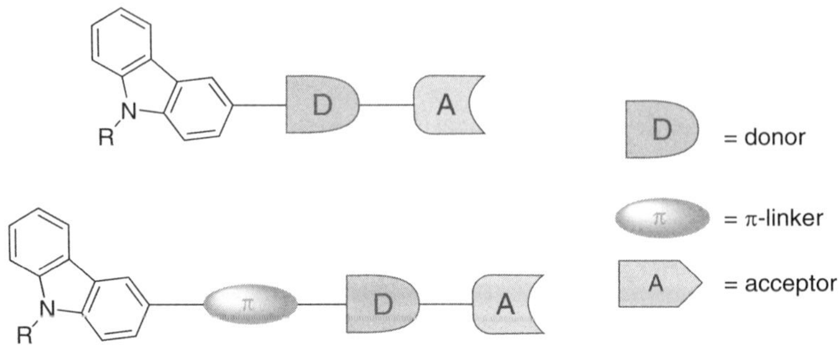

Figure 2.32 Molecular architecture of metal-free organic dyes utilizing carbazole as auxiliary donor via C3 linkage.

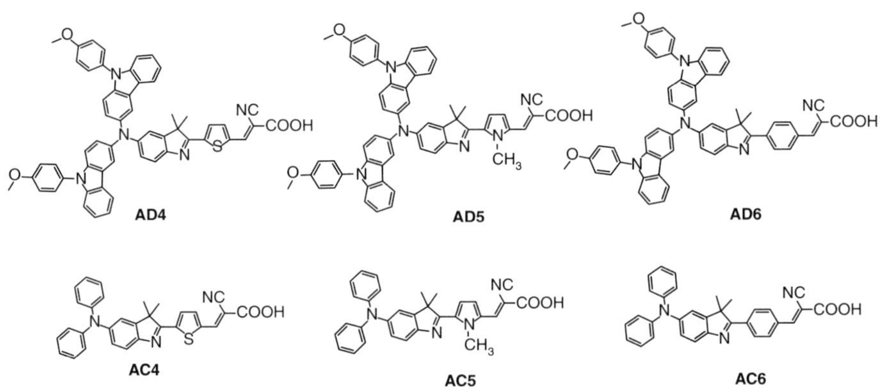

Figure 2.33 Dicarbazolylamine-based organic dyes.

red-shifted absorption (~ 40 nm) when compared to the diphenylamine analogs (**AC4–AC6**) [94] due to more planar geometry and high donor strength of amine. However, their performance in DSSC was affected by their aggregation tendency. The best performance obtained was for **AD4** with J_{sc} = 5.35 mA cm^{-2}, V_{oc} = 564 mV, ff = 0.68 corresponding to an overall conversion efficiency of 2.05%. The addition of CDCA improved the efficiency of the dyes marginally.

A novel strategy was adopted by Li *et al.* to suppress the aggregation at the surface of TiO_2 [95]. They have taken a dye with triphenylamine donor, pyrrole linker and cyanoacrylic acid acceptor and functionalized the pyrrole nitrogen with an auxiliary donor of carbazole, triphenylamine donor (Figure 2.34). The absorption of this exotic dye, **AD7**, showed blue-shifted absorption profile

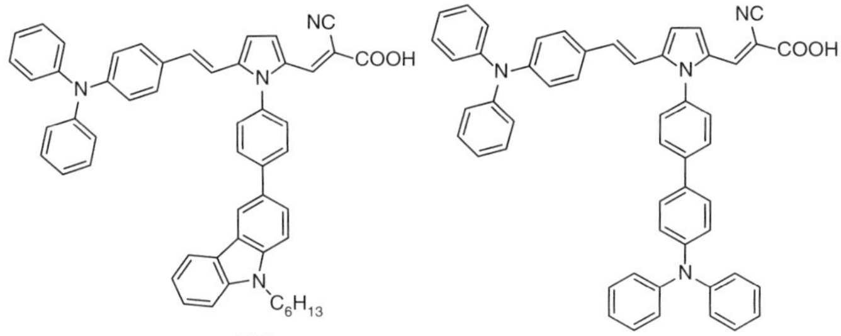

AD7 AC7

Figure 2.34 Dyes designed to impede dye aggregation.

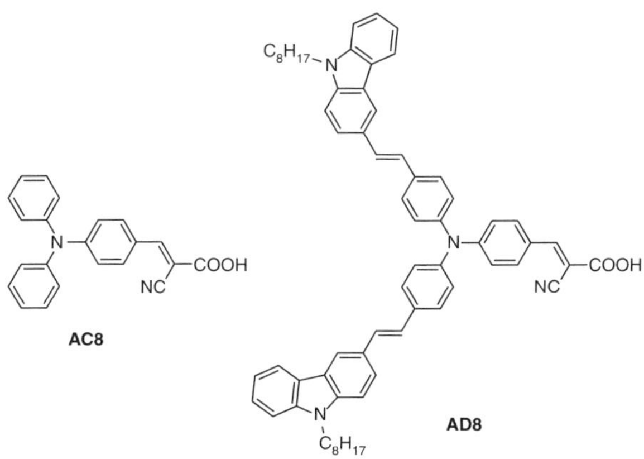

AC8 AD8

Figure 2.35 Dye possessing 3-vinylcarbazole chromophore and its parent dye.

when compared to the triphenylamine functionalized dye, **AC7**. This resulted in a lower J_{sc} and efficiency (4.62%). However, the dye deploying triphenylamine as auxiliary donor (**AC7**) exhibited high power conversion efficiency (7.21%).

The performance of the simple triphenylamine-based dye (*E*)-2-cyano-3-(4-(diphenylamino)phenyl)acrylic acid (**AC8**) was improved by extending the conjugation with vinylcarbazole chromophore (Figure 2.35). Lu *et al.* successfully made Y-shaped dyes based on triphenylamine containing 9-octyl-9*H*-carbazole-vinyl, triphenylamino-vinyl and 10-octyl-10*H*-phenothiazine-vinyl as arms cyanoacrylicacid as acceptor [96]. Generally the vinyl conjugated organic dyes show exceptional molar extinction coefficient for the absorption peaks and this increases the generation of photocurrent density. **AD8** showed overall efficiency (3.63%) with J_{sc} = 9.42 mA cm^{-2}, V_{oc} = 730 mV and *ff* = 0.53.

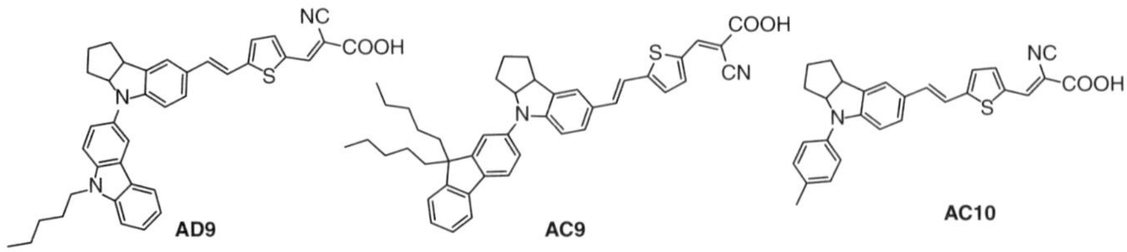

Figure 2.36 Organic dye containing *N*-carbazolyindoline donor and related compounds.

Zhu *et al.* achieved an excellent stability through molecular engineering of indoline dyes (Figure 2.36) [97]. They have introduced 4-tolyl, 2-fluorenyl and 3-carbazolyl units at indoline nitrogen to arrive at highly efficient organic dyes. The carbazole substituted dye **AD9** showed longer wavelength absorption in the series. This supports the strong electron releasing effect of carbazole. The LUMO of **AD9** is highly negative (–1.95 V) when compared to the CB of TiO_2 which facilitates the injection of electrons to CB of TiO_2. This thermodynamic driving force results in a high J_{sc} of 14.16 mA cm^{-2} and overall efficiency of 5.87%. Despite **AD9** showing broad IPCE spectra in the range of 350–800 nm, the low efficiency observed was attributed to aggregation of dyes at the surface of TiO_2. Supporting this hypothesis was the addition of CDCA to the dye bath which significantly improved the device efficiency (8.49%) with J_{sc} = 18.53 mA cm^{-2}, V_{oc} = 649 mV and *ff* = 0.71. The carbazole substitution increased the donor strength and shifted the LUMO more negatively.

2.4.3 Carbazole as Auxiliary Donor via *N*9-Position

The general structure of the dyes containing carbazole as auxiliary donor via *N*9 substitution is illustrated in Figure 2.37. This structural arrangement facilitates the delocalization of nitrogen lone pair in conjugation bridge and enriches the donor strength.

Promarak *et al.* utilized 3,6-di(*tert*-butyl)carbazole as an auxiliary donor and obtained three dyes, **AD10–AD12** (Figure 2.38) [98]. The dyes showed a progressive increase in the absorption wavelength on the introduction of additional thiophene units in the conjugation pathway between the main donor and the acceptor. Among the dyes, **AD12** displayed highest device efficiency (Table 2.3) owing to their light harvesting ability.

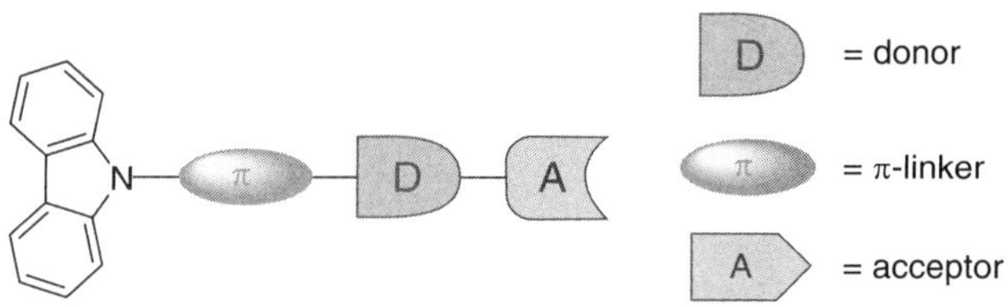

Figure 2.37 Molecular configuration of organic dyes derived using carbazole as auxiliary donor via N9-functionalization.

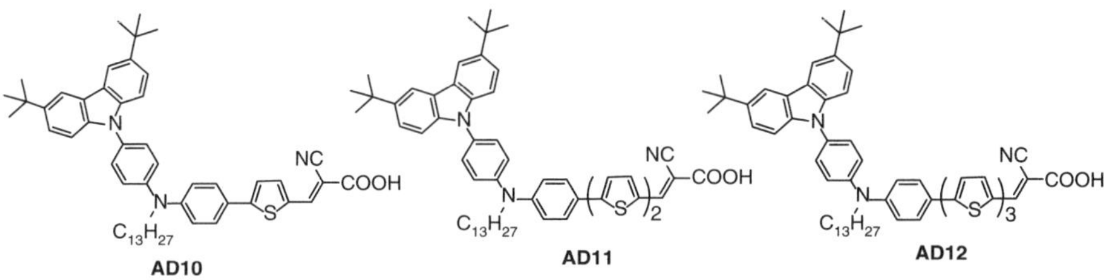

Figure 2.38 Structures of dyes with *N*-functionalized carbazole as auxiliary donor.

Kuang *et al.* found that the introduction of carbazole functionalized aggregation-induced emission (AIE) chromophore to the phenothiazine-based organic dye resulted in favorable absorption properties (Figure 2.39) [86]. When compared to (*E*)-2-cyano-3-(10-ethyl-10*H*-phenothiazin-3-yl)acrylic acid (**AC11**) [16] without chromophoric decoration, the absorption spectrum of **AD13** is significantly red-shifted (23 nm). The V_{oc} of **AD13** also improved from 709 mV to 793 mV and J_{sc} increased from 6.13 mA cm^{-2} to 10.76 mA cm^{-2}. The overall efficiency of **AD13** is 5.51%.

Lin *et al.* studied the influence of the carbazole- or phenothiazine-containing antennas in starburst triphenylamine-based organic dyes (Figure 2.40) [99]. When compared to (*E*)-2-cyano-3-(4-(diphenylamino)phenyl)acrylic acid (**AC8)** the absorption peak of **AD14** is bathochromically shifted and LUMO increased to 2.10 V. This increased the electron density in the CB of TiO_2 by facilitating the electron harvest and injection efficiencies. Consequently, the efficiency increased to 3.26% from 2.79% observed for **AC8**. But the drop in efficiency was observed when compared to phenothiazine auxiliary donor (**AC12**) owing to its low HOMO level not favorable for regeneration of dye which led to lower V_{oc}. Later Sun *et al.* improved the efficiency of the DSSC based on **AD14** by using Br^-/Br_3^- redox couple to 3.63% [58]. The hike in efficiency is mainly due

Table 2.3 Optical, electrochemical and photovoltaic performance parameters of dyes containing carbazole as auxiliary donor.

Dye	$\lambda_{max,}$ nm (ε_{max}, M^{-1} cm^{-1})	E_{ox}', V (*vs* NHE)	E_{ox}^{*}', V (*vs* NHE)	J_{SC} (mA cm^{-2})	V_{OC} (mV)	*ff*	η (%)	*Ref*
AD1	470 (35564)	0.85	–1.27	10.60	579	0.58	3.54	[91]
AD2	491 (28702)	0.88	–1.08	12.70	598	0.61	4.63	[91]
AD3	477 (20170)	0.97	–1.04	5.06	561	0.59	1.67	[91]
AC1	455 (35264)	0.81	–1.53	13.20	562	0.58	4.28	[91]
AC2	467 (44827)	0.81	–1.36	14.20	560	0.60	4.77	[91]
AC3	480 (54656)	0.81	–1.51	11.00	530	0.51	2.97	[91]
AD4	513 (26013)	1.08	–0.99	5.35	560	0.68	2.05	[93]
AD5	417 (19567)	0.86	–1.66	3.86	582	–	1.55	[93]
AD6	477 (26046)	1.10	–1.11	–	–	–	1.49	[93]
AC4	464 (25491)	1.39	–0.92	8.19	618	0.70	3.53	[94], [93]
AC5	395 (17062)	–	–	5.85	664	0.74	2.86	[94], [93]

Table 2.3 (*Cont.*)

Dye	$\lambda_{max,}$ nm (ε_{max}, M^{-1} cm^{-1})	E_{ox}, V (*vs* NHE)	E_{ox}^{*}, V (*vs* NHE)	J_{SC} (mA cm^{-2})	V_{OC} (mV)	*ff*	η (%)	*Ref*
AC6	435 (22905)	1.37	−1.06	6.89	627	0.72	3.09	[94], [93]
AD7	415 (27900)	0.80	−1.60	10.50	710	0.62	4.62	[95]
AC7	448 (40500)	0.80	−1.60	16.34	720	0.61	7.21	[95]
AD8	395 (47400)	0.64	−1.64	9.42	730	0.53	3.63	[96]
AC8	370 (9400)	0.98	−1.51	6.88	640	0.63	2.79	[96]
AD9	523 (27900)	0.83	−1.11	14.16	628	0.66	5.87	[97]
AC9	504 (27200)	0.98	−1.05	12.28	614	0.68	5.09	[97]
AC10	491 (22300)	1.04	−1.01	9.77	622	0.65	3.99	[97]
AD10	457 (25174)	1.07	−1.24	7.19	730	0.67	3.52	[98]
AD11	463 (27460)	1.03	−1.23	8.88	700	0.66	4.10	[98]
AD12	472 (30102)	0.98	−1.26	10.89	700	0.67	5.12	[98]
AD13	462 (13000)	–	–	10.76	793	0.64	5.51	[86]
AC11	439 (1600)	–	–	6.13	709	0.67	2.91	[16]

(*Continued*)

Table 2.3 (*Cont.*)

Dye	λ_{max}, nm (ε_{max}, $M^{-1} cm^{-1}$)	E_{ox}, V (*vs* NHE)	E_{ox}^{*}, V (*vs* NHE)	J_{SC} ($mA\ cm^{-2}$)	V_{OC} (mV)	*ff*	η (%)	*Ref*
AD14	424 (13834)	0.45	−2.10	7.30	600	0.74	3.26	[99]
AC12	422 (29367)	0.64	−1.99	9.20	625	0.79	4.54	[99]
AD14	418 (12800)	1.24	−1.30	4.41	1015	0.76	3.38	[58]
AD15	480 (25000)	-	-	13.80	630	0.69	6.02	[6]
AC13	487 (8850)	-	-	13.60	615	0.69	5.77	[6]
AC14	508 (33300)	-	-	8.02	543	0.64	2.79	[6]
AC15	522 (19800)	-	-	9.05	567	0.56	2.87	[6]
AD16	458 (25225)	1.65	−0.60	7.53	720	0.69	3.70	[100]
AD17	455 (29322)	1.60	−0.61	9.02	710	0.68	4.34	[100]
AD18	464 (33926)	1.55	−0.58	9.98	700	0.67	4.62	[100]
AD19	425	1.00	−1.34	11.29	710	0.65	5.22	[101]

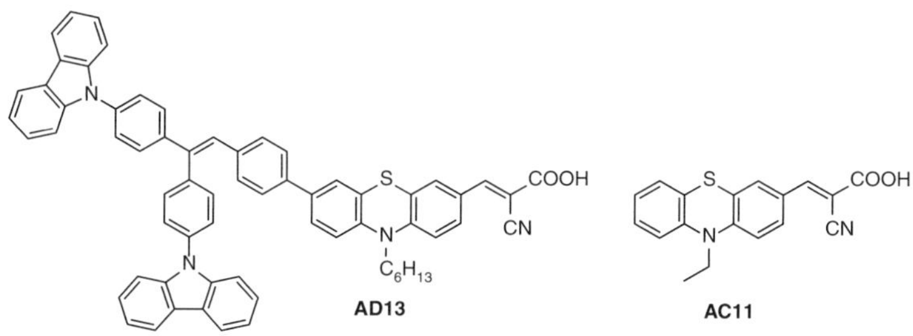

Figure 2.39 Phenothiazine-based organic dyes.

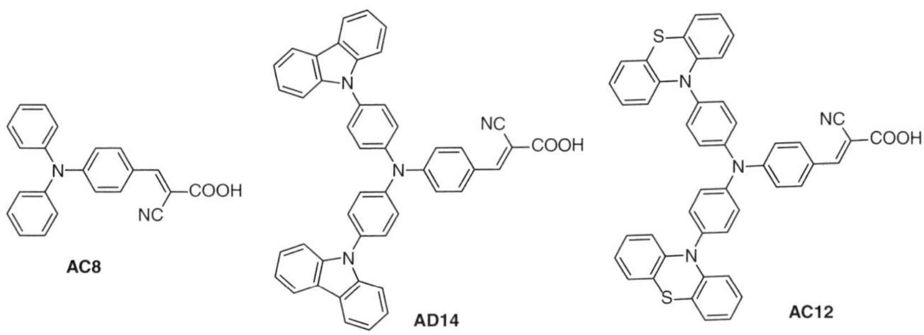

Figure 2.40 Starburst organic dyes featuring auxiliary donors.

to the tremendous increase in V_{oc} (1015 mV) with little sacrifice in *Jsc* (4.41 mA cm^{-2}).

Zhang *et al.* reported a series of dyes (Figure 2.41) based on starburst triarylamine dyes for efficient DSSC [6]. The authors demonstrated the fine tuning of HOMO and LUMO energy levels by the modification of auxiliary donors with triphenylamine and carbazole. The LUMO increased on introduction of carbazole in (*Z*)-2-(3-((*E*)-4-(diphenylamino)styryl)-5,5-di-methylcyclohex-2-enylidene)-2-cyanoacetic acid (**AC13**), while swapping it with triphenylamine reduced the LUMO level significantly. Due to the difference in the thermodynamic driving force for the electron injection, the carbazole substituted dye, **AD15**, showed higher efficiency (6.02%) than the triphenylamine analog (2.87%). Also when compared to **AD14**, **AD15** showed nearly double device efficiency due to the presence of rigid conjugation bridge in the form of 5,5-dimethyl-3-vinylcyclohex-2-enylidene.

Promarak *et al.* reported three dyes, **AD16–AD18** (Figure 2.42), containing 3,6-di-*tert*-butylcarbazole as auxiliary donor [100]. In

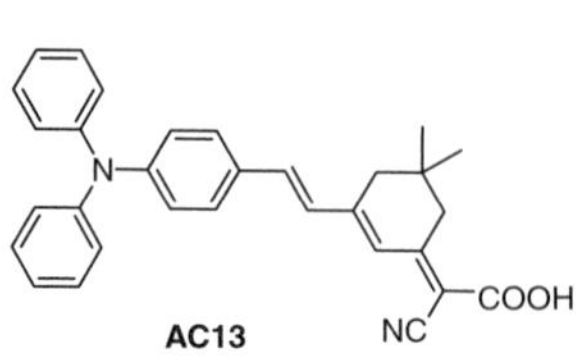

AC13

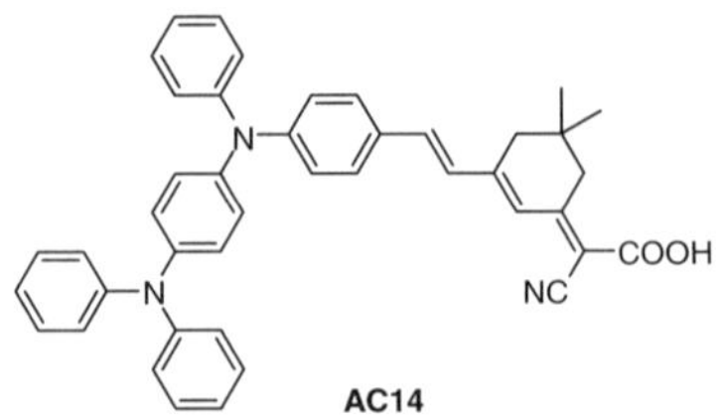

AC14

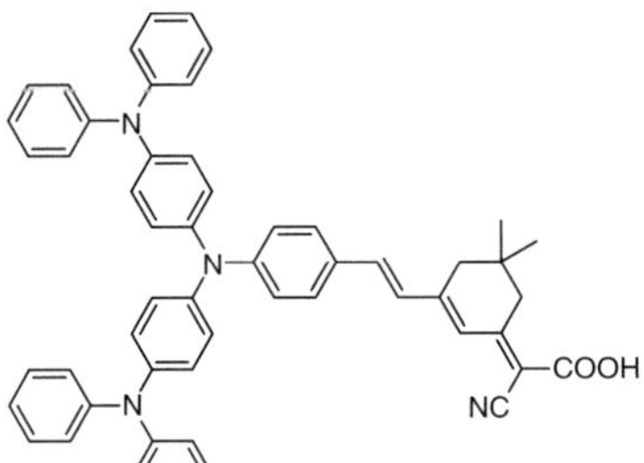

AC15

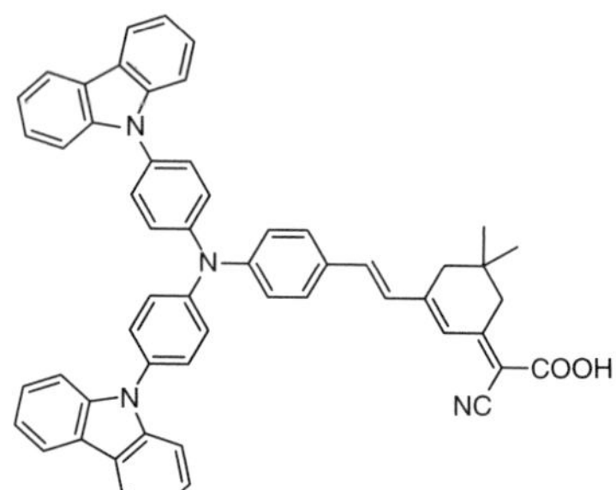

AD15

Figure 2.41 Starburst organic dyes featuring auxiliary donors.

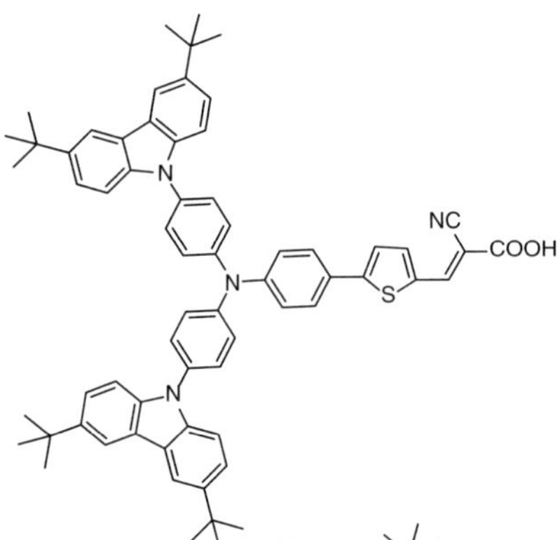

AD16

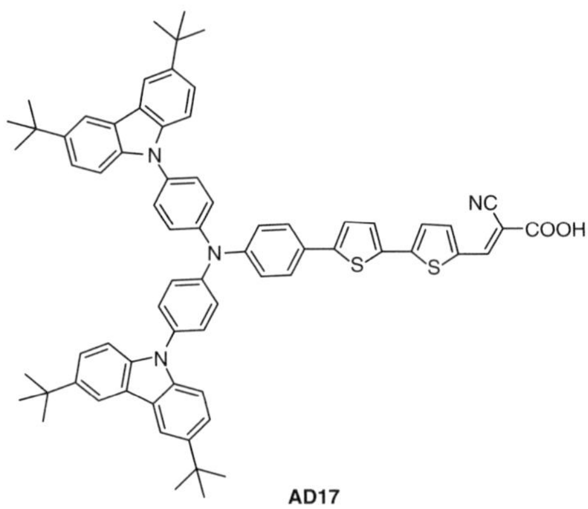

AD17

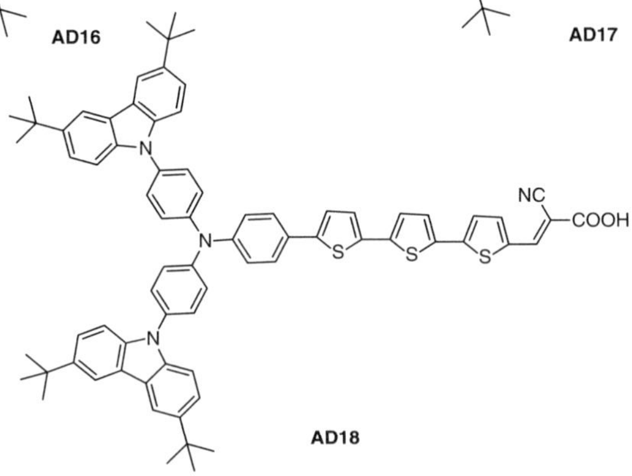

AD18

Figure 2.42 Starburst organic dyes with elongated conjugation.

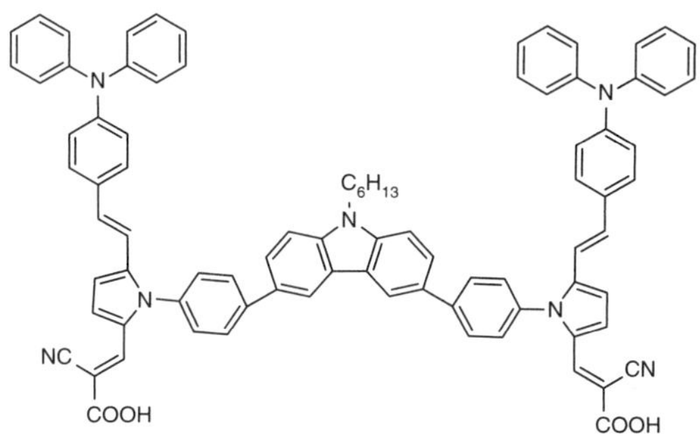

Figure 2.43 H-shaped dianchoring organic dyes.

these dyes, the conjugation extension with thiophene red-shifted the absorption maximum and absorption coefficient showed an upward hike when compared to the dyes without carbazole auxiliary donors. They exhibited efficiencies in the range of 3.70–4.62%.

2.4.4 Carbazole as Auxiliary Donor via C3, C6-positions

In 2012, Li *et al.* reported a new pyrrole-based dye (Figure 2.43) with H-shaped molecular architecture [101]. The authors have linked two D-π-A segments via a π-bridge featuring phenyl, fluorene or carbazole units. The carbazole-bridged dye **AD19** showed superior performance in the DSSC (η = 5.22%) over the fluorene- and phenylene-bridged dyes. The efficiency of the devices did not improve on addition of CDCA which indicated the absence of aggregation at the TiO_2 surface due to nonplanarity.

2.5 Carbazole as Donor as Well as Linker for DSSC

Different modes of chemical modifications possible for carbazole make it a versatile building block for organic dyes. As described above, carbazole was liberally used as a donor, auxiliary donor or linker in the construction of organic dyes. In some instances, carbazole served a dual function, namely, donor and linker. The structural compositions differ due to the point of attachment at carbazole with the rest of the molecule as shown in Figure 2.44.

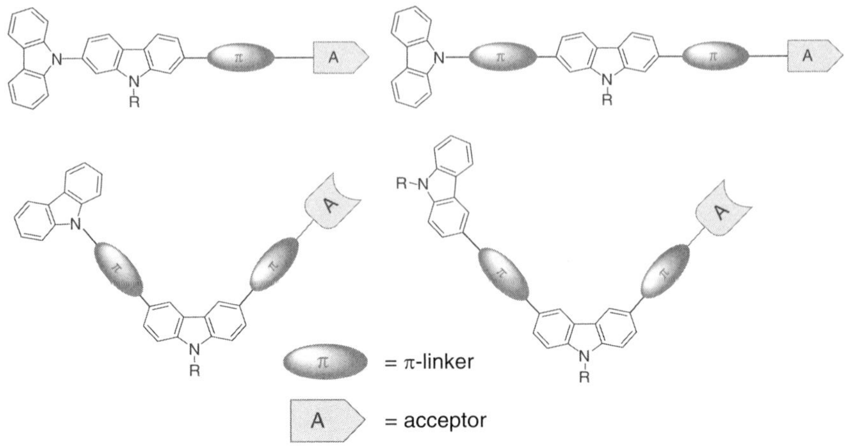

Figure 2.44 Molecular architectures of the organic dyes possessing carbazole as linker and donor.

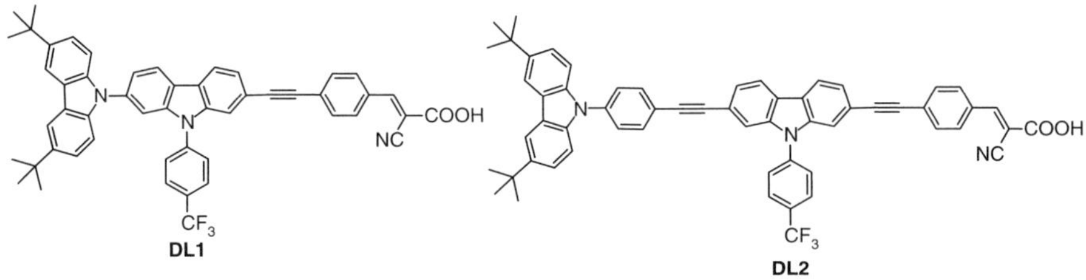

Figure 2.45 Organic dyes containing carbazole as *N*-donor and 2,7-disubstituted linker.

Castellano *et al.* reported dyes **DL1** and **DL2** (Figure 2.45) where carbazole served as *N*-donor and 2,7-disubstituted linker along with cyanoacrylic acid as acceptor [102]. **DL1** and **DL2** showed similar performance in DSSC with efficiency at 2.70%. It is interesting to compare with the dye **D43**, which contains bithiophene instead of carbazole linker. **D43** exhibited red-shifted absorption profile (Table 2.1) when compared to **DL2** (Table 2.4). However, the efficiency of **DL2** is better than that observed for **D43** (Figure 2.16). This clearly indicates that the carbazole acts as an efficient linker and facilitates charge separation in the molecule. The DSSCs lasted for over 1000 h at 55 °C without much efficiency loss.

Fang *et al.* designed three dyes, **DL3–DL5** (Figure 2.46) in an attempt to minimize the charge recombination and aggregation at the surface of photoanodes by using oligocarbazole as donor and

Table 2.4 Optical, electrochemical and photovoltaic performance parameters of dyes containing carbazole as donor and linker.

Dye	**λ_{max}, nm (ε_{max}, M^{-1} cm^{-1})**	**E_{ox}, V (*vs* NHE)**	**E_{ox}^{*}, V (*vs* NHE)**	**J_{SC} (mA cm^{-2})**	**V_{OC} (mV)**	***ff***	**η (%)**	***Ref***
DL1	395 (34120)	1.43	−1.28	6.24	617	0.70	2.70	[102]
DL2	385 (54740)	1.45	−1.18	5.96	642	0.71	2.70	[102]
DL3	448 (35700)	0.99	−1.42	11.47	767	0.67	5.87	[103]
DL4	444 (34400)	0.94	−1.48	11.67	796	0.68	6.33	[103]
DL5	456 (38700)	0.94	−1.47	11.46	792	0.68	6.16	[103]
DL6	412 (16000)	–	–	4.55	682	0.69	2.14	[86]

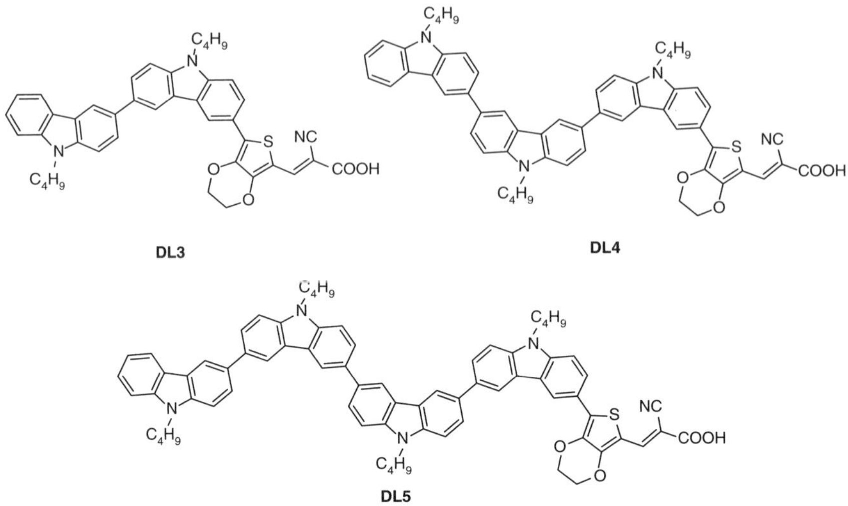

Figure 2.46 Organic dyes containing oligocarbazole units.

linker [103]. The number of carbazole units progressively increased from **DL3** to **DL5**. **DL3** possessed two carbazole units while **DL5** four. Irrespective of the variation in the number of carbazole units, all the dyes exhibited the same absorption wavelength. However, the molar extinction coefficient increased with the number of carbazole units in the chain. The presence of zigzag structure and alkyl chains played a major role in suppressing the formation of aggregates on TiO_2. The V_{oc} of the devices are high, in the range of 767–796 mV, and ascribed to the low propensity for charge recombination. Despite the twisted structure, **DL5** displayed slightly lower V_{oc} (792 mV) than **DL4** in DSSC. This is probably due to the loose packing of dye molecules creating voids through which I_3^- ions may diffuse and reach TiO_2 surface and scavenge the electrons. Among the dyes, **DL4** exhibited exceptionally high DSSC efficiency (6.33%). This clearly points out that a careful molecular engineering controlling the size and three-dimensional arrangement is necessary to realize efficient DSSCs.

Kuang *et al.* reported a dye (**DL6**, Figure 2.47) containing carbazole substituted triphenylethylene chromophore and 3,6-disubstituted carbazole linker [86]. On comparison with **D1** (0.42% in Table 2.1), it is obvious that the additional donor increases the efficiency of the DSSC based on **DL6** (2.15%). This is attributed to the red-shifted absorption observed for **DL6** (Table 2.4) when compared

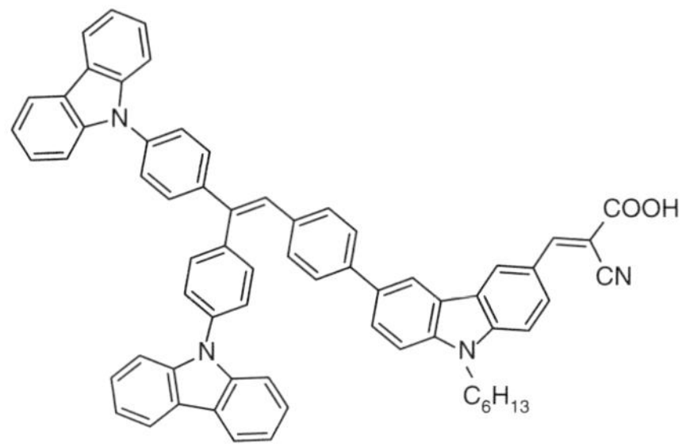

DL6

Figure 2.47 Organic dye featuring 3,6-disubstituted carbazole linker and triphenylethylene-based donor.

to **D1** (Table 2.1). This enhanced light-harvesting capability of **DL6** leads to large photocurrent density.

2.6 Conclusion and Outlook

In the past five years, carbazole-based organic sensitizers have drawn immense attention of researchers due to the chemical flexibility in the structural modifications and functional tuning. Many carbazole-based dyes have registered impressive power conversion efficiencies in DSSC. A survey of the available dyes showed the use of carbazole as a robust multifunctional building block. It has been demonstrated as donor, auxiliary donor and π-linker in the conventional donor-acceptor molecular architecture. The best efficiency reported for a carbazole-based dye is 8.30%, which is very close to that realized for the promising ruthenium dyes. For using the carbazole in organic dyes different connectivity on carbazole have been utilized. It is well established now that the C3 and C6 positions are useful for donor strength alterations, and connections through C2 and C7 positions are effective as π-linkers. The C3 and C6 functionalization helps to dissipate the charge and also pushes the electron density towards the acceptor. Among the organic dyes containing carbazole as donor the best efficiency is registered at 8.10%. By linking the amine donor and cyanoacrylic acid acceptor using a 2,7-disubstituted carbazole a best efficiency of 7.2% has been achieved. Usage of carbazole as auxiliary donor is beneficial to tune the LUMO energy and attain perfect packing at the surface of TiO_2. With carbazole as auxiliary donor efficiency as high as 8.50% has been reported. Though disubstitution on carbazole has been successfully translated to the design of organic dyes, tetra-substituted carbazole derivatives have not yet been utilized in the construction of organic dyes. We expect better performance in the future from carbazole based on various molecular engineering techniques and panchromatic design of dyes.

Acknowledgements

This work was supported by a grant-in-aid (Ref.: DST/TSG/ME/2010/27-C/G) from Department of Science and Technology, New Delhi to KRJT. AV acknowledges a receipt of research fellowship from University Grants Commission, New Delhi.

References

1. Y. Liang, and L. Yu, *Accounts of Chemical Research*, Vol. 43, 1227-1236, 2010.
2. B. O'Regan, and M. Gratzel, *Nature*, Vol. 353, 737-740, 1991.
3. M. Reisch, *Chemical & Engineering News Archive*, Vol. 85, 15, 2007.
4. A. Hagfeldt, G. Boschloo, L. Sun, L. Kloo, and H. Pettersson, *Chemical Reviews*, Vol. 110, 6595-6663, 2010.
5. D.P. Hagberg, T. Marinado, K.M. Karlsson, K. Nonomura, P. Qin, G. Boschloo, T. Brinck, A. Hagfeldt, and L. Sun, *The Journal of Organic Chemistry*, Vol. 72, 9550-9556, 2007.
6. Z. Ning, Q. Zhang, W. Wu, H. Pei, B. Liu, and H. Tian, *The Journal of Organic Chemistry*, Vol. 73, 3791-3797, 2008.
7. M. Cheng, X. Yang, F. Zhang, J. Zhao, and L. Sun, *The Journal of Physical Chemistry C*, Vol., 2013.
8. M. Liang, and J. Chen, *Chemical Society Reviews*, Vol. 42, 3453-3488, 2013.
9. K. Hara, Z.-S. Wang, T. Sato, A. Furube, R. Katoh, H. Sugihara, Y. Dan-oh, C. Kasada, A. Shinpo, and S. Suga, *The Journal of Physical Chemistry B*, Vol. 109, 15476-15482, 2005.
10. Z.S. Wang, Y. Cui, K. Hara, Y. Dan-oh, C. Kasada, and A. Shinpo, *Advanced Materials*, Vol. 19, 1138-1141, 2007.
11. K. Hara, K. Sayama, Y. Ohga, A. Shinpo, S. Suga, and H. Arakawa, *Chemical Communications*, Vol. 0, 569-570, 2001.
12. K.R.J. Thomas, J.T. Lin, Y.-C. Hsu, and K.-C. Ho, *Chemical Communications*, Vol. 4098-4100, 2005.
13. A. Baheti, P. Tyagi, K.R.J. Thomas, Y.-C. Hsu, and J.T.s. Lin, *The Journal of Physical Chemistry C*, Vol. 113, 8541-8547, 2009.
14. A. Baheti, P. Singh, C.-P. Lee, K.R.J. Thomas, and K.-C. Ho, *The Journal of Organic Chemistry*, Vol. 76, 4910-4920, 2011.

15. H. Tian, X. Yang, R. Chen, Y. Pan, L. Li, A. Hagfeldt, and L. Sun, *Chemical Communications*, Vol. 0, 3741-3743, 2007.
16. D. Cao, J. Peng, Y. Hong, X. Fang, L. Wang, and H. Meier, *Organic Letters*, Vol. 13, 1610-1613, 2011.
17. M. Velusamy, K.R.J. Thomas, J.T. Lin, Y.-C. Hsu, and K.-C. Ho, *Organic Letters*, Vol. 7, 1899-1902, 2005.
18. Y. Wu, and W. Zhu, *Chemical Society Reviews*, Vol. 42, 2039-2058, 2013.
19. Y. Cui, Y. Wu, X. Lu, X. Zhang, G. Zhou, F.B. Miapeh, W. Zhu, and Z.-S. Wang, *Chemistry of Materials*, Vol. 23, 4394-4401, 2011.
20. T. Horiuchi, H. Miura, K. Sumioka, and S. Uchida, *Journal of the American Chemical Society*, Vol. 126, 12218-12219, 2004.
21. S. Ito, H. Miura, S. Uchida, M. Takata, K. Sumioka, P. Liska, P. Comte, P. Pechy, and M. Gratzel, *Chemical Communications*, Vol. 0, 5194-5196, 2008.
22. K.R.J. Thomas, J.T. Lin, Y.-T. Tao, and C.-W. Ko, *Journal of the American Chemical Society*, Vol. 123, 9404-9411, 2001.
23. J. Li, C. Ma, J. Tang, C.-S. Lee, and S. Lee, *Chemistry of Materials*, Vol. 17, 615-619, 2005.
24. S.L. Lai, Q.X. Tong, M.Y. Chan, T.W. Ng, M.F. Lo, C.C. Ko, S.T. Lee, and C.S. Lee, *Organic Electronics*, Vol. 12, 541-546, 2011.
25. M.H. Tsai, Y.H. Hong, C.H. Chang, H.C. Su, C.C. Wu, A. Matoliukstyte, J. Simokaitiene, S. Grigalevicius, J.V. Grazulevicius, and C.P. Hsu, *Advanced Materials*, Vol. 19, 862-866, 2007.
26. K.R.J. Thomas, M. Velusamy, J.T. Lin, Y.T. Tao, and C.H. Chuen, *Advanced Functional Materials*, Vol. 14, 387-392, 2004.
27. K.R.J. Thomas, J.T. Lin, Y.-T. Tao, and C.-H. Chuen, *Chemistry of Materials*, Vol. 16, 5437-5444, 2004.
28. C.-W. Ko, Y.-T. Tao, J.T. Lin, and K.R.J. Thomas, *Chemistry of Materials*, Vol. 14, 357-361, 2001.
29. M. Sonntag, K. Kreger, D. Hanft, P. Strohriegl, S. Setayesh, and D. de Leeuw, *Chemistry of Materials*, Vol. 17, 3031-3039, 2005.
30. Y. Wang, L. Hou, K. Yang, J. Chen, F. Wang, and Y. Cao, *Macromolecular Chemistry and Physics*, Vol. 206, 2190-2198, 2005.
31. Y. Song, C.-a. Di, Z. Wei, T. Zhao, W. Xu, Y. Liu, D. Zhang, and D. Zhu, *Chemistry – A European Journal*, Vol. 14, 4731-4740, 2008.
32. J. Li, and A.C. Grimsdale, *Chemical Society Reviews*, Vol. 39, 2399-2410, 2010.
33. E. Wang, L. Hou, Z. Wang, Z. Ma, S. Hellström, W. Zhuang, F. Zhang, O. Inganäs, and M.R. Andersson, *Macromolecules*, Vol. 44, 2067-2073, 2011.
34. N. Koumura, Z.-S. Wang, S. Mori, M. Miyashita, E. Suzuki, and K. Hara, *Journal of the American Chemical Society*, Vol. 128, 14256-14257, 2006.
35. P. Singh, A. Baheti, and K.R.J. Thomas, *The Journal of Organic Chemistry*, Vol. 76, 6134-6145, 2011.
36. H. Sasabe, N. Toyota, H. Nakanishi, T. Ishizaka, Y.-J. Pu, and J. Kido, *Advanced Materials*, Vol. 24, 3212-3217, 2012.
37. A. Midya, Z. Xie, J.-X. Yang, Z.-K. Chen, D.J. Blackwood, J. Wang, S. Adams, and K.P. Loh, *Chemical Communications*, Vol. 46, 2091-2093, 2010.
38. S. Pansay, N. Prachumrak, S. Jungsuttiwong, T. Keawin, T. Sudyoadsuk, and V. Promarak, *Tetrahedron Letters*, Vol. 53, 4568-4572, 2012.

39. K. Smith, D.M. James, A.G. Mistry, M.R. Bye, and D.J. Faulkner, *Tetrahedron*, Vol. 48, 7479-7488, 1992.
40. T. Zhao, Z. Liu, Y. Song, W. Xu, D. Zhang, and D. Zhu, *The Journal of Organic Chemistry*, Vol. 71, 7422-7432, 2006.
41. Y. Zheng, A.S. Batsanov, V. Jankus, F.B. Dias, M.R. Bryce, and A.P. Monkman, *The Journal of Organic Chemistry*, Vol. 76, 8300-8310, 2011.
42. T. Kumchoo, V. Promarak, T. Sudyoadsuk, M. Sukwattanasinitt, and P. Rashatasakhon, *Chemistry – An Asian Journal*, Vol. 5, 2162-2167, 2010.
43. C. Teng, X. Yang, C. Yuan, C. Li, R. Chen, H. Tian, S. Li, A. Hagfeldt, and L. Sun, *Organic Letters*, Vol. 11, 5542-5545, 2009.
44. M.S. Kim, H.S. Yang, D.Y. Jung, Y.S. Han, and J.H. Kim, *Colloids and Surfaces A: Physicochemical and Engineering Aspects*, Vol. 420, 22-29, 2013.
45. D. Kim, J.K. Lee, S.O. Kang, and J. Ko, *Tetrahedron*, Vol. 63, 1913-1922, 2007.
46. Y.J. Chang, P.-T. Chou, S.-Y. Lin, M. Watanabe, Z.-Q. Liu, J.-L. Lin, K.-Y. Chen, S.-S. Sun, C.-Y. Liu, and T.J. Chow, *Chemistry – An Asian Journal*, Vol. 7, 572-581, 2012.
47. T. Duan, K. Fan, C. Zhong, T. Peng, J. Qin, and X. Chen, *RSC Advances*, Vol. 2, 7081-7086, 2012.
48. D. Liu, B. Zhao, P. Shen, H. Huang, L. Liu, and S. Tan, *Sci. China Ser. B-Chem.*, Vol. 52, 1198-1209, 2009.
49. M. Akhtaruzzaman, Menggenbateer, A. Islam, A. El-Shafei, N. Asao, T. Jin, L. Han, K.A. Alamry, S.A. Kosa, A.M. Asiri, and Y. Yamamoto, *Tetrahedron*, Vol. 69, 3444-3450, 2013.
50. M. Marszalek, S. Nagane, A. Ichake, R. Humphry-Baker, V. Paul, S.M. Zakeeruddin, and M. Gratzel, *RSC Advances*, Vol., 2013.
51. M. Miyashita, K. Sunahara, T. Nishikawa, Y. Uemura, N. Koumura, K. Hara, A. Mori, T. Abe, E. Suzuki, and S. Mori, *Journal of the American Chemical Society*, Vol. 130, 17874-17881, 2008.
52. K. Hara, Z.-S. Wang, Y. Cui, A. Furube, and N. Koumura, *Energy & Environmental Science*, Vol. 2, 1109-1114, 2009.
53. Z.-S. Wang, N. Koumura, Y. Cui, M. Takahashi, H. Sekiguchi, A. Mori, T. Kubo, A. Furube, and K. Hara, *Chemistry of Materials*, Vol. 20, 3993-4003, 2008.
54. N. Koumura, Z.-S. Wang, M. Miyashita, Y. Uemura, H. Sekiguchi, Y. Cui, A. Mori, S. Mori, and K. Hara, *Journal of Materials Chemistry*, Vol. 19, 4829-4836, 2009.
55. T.N. Murakami, N. Koumura, T. Uchiyama, Y. Uemura, K. Obuchi, N. Masaki, M. Kimura, and S. Mori, *Journal of Materials Chemistry A*, Vol. 1, 792-798, 2013.
56. S. Kajiyama, Y. Uemura, H. Miura, K. Hara, and N. Koumura, *Dyes and Pigments*, Vol. 92, 1250-1256, 2012.
57. X.H. Zhang, J. Ogawa, K. Sunahara, Y. Cui, Y. Uemura, T. Miyasaka, A. Furube, N. Koumura, K. Hara, and S. Mori, *The Journal of Physical Chemistry C*, Vol. 117, 2024-2031, 2013.
58. C. Teng, X. Yang, S. Li, M. Cheng, A. Hagfeldt, L.-z. Wu, and L. Sun, *Chemistry – A European Journal*, Vol. 16, 13127-13138, 2010.
59. E.M. Barea, C. Zafer, B. Gultekin, B. Aydin, S. Koyuncu, S. Icli, F.F. Santiago, and J. Bisquert, *The Journal of Physical Chemistry C*, Vol. 114, 19840-19848, 2010.

60. K. Srinivas, C.R. Kumar, M.A. Reddy, K. Bhanuprakash, V.J. Rao, and L. Giribabu, *Synthetic Metals*, Vol. 161, 96-105, 2011.
61. A.C. Onicha, K. Panthi, T.H. Kinstle, and F.N. Castellano, *Journal of Photochemistry and Photobiology A: Chemistry*, Vol. 223, 57-64, 2011.
62. W. Lee, N. Cho, J. Kwon, J. Ko, and J.-I. Hong, *Chemistry – An Asian Journal*, Vol. 7, 343-350, 2012.
63. J.-Y. Li, T.-H. Lin, S.-C. Chen, and C.-G. Wu, *Journal of the Chinese Chemical Society*, Vol. 59, 1337-1344, 2012.
64. C. Qin, A. Islam, and L. Han, *Journal of Materials Chemistry*, Vol. 22, 19236-19243, 2012.
65. T. Wang, J. Han, Z. Zhang, B. Xu, J. Huang, and J. Su, *Tetrahedron*, Vol. 68, 10372-10377, 2012.
66. C.-C. Yu, K.-J. Jiang, J.-H. Huang, F. Zhang, X. Bao, F.-W. Wang, L.-M. Yang, and Y. Song, *Organic Electronics*, Vol. 14, 445-450, 2013.
67. C. Olivier, F. Sauvage, L. Ducasse, F. Castet, M. Grätzel, and T. Toupance, *ChemSusChem*, Vol. 4, 731-736, 2011.
68. F. Yang, M. Akhtaruzzaman, A. Islam, T. Jin, A. El-Shafei, C. Qin, L. Han, K.A. Alamry, S.A. Kosa, M.A. Hussein, A.M. Asiri, and Y. Yamamoto, *Journal of Materials Chemistry*, Vol. 22, 22550-22557, 2012.
69. K. Srinivas, K. Yesudas, K. Bhanuprakash, V.J. Rao, and L. Giribabu, *The Journal of Physical Chemistry C*, Vol. 113, 20117-20126, 2009.
70. C. Teng, X. Yang, C. Yang, S. Li, M. Cheng, A. Hagfeldt, and L. Sun, *The Journal of Physical Chemistry C*, Vol. 114, 9101-9110, 2010.
71. K.R.J. Thomas, P. Singh, A. Baheti, Y.-C. Hsu, K.-C. Ho, and J.T.s. Lin, *Dyes and Pigments*, Vol. 91, 33-43, 2011.
72. S. Hwang, J.H. Lee, C. Park, H. Lee, C. Kim, C. Park, M.-H. Lee, W. Lee, J. Park, K. Kim, N.-G. Park, and C. Kim, *Chemical Communications*, Vol. 0, 4887-4889, 2007.
73. S.-B. Ko, A.-N. Cho, M.-J. Kim, C.-R. Lee, and N.-G. Park, *Dyes and Pigments*, Vol. 94, 88-98, 2012.
74. A. Mishra, M.K.R. Fischer, and P. Bäuerle, *Angewandte Chemie International Edition*, Vol. 48, 2474-2499, 2009.
75. R.Y.-Y. Lin, Y.-S. Yen, Y.-T. Cheng, C.-P. Lee, Y.-C. Hsu, H.-H. Chou, C.-Y. Hsu, Y.-C. Chen, J.T. Lin, K.-C. Ho, and C. Tsai, *Organic Letters*, Vol. 14, 3612-3615, 2012.
76. S. Cai, X. Hu, J. Han, Z. Zhang, X. Li, C. Wang, and J. Su, *Tetrahedron*, Vol. 69, 1970-1977, 2013.
77. H. Tian, X. Yang, J. Cong, R. Chen, J. Liu, Y. Hao, A. Hagfeldt, and L. Sun, *Chemical Communications*, Vol. 0, 6288-6290, 2009.
78. H. Tian, I. Bora, X. Jiang, E. Gabrielsson, K.M. Karlsson, A. Hagfeldt, and L. Sun, *Journal of Materials Chemistry*, Vol. 21, 12462-12472, 2011.
79. Y. Ooyama, S. Inoue, T. Nagano, K. Kushimoto, J. Ohshita, I. Imae, K. Komaguchi, and Y. Harima, *Angewandte Chemie International Edition*, Vol. 50, 7429-7433, 2011.
80. Y. Ooyama, T. Nagano, S. Inoue, I. Imae, K. Komaguchi, J. Ohshita, and Y. Harima, *Chemistry-a European Journal*, Vol. 17, 14837-14843, 2011.
81. A. Venkateswararao, K.R.J. Thomas, C.-P. Lee, and K.-C. Ho, *Tetrahedron Letters*, Vol. 54, 3985-3989, 2013.

82. A. Baheti, K.R.J. Thomas, C.-P. Lee, and K.-C. Ho, *Chemistry – An Asian Journal*, Vol. 7, 2942-2954, 2012.
83. H.-C. Chu, D. Sahu, Y.-C. Hsu, H. Padhy, D. Patra, J.-T.S. Lin, D. Bhattacharya, K.-L. Lu, K.-H. Wei, and H.-C. Lin, *Dyes and Pigments*, Vol. 93, 1488-1497, 2012.
84. Y.-S. Yen, Y.-C. Hsu, J.T. Lin, C.-W. Chang, C.-P. Hsu, and D.-J. Yin, *The Journal of Physical Chemistry C*, Vol. 112, 12557-12567, 2008.
85. T.-Y. Wu, M.-H. Tsao, S.-G. Su, H.P. Wang, Y.-C. Lin, F.-L. Chen, C.-W. Chang, and I.-W. Sun, *Journal of the Brazilian Chemical Society*, Vol. 22, 780-789, 2011.
86. C. Chen, J.-Y. Liao, Z. Chi, B. Xu, X. Zhang, D.-B. Kuang, Y. Zhang, S. Liu, and J. Xu, *Journal of Materials Chemistry*, Vol. 22, 8994-9005, 2012.
87. R. Grisorio, L. De Marco, G. Allegretta, R. Giannuzzi, G.P. Suranna, M. Manca, P. Mastrorilli, and G. Gigli, *Dyes and Pigments*, Vol. 98, 221-231, 2013.
88. S. Ramkumar, and S. Anandan, *Dyes and Pigments*, Vol. 97, 397-404, 2013.
89. M.-D. Zhang, H.-X. Xie, X.-H. Ju, L. Qin, Q.-X. Yang, H.-G. Zheng, and X.-F. Zhou, *Physical Chemistry Chemical Physics*, Vol. 15, 634-641, 2013.
90. Y. Ooyama, N. Yamaguchi, I. Imae, K. Komaguchi, J. Ohshita, and Y. Harima, *Chemical Communications*, Vol. 49, 2548-2550, 2013.
91. C.-H. Yang, S.-H. Liao, Y.-K. Sun, Y.-Y. Chuang, T.-L. Wang, Y.-T. Shieh, and W.-C. Lin, *The Journal of Physical Chemistry C*, Vol. 114, 21786-21794, 2010.
92. C.-H. Yang, H.-L. Chen, Y.-Y. Chuang, C.-G. Wu, C.-P. Chen, S.-H. Liao, and T.-L. Wang, *Journal of Power Sources*, Vol. 188, 627-634, 2009.
93. B.H. Kim, and H.S. Freeman, *Photochemical & Photobiological Sciences*, Vol. 12, 421-431, 2013.
94. B.H. Kim, and H.S. Freeman, *Journal of Materials Chemistry*, Vol. 22, 20403-20409, 2012.
95. Q. Li, L. Lu, C. Zhong, J. Huang, Q. Huang, J. Shi, X. Jin, T. Peng, J. Qin, and Z. Li, *Chemistry – A European Journal*, Vol. 15, 9664-9668, 2009.
96. J. Jia, K. Ca, P. Xue, Y. Zhang, H. Zhou, and R. Lu, *Tetrahedron*, Vol. 68, 3626-3632, 2012.
97. B. Liu, Q. Liu, D. You, X. Li, Y. Naruta, and W. Zhu, *Journal of Materials Chemistry*, Vol. 22, 13348-13356, 2012.
98. S. Namuangruk, R. Fukuda, M. Ehara, J. Meeprasert, T. Khanasa, S. Morada, T. Kaewin, S. Jungsuttiwong, T. Sudyoadsuk, and V. Promarak, *The Journal of Physical Chemistry C*, Vol. 116, 25653-25663, 2012.
99. Z. Wan, C. Jia, Y. Duan, L. Zhou, J. Zhang, Y. Lin, and Y. Shi, *RSC Advances*, Vol. 2, 4507-4514, 2012.
100. T. Khanasa, N. Jantasing, S. Morada, N. Leesakul, R. Tarsang, S. Namuangruk, T. Kaewin, S. Jungsuttiwong, T. Sudyoadsuk, and V. Promarak, *European Journal of Organic Chemistry*, Vol. 2013, 2608-2620, 2013.
101. Q. Li, J. Shi, H. Li, S. Li, C. Zhong, F. Guo, M. Peng, J. Hua, J. Qin, and Z. Li, *Journal of Materials Chemistry*, Vol. 22, 6689-6696, 2012.
102. Y. Ooyama, Y. Shimada, A. Ishii, G. Ito, Y. Kagawa, I. Imae, K. Komaguchi, and Y. Harima, *Journal of Photochemistry and Photobiology A: Chemistry*, Vol. 203, 177-185, 2009.
103. H. Lai, J. Hong, P. Liu, C. Yuan, Y. Li, and Q. Fang, *RSC Advances*, Vol. 2, 2427-2432, 2012.

3

Colloidal Synthesis of $CuInS_2$ and $CuInSe_2$ Nanocrystals for Photovoltaic Applications

Joanna Kolny-Olesiak

Energy and Semiconductor Research Laboratory, Department of Physics, Carl von Ossietzky University of Oldenburg, Oldenburg, Germany

Abstract

Semiconductor nanocrystals attract scientific attention due to their size- and shape-dependent properties. They are interesting candidates for photovoltaic or light-emitting applications. However, the best studied nanomaterials contain toxic elements like cadmium or lead, which severely restricts their industrial applications. That is why I-III-VI semiconductor nanocrystals, which are much less toxic, but possess similar optical properties, are gaining scientific interest. The number of reports about colloidal synthesis of I-III-VI nanocrystals is steadily increasing. However, controlling the size, shape, composition and optical properties of these materials still remains a challenge. This chapter summarizes the experimental methods of colloidal synthesis of $CuInS_2$ and $CuInSe_2$ nanocrystals, and the possibilities to control the properties of these nanomaterials by varying the conditions of their colloidal synthesis. In the last part of the chapter a short overview of the applications of $CuInS_2$ and $CuInSe_2$ nanocrystals in solar energy conversion is given.

Keywords: Colloidal synthesis, I-III-VI semiconductor nanocrystals, $CuInS_2$, $CuInSe_2$

3.1 Introduction

Optical and electrical properties of semiconductor nanocrystals depend not only on their composition and the crystallographic

**Corresponding author*: joanna.kolny@uni-oldenburg.de

Atul Tiwari, Rabah Boukherroub, and Maheshwar Sharon (eds.) Solar Cell Nanotechnology, (97–116) 2014 © Scrivener Publishing LLC

structure, but also to a great extent on their morphological parameters, such as size and shape [1]. This opens up new possibilities to control and design the properties of nanomaterials and pushes forward efforts to develop novel procedures to generate nanoscale particles. That is why colloidal synthesis of nanocrystalline materials, which allows for precise control of the morphological parameters of nanocrystals, is a rapidly developing field of research [2–4]. During the last two decades immense progress has been achieved in the synthesis of nanocrystals and, consequently, in controlling the properties of nanocrystalline materials. However, the main focus of the researchers laid for a long time on the II-VI or IV-VI semiconductor materials, which exhibit interesting optical properties, but have a severe drawback of containing highly toxic heavy metals, such as cadmium or lead. Therefore, for large-scale applications more environmentally friendly and less toxic alternatives are needed, which have similar properties. Appropriate candidates for applications in solar energy conversion were found among I-III-VI semiconductors. Especially the ternary copper-based materials, $CuInS_2$ (CIS) and $CuInSe_2$ (CISe), possess bandgaps, which match well the solar spectrum (1.4 eV and 1.04 eV respectively). Their optical properties can be further adjusted by alloying them with ZnS or exchanging some fraction of indium ions with gallium. Because of their optical properties, exceptional radiation stability and high extinction coefficients in the visible range of the solar spectrum, alloys such as $Cu(In,Ga)(S,Se)_2$ already find application in thin-film solar cells, where they exhibit power conversion efficiencies up to 20% [5]. The fabrication of this kind of photovoltaic device requires, however, expensive production techniques (high investment costs, vacuum, high temperature). Therefore, extensive research activities are focused on finding low-priced alternative manufacturing methods of thin-film solar cells. Here, the application of colloidal nanocrystals opens up new possibilities to generate thin layer devices using solution-based processes. Furthermore, new kinds of solar cells can be obtained from colloidal nanoparticles, which are not accessible with standard vacuum-based techniques, e.g., hybrid organic-inorganic solar cells (see Section 3.3.2).

Properties, which are important for successful application of inorganic nanocrystals in photovoltaic devices, can be adjusted by changing the parameters of the synthesis. The factors influencing the growth and shape control of binary materials have been extensively studied during the last years [2]. Much less is known, yet, about the growth of ternary and quaternary nanomaterials, such as I-III-VI semiconductor nanocrystals. The presence of two different cationic precursors with different reactivities makes planning the synthesis more complicated than in the case of binary materials. The formation of undesirable side products, which can have negative effects on the application of these materials, is frequently observed. On the other hand, in contrast to binary materials, the composition becomes another parameter, which can significantly influence the properties of ternary or quaternary nanocrystals. This chapter summarizes the recent achievements in the field of colloidal synthesis of CIS, CISe and their alloys. In particular, the possibilities of controlling parameters are discussed, which are relevant for the application of these materials in photovoltaics.

3.2 Synthesis of $CuInS_2$ and $CuInSe_2$ Nanocrystals

The synthesis of colloidal nanocrystals in organic high-boiling solvent turned out to be an effective way to produce a variety of highly crystalline semiconductor and metal nanostructures with controllable size and shape [2–4]. In order to obtain uniform particles, it is important to separate the nucleation from the further growth step. An effective way to achieve this is the so called hot-injection-technique, which was developed in the early nineties for the synthesis of CdSe nanoparticles [6], and meanwhile, adapted for many other materials. For more details on this kind of synthesis the reader is referred to review articles [7–10]. Briefly, in this method two solutions are prepared, which contain precursors for the reaction. One of them is heated to a high temperature (typically between 200°C and 300°C). Rapid oversaturation and, consequently, burst nucleation is achieved by the injection of the second solution into the first one. The nucleation step leads to a fast consumption of the monomers; therefore, the oversaturation of the solution lasts only for a short time period and all the seeds are formed approximately at the same time. This reduces the differences between the particles; they all form and, then, grow under nearly identical conditions. This method, however, is not easy to scale up. Heating-up procedures, on the other hand, in which all the precursors are mixed together at room temperature and subsequently heated to the required reaction temperature, can be scaled up without problems. There drawback is frequently the lack of precise control over the nucleation of the particles, and resulting broad size distribution of the nanoparticles. Both methods were already successfully applied for the synthesis of I-III-VI semiconductor materials.

Common starting materials for the formation of I-III-VI semiconductor nanoparticles are metal salts (e.g. chlorides, iodides, acetates, acetylacetonates) and chalcogene compounds, such as thiols, thiourea, selenourea, trioctylalkyl selenide, or elemental sulfur and selenium. When brought together at high temperature, these chalcogene and metal precursors (also referred to as monomers) react rapidly with each other to form binary or ternary compounds. This reaction can be slowed down and controlled by the use of ligand molecules. Their role will be discussed in the following section.

3.2.1 Ligand Shell and Colloidal Stability

One of the biggest advantages of colloidal nanocrystals as alternative materials for photovoltaic applications is their processability from solution. Because of their ability to form stable colloidal suspensions with common solvents, nanoparticles can be handled as "normal" chemicals: they can be dried to form powders and redispersed again. This is achieved by covering the inorganic core of the nanoparticles with an organic ligand shell during the synthesis.

In the colloidal synthesis of inorganic materials the formation of the nanocrystals takes place in presence of organic stabilizer molecules. Compounds such as thiols, carboxylic acids, phospine oxides, phosphonic acids or amines, are the most common ligands. All of them have a functional group with an electron pair that can bind to the metal ions on the surface of the nanocrystals. They play an important role during the formation and the growth of the nanocrystals, because they not only stabilize the crystallites in solution during the synthesis and prevent them from fast, uncontrolled growth, but also bind to the monomers and regulate their activity. Three kinds of effects influence the reactivity of the precursors. High concentration of ligands reduces the activity of the monomers, as well as ligands, which form strong bonds with the precursors; finally, steric effects play a role: bulky ligands, which reduce the mobility of the monomers, also lower their activity. All this strongly affects the whole nucleation and growth process as will be discussed in more detail in the following sections.

Another important function of the ligands is providing chemical and colloidal stability of the particles. The ligand shell prevents the nanocrystals from aggregation. In unpolar solvents this is mainly due to steric effects, i.e., the repulsion of the carbon chains of the ligand molecules. Because the surfactant layer is strongly responsible for the interaction of the nanocrystals in colloidal suspension, it also plays an important role in self-assembly processes of nanocrystals. From the point of view of colloidal stability, organic compounds with long carbon chains are especially suitable ligand molecules, because they effectively prevent aggregation of the nanoparticles. They also turn out to be the most suitable ligands for most synthetic procedures. However, a dense and thick ligand shell has disadvantages for the application of the particles in photovoltaics, where charge separation and charge transport are important steps of electrical power generation [11]. The organic ligand layer is an insulator, which severely interferes with these two elementary processes. Therefore, usually a ligand exchange is conducted after the synthesis, in which the original surfactant molecules are replaced with less bulky ones. Ideally, these ligands should still prevent the particles from aggregation, while enabling charge transport. In practice, small organic ligands frequently do not stabilize the particles in a sufficient manner, which leads to the formation of small aggregates in the colloidal suspension. This diminishes the processability of the particles. Another problem related to weak stabilization is the presence of defects on the surface, which act as traps for charge carriers and deteriorate the performance of the device [12]. The amount of surface defects, can, however, be reduced by covering the surface with a layer of another inorganic material (see Section 3.2.5).

Even in cases, where the organic ligand shell is removed prior to application, such as in thin films formed through thermal treatment of a layer of nanoparticles processed from colloidal suspensions, the presence of bulky ligands causes problems. Carbon usually cannot be removed completely from the inorganic film; furthermore, cracks form because of the change in the volume resulting from the removal of the organic material.

During the last years, fully inorganic surfactants have also been developed, which can also provide colloidal stability of nanocrystalline materials in polar solvents and have beneficial effects on the charge transport in superlattices composed of these semiconductor nanoparticles [13–16]. Even though the strategy of producing nanomaterials with inorganic ligands has a lot of advantages, only a few examples can be found until now of the use of these kinds of ligands for I-III-VI semiconductor nanocrystals [17].

3.2.2 Adjusting the Reactivity of the Precursors

The synthesis of ternary I-III-VI semiconductor nanocrystals requires a reaction between a chalcogen source and two different cations, e.g., Cu^+ and In^{3+}. A challenge for designing synthetic procedures for these particles arises from the fact that the two cations have different chemical properties: copper is a soft Lewis acid, while indium and gallium are hard Lewis acids. Selenium and sulfur are soft Lewis bases. Therefore, both chalcogens preferentially react with copper, which leads to the formation of copper chalcogenides. To avoid this, the reactivity of the cations has to be adjusted by using appropriate ligands. Among the common ligands used in the colloidal synthesis, thiols or phosphines are soft Lewis bases, while amines or carboxylic acids are hard ones. Thus, the reactivity of the cations cannot be controlled easily with one kind of organic stabilizer only, unless it is applied in a very high concentration [18]. A precise control of the activity of both cations can be achieved by using more than one kind of stabilizer, e.g., a thiol (a soft Lewis base) for copper and a carboxylic acid (a hard Lewis base) for indium [19]. Another possibility for overcoming the problems connected with the different reactivities of copper and indium is the use of single molecule precursors [20–23] containing both metals, and releasing them simultaneously during the reaction. In some reactions beneficial effects on the stoichiometry and phase purity of the resulting nanoparticles were observed, if the indium and the sulfur source were mixed together prior to addition of copper precursor [24]. In this way an In-S bond could be formed before the reaction with copper. This shows that not only the right choice of the precursors, but also the way in which the synthesis is carried out, has a strong influence on the properties of the resulting particles.

Copper reacts faster with sulfur or selenium than indium or gallium, which frequently results in the formation of copper sulfide or copper-rich CIS nanocrystals in the beginning of the reaction [19, 24–28]. At later growth stages the element ratio in the nanoparticles can change towards higher indium contents, due to the copper depletion of the solution [25]. Thus, controlling the size and the composition of the particles independently from each other is not always possible. Obtaining stoichiometries close to the ideal 1:1:2 atomic ratios requires a precise control of the activities of all the precursors. This was achieved by Li *et al.* [18] in a reaction taking place in dodecanethiol as solvent and sulfur source. The large excess of the anionic precursor prevented the formation of copper interstitials, which seem to be responsible for the changes in the stoichiometry during the growth process.

The formation of copper sulfide particles in the beginning of the synthesis is usually considered a problem, and in most of the synthetic procedures the conditions are chosen so that this can be avoided. However, the formation of copper sulfide particles may also have advantages and was successfully used for shape control of CIS nanorods (see next section) [26–29].

3.2.3 Shape Control

The shape of semiconductor nanocrystals depends on many different factors. Particularly essential are the ligands used during the synthesis, because they can delay the growth of some facets of the crystals, and, thus, influence their shape. Furthermore, they influence the crystallographic structure, and consequently the symmetry of the crystallites (see Section 3.2.4), which strongly affects the final shape. Also, the temperature plays an important role in the shape control because it can influence the crystallographic structure of the nanocrystals. Furthermore, it depends on the temperature if the growth takes place in the thermodynamic or in the kinetic regime. Uniform facetted crystals formed under thermodynamic control (high temperature, low monomer flux towards the surface of growing nanocrystals) can serve as building blocks for ordered superstructures of nanocrystals. They are, therefore, suitable for applications which require the formation of layers of one kind of material.

The first report about successful size control of CIS nanocrystals was published in 2009 by Xie *et al.* [19]. Quasi-spherical particles with sizes between 2 nm and 16 nm were synthesized and the stoichiometry of the particles was controlled by adjusting the reactivities of the cationic precursors. In the same year, uniform CIS nanodiscs with wurtzite structure were synthesized by heating metal chlorides and thiourea in oleylamine [30]. Analog reaction with selenourea results in the formation of uniform CISe trigonal pyramides (Figure

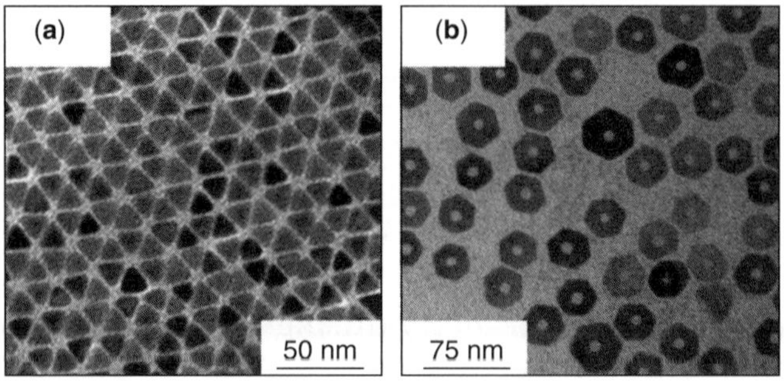

Figure 3.1 Shape control of CISe nanocrystals: a) trigonal pyramids (adapted with permission from: *J. Am. Chem. Soc.*, 2009, 131 (9), pp 3134–3135) Copyright (2009) American Chemical Society. and b) nanorings (adapted with permission from *Nano Lett.*, 2008, 8 (9), pp 2982–2987) Copyright (2008) American Chemical Society.

3.1a) [31]. Uniform CISe nanorings (Figure 3.1b) were obtained in a reaction of CuCl, $InCl_3$ and Se in oleylamine in the presence of trioctylphosphine [32]. Trioctylphosphine proved necessary for the shape control in this case; however the underlying mechanism is not clear. A non-injection method for the synthesis of CIS pyramids with sizes from 3.5 nm to 7.3 nm with chalcopyrite structure was developed by Zhong *et al.* [25]; two surfactants, dodecanethiol and oleic acid, were applied in this synthesis to control the activity of the monomers, however, together with the variation of the size, changes in the composition of the particles were observed. Another non-injection method, which applies dodecanethiol as solvent and sulfur source, yields particles with pyramidal shape, and without changes in the composition during the growth process [18].

Under kinetic control (low temperature, high monomer flux), shapes are accessible, which are thermodynamically less stable, such as elongated nanocrystals. These shapes are particularly suitable for application in hybrid solar cells, which are composed of a conductive polymer and inorganic semiconductor nanocrystals. Conductive polymers have a good hole conductivity; blending them with inorganic nanocrystals increases the electron conductivity of the active layer. This is especially efficient if nanorods or nanowires are applied. However, nanorods tend to form aggregates, which, in turn, have a negative influence on the morphology of the polymer-nanorod-blend and impair the device performance. Branched nanocrystals (e.g. tetrapods) do not have this drawback; in general, they are better soluble than nanorods, while still offering pathways for effective electron transport through the active layer.

In the synthesis of CIS nanoparticles a new strategy of shape control has been developed in which the formation of $Cu_{2-x}S$-CIS hybrid nanostructure plays a central role [26–28]. If the reaction conditions are chosen so that $Cu_{2-x}S$ nanocrystals are formed in the beginning of the reaction, these particles can serve as a starting point for the growth of CIS nanostructures (Figure 3.2). Hexagonal CIS and $Cu_{2-x}S$ have the same anion lattice; furthermore, the mobility of copper within the particles is relatively high at elevated temperatures. This facilitates the formation of $Cu_{2-x}S$-CIS hybrid nanostructures. During their growth the nucleation of CIS takes place within the $Cu_{2-x}S$ nanocrystals. When the CIS seeds reach a certain size, the $Cu_{2-x}S$-CIS hybrid structure reorganizes to form two discs connected together. During the next growth stage both phases grow; however, the growth of the CIS phase takes place only at the interface between the two phases. Thus, the shape of the CIS particle can be controlled by changing the size and the position of the $Cu_{2-x}S$ part of the hybrid nanostructure. In the last stage of the reaction the size of the $Cu_{2-x}S$ particle decreases, until, finally, pure CIS nanocrystals are left (Figure 3.2d) [27].

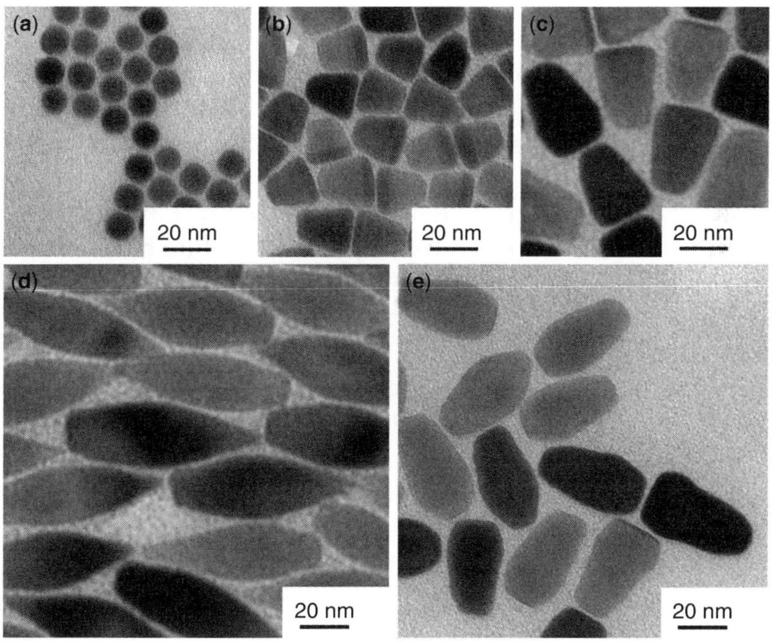

Figure 3.2 Growth process of CIS nanorods (adapted with permission from *J. Am. Chem. Soc.*, 2010, 132 (45), pp 15976–15986. Copyright (2010) American Chemical Society).

Controlled aggregation resulting in the formation of multipods or nano-networks can be induced by destabilizing nanorods at later growth stages. This can be obtained if the ligands and the solvent react with each other to form a compound which cannot stabilize the particles. This reaction must be, however, slower than the formation and the growth of the nanocrystals to avoid their uncontrolled growth and to maintain shape control. Also, other shapes such as flower-like particles can be obtained if the stabilization of the particles is not sufficient. To minimize the surface energy, these particles attach together to form small, flower-shaped aggregates [33].

3.2.4 Crystallographic Structure

Bulk CIS and CISe both crystallize in chalcopyrite structure at room temperature. Cation disordered polymorphs, wurtzite (WZ) and zincblende (ZB) are observed in bulk only at high temperatures. In contrast to this, nanocrystalline CIS and CISe are stable at room temperature in all three modifications. Structures with both cations sharing one lattice site allow a flexibility of stoichiometry, which enables tuning the Fermi energy of the nanostructure in a wide range; this is beneficial for device fabrication. Theoretical calculation for wurtzite CISe particles show that this structure has advantages for being used as active layer of photovoltaic devices with improved absorption and electron transport, compared to chalcopyrite structure [34].

In colloidal synthesis of I-III-VI semiconductor particles the crystallographic phase of the product can be controlled by the presence of specific ligand molecules and by the temperature of the reaction. Some ligands can stabilize structures, which are otherwise not thermodynamically stable. This has been observed, e.g., in II-VI semiconductor nanoparticles and can also be applied to generate I-III-VI semiconductor nanoparticles with the desired crystallographic structure.

The first synthesis of zincblende and wurtzite CIS nanoparticles was described by Pan *et al.* The crystallographic structure was controlled by the use of oleic acid (ZB) or dodecanethiol (WZ) as capping agent, under otherwise identical experimental conditions. Other examples of controlling the crystallographic structure followed in the next years, which not only describe the synthetic procedures, but also attempt to give an explanation for the observed results [23, 28, 35]. The formation of thermodynamically more stable chalcopyrite and zincblende particles was usually observed in slower reactions, taking place under thermodynamically controlled growth conditions. Fast reactions and the presence of thiols in the reaction solution favor the generation of wurtzite particles [28].

Also, in the synthesis of CISe nanoparticles the crystallographic structure could be controlled. In 2010 Norako and Brutchey reported the first synthesis of wurtzite CISe nanoparticles [36], which had a rather broad size distribution and sizes around 30 nm. The presence of oleylamine and the use of diphenyl diselenide as selenium source turned out as critical for the formation of wurtzite NPs. Monodisperse CISe nanocrystals with hexagonal shape and a diameter of about 21.3 nm were later obtained with a similar procedure by Wang *et al.* [37].

3.2.5 Composition

The possibility of adjusting the positions of the energy levels is one of the essential requirements for successful application of a material in photovoltaic devices. This can be achieved by reducing the size semiconductor nanocrystals below the value of the Bohr exciton radius (4.2 nm and 10.2 nm in CIS and CISe, respectively). Thus, this method can be applied effectively only in the size range of a few nanometers. However, small particles are less stable and have more defects than larger ones. An alternative approach for changing the optical properties without changing the size of the particles uses alloying with another semiconductor material, e.g., $CuInSe_2$-$CuGaSe_2$ [38, 39], $CuInS_2$-$CuInSe_2$ [40, 41], or $CuInS_2$-ZnS [42–45]. This is also the only method for adjusting the position of energy levels in materials, which are not going to maintain their nanocrystalline character in the final device (thin-film solar cells processed from colloidal solution).

The bulk bandgap of CISe (1.04 eV) lies below the optimum value of a single junction solar cell (1.3 eV); therefore, Cu(In, Ga)Se_2 with bandgap adjustable between 1.04 eV and 1.68 eV (bulk bandgap of $CuGaSe_2$) is applied in thin-film solar cells. Also, methods to synthesize Cu(In, Ga)Se_2 nanoparticles were studied; mainly with the goal of using the colloidal solution as ink for printable photovoltaic devices [39, 46, 47]. Size and composition control was achieved by adjusting the parameters of the synthesis, such as concentrations, ligands and temperature. Unfortunately, gallium is a rare and expensive element; therefore, it would be beneficial in view of reducing the costs of photovoltaic devices to find cheaper alternatives. Alloying of zincblende CISe with ZnSe offers the possibility to vary the bandgap in the range between 1.04 eV and 2.7 eV. Because of the small lattice mismatch between the two materials, alloyed nanocrystals can be formed in the whole composition range [48].

Also, the crystallographic structures of $CuInS_2$ and ZnS are similar – the lattice mismatch between the two materials is only 2.2%; therefore copper and indium ions can be easily replaced by zinc ions and homogenous alloys can be produced within the whole composition range (between 0% and 100% ZnS). The bandgap of $CuInS_2$-ZnS alloys can vary between 1.55 and 3.7 eV (bulk bandgap of ZnS), while the positions of the HOMO and the LUMO should lie between those of pure $CuInS_2$ and ZnS. DFT calculations [49] show that the valence band of $CuInS_2$-ZnS alloys is formed from Cu 3d and S 3p orbitals, while the In 5s5p and Zn 4s4p orbitals contribute to the formation of the conduction band. Thus, only by varying the ZnS content in the alloy material can the position of the energy levels be influenced and nanocrystals with absorption covering the whole visible spectrum be obtained. Several examples of a direct synthesis of $CuInS_2$-ZnS alloyed nanocrystals were described in the literature recently [43, 45, 50–52]; e.g., Pan *et al.* [51] applied thermolysis of complexes of dithiocarbamate with Zn^{2+}, Cu^{2+} and In^{3+} in the presence of oleylamine as activating agent, which facilitated the formation of homogenous nanocrystals. Another approach is a two-step synthesis, in which $CuInS_2$ nanocrystals are synthesized first and alloy particles are obtained by cation exchange [42, 45]. Some of these methods allow for a precise control of the composition, and, consequently, of the optical properties of quasi-spherical [43, 51] or cube-shaped [53] alloyed nanocrystals. Growth of elongated structures can be obtained in a two-step method, using copper sulfide particles as seeds [27, 54].

Combining two materials in a core-shell nanostructure can also be beneficial for solar cell applications. Defects in CIS and CISe nanocrystals can trap the charge carriers in photovoltaic devices, and, thus, reduce the power conversion efficiency. Surface defects can only partly be reduced by using strong capping agents. Another effective method for reducing the defects related to the surface is covering CIS or CISe with a shell of a material with larger bandgap, such as ZnS [18, 55, 56] or CdS [57–59]. The charge carriers are localized in the core of such a structure; the probability density of both the electron and the hole on the surface of the shell material is relatively low, which reduces the impact of the surface defects on the properties of a core-shell structure. Furthermore, CIS-ZnS core shell nanocrystals exhibit longer lifetimes of the excited charge carriers, which is beneficial for their application in solar cells [56].

3.3 Application of Colloidal $CuInS_2$ and $CuInSe_2$ Nanoparticles in Solar Energy Conversion

Several different solar cell geometries can be realized using nanocrystalline materials. Positive and negative charge carriers can be separated in a planar p-n junction, a Schottky barrier, an interpenetrating network of a donor and an acceptor material, or at a semiconductor-liquid interface. The application of nanocrystals in all these kinds of devices offers a multitude of advantages (variable bandgaps, radiation stability, high extinction coefficients, and the possibility of multiple exciton generation); however, there are still many problems related mainly to the presence of defects and stability issues which have to be solved before the potential of nanocrystaline building blocks for photovoltaic devices can be fully exploited.

3.3.1 All-Inorganic Solar Cells

$Cu(In,Ga)Se_2$ solar cells reached 20% efficiency, which is the current record value among the thin-film devices [5]. However, the high production costs (investments and power consumption) hinder their widespread utilization; thus, less expensive alternative production methods are needed. Colloidal solutions of semiconductor nanocrystals, which can be used as inks, are seen as one of the possible options to avoid the power consuming vacuum and high temperature processes and to reduce the investment costs [60]. As has been shown by Guo *et al.* [32, 47], $Cu(In,Ga)Se_2$ thin film composed of large crystalline domains can indeed be obtained by sintering colloidal nanocrystals. When thermal treatment is applied to $Cu(In,Ga)Se_2$ nanoparticles, devices can be obtained exhibiting a power conversion efficiency of 3.2% under AM1.5 illumination [32]. Because of the removal of the organic ligands during the sintering, the film contains some void space inclusions. This can be avoided when using Cu(In,Ga)S_2 particles for the film formation, which are heated in Se-atmosphere. The exchange of sulfur for selenium leads to an increase of the volume of the inorganic material, which compensates the loss of organic ligands. Therefore, films with a better morphology are obtained, which yield higher power conversion efficiencies (4.76%) [47].

The sintering step, applied to obtain a polycrystalline film from nanoparticles, is, however, also power consuming; therefore, active layers, which do not require sintering, are also investigated. The highest efficiency, 6% in nanocrystalline devices without thermal treatment, was obtained for PbS nanocrystals covered with inorganic ligands [61]. Devices produced with oleylamine-capped CISe nanocrystal inks reached 3% power conversion efficiency under AM1.5 [62]. However, a Mott-Schottky analysis of the space-charge capacitance showed that the active region of these devices was only 50 nm; thus, only a fraction of the incident light absorbed in thicker active layers was effectively converted to electrical power. The efficiency of these devices could be enhanced if the extraction of photogenerated charge carriers from the CISe film could be improved. CISe nanocrystals covered with inorganic ligands show lower power conversions efficiencies: when a metal chalcogene-hydrazinium complex was applied as capping agent, efficiencies up to 1.7% were obtained [17]. However, because of the need to use hydrazine as solvent, these kinds of nanocrystals cannot find industrial application. Hydrazine-free ligands, such as S^{2-} are also suitable capping agents for CISe nanocrystals, but the efficiency of devices using them as the active layer are substantially lower [17].

3.3.2 Organic-Inorganic Hybrid Solar Cells

Solar cells composed of conducting polymers possess several advantages compared to thin-film solar cells. They are low cost devices because they can be fabricated from solution, e.g., by printing, and if needed, on flexible substrates. In a single-junction low-bandgap polymer cell an efficiency of 8.4% has been reached [63]. A drawback of organic polymers is their low chemical and photochemical stability, and relatively low electron mobility. Furthermore, most semiconducting polymers have wide energy gaps, and consequently, can utilize only a small fraction of the solar spectrum. The absorption of semiconductor nanocrystals can reach the near-IR spectral range; the mobility of electrons in inorganic semiconductor materials is much higher than in polymers. Therefore, combining semiconducting polymers with inorganic nanomaterials seems to be a suitable way to overcome the limitations of the organic materials. In spite of the theoretical advantages of organic-inorganic bulk heterjunction solar cells, and vivid research activity in this field, these kinds of devices did not bring an increase of efficiency, compared with purely organic ones. Hybrid solar cells suffer from problems related to insufficient passivation of the surface of the nanoparticles, trap states on the surface of the inorganic particles and aggregation, which deteriorates the morphology of the active layer, as discussed in Section 3.2.1. With materials such as PbS, PbSe or CdSe, efficiencies up to 5% could be achieved [64], so far, with optimized morphology and surface properties of the particles. Blends of CIS or CISe with conducting polymers such as poly hexylthiophene (P3HT), poly(2-methoxy-5-(2-ethylhexyloxy)-1,4-phenylene vinylene) (MEH-PPV) or poly(3-(ethyl-4- butanoate) thiophene) (P3EBT) show photovoltaic response, however, hybrid solar cells based on these blends usually exhibit power conversion efficiencies far below 1% [65–70].

3.3.3 Nanocrystal Sensitized Solar Cells

In dye-sensitized solar cells light is converted photoelectrochemically into electrical power: a TiO_2 electrode is covered with a dye, which absorbs light. The photogenerated electrons are transferred to the electrode, subsequently, the positively-charged dye reacts with a redox couple (typically I^- and I_3^-), which is dissolved in an organic solvent, and which transports the positive charge to the counter electrode. By using a co-sensitizer together with a black ruthenium dye an efficiency of 11.4% was achieved in dye-sensitized solar cells [71]. Instead of the dye, nanoparticles can be applied as sensitizer. Nanocrystals have higher extinction coefficients and are more stable under illumination than conventional dye molecules. The relative positions of the energy levels of TiO_2 and CIS nanoparticles allow for a charge separation, therefore combinations of this material show photocatalytic activity [59, 72–74] and can also be applied in nanocrystal-sensitized solar cells. Devices in which ZnSe-coated CIS particles were attached to TiO_2 covered with a Cu_2S buffer layer reach 2.52% efficiency. Alloying CIS with selenium further extends the absorption into the near-IR spectral region, while the formation of a thin Cd(S, Se) shell can reduce surface trapping and increase the stability of the particles. The resulting quantum-dot-sensitized solar cells exhibit up to 4.2% power conversion efficiency [41, 58].

3.4 Conclusion and Outlook

Colloidal synthesis offers excellent possibilities for controlling the properties of CIS and CISe nanocrystals. Much progress has been achieved in the control of size, shape and composition of these nanomaterials in the past few years. This has provided an opportunity to apply nanocrystalline CIS and CISe in solar energy conversion, and to benefit from the unique optical properties and stability of these materials together with the possibility of fabricating photovoltaic devices from solution, which could significantly reduce their production costs and enable large-scale application. Some of the studies devoted to the application of colloidal I-III-VI nanomaterials in photovoltaics show promising results, especially when the particles are used as inks for the generation of thin-layer devices or in nanocrystal-sensitized solar cells. However, in hybrid solar cells and in all-inorganic devices without a sintering step CIS and CISe do not reach the efficiencies which have already been demonstrated for other materials, such as PbS or CdSe. Here, successful application of CIS and CISe requires further research; the surface properties and the stability of these materials have to be investigated in more detail and improved. Furthermore, synthetic procedures for particles with a smaller amount of internal defects have to be developed. To achieve this goal, the growth process and the underlying reaction mechanisms have to be extensively studied to enhance our understanding of the molecular processes taking place during the synthesis.

References

1. C. Burda, X. Chen, R. Narayanan, and M. a. El-Sayed, *Chemical Reviews*, Vol. 105, p. 1025, 2005.
2. S. Kumar and T. Nann, *Small*, Vol. 2, p. 316, 2006.
3. Y. Jun, J. Choi, and J. Cheon, *Angewandte Chemie International Edition*, Vol. 45, p. 3414, 2006.
4. A. R. Tao, S. Habas, and P. Yang, *Small*, Vol. 4, p. 310, 2008.
5. M. A. Green, K. Emery, Y. Hishikawa, W. Warta, and E. D. Dunlop, *Progress in Photovoltaics: Research and Applications*, Vol. 21, p. 1, 2013.

6. C. B. Murray, D. J. D. Norris, and M. G. Bawendi, *Journal of the American Chemical Society*, Vol. 115, p. 8706, 1993.
7. B. L. Cushing, V. L. Kolesnichenko, and C. J. O'Connor, *Chemical Reviews*, Vol. 104, p. 3893, 2004.
8. J. Park, J. Joo, S. G. G. Kwon, Y. Jang, and T. Hyeon, *Angewandte Chemie International Edition*, Vol. 46, p. 4630, 2007.
9. C. de Mello Donegá, P. Liljeroth, and D. Vanmaekelbergh, *Small*, Vol. 1, p. 1152, 2005.
10. C. B. Murray, C. R. Kagan, and M. G. Bawendi, *Annual Review of Materials Science*, Vol. 60, p. 4300, 2000.
11. F. Hetsch, X. Xu, H. Wang, S. V Kershaw, and A. L. Rogach, *The Journal of Physical Chemistry Letters*, Vol. 2, p. 1879, 2011.
12. I. Lokteva, N. Radychev, F. Witt, H. Borchert, J. Parisi, and J. Kolny-Olesiak, *The Journal of Physical Chemistry C*, Vol. 114, p. 305, 2010.
13. M. V Kovalenko, M. I. Bodnarchuk, and D. V. Talapin, *Journal of the American Chemical Society*, Vol. 132, p. 15124, 2010.
14. A. Nag, M. V Kovalenko, J.-S. Lee, W. Liu, B. Spokoyny, and D. V Talapin, *Journal of the American Chemical Society*, Vol. 133, p. 10612, 2011.
15. M. V Kovalenko, M. Scheele, and D. V Talapin, *Science*, Vol. 324, p. 1417, 2009.
16. A. T. Fafarman, W. Koh, B. T. Diroll, D. K. Kim, D.-K. Ko, S. J. Oh, X. Ye, V. Doan-Nguyen, M. R. Crump, D. C. Reifsnyder, C. B. Murray, and C. R. Kagan, *Journal of the American Chemical Society*, Vol. 133, p. 15753, 2011.
17. C. J. Stolle, M. G. Panthani, T. B. Harvey, V. a Akhavan, and B. a Korgel, *ACS Applied Materials & Interfaces*, Vol. 4, p. 2757, 2012.
18. L. Li, A. Pandey, D. J. Werder, B. P. Khanal, J. M. Pietryga, and V. I. Klimov, *Journal of the American Chemical Society*, Vol. 133, p. 1176, 2011.
19. R. Xie, M. Rutherford, and X. Peng, *Journal of the American Chemical Society*, Vol. 131, p. 5691, 2009.
20. S. L. Castro, S. G. Bailey, R. P. Raffaelle, K. K. Banger, and A. F. Hepp, *The Journal of Physical Chemistry B*, Vol. 108, p. 12429, 2004.
21. S. L. Castro, S. G. Bailey, R. P. Raffaelle, K. K. Banger, and A. F. Hepp, *Chemistry of Materials*, Vol. 15, p. 3142, 2003.
22. J. J. Nairn, P. J. Shapiro, B. Twamley, T. Pounds, R. von Wandruszka, T. R. Fletcher, M. Williams, C. Wang, and M. G. Norton, *Nano Letters*, Vol. 6, p. 1218, 2006.
23. S. K. Batabyal, L. Tian, N. Venkatram, W. Ji, and J. J. Vittal, *The Journal of Physical Chemistry C*, Vol. 113, p. 15037, 2009.
24. F. M. Courtel, A. Hammami, R. Imbeault, G. Hersant, R. W. Paynter, B. Marsan, and M. Morin, *Chemistry of Materials*, Vol. 22, p. 3752, 2010.
25. H. Zhong, S. S. Lo, T. Mirkovic, Y. Li, Y. Ding, Y. Li, and G. D. Scholes, *ACS Nano*, Vol. 4, p. 5253, 2010.
26. S. T. Connor, C.-M. Hsu, B. D. Weil, S. Aloni, and Y. Cui, *Journal of the American Chemical Society*, Vol. 131, p. 4962, 2009.
27. M. Kruszynska, H. Borchert, J. Parisi, and J. Kolny-Olesiak, *Journal of the American Chemical Society*, Vol. 132, p. 15976, 2010.
28. M. Kruszynska, H. Borchert, J. Parisi, and J. Kolny-Olesiak, *Journal of Nanoparticle Research*, Vol. 13, p. 5815, 2011.
29. X. Lu, Z. Zhuang, Q. Peng, and Y. Li, *CrystEngComm*, Vol. 13, 2011.

30. B. Koo, R. N. Patel, and B. a. Korgel, *Chemistry of Materials*, Vol. 21, p. 1962, 2009.
31. B. Koo, R. N. R. N. Patel, and B. a B. A. Korgel, *Journal of the American Chemical Society*, Vol. 131, p. 3134, 2009.
32. Q. Guo, S. Kim, M. Kar, W. Shafarman, W. Birkmire, Robert, E. A. Stach, R. Agrawal, and H. W. Hillhouse, *Nano Letters*, Vol. 8, p. 2982, 2008.
33. J.-Y. Chang, M.-H. Tsai, K.-L. Ou, C.-H. Yang, and J.-C. Fan, *CrystEngComm*, Vol. 13, p. 4236, 2011.
34. L.-C. Xu, R.-Z. Wang, L.-M. Liu, Y.-P. Chen, X.-L. Wei, H. Yan, and W.-M. Lau, *Journal of Materials Chemistry*, Vol. 22, p. 21662, 2012.
35. K. Nose, Y. Soma, T. Omata, and S. Otsuka-Yao-Matsuo, *Chemistry of Materials*, Vol. 21, p. 2607, 2009.
36. M. E. Norako and R. L. Brutchey, *Chemistry of Materials*, Vol. 22, p. 1613, 2010.
37. J.-J. Wang, Y.-Q. Wang, F.-F. Cao, Y.-G. Guo, and L.-J. Wan, *Journal of the American Chemical Society*, Vol. 132, p. 12218, 2010.
38. Y. Luo, *Colloid Journal*, Vol. 71, p. 375, 2009.
39. J. Tang, S. Hinds, S. O. Kelley, and E. H. Sargent, *Chemistry of Materials*, Vol. 20, p. 6906, 2008.
40. M. Y. Chiang, S. Chang, C. Y. Chen, F. W. Yuan, and H. Y. Tuan, *The Journal of Physical Chemistry C*, Vol. 115, p. 1592, 2011.
41. H. McDaniel, N. Fuke, J. M. Pietryga, and V. I. Klimov, *The Journal of Physical Chemistry Letters*, Vol. 4, p. 355, 2013.
42. L. De Trizio, M. Prato, A. Genovese, A. Casu, M. Povia, R. Simonutti, M. J. P. Alcocer, C. D. Andrea, F. Tassone, and L. Manna, *Chemistry of Materials*, Vol. 24, p. 2400–2406, 2012.
43. X. Wang, D. Pan, D. Weng, C. Y. C.-Y. Low, L. Rice, J. Han, and Y. Lu, *The Journal of Physical Chemistry C*, Vol. 114, p. 13406, 2010.
44. X. Yuan, J. Zhao, P. Jing, W. Zhang, H. Li, L. Zhang, X. Zhong, and Y. Masumoto, *The Journal of Physical Chemistry C*, Vol. 116, p. 11973, 2012.
45. W. Zhang and X. Zhong, *Inorganic Chemistry*, Vol. 50, p. 4065, 2011.
46. M. G. Panthani, V. Akhavan, B. Goodfellow, J. P. Schmidtke, L. Dunn, A. Dodabalapur, P. F. Barbara, and B. A. Korgel, *Journal of the American Chemical Society*, Vol. 130, p. 16770, 2008.
47. Q. Guo, G. M. Ford, H. W. Hillhouse, and R. Agrawal, *Nano Letters*, Vol. 9, p. 3060, 2009.
48. S. Li, Z. Zhao, Q. Liu, L. Huang, G. Wang, D. Pan, H. Zhang, and X. He, *Inorganic Chemistry*, Vol. 50, p. 11958, 2011.
49. I. Tsuji, H. Kato, H. Kobayashi, and A. Kudo, *The Journal of Physical Chemistry B*, Vol. 2, p. 7323, 2005.
50. J. Feng, M. Sun, F. Yang, and X. Yang, *Chemical Communications*, Vol. 47, p. 6422, 2011.
51. D. Pan, D. Weng, X. Wang, Q. Xiao, W. Chen, C. Xu, Z. Yang, and Y. Lu, *Chemical Communications*, p. 4221, 2009.
52. H. Nakamura, W. Kato, M. Uehara, K. Nose, T. Omata, S. Otsuka-Yao-Matsuo, M. Miyazaki, and H. Maeda, *Chemistry of Materials*, Vol. 18, p. 3330, 2006.
53. X. Tang, W. Cheng, E. S. G. Choo, and J. Xue, *Chemical Communications*, p. 5217, 2011.
54. J.-Y. Chang and C.-Y. Cheng, *Chemical Communications*, p. 9089, 2011.

55. J. Park and S.-W. Kim, *Journal of Materials Chemistry*, Vol. 21, p. 3745, 2011.
56. V. K. Komarala, C. Xie, Y. Wang, J. Xu, and M. Xiao, *Journal of Applied Physics*, Vol. 111, p. 124314, 2012.
57. Q. Wei and J. Mu, *Journal of Dispersion Science and Technology*, Vol. 28, p. 916, 2007.
58. T.-L. Li, Y.-L. Lee, and H. Teng, *Energy & Environmental Science*, Vol. 5, p. 5315, 2012.
59. T.-L. Li, Y.-L. Lee, and H. Teng, *Journal of Materials Chemistry*, Vol. 21, p. 5089, 2011.
60. S. E. Habas, H. a S. Platt, M. F. a M. van Hest, and D. S. Ginley, *Chemical Reviews*, Vol. 110, p. 6571, 2010.
61. J. Tang, K. W. Kemp, S. Hoogland, K. S. Jeong, H. Liu, L. Levina, M. Furukawa, X. Wang, R. Debnath, D. Cha, K. W. Chou, A. Fischer, A. Amassian, J. B. Asbury, and E. H. Sargent, *Nature Materials*, Vol. 10, p. 765, 2011.
62. V. a Akhavan, M. G. Panthani, B. W. Goodfellow, D. K. Reid, and B. a Korgel, *Optics Express*, Vol. 18 Suppl 3, p. A411, 2010.
63. Z. He, C. Zhong, X. Huang, W.-Y. Wong, H. Wu, L. Chen, S. Su, and Y. Cao, *Advanced Materials*, Vol. 23, p. 4636, 2011.
64. R. Zhou and J. Xue, *Chemphyschem*, Vol. 13, p. 2471, 2012.
65. A. Takeda, T. Oku, A. Suzuki, K. Kikuchi, and S. Kikuchi, *Journal of the Ceramic Society of Japan*, Vol. 117, p. 967, 2009.
66. W. Yue, S. Han, R. Peng, W. Shen, H. Geng, F. Wu, S. Tao, and M. Wang, *Journal of Materials Chemistry*, Vol. 20, p. 7570, 2010.
67. E. Maier, T. Rath, W. Haas, O. Werzer, R. Saf, F. Hofer, D. Meissner, O. Volobujeva, S. Bereznev, E. Mellikov, H. Amenitsch, R. Resel, and G. Trimmel, *Solar Energy Materials and Solar Cells*, Vol. 95, p. 1354, 2011.
68. N. Radychev, D. Scheunemann, M. Kruszynska, K. Frevert, R. Miranti, J. Kolny-Olesiak, H. Borchert, and J. Parisi, *Organic Electronics*, Vol. 13, p. 3154, 2012.
69. E. Arici, H. Hoppe, F. Schäffler, D. Meissner, M. a. Malik, and N. S. Sariciftci, *Applied Physics A*, Vol. 79, p. 59, 2004.
70. Y. Yang, H. Zhong, Z. Bai, B. Zou, Y. Li, and G. D. Scholes, *The Journal of Physical Chemistry C*, Vol. 116, p. 7280–7286, 2012.
71. L. Han, A. Islam, H. Chen, C. Malapaka, B. Chiranjeevi, S. Zhang, X. Yang, and M. Yanagida, *Energy & Environmental Science*, Vol. 5, p. 6057, 2012.
72. Y. Lin, F. Zhang, D. Pan, H. Li, and Y. Lu, *Journal of Materials Chemistry*, Vol. 22, p. 8759, 2012.
73. F. Shen, W. Que, Y. Liao, and X. Yin, *Industrial & Engineering Chemistry Research*, Vol. 50, p. 9131, 2011.
74. T.-L. Li and H. Teng, *Journal of Materials Chemistry*, Vol. 20, p. 3656, 2010.

4

Two Dimensional Layered Semiconductors: Emerging Materials for Solar Photovoltaics

Mariyappan Shanmugam and Bin Yu*

College of Nanoscale Science and Engineering, State University of New York, Albany, New York, USA

Abstract

Graphene-like, two-dimensional (2D), lattice-structured transition metal dichalcogenides are layered semiconductors consisting of covalent bonded atomic planes coordinated by van der Waals forces. This chapter presents two important layered semiconductors, molybdenum disulfide (MoS_2) and tungsten disulfide (WS_2), as photoactive material candidates for bulk heterojunction (BHJ) and Schottky barrier solar cells. Development of functional materials by chemical exfoliation and chemical vapor deposition are presented with experimental processing steps. Microstructural and Raman spectroscopic studies were performed on the developed MoS_2 and WS_2 semiconductors. Fabrication and characterization of functional photovoltaic devices, performance evaluation, electronic transport and interfacial recombination mechanisms in BHJ and Schottky barrier solar cells are presented. Nanoflakes and nanofilms of 2D-layered semiconductors employed in BHJ and Schottky barrier solar cells emphasize that the absorption limit can be tuned by controlling thickness of the material confirmed by solar cell performance.

Keywords: Layered semiconductors, chemical vapor deposition, solar cell, bulk heterojunction, Schottky barrier, electronic transport, interfacial recombination

**Corresponding author*: byu@albany.edu

Atul Tiwari, Rabah Boukherroub, and Maheshwar Sharon (eds.) Solar Cell Nanotechnology, (117–134) 2014 © Scrivener Publishing LLC

4.1 Introduction

Bulk heterojunction (BHJ) solar cells, employing both organic and inorganic semiconductors, show promise in emerging photovoltaic technology due to their improved stability and performance. Titanium dioxide (TiO_2) has been widely used as an electron acceptor in bulk BHJ solar cells and has shown promising photoconversion efficiencies [1–5]. A porous TiO_2 nanoparticle-based electron acceptor is preferred for BHJ solar cell application due to the fact that a large amount of surface is required to load a hole-conducting polymer. The porous TiO_2 offers a potential opportunity to form heterojunction with polymer hole conductors in a large area to enhance exciton generation and separation. Various nanostructures of TiO_2 such as nanoparticles, nanorods and nanotubes have been employed in BHJ solar cells [6–8]. Poly(3-hexylthiophene) (P3HT) is widely used as a hole conductor with TiO_2 as an electron acceptor in BHJ solar cell technology. The blend morphology between electron acceptor and hole conductor is an essential factor in determining the charge transport, recombination and performance of the BHJ solar cells [9, 10].

Shottky barrier solar cells are relatively simple structures compared to heterojunction and multi-junction solar cells and exhibit promising photovoltaic performance. Both p-type and n-type doped semiconductors can be used in Schottky barrier solar cells with a metal of appropriate work function. Bulk and thin films of various semiconductor materials have been employed in Schottky barrier solar cells, and the results have shown the possible potential opportunities of Schottky barrier solar cells in renewable energy technology [11–15]. Allotropes of carbon, a low cost material candidate compared to expensive bulk and thin-film semiconductors and metals, have recently been employed in Schottky barrier solar cells as metal contact [16–20]. Major disadvantage with conventional semiconductors such as Si, Ge and GaAs are the surface states due to the unsaturated dangling bonds which play a role of recombination centers in solar cells [21–23]. Surface states adversely affect electron lifetime in solar cells by increasing the surface recombination velocity [24]. Semiconductor surface is a critical part in Schottky barrier solar cells at which the barrier for electrons and holes is formed due to the work function difference between semiconductors and metal. Surface states on the semiconductor surface deteriorate the formation of Schottky barrier height, and thus affect the built-in potential at the metal/semiconductor interface. Surface states on a semiconductor can be passivated by high-k dielectric materials to enhance the solar cell performance [25, 26]. The surface states on the active semiconductors employed in Schottky barrier solar cell become interfacial states while the metal contact is deposited on the semiconductor. Metal/semiconductor junction with interfacial defect states experience Fermi level pinning which lowers the barrier height and hence lowers solar cell performance.

Graphite-like layered semiconductors with two-dimensional (2D) lattice structure belong to transition metal dichalcogenides (TMDs), and have been employed as active semiconductor layers in nano/opto-electronics. Optical properties of the 2D layered semiconductors such as molybdenum disulfide (MoS_2) and tungsten disulfide (WS_2) can be tuned from 1.2 eV to 1.8 eV by changing layer thickness. Both direct and indirect bandgap nature can be achieved in 2D layered semiconductors by controlling the number of layers in a stack. Such flexibility in tuning optical bandgap of the active layer semiconductors offers unique opportunities in opto-electronics and photovoltaic technology. MoS_2 and WS_2 have recently been studied for nanoelectronics applications [27–30]. Dependency of electrical conductivity, optical bandgap and carrier mobility of 2D layered semiconductors are the major key factors influencing them in opto-electronic device applications [31–36].

In this chapter, we present functional 2D layered semiconductor material development and characterization for BHJ and Schottky barrier solar cells. Chemically exfoliated 2D layered semiconductor nanoflakes-based BHJ solar cells and CVD-synthesized, 2D layered semiconductors-based Schottky barrier solar cells are presented. Photovoltaic performance, electron transport and recombination processes at interfaces are studied to demonstrate 2D layered semiconductors as potential candidates for emerging photovoltaic technology.

4.2 Material Synthesis

4.2.1 Chemical Exfoliation

Bulk crystals of 2D layered semiconductors MoS_2 and WS_2 were used to exfoliate nanoflakes by ultra-sonication using 2-propanol as a solvent. The 2D layered semiconductor crystals were used as purchased from US Research Nanomaterials, Inc. The bulk crystals

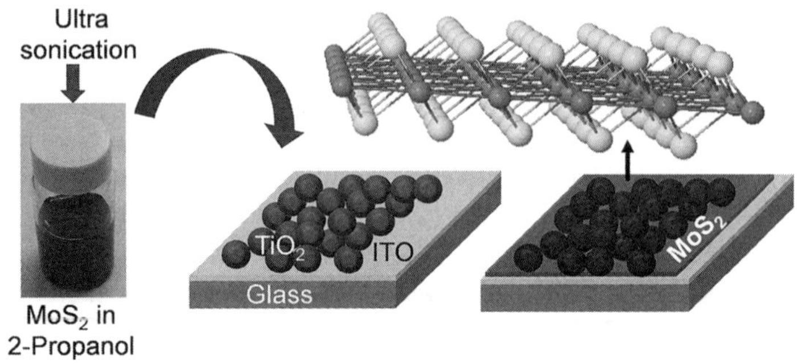

Figure 4.1 Schematic of ultra-sonication-based chemical exfoliation of 2D-layered semiconductors (in this case MoS_2), mixing with TiO_2 nanoparticle layer coated on ITO transparent electrode used for BHJ solar cell fabrication.

were manually grinded further to achieve smaller size particles and dissolved in 2-propanol. This solution was ultra-sonicated for 15 minutes to produce ultrathin nanoflakes of the 2D layered semiconductors MoS_2 and WS_2. After solvent annealing at 100°C, colloidal TiO_2 nanoparticles were mixed with 2D layered semiconductor nanoflakes in a ratio of 2:1, followed by a thermal annealing at 450°C for 30 minutes in atmosphere. Figure 4.1 shows key processing steps including ultra-sonicated 2D layered semiconductor in 2-propanol solution (in this case MoS_2), TiO_2 nanoparticle layer on ITO electrode, mixing the nanoflakes of MoS_2 with TiO_2 and a crystal structure of MoS_2 semiconductor. Similarly, WS_2 nanocrystals also were exfoliated by ultra-sonication process and mixed with TiO_2 nanoparticle layer as a photoactive material for BHJ solar cell application.

4.2.2 CVD Synthesis of 2D Layered Semiconductors MoS_2 and WS_2

Two Si wafers, coated with 100 nm of SiO_2, were used as substrates for Molybdenum (Mo) and Tungsten (W) metal depositions. Mo and W metal thin films were deposited on Si/SiO_2 substrates with an approximate thickness of 50 nm by electron beam evaporation method. The CVD growth process started with flowing Argon of 200 sccm in the quartz tube chamber leading to a pressure level of ~1 Torr. The furnace was then heated to 750°C and stabilized over one hour. Sulfur powder, placed up-stream in the quartz tube chamber, was heated to just above its melting temperature (~115°C) to vaporize in Ar gas environment. The sulfur vapor flowed over the Mo and W thin films coated on SiO_2/Si substrates for approximately 30 minutes. Vapor-solid reaction takes place at an elevated temperature (~750°C), leading to MoS_2 and WS_2 nanofilms growth driven by increase in enthalpy. After the completion of MoS_2 and WS_2 growth, the samples were annealed at 1000°C for one hour to improve the crystalline quality and uniformity of the resulting MoS_2 and WS_2 nanofilms. Figure 4.2(a) shows a schematic diagram of the CVD set-up used for 2D layered semiconductors growth illustrating Ar carrier gas, sulfur powder evaporation onto Mo metal-coated Si/SiO_2 substrate leading to MoS_2.

Indium tin oxide (ITO)-coated glass substrates were used as substrates for Schottky barrier solar cells. So, the 2D layered semiconductor nanofilms need to be transferred from the growth substrates to device substrates. A 15 M concentrated Pottasium hydroxide (KOH) solution was prepared in which the MoS_2 and WS_2 grown Si/SiO_2 substrates were immersed to etch the SiO_2 layer to lift-off the grown 2D layered semiconductor nanofilms. SiO_2 was etched by KOH and free-floating nanofilms of MoS_2 and WS_2 were achieved

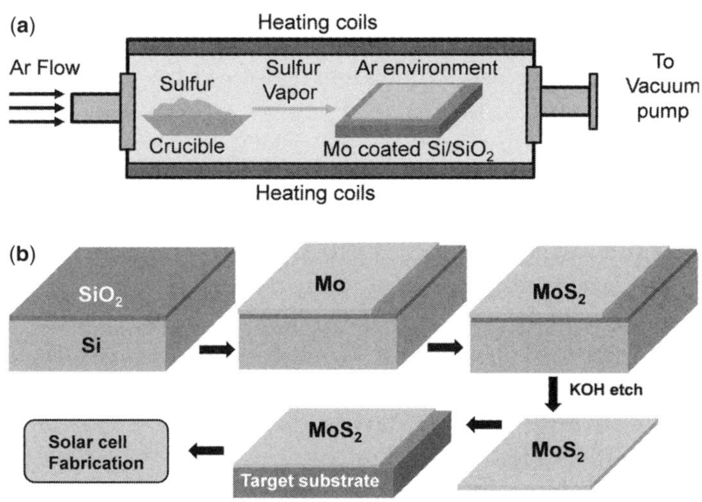

Figure 4.2 Schematic diagram showing (a) CVD growth set-up and (b) key processing steps involved in making 2D-layered semiconductor nanofilms (in this case MoS_2) by CVD method showing metal film deposition on Si/SiO_2 substrate, evaporation of sulfur on metal film leading to 2D-layered semiconductor growth, layer transfer onto target substrate used for Schottky barrier solar cell fabrication.

in KOH solution. Figure 4.2(b) shows key process steps involved in CVD synthesis of 2D layered semiconductor growth and transfer (in this case MoS_2 nanofilm) from SiO_2-coated Si substrate and layer transfer after the KOH etch to the target substrate used for Schottky barrier solar cell fabrication. The process steps are the same for WS_2 nanofilm growth and transfer.

4.2.3 Material Characterization

A HORIBA Scientific Raman spectrometer with a laser of 514.5 nm excitation wavelength was used to study the MoS_2 and WS_2 nanoflakes-coated TiO_2 electron acceptor used for the BHJ solar cell fabrication, as reported in this chapter. A LEO 1550 scanning electron microscope was used to study the thickness and surface morphology of the 2D layered semiconductor nanoflakes-coated TiO_2.

4.3 Photovoltaic Device Fabrication

4.3.1 Bulk Heterojunction Solar Cells

Colloidal TiO_2 nanoparticle film was deposited onto indium tin oxide (ITO)-coated glass substrates, followed by a thermal annealing at 450°C for 30 minutes in atmosphere. Two BHJ solar cells were fabricated employing MoS_2 and WS_2 as photoactive material candidates. P3HT solution, prepared by dissolving 1 wt% of P3HT in chlorobenzene, was spin-coated onto the 2D layered semiconductor nanoflakes-coated TiO_2 film. The spin-coated P3HT layer on TiO_2/2D layered semiconductor film diffuse into the porous TiO_2 nanoparticle network and forms a large surface BHJ structure to facilitate effective charge separation. A 50 nm gold thin film was deposited by electron-beam evaporation as the back contact for the BHJ solar cells. Figure 4.3(a) shows a schematic functional diagram of the fabricated BHJ solar cell employing 2D layered semiconductor as a photoactive material.

4.3.2 Schottky Barrier Solar Cells

Two Schottky barrier solar cells were fabricated employing MoS_2 and WS_2 nanofilms. The MoS_2 and WS_2 nanofilms were synthesized

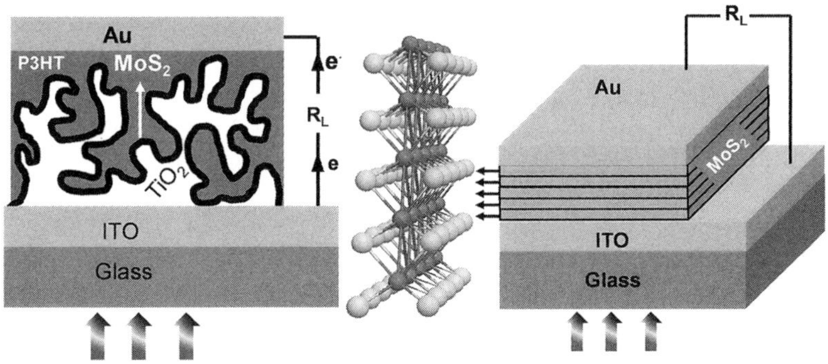

Figure 4.3 Schematic of cross sectional views of the fabricated (a) BHJ solar cell employing 2D-layered semiconductor nanoflakes as photoactive materials (in this case MoS_2) and (b) Schottky barrier solar cell with layer transferred 2D-layered semiconductor nanofilm stack (in this case MoS_2).

by CVD method as explained in Section 4.2.2. The nanofilms of 2D layered semiconductor MoS_2 and WS_2 were transferred onto the ITO-coated substrate as the photoactive material for the Schottky barrier solar cell. A 50 nm of Au metal film was deposited by electron-beam evaporation on the 2D layered semiconductor nanofilms, forming Schottky junction with the nanofilms. The transparent ITO electrode was used as the front contact of the Schottky barrier solar cell through which the photons incident on the 2D layered semiconductors. Multiple transfers of nanofilms were made onto ITO to yield the total active layer thickness of ~200 nm. Figure 4.3(b) shows a schematic diagram of the fabricated Schottky barrier solar cells employing CVD-synthesized 2D layered semiconductor nanofilms of MoS_2 and WS_2.

4.3.3 Device Characterization

The current density-voltage (J-V) characteristics of the fabricated BHJ and Schottky barrier solar cells were measured under both dark and standard AM 1.5 illumination conditions calibrated using a standard silicon solar cell by Agilent B1500A semiconductor device analyzer. A Xe arc lamp was used to simulate the AM 1.5 photon flux for the J-V characteristic measurements. The external quantum efficiency (EQE) spectra of the BHJ and Schottky barrier solar cells were measured in the wavelength range of 350 nm ~ 900 nm using a Newport monochromator equipped with the same Xe arc lamp that was used to measure the J-V characteristics of the BHJ solar cells. Capacitance-voltage (C-V) characteristics of the Schottky barrier solar cells were measured by C-V module assembled with Agilent B1500 semiconductor device analyzer.

4.4 Microstructural and Raman Spectroscopic Studies of MoS_2 and WS_2

Figure 4.4 shows the scanning electron microscopic (SEM) images of the 2D layered semiconductor nanoflakes (in this case WS_2 nanoflakes) coated TiO_2 nanoparticle film on glass substrate (a) cross sectional view and (b) surface morphology. The cross-sectional SEM image shows the thicknesses of the 2D layered semiconductor WS_2-coated TiO_2 was about 12.7 μm as shown. A clear difference between the WS_2-coated porous TiO_2 films and the glass substrate is observed. Surface morphology of the WS_2-coated porous TiO_2 film is shown in Figure 4.4(b). It is observed that randomly oriented TiO_2 nanoparticles with an average size of ~20 nm are well interconnected to each other which facilitate effective electron transport in BHJ solar cells as an electron acceptor. The film is highly porous and clusters of TiO_2 nanoparticles are formed as can be seen in the surface morphology. Since the nanoflakes of WS_2 obtained by the ultrasonication process are a combination of monolayer and a few layer stacks, it is difficult to visualize them separately by SEM.

Figure 4.4 SEM images of 2D-layered semiconductor nanoflakes coated TiO_2 electron acceptor used for BHJ solar cell fabrication showing (a) cross sectional view of 12.7 μm thick layer and (b) surface morphology of highly porous TiO_2 network coated by 2D-layered semiconductor nanoflakes.

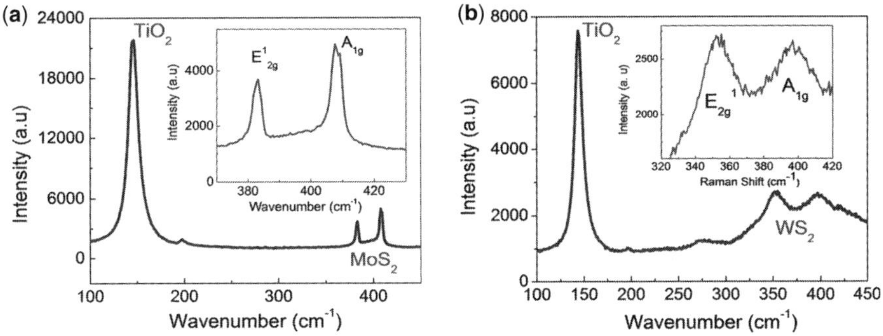

Figure 4.5 Raman spectroscopic measurements performed on 2D-layered semiconductor nanoflakes coated TiO_2 electron acceptors showing (a) MoS_2 with TiO_2 and (b) WS_2 with TiO_2. The insets show the closer view of the active vibrational modes of 2D-layered semiconductors with respect to their wavenumbers.

Figure 4.5 shows the Raman spectrum obtained on the 2D layered semiconductor-coated TiO_2 nanoparticle films. Figure 4.5(a) represents Raman spectrum obtained on MoS_2 nanoflakes-coated TiO_2 and (b) represents WS_2 nanoflakes-coated TiO_2. The presence of TiO_2 nanoparticles in the film was confirmed by the major Raman peak at 146 cm^{-1}, while the other two peaks observed at 383 cm^{-1} and 407 cm^{-1} represent the signature of MoS_2. The bulk MoS_2 film shows Raman peaks at 383 cm^{-1} and 409 cm^{-1}. As seen in the inset of Figure 4.5(a), one of the representative peaks of MoS_2 nanoflakes was at 383 cm^{-1}, but the other one was 407 cm^{-1}. While 405 cm^{-1} is the known signature of MoS_2 monolayer, the wave-number shift suggests the mixed presence of monolayer and multilayer MoS_2 flakes in the film. Figure 4.5(b) shows the Raman spectrum of the WS_2 nanoflakes-coated TiO_2 nanoparticle film. The dominant Raman peak at 143 cm^{-1} represents anatase phase of the TiO_2. Two other Raman peaks at 353 cm^{-1} and 397 cm^{-1} confirm the WS_2. The closer view of the two signature peaks of WS_2 is shown in the figure inset. The peaks at 353 cm^{-1} and 397 cm^{-1} correspond to the E^1_{2g} (in-plane vibrations) and A_{1g} (out-of-plane vibrations) modes of the WS_2 nanoflakes. Raman peaks for multilayers of WS_2 exhibit E^1_{2g} at ~356 cm^{-1} and A_{1g} at ~ 421 cm^{-1}. Compared to the multilayers of WS_2, the nanoflakes used in our studies have shown small shift in A_{1g} (from 356 cm^{-1} to 353 cm^{-1}) and a major shift in A_{1g} (from 421 cm^{-1} to 397 cm^{-1}), confirming that the nanoflakes of WS_2 consist of a combination of monolayer/few layers.

4.5 Photovoltaic Performance Evaluation

4.5.1 BHJ Solar Cells

Figure 4.6(a) and 4.6(b) show the dark and illuminated J-V characteristics of the BHJ solar cells employing MoS_2 and WS_2 nanoflakes as photoactive materials on TiO_2 nanoparticles respectively. The BHJ solar cell employing MoS_2 exhibit a short-circuit current density (J_{SC}) of 4.7 mA/cm², an open-circuit voltage (V_{OC}) of 560 mV, and photoconversion efficiency (η) of 1.7% under 100 mW/cm² illumination level. The photovoltaic performance of the solar cell suggests that photo-excited electron transfer from MoS_2 and P3HT to the conduction band of TiO_2 could be suffered due to interfacial defects, a possible reason for degraded performance. The BHJ solar cell with WS_2 shows J_{SC}, V_{OC}, P_{MAX} and photoconversion efficiency of 7.1 mA/cm², 611 mV, $3.322 * 10^{-4}$ W and 2.6%, respectively. All measured solar cell parameters were estimated under an illumination level of 100 mW/cm².

Figure 4.7 shows the EQE spectra of the BHJ solar cells employing MoS_2 and WS_2 as photoactive materials in a wavelength region of 350 nm ~ 750 nm. The BHJ solar cell with MoS_2 photoactive material showed a maximum value of the EQE was ~ 61% at 485 nm. The optical bandgap of P3HT and a few layers of MoS_2 is very close and the measured EQE is a combined effect of both MoS_2 and P3HT. We observed that maximum value of the EQE for the BHJ solar cell with WS_2 photoactive material is nearly the same as in the BHJ solar cell

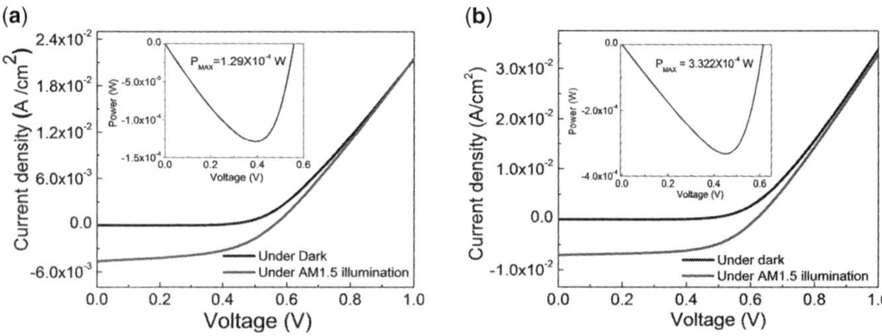

Figure 4.6 Dark and illuminated J-V characteristics of the BHJ solar cells employing (a) nanoflakes of MoS_2-coated TiO_2 and (b) WS_2-coated TiO_2. The insets show maximum power output from the BHJ solar cells.

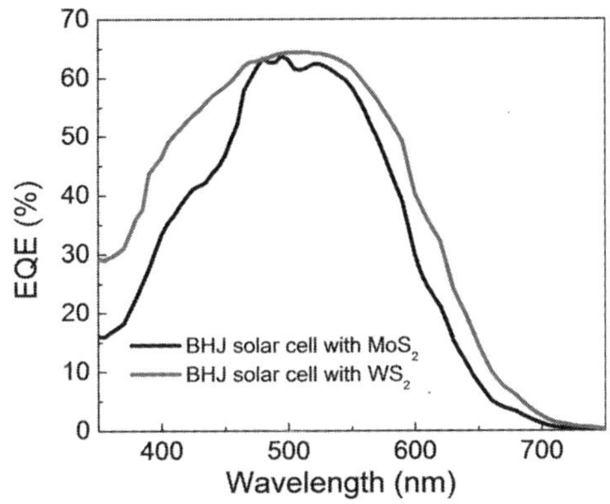

Figure 4.7 EQE spectra of the BHJ solar cells with MoS_2- and WS_2-coated TiO_2 nanoparticle-based electrodes showing effective electron collection in a wavelength window of 350 nm ~ 750 nm.

employing MoS_2. However, in the shorter and longer wavelength regions, EQE of the BHJ solar cell with WS_2, was ~10% more than that of BHJ solar cell with MoS_2.

4.5.2 Schottky Barrier Solar Cells

Figure 4.8(a) shows dark and illuminated J-V characteristics of the Schottky barrier solar cell employing MoS_2 nanofilm, and (b) WS_2 nanofilm as a photoactive material in the forward bias region. The Schottky barrier solar cell with MoS_2 as a photoactive material exhibits J_{SC}, V_{OC}, P_{MAX}, and η are 5.37 mA/cm^2, 597 mV, 2.99*10^{-4} W, and 1.8 % respectively. The J_{SC}, V_{OC}, and η of the Schottky barrier solar cell with WS_2 nanofilm were measured to be 5.5 mA/cm^2, 572 mV, and 1.7%, respectively. The figure inset shows the power output of solar cell with the maximum value of 2.9*10^{-4} W.

Figure 4.9 shows the EQE spectra of the two Schottky barrier solar cells employing MoS_2 and WS_2 nanofilms in the wavelength window of 350 nm ~ 950 nm. Both MoS_2- and WS_2-based Schottky barrier solar cells showed similar EQE spectra confirming that the thickness of the materials is nearly the same. The maximum EQE value (50.7%) was obtained at 585 nm. The effective photo-carrier generation in the wavelength range of 350 nm ~ 950 nm implies that MoS_2 and WS_2 nanofilm absorbs photons in the respective spectral region.

Figure 4.10 shows the Mott-Schottky (C^{-2} vs. V) characteristics with the inset showing the measured C–V curves of (a) MoS_2- and

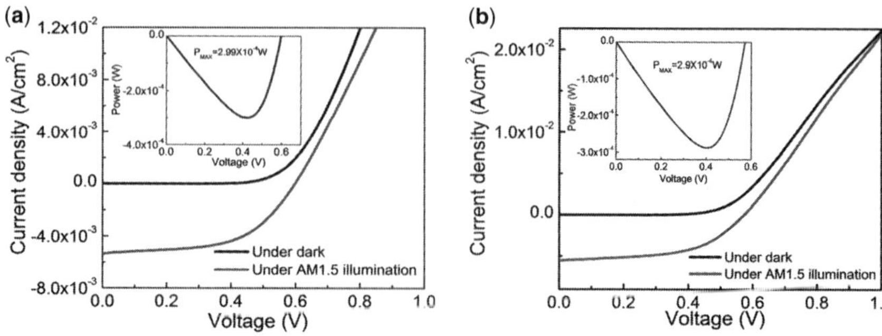

Figure 4.8 Dark and illuminated J-V characteristics of the Schottky barrier solar cells employing (a) nanofilms of MoS_2 and (b) WS_2 as photoactive materials developed by layer transfer process. The insets show maximum power output from the Schottky barrier solar cells.

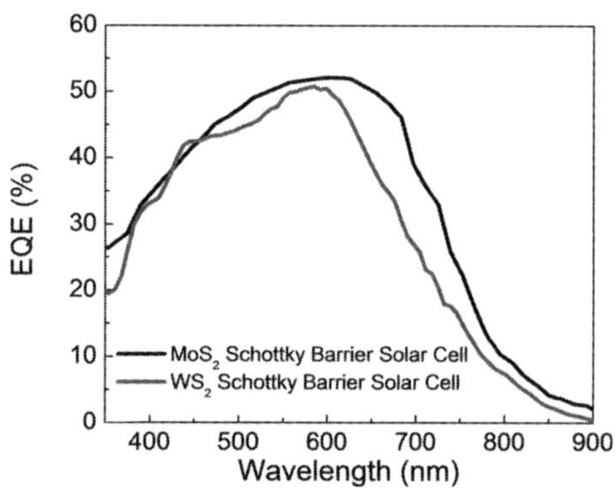

Figure 4.9 EQE spectra of the Schottky barrier solar cells with MoS_2 and WS_2 nanofilms as photoactive materials showing effective electron collection in a wavelength window of 350 nm ~ 900 nm.

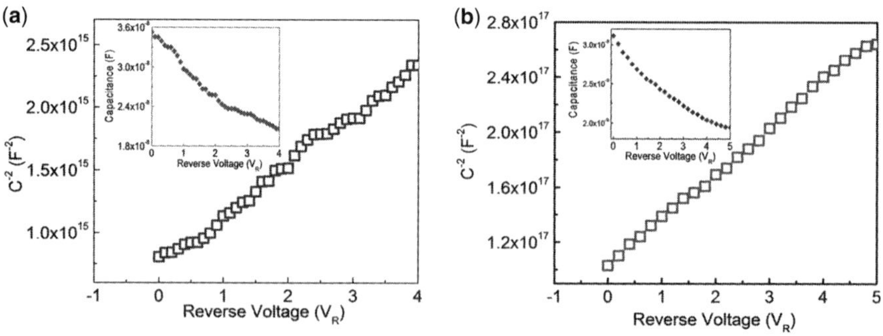

Figure 4.10 Mott-Schottky characteristics of the Schottky barrier solar cells employing (a) MoS_2 and (b) WS_2 nanofilms as photoactive materials. The insets show the C-V characteristics of the Schottky barrier solar cells.

(b) WS_2-based Schottky barrier solar cells. Linear dependency of C^{-2} on V behavior was observed in the reverse voltage bias region confirming that the CVD-grown MoS_2 and WS_2 nanofilms are n-type doped. The C-V characteristic of the Schottky barrier solar cell is given by Eq. 4.1,

$$\frac{1}{C^2} = \frac{2(V_R + V_{bi})}{q\varepsilon_s N_D A^2} \tag{4.1}$$

where V_R is reverse voltage bias, V_{bi} is built-in-potential, q is electronic charge, ε_s is permittivity, N_D is doping density in the MoS_2 and WS_2 nanofilms, and A is the junction area. The value of Schottky barrier height was acquired by extrapolating the linear C^{-2} region to the abscissa, yielding about ~0.5 eV in both MoS_2- and WS_2-based Schottky barrier solar cells. Under forward applied bias, Fermi levels of 2D layered semiconductors and Au metal contact shift upward and downward, respectively, with respect to the value at equilibrium condition (zero applied bias). Under reverse applied bias, Fermi levels shift in the opposite direction, forming maximum Schottky barrier height. The layer-structured MoS_2 and WS_2 nanofilms possess dangling-bond-free surface on which the recombination velocity of photogenerated carriers would be much lower than that of conventional semiconductors. However, since each MoS_2 and WS_2 nanofilm in the active layer stack of the Schottky barrier solar cell was manually transferred, we expect that microstructural deformation might have been introduced into the stack. This could be a major factor that deteriorates the Schottky barrier height in both solar cells.

4.6 Electronic Transport and Interfacial Recombination

4.6.1 BHJ Solar Cells

Figure 4.11(a) shows the energy band diagram of the BHJ solar cells illustrating that various physical processes associated with photoexcited electrons and holes occur at the TiO_2/2D layered semiconductor/P3HT interfaces under solar cell operating condition. Incident photons with an energy range of ~1.7 eV to 1.9 eV excite electrons simultaneously from both the 2D layered semiconductors and P3HT.

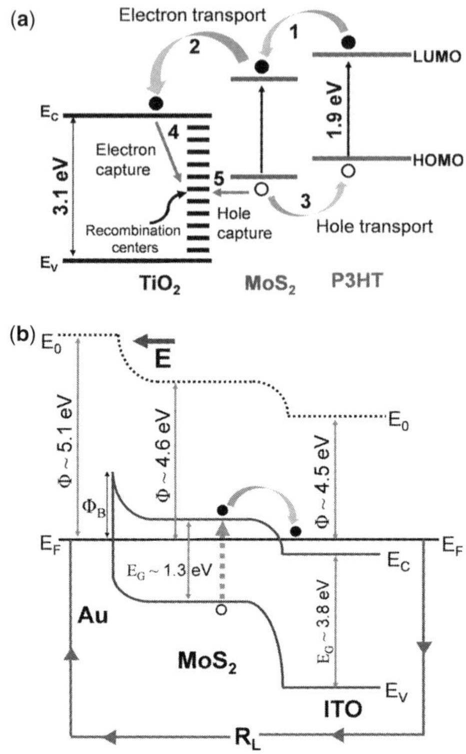

Figure 4.11 Energy level alignment in the (a) BHJ solar cells showing major electron transport and recombination processes occuring at TiO_2/2D-layered semiconductor/P3HT interfaces (b) Schottky barrier solar cell with ITO/2D-layered semiconductor/metal interfaces.

The photogenerated electrons are injected into the conduction band of the TiO_2 in which electrons diffuse to reach the ITO electrode. Processes (1) and (2) in Figure 4.11(a) represent electron injection from the lowest unoccupied molecular orbital (LUMO) of P3HT to the conduction band of 2D layered semiconductor (in this case MoS_2). Process (3) represents hole transport from the valence band of the MoS_2 to the highest occupied molecular orbital (HOMO) of the P3HT and then to the Au back contact. These three processes (1 to 3) are favorable to enhance the solar cell performance. However, due to the large surface area of TiO_2, surface states (recombination centers) exist in the bandgap of TiO_2 in which photoinjected electrons can be trapped from the conduction band and recombine with holes present at the valence band of MoS_2. Process (4) represents electron capture by TiO_2 surface state followed by hole capture (process (5)) from the valence band of

MoS_2-coated on TiO_2. We propose that it is these two interfacial processes (electron and hole capture) which eventually lead to recombination and deteriorate the solar cell performance significantly.

The Shockley-Read-Hall (SRH) recombination model [37] can explain the major interfacial recombination processes and gives electron lifetime (Eq. 4.2),

$$\tau_{SRH} = \frac{\tau_p\left(n_0 + n_1 + \Delta_n\right) + \tau_n\left(p_0 + p_1 + \Delta_p\right)}{p_0 + n_0 + \Delta_n} \tag{4.2}$$

where τ_p and τ_n are lifetimes of hole and electron, respectively. The n_0 and p_0 are electron and hole densities at thermal equilibrium condition (no excess electron-hole pair generation). The Δ_n and Δ_p are the excess electrons and holes generated by the BHJ solar cell under illumination condition. Further, the electron and hole lifetimes can be defined by $\tau_n = \frac{1}{\sigma_n V_{Th} N_T}$ and $\tau_p = \frac{1}{\sigma_p V_{Th} N_T}$ where σ_n and σ_p are capture cross-sections for electrons and holes, respectively. The quantities V_{Th} and N_T represent thermal velocity and density of TiO_2 surface states often called as trap states. As the density of TiO_2 surface states increases in the large surface area TiO_2 nanoparticle layer, electron lifetime decreases significantly due to an increase in the probability of recombination via TiO_2 surface states. In Eq. 4.2 the quantities n_1 and p_1 are electrons and holes that participate in the transport and recombination process at TiO_2/2D layered semiconductors interfaces under illumination condition. The electrons and holes take part in the transport process caused by the presence of TiO_2 surface states that are distributed all over the bandgap. However, only those surface states close to valence band of the 2D layered semiconductors would actively participate in the capture and recombination process due to the presence of free holes at valence band. The same analysis for electronic transport and recombination processes is valid in the case of BHJ solar cell with WS_2 as a photoactive material candidate. Both MoS_2 and WS_2 behave very similarly in photonic absorption as they possess very similar optical bandgap values.

4.6.2 Schottky Barrier Solar Cells

Figure 4.11(b) shows the energy band diagram of the fabricated Schottky barrier solar cells having a stacked structure of ITO/2D layered semiconductors/Au. If the metal work function is higher than that of n-type semiconductor, then an ideal Schottky barrier could be formed. Au metal film with a work function: ~5.1 eV forms a Schottky barrier on 2D layered semiconductors MoS_2 and WS_2 (work function: ~4.6 eV). Work function of ITO is ~4.5 eV which is suitable to form a good Ohmic contact with n-type doped MoS_2 and WS_2 nanofilms. Since the ITO is heavily doped material, its Fermi level enters into the conduction band as shown in Figure 4.11(b). Optical bandgap of the 2D layered semiconductors changes from monolayer to six layers (in the case of MoS_2) and becomes constant. Therefore layer dependency of optical bandgap becomes negligible while bulk material is employed in the solar cells. The MoS_2- and WS_2-stacked nanofilms used for Schottky barrier solar cells with an optical bandgap of ~1.3 eV effectively absorb photons in the wavelength region of 350 nm ~ 950 nm. ITO does not absorb solar photons in visible spectral window due to the wide bandgap through which the solar cells were illuminated. Incident photons generate electron-hole pairs in the 2D layered semiconductors, and electrons are subsequently excited to the conduction band of 2D layered semiconductors. As the work function difference between Au metal contact and n-type 2D layered semiconductors results in a Schottky barrier, the photoexcited electrons move from the conduction band of 2D layered semiconductors towards the Fermi level of ITO where they can be collected to the external load. A larger work function difference between Au and n-type 2D layered semiconductors would generate higher electric field in the depletion region. The photogenerated electron-hole pairs are separated and driven by the built-in electric field developed at the 2D layered semiconductor/Au interface. The processes of optical excitation in the 2D layered semiconductor (in this case MoS_2) and electron transport from MoS_2 to ITO electrode are illustrated in Figure 4.11(b). The same analysis and band energy level alignment can be generalized to the Schottky barrier solar cell employing WS_2 as active semiconductor material candidate.

4.7 Conclusion and Outlook

While electronic transport in conventional bulk and thin-film semiconductors-based solar cells is limited by electronically active interfacial defects, 2D layered semiconductors offer self-terminated, dangling, bond-free surfaces to form interfaces with other active layers in solar cells to achieve effective electronic transport with better control on interfacial recombination. Large area 2D layered semiconductor material development is viable by the CVD method and hence useful for large area solar cell applications. Thickness-dependent physical properties are attractive for photovoltaic technology through which material parameters can be tuned in a simple way. The demonstrated BHJ and Schottky barrier solar cells elucidated 2D layered semiconductors can be considered as potential material candidates for emerging photovoltaic technology.

References

1. Y. Li, C.-Wei Wang, Y. Zhao, J. Wang, F. Zhou, *Journal of Solid State Chemistry*, Vol. 196, P. 349, 2012.
2. S. Lu, L. Zeng, T. Wu, B. Ren, J. Niu, H. Liu, X. Zhao, *J. Mao, Solar Energy*, Vol. 85, p,1967, 2011.
3. M. Wang, X. Wang, *Solar Energy Materials and Solar Cells*, Vol. 91, p.178223, 2007.
4. J. A. Mikroyannidis, M. M. Stylianakis, P. Suresh, G.D. Sharma, *Solar Energy Materials and Solar Cells*, Vol. 93, p. 1792,2009.
5. Y. Wang, W. Wei, X. Liu, Y. Gu, *Solar Energy Materials and Solar Cells*, Vol.98, p. 129, 2012.
6. J. Lee, J. Y. Jho, *Solar Energy Materials and Solar Cells*, Vol. 95, p. 3152, 2011.
7. J. Bouclé, S. Chyla, M. S.P. Shaffer, J. R. Durrant, D. D.C. Bradley, J. Nelson, *Comptes Rendus Physique*, Vol.9, p.110, 2008.
8. Z.J. Wang, S.C. Qu, X.B. Zeng, J.P. Liu, C.S. Zhang, F.R. Tan, L. Jin, Z.G. Wang, *Applied Surface Science*, Vol. 255, p.1916, 2008.
9. Y.Y. Yu, W.C. Chien, Y.H. Ko, Y.C. Chan, S. C. Liao, *Current Applied Physics*, Vol.12, p.S7, 2012.
10. H. Flügge, M. Pfannmöller, G. Benner, I. Wacker, H. Schmidt, T. Rabe, R. R. Schröder, W. Kowalsky, *Energy Procedia*, Vol. 31, p.46, 2012.
11. M. Bhushan, *Appl. Phys. Lett.*, Vol. 40, p. 51,1982.
12. M. Vanecek, A. Abrahám, O. Štika, J. Stuchlík, J. Koka, *phys. stat. sol.* (a), Vol.83, p.617,1984.
13. Z. B. Zhou, R. Q. Cui, Q. J. Pang, G. M. Hadi, Z. M. Ding, W. Y. Li, *Sol. Energy Mater. Sol. Cells*, Vol. 70, p. 487, 2002.
14. B. Ghosh, M. Das, P. Banerjee, S. Das, *Solid State Sci.* Vol. 11, p. 461, 2009.
15. M. Soylu, F. Yakuphanoglu, *Thin Solid Films* Vol. 519, p. 1950, 2011.
16. J. Wei, Y. Jia, Q. Shu, Z. Gu, K. Wang, D. Zhuang, G. Zhang, Z. Wang, J. Luo, A. Cao, and D. Wu, *Nano Lett.*, Vol. 7, p. 2317, 2007.
17. C. Xie, P. Lv, B. Nie, J. Jie, X. Zhang, Z. Wang, P. Jiang, Z. Hu, L. Luo, Z. Zhu, L. Wang, and C. Wu, *Appl. Phys. Lett.*, Vol. 99, p. 133113, 2011.

18. Y. Ye, Y. Dai, L. Dai, Z. Shi, N. Liu, F. Wang, L. Fu, R. Peng, X. Wen, Z. Chen, Z. Liu, and G, Qin, *Appl. Mater. Interfaces*, Vol. 2, p. 3406, 2010.
19. L. Zhang, Y. Jia, S. Wang, Z. Li, C. Ji, J. Wei, H. Zhu, K. Wang, D. Wu, E. Shi, Y. Fang,and A. Cao, *Nano Lett.*, Vol. 10, p. 3583, 2010.
20. C. C. Chen, M. Aykol, C. C. Chang, A. F. J. Levi, and S. B. Cronin, *Nano Lett.*, Vol. 11, p. 1863, 2011.
21. P. Broqvist, A. Alkauskas, and A. Pasquarello, *Phys. Rev. B*, Vol. 78, p. 075203, 2008.
22. R. Brüggemann, G.H Bauer, *J. Non-Cryst. Solids*, Vol. 227, p.197, 1998.
23. E. Kamiyama, K. Sueoka, *Appl. Surf. Sci.*, Vol. 258, p. 5265, 2012.
24. C. Boehme, K. Lips, *J. Non-Cryst. Solids*, Vol. 338, p. 434, 2004.
25. R. Barrio, J. J. Gandía, J. Cárabe, N. González, I. Torres, D. Muñoz, C. Voz, *Sol. Energy Mater. Sol. Cells*, Vol. 94, p. 282, 2010.
26. P. Repo, H. Talvitie, S.Li, J. Skarp, H. Savin, *Energy Procedia*, Vol. 8, p. 681, 2011.
27. A. Splendiani, L. Sun, Y. Zhang, T. Li, J. Kim, C. Y. Chim, G. Galli, and F. Wang, *Nano Lett.*, Vol. 10 (4), p. 1271, 2010.
28. K. K. Liu, W. Zhang, Y. H. Lee, Y. C. Lin, M. T. Chang, C. Y. Su, C. S. Chang, H. Li, Y. Shi, H. Zhang, C. S. Lai, and L. J. Li, *Nano Lett.*, Vol. 12 (3), p.1538, 2012.
29. K. F. Mak, C. Lee, J. Hone, J. Shan, T. F. Heinz, *Phys. Rev. Lett.*, Vol. 105, p. 136805, 2010.
30. S. Najmaei, Z. Liu, P. M. Ajayan, and J. Lou, *Appl. Phys. Lett.* Vol.100, p.013106, 2010.
31. B. Radisavljevic, A. Radenovic, J. Brivio, V. Giacometti, and A. Kis, *Nature Nanotechnol.* Vol. 6, p.147, 2011.
32. Y. Yoon, K .Ganapathi, and S. Salahuddin, *Nano Lett.* Vol. 11, p.3768, 2011.
33. Y. Zhang, J. Ye, Y. Matsuhashi, and Y. Iwasa, *Nano Lett.* Vol. 12, p. 1136, 2012.
34. H. Qiu, L. Pan, Z. Yao, J. Li, Y. Shi, and X. Wang, *Appl. Phys. Lett.* Vol. 100, 123104, 2012.
35. H. Hwang, H. Kim, and J. Cho, *Nano Lett.* Vol. 11, p. 4826, 2011.
36. M. Shanmugam, T. Bansal, C. A. Durcan, B. Yu, *Appl. Phys. Lett.* Vol. 100, p. 153901, 2012.
37. W. Shockley, W.T. Read, *Phys. Rev.*, Vol. 87, p. 835, 1952.

5

Control of ZnO Nanorods for Polymer Solar Cells

Hsin-Yi Chen[1], Ching-Fuh Lin[1,2,*]

[1]*Graduate Institute of Photonics and Optoelectronics Engineering, National Taiwan University, Taipei, Taiwan (R.O.C.)*
[2]*Graduate Institute of Electronics Engineering, National Taiwan University, Taipei, Taiwan (R.O.C.)*

Abstract

Zinc oxide (ZnO), an n-type inorganic semiconductor, has been extensively utilized in polymer solar cells due to its salient characteristics such as low cost, non-toxicity, high stability, and good optoelectronic properties. Their one-dimensional counterparts, ZnO nanorods (NRs), are particularly attracting more and more attention in recent years. Compared with devices in a planar configuration, this device architecture has several advantages, including that the morphology of the composite can be controlled in a systematic and reproducible way; devices using vertically-aligned ZnO NRs can increase electron donor/acceptor interface areas; and this device structure provides direct electron transport pathways toward the electrode that possesses high electron mobility. Therefore, the understanding of ZnO NRs for the control of polymer solar cells is critical. In this chapter, we will review the preparation, characterization and morphology control of ZnO NRs. The latest technology and principles of utilizing ZnO NRs to enhance device performance in polymer solar cells will also be reported in detail.

Keywords: Zinc oxide (ZnO), nanorod (NR), morphology, polymer solar cell

**Corresponding author*: cflin@cc.ee.ntu.edu.tw

Atul Tiwari, Rabah Boukherroub, and Maheshwar Sharon (eds.) Solar Cell Nanotechnology, (135–158) 2014 © Scrivener Publishing LLC

5.1 Introduction

Organic solar cells (OSCs) are good alternatives compared to conventional inorganic photovoltaic devices because they can be fabricated on flexible substrates by solution process. However, in order to overcome the short exciton diffusion length in most organic semiconductors, a bulk heterojunction (BHJ) structure is widely utilized in organic solar cells. In terms of BHJ cells, poly(3-hexylthiophene) (P3HT) and (6, 6)-phenyl C61 butyric acid methyl ester (PCBM) are the most commonly utilized materials for electron donor and acceptor with a high power conversion efficiency (PCE) of 5% [1–3]. A number of inorganic semiconductors, such as TiO_2 [4, 5], ZnO [6, 7] and CdSe [8] have also been applied as electron acceptors. Although efficient charge separation is achieved by a large interfacial area between electron donor and acceptor, BHJ devices usually suffer from discontinuous charge transport pathways [9]. In addition, the percolation network is susceptible for process, which makes it difficult to achieve a reliable performance. Consequently, creation of vertically-aligned inorganic nanostructures with respect to the substrate is thus considered to be a potential approach to overcome the above disadvantages.

In an ideal case, vertically-aligned nanostructures could be synthesized with an optimized dimension which is short enough for the most excitons to diffuse through the interface between the inorganic and organic materials. By controlling the dimension of the vertically-aligned nanostructure, polymer chains are also allowed to align themselves perpendicular to the substrate. The inter-chain coupling is thus improved, causing better exciton diffusion and charge transport in conjugated polymers [10–12]. On the other hand, for electrons, there is also a direct transport pathway to reach the electrode in vertically-aligned nanostructures. Among those nanostructures, ZnO NRs are the most extensively utilized nanostructures for organic solar cells in the past few years due to their excellent optical characterization, good structural property, large-area availability and low cost synthesis process. As a result, in this chapter, we will review the recent work focusing on the control of ZnO NRs for polymer solar cells. The preparation, characterization and morphology control of ZnO NRs will be demonstrated. The latest technology and principle by utilizing ZnO NRs to improve the surface morphology and enhance charge separation in polymer solar cells (PSC) will also be addressed in detail in this chapter.

5.2 Preparation and Characterization of ZnO NRs

ZnO are receiving considerable interest because of their excellent photonic properties, as well as a wide bandgap of 3.4 eV that is beneficial to short wavelength optoelectronic applications. Its transparent property is also good for visible light transmission. Its nanostructure counterparts, ZnO NRs, are able to facilitate more efficient carrier transport because of decreased grain boundaries, surface defects, disorders, and discontinuous interfaces [13, 14]. Conventionally, various methods of growing ZnO NRs include vapor liquid-solid (VLS) deposition and chemical vapor deposition (CVD) [15]. So far, the simplest, most energy-efficient method for synthesizing ZnO NRs is using the hydrothermal method [15, 16]. This technique has been shown to be the most economical synthesis process compared to the conventional growth method [16]. Consequently, in terms of the preparation of ZnO NRs, we will focus on the ZnO NRs preparation and characterization by hydrothermal method in the following section.

5.2.1 ZnO NRs Prepared by Hydrothermal Method

The hydrothermal method is generally regarded as a solution-phase synthesis compared to vapor-phase synthesis. The growth process is carried out in a liquid. Normally aqueous solutions are introduced and the process is then referred to as "hydrothermal" growth process. Some of the solution-phase synthesis processes are Zinc Acetate Hydrate (ZAH) derived nanocolloidal sol-gel route [17], ZAH in alcoholic solutions with sodium hydroxide (NaOH) or tetra methyl ammonium hydroxide (TMAH) [18–20] and template-assisted growth [21].

So far, the general hydrothermal process to synthesize ZnO NRs is as follows: Si (or alternative substrate) is cleaned by a standard cleaning method. Approximately 200–500 nm ZnO thin film is deposited on the substrate using RF magnetron sputtering at room temperature. The samples are then heated in air at various temperatures (250–450°C) for 10–30 min to obtain ZnO seeds for hydrothermal growth. Other method like sol-gel processes can also be applied to form this thin film of seed layer [17]. ZnO nanoparticles have also been used for the seed layer by directly coating the solution that contains the suspended colloidal nanoparticles on the substrates, followed by a simple annealing step to remove the solvent [21]. The ZnO-seeded samples are then placed in a solution containing zinc nitrate ($Zn(NO_3)_2$) and hexamethylamine (HMT) as a precursor. The hydrothermal reaction is usually performed in a preheated oven at a certain temperature (usually < 100°C) for several hours. In the hydrothermal process, several factors are often considered to affect the synthesis of the ZnO NRs, including control of HMT and $Zn(NO_3)_2$, control of seed layer synthesis and heating temperature, as well as morphology control of ZnO NRs. Herein we will discuss the above three important parameters to obtain excellent ZnO NRs.

5.2.1.1 *Control of HMT and $Zn(NO_3)_2$*

Vayssieres *et al.* [22] first utilized the hydrothermal method for the controlled fabrication of ZnO NRs on glass and Si substrates by the thermal decomposition of methenamine and zinc nitrate. To initiate the growth from the substrate, a thin layer of ZnO nanoparticles is put on the substrate. Methenamine, also known as hexamethylenetetramine (HMT), is a highly water soluble, non-ionic tetradentate cyclic tertiary amine. Thermal degradation of HMT releases hydroxyl ions which react with Zn^{2+} ions to form ZnO [23]. This can be described in the following equations:

$$(CH_2)6N_4 + 6H_2O \longrightarrow 6HCHO + 4NH_3, \tag{5.1}$$

$$NH_3 + H_2O \longrightarrow NH^+_4 + OH^-, \tag{5.2}$$

$$2OH^- + Zn^{2+} \longrightarrow ZnO(s) + H_2O. \tag{5.3}$$

It is generally known that the role of HMT is to supply the hydroxyl ions to drive the precipitation reaction [24]. Besides, HMT also acts as a buffer as the rate of its hydrolysis decreases with an increase in the pH value [25]. Ashfold *et al.* [24] have demonstrated that the rate of decomposition of HMT is independent of the reaction and that HMT does act as a kinetic buffer. In the initial growth stage, the pH value and the concentration of Zn^{2+} ions are higher so that the ZnO growth will be through $Zn(OH)_2$. With the gradual increase in the pH value and the concentration of the Zn ions, $Zn(OH)_2$ becomes thermodynamically unstable and the $Zn(OH)_2$ formed on the substrate begins dissolving. Further growth of the NRs will be through direct deposition of ZnO [24]. The contribution of HMT in the growth process of ZnO NRs has also been discussed by Sugunan *et al.* [26]. It was proposed that HMT, a long-chain polymer and a nonpolar chelating agent, will preferentially attach to the nonpolar facets of the zincite crystal. Therefore, it cuts off the access of Zn^{2+} ions to them, leaving only the polar (001) face for epitaxial growth. That is, HMT acts more like a shape-inducing polymer surfactant rather than as a buffer.

There is a lot of research investigating the growth mechanism of ZnO NRs synthesized using $Zn(NO_3)_2$ and HMT, such as the effect of substrates and seed layers on the morphology of nanorods [27–29], as well as the control of aspect ratio through the addition of citrate anions [30]. Pal and Santiago [31] have controlled the morphology of ZnO nanostructures by varying the amount of a soft surfactant, ethylenediamine and the pH value of the reaction mixture composed of sodium hydroxide and the surfactant. Homogenous growth was observed at a pH value of 12 with inhomogeneity creeping as the pH decreases [31]. A 10% concentration of ethylenediamine obtained higher aspect ratio than 5% concentration. ZnO NRs have been successfully synthesized using different surfactants which are available from numerous reports. Tang *et al.* [32] synthesized ZnO NRs by using zinc acetylacetonate $Zn(acac)_2 \cdot H_2O$ as a single source precursor and investigated the growth condition of the nanorods in the presence of four different surfactants: polyvinyl alcohol (PVA), polyethylene glycol (PEG), sodium dodecyl sulphate (SDS) and cetyltrimethyl ammonium bromide (CTAB). The usage of PVA resulted in more regular and defect-free rods than PEG, SDS and CTAB. Li *et al.* [33] have reported the growth of tapered ZnO NRs with diameter decreasing from 400 nm at the body to about 80 nm at the tip from CTAB-assisted hydrothermal growth. Chen *et al.* [34] studied the effect of potassium iodide (KI) as a surfactant in the crystallization of ZnO nanorod clusters from a chemical bath containing zinc nitrate hexahydrate and hydrazine hydrate. They obtained step growth of hexagonal nanorod clusters, which is attributed to the presence of iodine ions in the growth bath. The further results clearly demonstrated that the source of Zn ions could also affect the morphology. Ni *et al.* [35] obtained ZnO NRs in place of stepped columns simply by changing the source of Zn ions from $Zn(Ac)_2$ to $ZnCl_2$, and all other conditions remain. It is noted that the presence of Zn^{2+} containing salts is not a necessity for the growth of ZnO nanostructures through the hydrothermal growth method. Li *et al.* [36] have successfully grown large-scale arrays of

ZnO NRs on zinc foil without the assistance of any template, oxidant or coating of metal oxide layers. It is simply synthesized by dipping the foil into a 25% aqueous solution of ammonia and heating at a temperature 80°C in a Teflon-lined stainless steel autoclave. From the above research, it was observed that the thickness, density and morphology of the ZnO NRs are affected by the pH value and the concentration of HMT and $Zn(NO_3)_2$ in the growth bath. A good understanding of control of HMT and $Zn(NO_3)_2$ can promote high quality ZnO NR growth.

5.2.1.2 *Control of Seed Layer Synthesis and Heating Temperature*

In this hydrothermal method technique, the seeded substrates are utilized to achieve well-controlled morphology and growth orientation of ZnO NRs. Heterogeneous nucleation can favor the ZnO NR growth because the interfacial energy between crystals and substrate is usually smaller than the interfacial energy between crystals and solution [37]. Therefore, pre-coating the substrate with seed materials that are similar to nanocrystals can effectively control nanorod orientation and morphology.

A variety of methods have been employed to produce seeds such as thermal evaporation [14, 15], metal organic chemical vapor deposition (MOCVD) [38–40], spin coating [41], and sputtering [42–46]. Li *et al.* [42] and Song *et al.* [43] deposited ZnO on silicon substrate by using RF sputtering followed by hydrothermal reaction without any heat treatment and obtained columnar ZnO NRs. The above results show the importance of the heat treatment of ZnO seed layers in forming ZnO NRs using the hydrothermal method. ZnO NRs have been grown on a seeded Zn sheet, and the optimum ZnO seed formation was obtained after oxidation at 300°C [15]. Above 300°C, ZnO flakes other than spherical ZnO seeds form on the Zn sheet. For a Zn thin film sputtered on PTFE substrate, thermal oxidation was performed at 300°C followed by hydrothermal reaction to grow ZnO NRs. Liu *et al.* [46] used ZnO deposited on Si substrate by ion beam sputtering followed by annealing temperature at 200–600°C. The high annealing temperature could limit future applications of ZnO NRs because the ZnO buffer layer would degrade. This degradation can be caused by interdiffusion between the buffer layer and the Si substrate, and by the thinning of the buffer layer because of heating at high temperature [47].

As illustrated above, in addition to seed layer synthesis, heating temperature is also an important factor that affects ZnO NR morphology. There are several groups investigating the heating temperature effect on ZnO NR morphology by hydrothermal method. In 2012, Ridhuan *et al.* [48] reported the surface morphology of ZnO films deposited on Si substrates after different heat-treatment temperatures, as depicted in Figure 5.1. After heat treatment at the temperature of 250°C, the surface of ZnO films is covered with

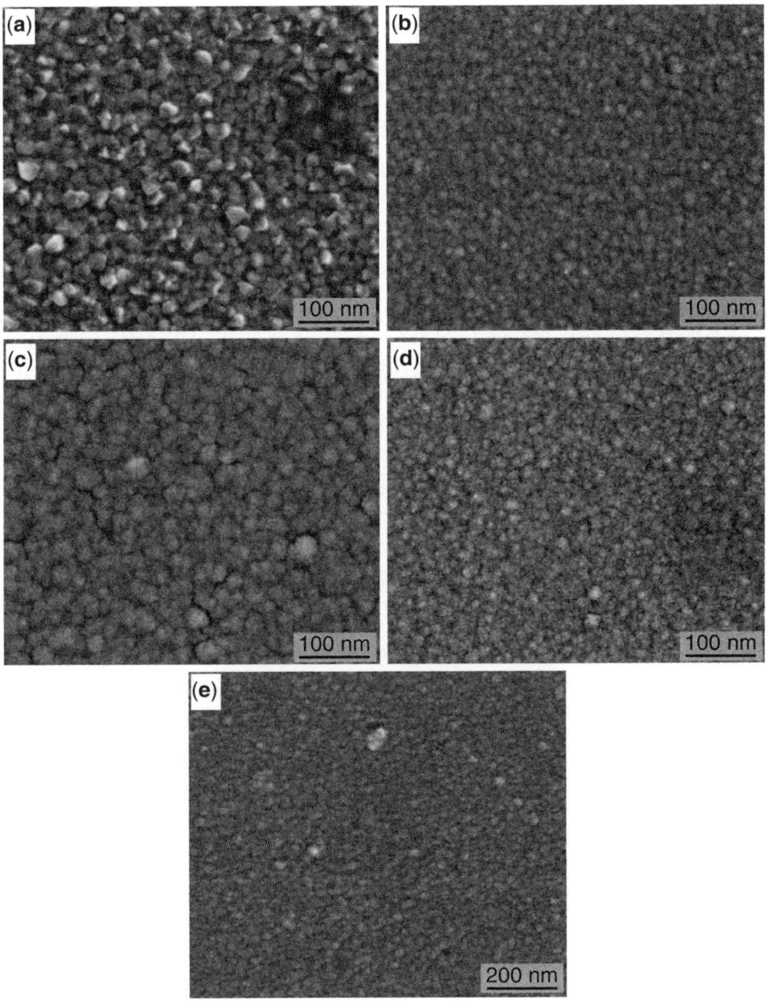

Figure 5.1 Surface morphologies of ZnO seeds heat treated at different temperatures: a) 250, b) 300, c) 350, d) 400, and e) 450C [48].

dispersed particulate grain size and an average diameter of 90 nm is obtained. The surface morphology of ZnO seed layer transforms from a flaky structure at 250°C to a circular structure with heating temperature increasing from 300°C to 450°C [48]. Obviously, the ZnO seed size decreases with increased heating temperature. With high heating temperature treatment, the seed layer particles tend to dislocate and pile up. At high heating temperatures, atoms vibrate at their lattice positions and exchange energy with neighboring atoms. As the atoms reach sufficient energy, they diffuse to achieve the lowest strain energy and form small seeds. That is, smaller grains form with increasing heating temperature. On the other hand, to achieve excellent nanorod formation by the hydrothermal method, ZnO seeds need to have uniform size and distribution. The ZnO seeds are required to provide enough spaces for ZnO nanorod growth during a hydrothermal reaction. From the FESEM images in Figure 5.2, the ZnO seeds which are treated at 400°C depict the best morphology for ZnO NRs formation.

The heat-treated seeded substrates are then subjected to hydrothermal growth at 80°C for 4 h to investigate the ZnO seeds effect on ZnO NRs formation. Figures 5.2(a)–5.2(e) [48] illustrate ZnO NRs growth on ZnO seed substrates at different heating temperatures corresponding to Figure 5.1. The average lengths of ZnO NRs are 105, 184, 231, 742, and 605 nm for ZnO seeds heat treated at 250, 300, 350, 400, and 450°C, respectively [48]. The average diameter of the ZnO NRs gradually increases from 64 nm to 80 nm with increased heating temperature of the seed layer. The increase in ZnO nanorod length is due to increasing reactivity of the seed layer surface during the nucleation stage [46]. Generally, the ZnO crystal structure consists of a number of alternating planes composed of coordinated Zn and O atoms along the c-axis direction. The Zn-terminated [0001] is at the top surface of the ZnO crystal structure, whereas the O-terminated [000)] is at the bottom surface, causing a dipole moment and a polar surface of ZnO seed layer [46]. As a result, the presence of two charges (Zn^{2+} and OH^- ions) results in the polar surface of the ZnO seed layer.

On the other hand, because $Zn(OH)_2$ forms and produces ZnO upon dehydration, the increased heating temperature can effectively improve the reactivity of the seed layer. This reactivity is stronger than the electrostatic interaction of the polar surface of the ZnO seed layer with the charges (Zn^{2+} and OH^-) because of the energy supplied during heat treatment [46]. Therefore, more

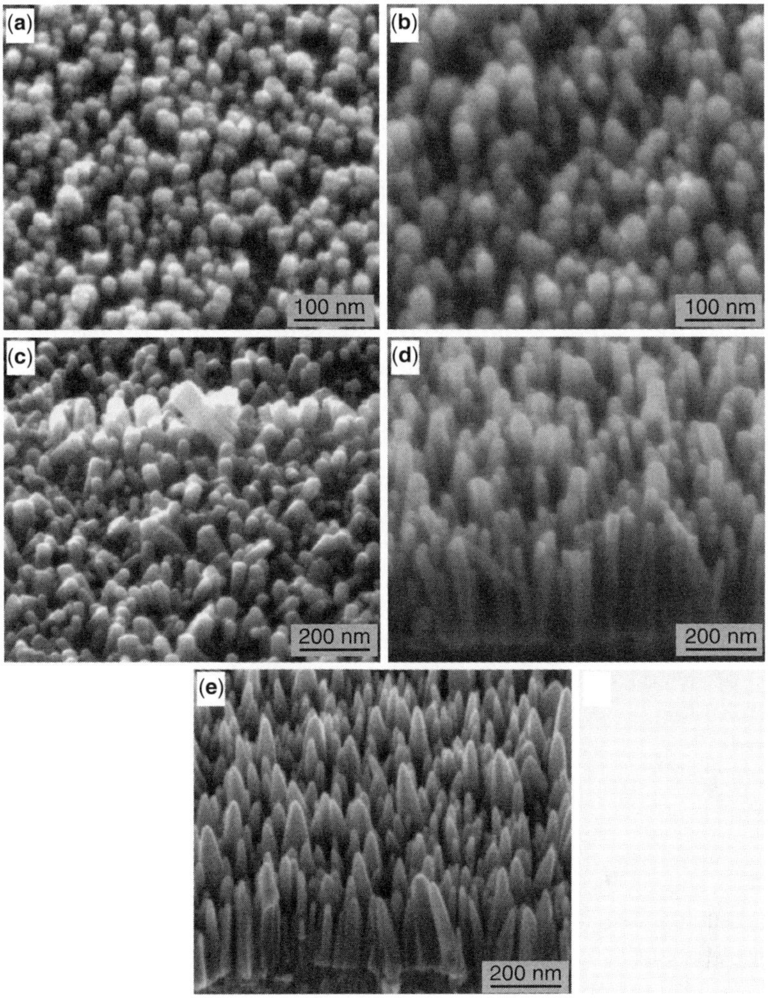

Figure 5.2 FESEM images of ZnO NRs grown on seeds heat treated at different temperatures: a) 250, b) 300, c) 350, d) 400, and e) 450C [48].

Zn^{2+} and OH^- ions are diffused on the ZnO seeds, which bring out increased ZnO nanorods length with heating temperature from 250°C to 400°C. However, further increase annealing temperature to 450°C results in decreased ZnO nanorod length. It clearly demonstrates that there is a threshold annealing temperature for the ZnO seed layer which leads to the structural deterioration.

The X-ray diffraction spectra of ZnO NRs with different heating temperatures are also shown to identify the structural property [48].

It illustrates two peaks corresponding to wurtzite ZnO with a strong (002) peak and a small (103) peak at 2θ= 34.4° and 61.7°, respectively. It is depicted that the (002) peak intensity increases with the rising heating temperature. This is because, when the annealing temperature increases, more energy is supplied, thus causing Zn and O atoms to move into the proper sites. As a result, the crystal quality is improved to reduce the Gibbs free energy and high-quality ZnO crystals are formed under thermal equilibrium [47]. As mentioned above, Figure 5.1 demonstrates that the increased heating temperature produces smaller ZnO seeds. A larger ratio of the surface area to volume also increases the surface area, thus increasing the ZnO growth rate. On the other hand, it demonstrates that the peak shifts to a slightly higher 2θ from 34.2° to 34.6° at higher temperature [48]. This may be due to the release of intrinsic strain induced by the presence of residual strain from structural imperfections, such as vacancies, stacking faults, and interstitials. According to the research investigation by Shin *et al.* [39], the lattice mismatch between ZnO film and Si-substrate could cause structural defects during the growth. Thus, an annealing treatment of the ZnO film deposited on Si-substrate could remove the intrinsic strain because the Zn and O atoms would be stacked in order. From the above results and discussions, the good control of seed layer synthesis and heating temperature for ZnO NRs growth by hydrothermal method is shown to be of great importance.

5.2.2 Morphology Control of ZnO NRs

As generally known, ZnO is regarded as a polar crystal. In its crystal structure, O^{2-} is in hexagonal closest packing and each Zn^{2+} lies within a tetrahedral group of four oxygen ions [49]. Zn and O atoms are stacked alternatively along the *c*-axis. The formation of hexagonal prism and pyramid of ZnO crystals is attributed to the difference in the growth velocities of various crystal facets. The growth velocities under hydrothermal conditions along the different directions are known to follow the relation $V(0001) > V(1011) > V(1010)$ [50]. Therefore, the relative growth rate of these crystal faces will determine the final shape and aspect ratio of the ZnO NRs.

Besides the ZnO NRs morphology affected by growth velocity in different directions, it is also significant to address the mechanism to bring out different morphology of ZnO NRs influenced by different $Zn(NO_3)_2$ concentration and reaction time. Li *et al.* [51] proposed the formation mechanism of ZnO NRs effected by different $Zn(NO_3)_2$

concentration and reaction time. In their experiment, the ZnO NRs are synthesized by a simple solution-phase method in two different kinds of solution [51]. The first solution contained 1.5 g ammonia (25.0%) and 5 g $Zn(NO_3)_2 \cdot 6H_2O$ (99.0%), and the second one contained 1.5 g ammonia and 3 g $Zn(NO_3)_2 \cdot 6H_2O$. Then two 2×3 cm^2 zinc foils were put into the above two kinds of aqueous solution, respectively, followed by 12 hours heating at a constant temperature of 80°C.

The formation mechanism and the schemes have been illustrated in Figure 5.3 and Figure 5.4, respectively [51]. The only difference in the two conditions is that the added $Zn(NO_3)_2 \cdot 6H_2O$ is less in the second solution. Therefore, the zinc precursor for the growth of ZnO crystals in the second solution mainly comes from the metal zinc, which leads to the supply of Zn^{2+} from $Zn(NO_3)_2 \cdot 6H_2O$ being very little. Then the zinc precursor is transported from the root regions of NRs to the top regions because of the gradient of concentration. As a result, the lower concentration region is around the tops of NRs. The growth of ZnO crystals is relatively slow compared with the root of NRs, thus causing the formation of ZnO NRs with sharp tops (Figure 5.3(b)). On the other hand, for the first solution with more $Zn(NO_3)_2 \cdot 6H_2O$, there is almost no gradient of concentration.

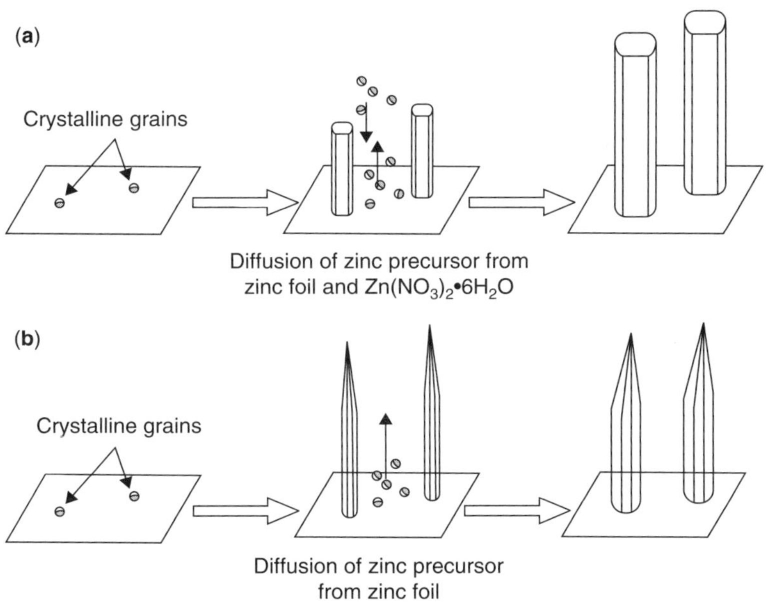

Figure 5.3 Growth diagram of the ZnO NRs (a) with flat-top for 12 h in the first solution; (b) with sharp-top for 12 h in the second solution. [51].

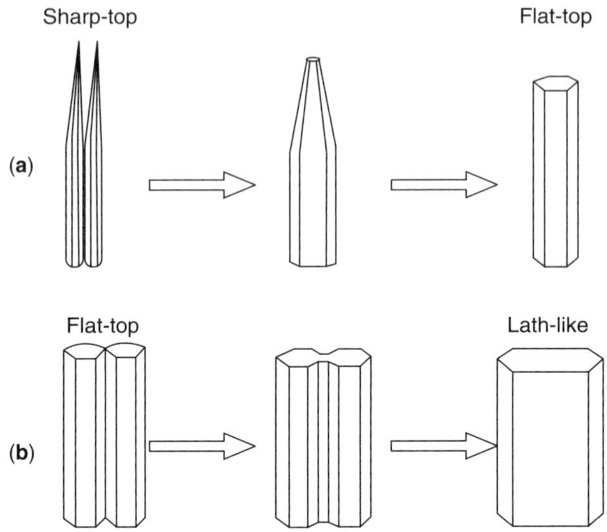

Figure 5.4 Mechanism of the morphology transition of ZnO NRs with increasing reaction time in the second solution. [51].

The concentration of zinc precursor is uniform throughout, which results in the formation of ZnO NRs with flat tops (Figure 5.3(a)).

Furthermore, with increasing reaction time, it also brings out different morphology as shown in Figure 5.4. As two single NR contact with each other, the local supersaturation of the solution is enhanced, which results in the secondary growth of crystals around the interface between the two NRs [52], and then the flat-top nanorods are formed reasonably well. On the other hand, the lath-like nanorods are also found during the growth, as shown in Figure 5.4(b). From the viewpoint of crystal growth, the six crystal planes have the same growth speed and it can not extend two crystal planes in one direction [53]. However, as shown in Figure 5.4(b), when two flat-top nanorods are closely parallel to each other, the secondary growth of crystals will happen in the interface regions and carry out the coalescence. If this situation continues, a lath-like crystal will be obtained thereby. It can elucidate the lath-like crystal which happens with the increasing reaction time.

5.2.3 Summary of ZnO NR Growth

The hydrothermal method is so far the simplest and most efficient process to synthesize ZnO NRs. Various additives are used in aqueous medium to successfully synthesize ZnO NRs of different morphologies. A mixture of $Zn(NO_3)_2$ and HMT is the most popular solution in the growth process. Based on the investigations above, ZnO NRs are attractive candidates for applications in optoelectronic devices such as solar cells, sensors, and energy generators. Good control of HMT and $Zn(NO_3)_2$, seed layer synthesis and heating temperature is beneficial to synthesize excellent vertically-aligned ZnO NRs, which can promote higher device efficiency in organic solar cells.

5.3 Application of ZnO NR in Polymer Solar Cells

In the past decade, rapid progress in the development of polymer solar cells has raised the power conversion efficiency from 3% to almost 9%, reflecting a promising future for production possibility. Polymer solar cells are processed from solution in organic solvents, whereas small molecule solar cells are usually processed using thermal evaporation deposition that requires vacuum equipment. Besides, PSCs are considerably attractive owing to a number of advantageous features, including thin-film architecture, less material consumption, and low manufacturing energy demands. On the other hand, as mentioned previously, vertically-aligned inorganic nanostructures are considered to be a practical method for promoting efficient charge transport in organic solar cells. As a result, over the recent years, ZnO NRs are extensively utilized in polymer solar cells to form organic-inorganic hybrid solar cells (HSC).

Organic-inorganic HSCs, which use a combination of organic and inorganic semiconductors, have been widely studied as an alternative to fully OSCs. This type of solar cell has been achieved by combining organic materials and inorganic semiconductors together. In general cases, a p-type organic semiconductor usually serves as both light harvester and electron donor, and the n-type inorganic semiconductor serves as electron acceptor. HSCs device structure possesses some advantages compared to OSCs. For example, the inorganic semiconductor of HSCs usually has higher electron mobility than the counterpart of organic materials. Besides, inorganic semiconductors usually have higher physical and chemical stability compared to that of organic semiconductors. Among these n-type inorganic semiconductors, ZnO has attracted a lot of attention in HSCs because it is less toxic than many other II–VI semiconductors and is relatively easy to synthesize in large quantities with low cost techniques. As an n-type semiconductor with a wide bandgap, ZnO also has very good optical property, which makes ZnO an excellent candidate as a good cathode buffer layer (CBL) or a transparent electrode in HSCs. In this section, we will report on the recent development of solar cells containing ZnO NRs and organic semiconductors. We will focus on the research of photovoltaic devices where ZnO acts as an electron acceptor or electrode buffer layer and transparent electrode.

5.3.1 ZnO-NR/Polymer Solar Cells Based on Vertically-Aligned Zno NRs

Compared with photovoltaic devices with a planar configuration, this kind of device configuration has several advantages. For instance, the morphology of the composite can be controlled in a reproducible way through controlling ZnO NR growth. The devices using vertically-aligned ZnO NRs can increase electron donor/acceptor interface areas, which can create direct electron transport pathways toward the electrode. Olson *et al.* first reported the study on ZnO-NR/P3HT HSCs based on vertically-aligned ZnO NRs in 2006 [54]. ZnO/P3HT HSCs with this device structure produced a V_{OC} of 0.44 V, a J_{SC} of 2.2 mA/cm^2, a fill factor (FF) of 0.56, and a PCE of 0.53%. They also investigated the effect of polymer on ZnO-NR/P3HT solar cell performance. The device performance was improved when dichlorobenzene was used as the solvent instead of chloroform, due to the enhanced infiltration. Furthermore, they reported that thermal annealing can improve the infiltration of the P3HT into the ZnO NRs, which is one of the most important issues for solar cells with this vertically-aligned NR structure [55]. Figure 5.5 depicts the cross-sectional SEM images of ZnO-NR/P3HT solar cells with and without annealing. It clearly indicates that the infiltration of the P3HT polymer is improved with increasing annealing temperature. It is also observed that more P3HT polymer adheres to the ZnO NRs for annealed cases after removing the polymer on top of the NRs.

In addition to P3HT, another polymer semiconductor poly (1-methoxy-4-(2ethylhexyloxy)-p-phenylenevinylene) (MEH-PPV) has also been utilized in this kind of device. ZnO-NR/MEH-PPV with vertically-aligned ZnO NRs has been reported by Wang *et al.* They prepared high quality ZnO NRs with low defect concentrations by electrodeposition method [56]. In order to further improve device

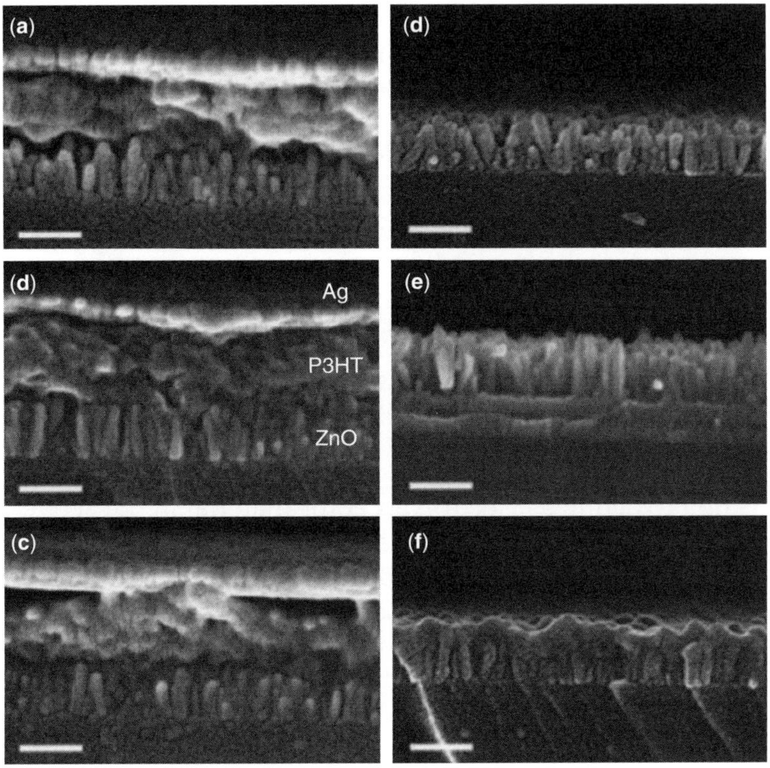

Figure 5.5 Cross-sectional SEM images of P3HT/ZnO NRs composite PV devices: (a) without annealing; (b) with annealing at 150 C; (c) with annealing at 225 C. Also shown are images after removing the polymer on top of the rods: (d) no anneal; (e) annealing at 150 C; (f) annealing at 225 C. Scale bars: 200 nm [55].

performance of ZnO-NR/organic HSCs, a lot of approaches have been investigated to modify and optimize the ZnO-NR/organic interface [57–59]. Furthermore, the ZnO surface can be modified by coating with a thin layer of other materials. The electrical properties of ZnO can be well controlled by the different doping level or coating composition. Various doping and surface treatment methods are demonstrated to modify the performance of HSCs with this device structure. For example, Ravirajan *et al.* [60] demonstrated the characterization of dye modified ZnO-NR/polymer solar cells with vertically-aligned NRs acting as electron acceptor materials. Three different solution chemicals were investigated for the synthesis of ZnO. An amphiphilic polypyridyl ruthenium complex dye was used to modify interfacial electron transfer between P3HT and the ZnO nanostructure surface, resulting in improved device performance compared to devices without dye treatment. The improvement is probably due to slower charge recombination after the amphiphilic dye modification treatment [60]. Other than these organic materials, inorganic coating has also been utilized to improve the device performance of ZnO-NR/organic HSCs. For example, Xu *et al.* [61] reported a simple and low-cost electrodeposition method to deposit n-type CdSe on the ZnO surface. ZnO-NR/P3HT solar cells with vertically-aligned ZnO NRs coated by CdSe exhibited an improved PCE of 0.88%, with short-circuit current density (Jsc) 4.89 mA/cm^2 and open-circuit voltage (Voc) of 0.51 V [61].

5.3.2 ZnO NR as a Cathode Buffer Layer in Polymer Solar Cells

In addition to acting as electron acceptor material in the active layer of ZnO/organic solar cells, ZnO can also be used as an cathode buffer layer (CBL) as well as a transparent electrode, which has attracted increasing attention during the past five years. As mentioned previously, ZnO is one of the n-type semiconductors with a wide bandgap of 3.4 eV, exhibiting attractive features such as high electron mobility, good optical transmittance, and excellent stability [62]. It also has attractive features such as non-toxicity, low cost process and large-scale area availability. Using ZnO as a CBL in solar cells can enhance the PCE by collecting and extracting electron carriers, as well as blocking positive hole carriers. Therefore, the ZnO is sometimes regarded as electron transport layer or hole blocking layer besides CBL. Furthermore, such ZnO CBL is considered as the optical spacer and oxygen barrier in solar cell devices to enhance their stabilities and lifetimes, which effectively improves the overall device performance [63, 64]. As a result, great efforts have been carried out to improve the performance of organic/inorganic HSCs by using ZnO NRs [65–67].

ZnO NRs were developed with an important role to construct an efficient CBL in polymer solar cells. Using ZnO NRs layer instead of a ZnO planar structure brings out the rapid collection of generated carriers because the ZnO NRs provide a directly conducting and ordered path from the point of photogeneration to the electrode, thus effectively reducing the electron recombination losses and improving the device performance [68]. It is particularly noted that, so far, ZnO NR are applied more often in inverted structure solar cells than conventional solar cells because of the stability issue. In conventional solar cells, it has been proven that acidic PEDOT:PSS layer is detrimental to the organic layer [69–71], and the low work function metal cathode is easily oxidized in air even with a delicate encapsulation [69, 72]. To effectively ameliorate the stability of conventional structure, a new device design of inverted HSC is proposed [73, 74]. Therefore, we will focus here on the recent research of ZnO NR utilized in inverted polymer solar cells.

In a typical case of ZnO NRs CBL-based solar cells, the inverted device structure has three main components: a thin film ITO cathode, a ZnO NRs CBL, and P3HT:PCBM BHJ active layer, as well as an anode layer of Ag [75]. At present, the FF of the solar cells increased from 40% to 60% by the incorporation of ZnO NRs, which contributed to the improvement of the PCEs that approach 5% for P3HT:PCBM system. Additionally, their photovoltaic performance was found to improve with the increase of the ZnO NR average length [75], similar to the reported hybrid P3HT/CdS solar cells with the FTO-coated ZnO NRs electrode [76]. Besides, solution-processed vanadium oxide (V_2O_5) as an anode interlayer is introduced between the organic layer and the Ag electrode for improving the performance of inverted polymer solar cells hybridized with ZnO NRs. In 2009, C. F. Lin [77] reported that the solution-processed V_2O_5 interlayer as an electron-blocking layer can effectively prevent the leakage current at the organic/Ag interface. The device schematic and its energy diagram are shown in Figure 5.6(a) and

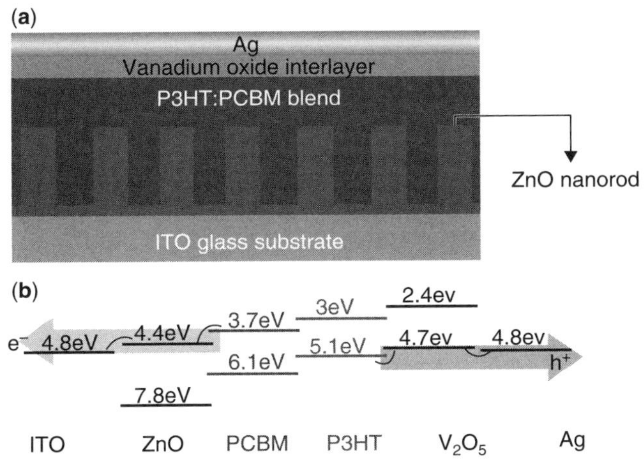

Figure 5.6 (a) Device structure of the hybrid solar cells formed by ITO/ZnO NR/PCBM:P3HT/V_2O_5/Ag. (b) Corresponding energy band diagram for the solar cells formed by ITO/ZnO NR/PCBM:P3HT/V_2O_5/Ag [77].

(b), respectively. The power conversion efficiency is improved from 2.5% to 3.56% by the introduction of the V_2O_5 interlayer [77]. The interlayer also serves as an optical spacer to enhance light absorption, and thereby increases the photocurrent. Moreover, the mixture structure by incorporating different transition metal oxides could also efficiently enhance the device efficiency. In 2010, C. F. Lin first demonstrated mixed oxides as anodic modification through simple solution processing to improve the efficiency and stability of inverted polymer photovoltaic devices [78]. The measured device performance is shown in Figure 5.7. The control device exhibits a Jsc of 10.67 mA/cm^2, a Voc of 0.53 V, and an FF of 54.82%, resulting in a PCE of 3.1%. Either V_2O_5 or WO_3 inserted, the device performance is improved (PCE of 3.74% for WO_3 and 3.81% for V_2O_5). When the mixed oxides are introduced, the device has a significant improvement, so Jsc increases to 11.97 mA/cm^2, Voc to 0.56 V, and FF to 62.06%. Therefore, the PCE is notably improved to 4.16%, a 34% enhancement. These investigations show that the mixed oxides can suppress the leakage current, enhance the light absorption, and provide a temporally durable device.

It was found that the solidification time of the active layer could also influence the device performance of inverted polymer/ZnO NRs solar cells. By slow drying the active layer, a device efficiency

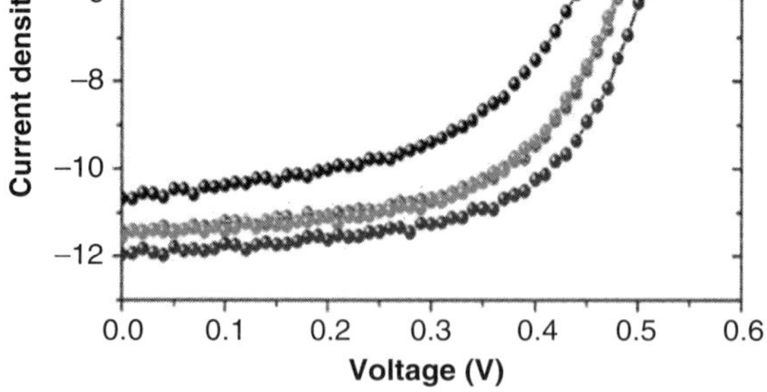

Figure 5.7 J-V curves of the photovoltaic devices shown in Figure. 5.6 with the different anode modification under 100 mW/cm^2 AM 1.5G irradiation [77].

of 3.58% was achieved, which could be attributed to the enhanced crystallinity of the polymer and the infiltration of the active materials [79]. The improved infiltration situation led to the increased interface region and good contact property, resulting in an improvement of the charge transport and better charge collection. Furthermore, the vertical phase separation effect with the slow drying conditions is beneficial to the enhanced electrode selectivity in favor of improved device performance [80].

More importantly, these kinds of solar cells exhibited an obvious improvement in performance after storage in air [73, 74], which is always an important issue in OSCs. A good stability of OSCs by applying ZnO NR as CBL and mixed oxides as anodic modification has been investigated [78]. Figure 5.8 illustrates the normalized PCEs as a function of storage time in ambient. These devices had no encapsulation and were periodically tested over 1000 h. After 1000 h, the control device PCE reduces from 3.1% to 1.9%, a 40% degradation. In contrast, with the mixed oxides, the PCE slowly declines from 4.16% to 3.75%; only 10% degradation is obtained [78]. The FF and Voc over this time remained relatively constant with the Jsc slightly decreasing, leading to normalized PCE of 90%. Firstly it is established that OSC is able to exhibit a long lifetime without apparent deterioration. This result also greatly implies that the ZnO NRs not only promote good carrier transport but also raise the stability in hybrid solar cells.

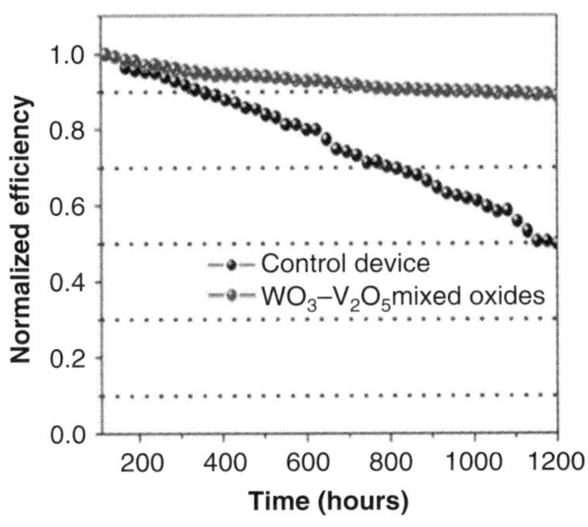

Figure 5.8 Normalized device PCE as a function of time with and without the mixed oxides over 1000 hrs in air. The device is a inverted PSC with the structure of ITO/ZnO NR/PCBM:P3HT/WO_3:V_2O_5/Ag [78].

5.4 Conclusion and Outlook

This chapter provides an overview of the preparation and characterization of ZnO NR, as well as the most recent research developments in the application of ZnO NRs in polymer solar cells. Vertically-aligned nanostructures have been exhaustively applied as electron transport material in polymer solar cells. They could introduce a high surface area for higher interface between the light-harvesting materials (polymer) and increase the photon absorption. Over the past few years, outstanding progress has been observed on the synthesis of vertically-aligned ZnO NRs. The good electron transport by applying ZnO NRs is a reality. The current densities have been effectively raised by applying vertical nanostructures compared to their planar counterparts. So far, the best solar cell performance by applying ZnO nanostructures in PSCs can achieve an efficiency of 7.2%. In order to further increase solar cell efficiency, the surface area of the nanostructured ZnO electrode must be enriched. In an effort to increase surface area and the interface between the semiconductor and the organic material (polymer), a synthesis technique that is able to control the vertically-aligned nanostructures at the 3D level will be in great demand in the future. Therefore, the vertical ZnO nanostructures such as branched nanorods or nanosheets, appear to be promising candidates to carry out high efficiency polymer solar cells. The ZnO nanostructures, which demonstrate most of the characteristics required for high electron transport and high solar cell efficiency of high density nanostructure, high surface area, low voids concentration, and single-crystalline structure for electron transport without grain boundaries, will still be a critical material applied in PSCs in the future.

References

1. F. Padinger, R. S. Rittberger and N. S. Sariciftci, *Advanced Functional Materials*, Vol. 13, p. 85, 2003.
2. G. Li, V. Shrotriya, J. S. Huang, Y. Yao, T. Moriarty, K. Emery and Y. Yang, *Nature Materials*, Vol. 4, p. 864, 2005.
3. M. Reyse-Reyes, K. Kim and D. L Carroll, *Applied Physical Letters*, Vol. 87, 083506, 2005.
4. C. Y. Kwong, A. B. Djurisic, P. C. Chui, K. W. Cheng and W. K. Chan, *Chemical Physics Letters*, Vol. 384, p. 372, 2004.

5. P. A. van Hal, M. M. Wienk, J. M. Kroon, W. J. H. Verhees, L. H. Slooff, W. J. H. van Gennip, P. Jonkheijm and R. A. J. Janssen, *Advanced Materials*, Vol. 15, p. 118, 2003.
6. W. J. E. Beek, M. M. Wienk and R. A. J. Janssen, *Advanced Materials*, Vol. 16, p. 1009, 2004.
7. W. J. E. Beek, L. H. Slooff, M. M. Wienk, J. M. Kroon and R. A. J. Janssen, *Advanced Functional Materials*, Vol. 15, p. 1703, 2005.
8. W. U. Huynh, J. J. Dittmer and A. P. Alivisatos, *Science*, Vol. 295, p. 2425, 2002.
9. S. D. Oosterhout, M. M. Wienk, S. S. van Bavel, R. Thiedmann, L. J. A. Koster, J. Gilot, J. Loos, V. Schmidt and R. A. J. Janssen, *Nature Materials*, Vol. 8, p. 818, 2009.
10. H. Sirringhaus *et al*, *Nature*, Vol. 401, p. 685, 1999.
11. K. M. Coakley, Y. X. Liu, M. D. McGehee, K. L. Frindell and G. D. Stucky, *Advanced Functional Materials*, Vol. 13, p. 301, 2003.
12. K. M. Coakley, B. S. Srinivasan, J. M. Ziebarth, C. Goh, X. Y. Liu and M. D. McGehee, *Advanced Functional Materials*, Vol. 15, p. 1927, 2005.
13. J. Chen, J. Li, J. Li, G. Xiao and X. Yang, *Journal of Alloys and Compounds*, Vol. 509, p. 740, 2010.
14. W. K. Tan, K. A. Razak, K. Ibrahim and Z. Lockman, *Journal of Alloys and Compounds*, Vol. 509, p. 820, 2010.
15. Z. Lockman, Y. P. Fong, T. W. Kian, K. Ibrahim and K. A. Razak, *Journal of Alloys and Compounds*, Vol. 493, p. 699, 2010.
16. S. Baruah and J. Dutta, *Journal of Sol-gel Science and Technology*, Vol. 50, p. 456, 2009.
17. L. Spanhel, *Journal of Sol-gel Science and Technology*, Vol. 39, p. 7, 2006.
18. X. Ma, H. Zhang, Y. Ji, J. Xu and D. Yang, *Materials Letters*, Vol. 59, p. 3393, 2005.
19. M. Kohls, M. Bonnani and L. Spanhel, *Applied Physical Letters*, Vol. 81, p. 3858, 2002.
20. H. Y. Xu, H. Wang, Y. C. Zhang, S. Wang, M. Zhu and H. Yan, *Crystal Research* and *Technology*, Vol. 38, p. 429, 2003.
21. S. Shingubara, *Journal of Nanoparticle Research*, Vol. 5, p. 17, 2003.
22. L. Vayssieres, K. Keis and S. E. Lindquist, *Journal of Physical Chemistry B*, Vol. 105, p. 3350, 2001.
23. L. Schmidt-Mende and J. L. MacManus-Driscoll, *Materials Today*, Vol. 10, p. 40, 2007.
24. M. N. R. Ashfold, R. P. Doherty, N. G. Ndifor-Angwafor, D. J. Riley and Y. Sun, *Thin Solid Films*, Vol. 515, p. 8679, 2007.
25. K. Govender, D. S. Boyle, P. B. Kenway and P. O'Brien, *Journal of Materials Chemistry*, Vol. 14, p. 2575, 2004.
26. A. Sugunan, H. C. Warad,M. Boman and J. Dutta, *Journal of Sol-gel Science and Technology*, Vol. 39, p. 49, 2006.
27. L. Vayssieres, K. Keis, S. E. Lindquist and A. Hagfeldt, *Journal of Physical Chemistry B*, Vol. 105, p. 3350, 2001.
28. L. Vayssieres, *Advanced Materials*, Vol. 15, p. 464, 2003.
29. Y. Sun, D. J. Riley and M. N. R. Ashfold, *Journal of Physical Chemistry B*, Vol. 110, p. 15186, 2006.
30. Z. R. Tian, J. A. Voigt , J. Liu, B. Mckenzie, M. J. Mcdermott, M. A. Rodriguez, H. Konishi and H. Xu, *Nature Materials*, Vol. 2, p. 821, 2003.
31. U. Pal and P. Santiago, *Journal of Physical Chemistry B*, Vol. 109, p. 15317, 2005.

32. L. Tang, X. B. Bao, H. Zhou and A. H. Yuan, *Physica E*, Vol. 40, p. 924, 2008.
33. Li F, Hu L, Li Z and Huang X, *Journal of Alloys and Compounds*, Vol. 465, p. L14, 2007.
34. Y. Chen, R. Yu, Q. Shi, J. Qin and F. Zheng, *Materials Letters*, Vol. 61, p. 4438, 2007.
35. Y. H. Ni, X. W. Wei, X. Ma and J. M. Hong, *Journal of Crystal Growth*, Vol. 283, p. 48, 2005.
36. Z. Li, X. Huang, J. Liu, Y. Li, X. Ji and G. Li, *Materials Letters*, Vol. 61, p. 4362, 2007.
37. N. S. Ridhuan, Y. P. Fong, Z. Lockman, K. A. Razak, *Applied Mechanics and Materials*, Vol. 83, p.116, 2011.
38. X. Zhao, J. Y. Lee, C. R. Kim, J. Heo, C. M. Shin et al., *Physica E*, Vol. p. 41, p. 1423, 2009.
39. C. M. Shin, J. Y. Lee, J. H. Heo, J. H. Park, C. R. Kim et al., *Applied Surface Science*, Vol. 255, p. 8501, 2009.
40. X. Q. Zhao, C. R. Kim, J. Y. Lee, C. M. Shin, J. H. Heo et al., *Applied Surface Science*, Vol. 255, p. 5861, 2009.
41. M. Guo, P. Diao and S. Cai, *Journal of Solid State Chemistry*, Vol. 178, p. 1864, 2005.
42. Q. Li, J. Bian, J. Sun, J. Wang, Y. Luo et al., *Applied Surface Science*, Vol. 256, p. 1698, 2010.
43. J. Song, S. Baek, J. Lee, S. Lim, *Journal of Chemical Technology and Biotechnology*, Vol. 83, p. 345, 2008.
44. S. F. Wang, T. Y. Tseng, Y. R. Wang, C. Y. Wang, H. C. Lu et al., *International Journal of Applied Ceramic Technology*, Vol. 5, p. 419, 2008.
45. N. Liu, G. Fang, W. Zeng, H. Zhou, F. Cheng et al., *ACS Applied Materials and Interfaces*, Vol. 2, p. 1973, 2010.
46. S. Y. Liu, T. Chen, J. Wan, G. P. Ru, B. Z. Li et al., *Applied Physics A*, Vol. 94, p. 775, 2009.
47. X. Q. Zhao, C. R. Kim, J. Y. Lee, J. H. Heo, C. M. Shin et al., *Applied Surface Science*, Vol. 255, p. 4461, 2009.
48. N. S. Ridhuan, K. A. Razak, Z. Lockman and A. A. Aziz, *PLoS one*, Vol. 7, e50405, 2012.
49. R. B. Peterson, C. L. Fields and B. A. Gregg, *Langmuir*, Vol. 20, p. 5114, 2004.
50. Z. Zhang and J. Mu, *Journal of Colloid Interface Scien*ce, Vol. 307, p. 79, 2007.
51. Z. Li, X. Huang, J. Liu, Y. Li, G. Li, *Materials Letters*, Vol. 62, p. 1503, 2008.
52. X. X. Liu, Z. G. Jin, S. J. Bu, J. Zhao, K. Yu, *Materials Science Engineering B*, Vol. 129, p. 139, 2006.
53. C. Herring, *Structure and Properties of Solid Surfaces*, University Chicago Press, Chicago, IL, 1953.
54. D. C. Olson, J. Piris, R. T. Collins, S. E. Shaheen and D. S. Ginley, *Thin Solid Films*, Vol. 496, p. 26, 2006.
55. D. C. Olson, Y. J. Lee, M. S. White, N. Kopidakis, S. E. Shaheen, D. S. Ginley, J. A. Voigt and J. W. P. Hsu, *Journal of Physical Chemistry C*, Vol. 111, p. 16640, 2007.
56. D. Q. Bi, F. Wu, W. J. Yue, Y. Guo, W. Shen, R. X. Peng, H. A. Wu, X. K. Wang and M. T. Wang, *Journal of Physical Chemistry C*, Vol. 114, p. 13846, 2010.
57. P. Atienzar, T. Ishwara, B. N. Illy, M. P. Ryan, B. C. O'Regan, J. R. Durrant and J. Nelson, *Journal of Physical Chemical Letters*, Vol. 1, p. 708, 2010.

58. D. C. Olson, Y. J. Lee, M. S. White, N. Kopidakis, S. E. Shaheen, D. S. Ginley, J. A. Voigt and J. W. P. Hsu, *Journal of Physical Chemistry C*, Vol. 112, p. 9544, 2008.
59. Y. Z. Hao, J. Pei, Y. Wei, Y. H. Cao, S. H. Jiao, F. Zhu, J. J. Li and D. H. Xu, *Journal of Physical Chemistry C*, Vol. 114, p. 8622, 2010.
60. A. M. Peiro, P. Ravirajan, K. Govender, D. S. Boyle, P. O'Brien, D. D. C. Bradley, J. Nelson and J. R. Durrant, *Journal of Materials Chemistry*, Vol. 16, p. 2088, 2006.
61. Y. Z. Hao, J. Pei, Y. Wei, Y. H. Cao, S. H. Jiao, F. Zhu, J. J. Li and D. H. Xu, *Journal of Physical Chemistry C*, Vol. 114, p. 8622, 2010.
62. U. Ozgur, Y. I. Alivov, C. Liu, A. Teke, M. A. Reshchikov, S. Dogan, V. Avrutin, S. J. Cho and H. Morkoc, *Journal of Applied Physics*, Vol. 98, 041301, 2005.
63. W. U. Huynh, J. J. Dittmer and A. P. Alivisatos, *Science*, Vol. 295, p. 2425, 2002.
64. C. Y. Kwong, A. B. Djurisic, P. C. Chui, K. W. Cheng and W. K. Chan, *Chemical Physics Letters*, Vol. 384, p. 372, 2004.
65. R. Po, C. Carbonera, A. Bernardi and N. Camaioni, *Energy Environmental Science*, Vol. 4, p. 285, 2011.
66. A. B. Djurisic, A. M. C. Ng and X. Y. Chen, *Progress in Quantum Electronics*, Vol. 34, p. 191, 2010.
67. S. K. Hau, H. L. Yip and A. K. Y. Jen, *Polymer Reviews*, Vol. 50, p. 474, 2010.
68. U. Ozgur, D. Hofstetter and H. Morkoc, *Proceedings of the IEEE*, Vol. 98, p. 1255, 2010.
69. M. Jørgensen, K. Norrman and F. C. Krebs, *Solar Energy Materials and Solar Cells*, Vol. 92, p. 686, 2008.
70. F. C. Krebs and K. Norrman, *Progress in Photovoltaics: Research and Applications*, Vol. 15, p. 697, 2007.
71. M. P. de Jong, L. J. van Ijzendoorn and M. J. A. de Voigt, *Applied Physical Letters*, Vol. 77, p. 2255, 2000.
72. M. Jørgensen, K. Norrman, S. A. Gevorgyan, T. Tromholt, B. Anderreasen and F. C. Krebs, *Advanced Materials*, Vol. 24, p. 580, 2012.
73. G. Li, C. W. Chu, V. Shrotriya, J. Huang and Y. Yang, *Applied Physical Letters*, Vol. 88, 253503, 2006.
74. M. S. White, D. C. Olson, S. E. Shaheen, N. Kopidakis, D. S. Ginley, *Applied Physical Letters*, Vol. 89, 143517, 2006.
75. K. Takanezawa, K. Hirota, Q. S. Wei, K. Tajima and K. Hashimoto, *Journal of Physical Chemistry C*, Vol. 111, p. 7218, 2007.
76. H. C. Liao, C. C. Lin, Y. W. Chen, T. C. Liu and S. Y. Chen, *Journal of Materials Chemistry*, Vol. 20, p. 5429, 2010.
77. J. S. Huang, C. Y. Chou, M. Y. Liu, K. H. Tsai, W. H. Lin and C. F. Lin, *Organic Electronics*, Vol. 10, p. 1060, 2009,
78. J. S. Huang, C. Y. Chou and C. F. Lin, *IEEE Electron Device Letters*, Vol. 31, p. 332, 2010.
79. C. Y. Chou, J. S. Huang, C. H. Wu, C. Y. Lee and C. F. Lin, *Solar Energy Materials and Solar Cells*, Vol. 93, p. 1608, 2009.
80. M. Campoy-Quiles, T. Ferenczi, T. Agostinelli, P. G. Etchegoin, Y. Kim, T. D. Anthopoulos, P. N. Stavrinou, D. D. C. Bradley and J. Nelson, *Nature Materials*, Vol. 7, p. 158, 2008.

Part 2

NOBLE APPROACHES

6

Dye-Sensitized Solar Cells

Lakshmi V. Munukutla, Aung Htun, Sailaja Radhakrishanan, Laura Main, and Arunachala M. Kannan*

Department of Engineering and Computing Systems, Arizona State University at the Polytechnic Campus, Mesa, Arizona, USA

Abstract
Dye-sensitized solar cells (DSSCs) are an inexpensive alternative to the common silicon (Si) solar cell technology. The DSSC technology is considered as a third generation photovoltaic technology, which is demonstrating tremendous potential to become a revolutionary technology that has several indoor applications (window panels and doors) in low light environment. Additionally, the DSSC fabrication process is simple and cost effective, and does not require clean-room facility and highly pure materials. The efficiency of the DSSCs is around 12% based on today's technology, which is higher than the organic solar cells and lower than the traditional silicon-based solar cells. There are several challenges yet to be solved associated with DSSCs for commercialization, and one of those challenges is related to its longevity and stability.

Keywords: Dye, semiconductor, titanium dioxide, polymer, solar cell

6.1 Introduction

The present dominance of photovoltaic field by the inorganic solid-state junction is challenged by the DSSCs as Michäel Gretzel first demonstrated in 1991 [1]. The key departure from the silicon junction devices is that the contact phase with the semiconductor

**Corresponding author*: amk@asu.edu

Atul Tiwari, Rabah Boukherroub, and Maheshwar Sharon (eds.) Solar Cell Nanotechnology, (161–184) 2014 © Scrivener Publishing LLC

is replaced by an electrolyte (liquid, gel and quasi-solid) in DSSC, which makes it a photochemical cell. Advances in nanocrystalline materials fabrication and characterization made it conceivable to attain systems with interpenetrating networks of mesoscopic semiconductors exhibiting higher conversion efficiencies which is one of the critical features for development of the DSSC technology. Secondly, achieving optical absorption and charge separation processes using a combination of a sensitizer as a light-absorbing material with a nanocrystaline wide bandgap semiconducting material, is directed to DSSC fabrication.

The best research cell efficiency between different solar cell technologies is shown in Figure 6.1 [2]. Multi-junction concentration solar cell has achieved record high efficiency of 40.7% in the year 2010, as compared to 20–27% efficiency in crystalline solar cells. DSSCs, which possess a simple fabrication procedure and promising applications on flexible substrates, are considered as third generation solar cells and their cost is predicted to be five times less than that of the second generation solar cells (silicon solar cells) as depicted in Figure 6.2. The recorded efficiency of 11.1% for the DSSCs has been reported by Y. Chiba *et al.* [3].

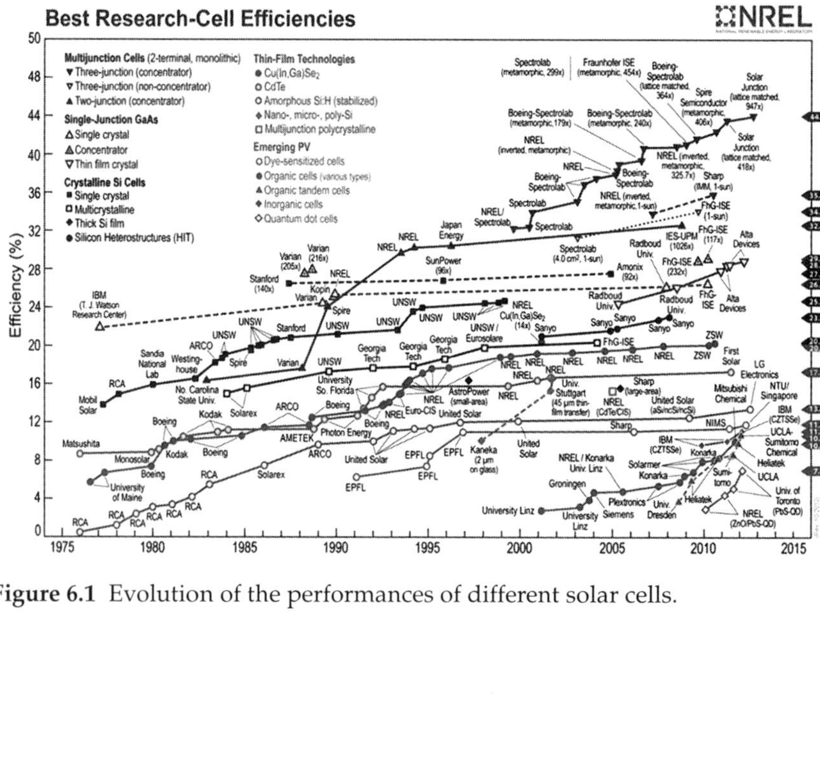

Figure 6.1 Evolution of the performances of different solar cells.

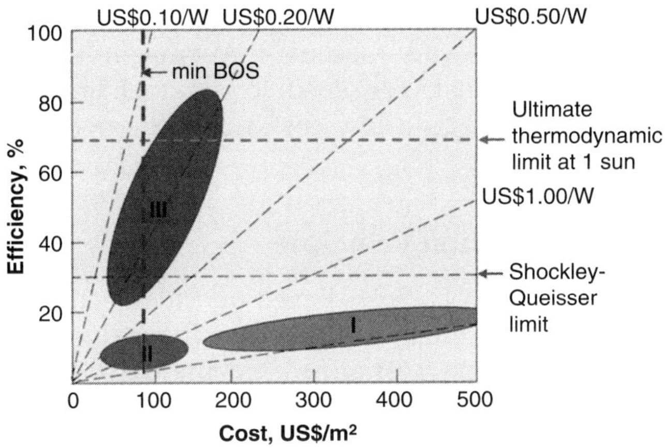

Figure 6.2 Cost comparison of different generation solar cells.

6.2 Background

The phenomenon of sensitizing a semiconductor with light having longer wavelength than its bandgap energy has been demonstrated earlier [4, 5]. This resembles convergence of photography and photoelectrochemisry based on the principle of photoinduced charge separation at a semiconductor-liquid interface. However, silver halides used in the photography had higher bandgap energies and so they became insensitive to most of the visible range of the electromagnetic spectrum. But in DSSC fabrication metal oxide films are used with much smaller bandgap energy to facilitate light absorption from the visible spectrum.

In 1839 Becquerel discovered the emergence of an electric potential between two electrodes connected to a solid or liquid system upon light irradiation [6]. According to Michael Gratzel, the first sensitization of a photoelectrode followed immediately after the rendering of a black and white image in the 1870s [7]. However, the realization that the two procedures work likewise through the injection of electrons from photoexcited dye molecules into the conduction band of the n-type semiconductor substrate did not occur until the late 1960s [8]. In subsequent years, the idea that the dye could function most efficiently if it is chemisorbed on the surface of the semiconductor was developed. The idea to use dispersed particles to provide a sufficient interface for chemisorption emerged [9], and then photoelectrodes were developed based on this concept [10]. Titanium dioxide (TiO_2) became the semiconductor of choice due to its many advantages for sensitized photochemistry and photoelectrochemistry. TiO_2 is a low cost, non-toxic, widely available and biocompatible material.

6.2.1 DSCC Operation Principle

Energy conversion in a DSSC is based on the injection of an electron from a photoexcited state of the sensitizer dye into the conduction band of the nanocrystalline semiconductor TiO_2. Titanium dioxide in antase phase is by far the most employed oxide semiconductor, as depicted in Figure 6.3 [11]. The original state of the dye is subsequently restored by electron donation from the electrolyte, usually an organic solvent containing redox system, such as the iodide/triiodide couple. The regeneration of the sensitizer by iodide intercepts the recapture of the conduction band electron by the oxidized dye. The iodide is regenerated in turn by the reduction of triiodide at the counter electrode, the circuit being completed via electron migration through the external load as shown in Figure 6.4.

The light-to-electron conversion process in DSSC differs fundamentally from the one in the conventional solar cells. Figure 6.5 shows the energy band diagram of a pn-junction after thermal

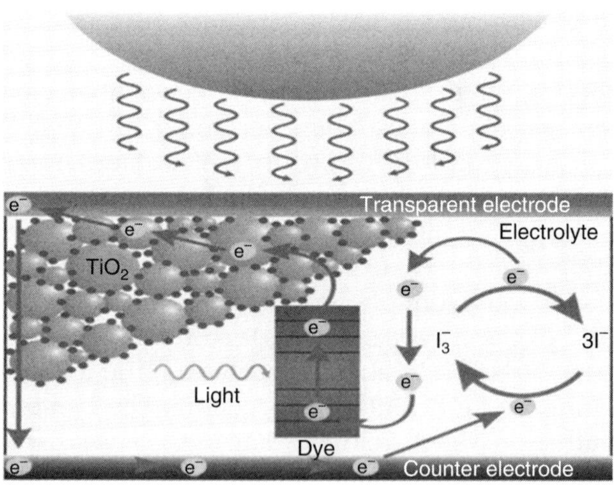

Figure 6.3 DSCC operating principle.

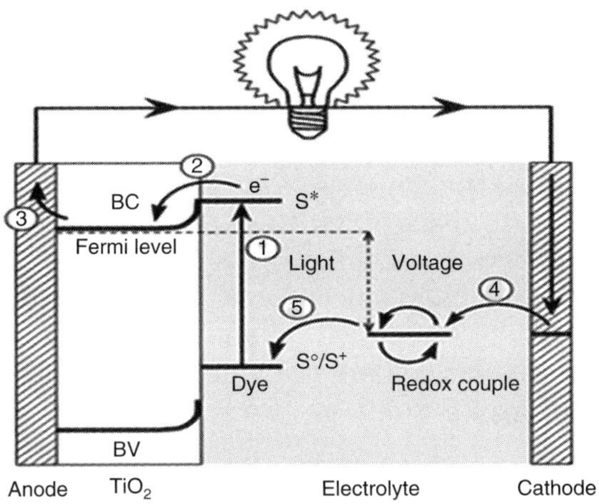

Figure 6.4 Energy Band structure of DSSC.

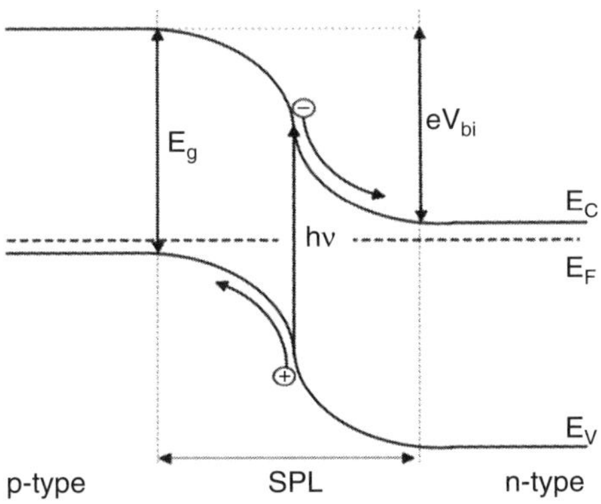

Figure 6.5 Energy band diagram of a conventional pn-junction solar cell under Short-circuit conditions [1].

equilibration of charge carriers [12]. The uncompensated charge caused by the diffusion of holes into the n-region, and electrons to the p-region, forms an electric field at the interface between the n and the p type semiconductors (space charge layer), which is characterized by the built in potential V_{bi}. Absorption of photons with an energy higher than the bandgap results in the generation of excitons, which interact via columbic forces. The excitons recombine after a certain time under emission of photons or heat, unless they are separated by an electric field. Therefore, only excitons created in the space charge layer or close to it will contribute to the photocurrent.

The distinctive features between the traditional silicon-based solar cell and DSSC are as follows:

- Light adsorption and charge carrier transport are separated in dye-sensitized solar cell, whereas the semiconductor in the conventional cell establishes both of these processes.
- An electric field is necessary for charge separation in the pn-junction cell. Nanoparticles in the dye-sensitized solar cell are too small to sustain a build-in field; accordingly charge transport occurs mainly via diffusion.
- Inside a pn-junction minority and majority charge carriers coexist in the same bulk volume. This makes conventional solar cells sensitive to bulk recombination and demands the absence of any recombination centers such as trace impurities. Dye-sensitized solar cells are majority charge carrier devices in which the electron transport occurs in the TiO_2 and the hole transport in the electrolyte.

Important challenges of the dye-sensitized solar cell are the volatility of the electrolyte and the corrosive action of the iodide redox couple. Correspondingly, long-term application of dye-sensitized solar cell is limited by the electrolyte permeability of the encapsulation and its inertness towards the iodine. Therefore, a lot of effort has been invested in finding less volatile electrolytes while maintaining sufficient hole transport. Approaches to replace the liquid electrolyte include application of ionic conducting polymer gels [13–15], electrolyte gellated with amino-acid derivatives [16] or room temperature molten salts [17]. Complete replacement of the liquid electrolyte system has been established using inorganic p-type semiconductors, such as CuI [18–21] or CuSCN [22, 23] as well as organic hole conductors, including low molecular weight charge transport materials [24] and semiconducting polymers [25, 26].

6.2.2 DSSC Structure

The DSSC comprises two electrochemical electrodes with a sandwiched iodide-based electrolyte. The working electrode of

DSSC generally consists of mesoporus titanium dioxide (TiO_2) semiconductor, which is produced by sputtering, doctor bladed, sol-gel spraying, or electrophoretic deposition [27–31]. The working electrode of a DSSC is made of dye-sensitized TiO_2 nanoparticles fabricated onto a transparent conducting oxide (TCO), a Pt counter electrode, and an electrolyte containing iodide/trididode (I^-/I_3^-) redox couple are the remaining components of the DSSC structure. The iodide is regenerated in turn by the reduction of triiodide at the counter electrode. Many researchers are working on improving the solar conversion efficiency by optimizing different components of DSSCs, especially the development of new materials. The DSSC structure is depicted in Figures 6.6 and 6.7 [32].

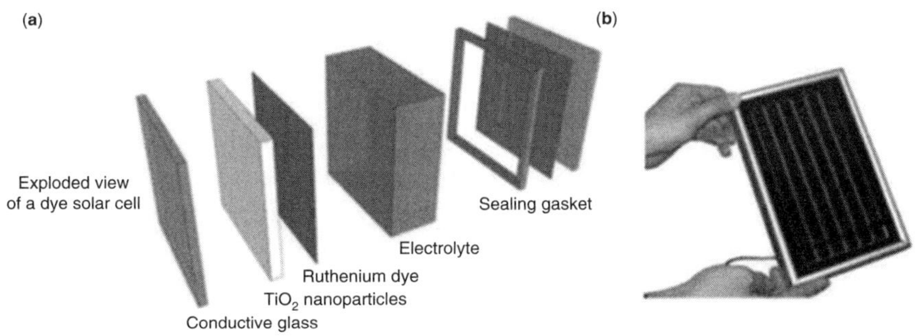

Figure 6.6 (a) Schematic structure; (b) outlook of DSSCs [1].

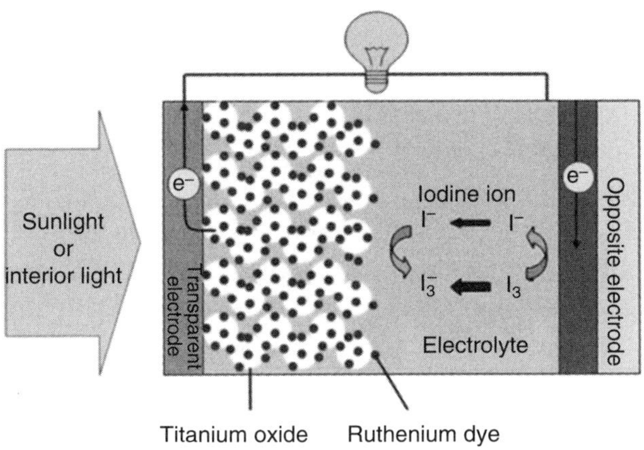

Figure 6.7 Assembled structure of DSSC [1].

6.2.3 DSSC Challenges

The DSSC performance improvement paths for exploration are; develop dyes that absorb more photons, create new electrolytes that provide higher voltages with stability, and finally develop DSSCs that can last for 30 years to compete with the conventional silicon technology. There are several challenges to be addressed to enhance DSSC efficiency beyond 12%. One path is increasing absorption light wavelength to 900 nm by designing near-infrared dyes. The second challenge is improving the DSSC performance and longevity which could be achieved by successfully replacing the liquid electrolyte with quasi-solid electrolyte and/or nanoclay-based electrolytes and exploring better sealing materials and techniques to improve the lifetime of the DSSCs to close to 30 years. Significant research is going on in these specified areas to reach theoretical efficiency range of 14–20% at the cell level and 6–9% at module level.

The aesthetic advantages of DSSCs over conventional solar cells are: dye selection to determine the device color and devices that can be transparent and also flexible with simple fabrication procedures. Michael Gratzel and others documented the DSSC fabrication details with complete instructions [33–35].

6.2.4 DSSC Components

6.2.4.1 *Working Electrode*

The need for dye-sensitized solar cells to absorb far more of the incident light was the driving force for the development of mesoscopic semiconductor materials [36], which have attracted great attention during recent years. The dye-sensitized, TiO_2 film is typically deposited on a highly doped, transparent conducting oxide (TCO), which allows light transmission while providing sufficient conductivity for current collection. The surface of TCO should make good mechanical and electrical contact with the porous TiO_2 film. The TiO_2 semiconductor has three functions in the dye-sensitized solar cell: it provides the surface for the dye adsorption, functions as an electron acceptor for the excited dye and serves as an electron conductor. Therefore, preparation of the working electrode (in particular TiO_2 film preparation) is crucial for the DSSC performance. The most important process step for the performance of dye-sensitized solar cells is the development of a mesoporous semiconductor structure. TiO_2 layer with mesoporous structures allows superior surface areas of 1000 times larger than the apparent geometric one based on particle size (10–20 nm), film thickness and porosity. Sintering the nanoparticles together enables the entire surface-absorbed molecular layer to be retrieved electronically to achieve the electronic contact between the nanoparticles. The interconnection of the nanoparticles by the sintering process allows the deposition of a mechanically stable, transparent film thickness of typically a few microns.

Semiconductors such as TiO_2, ZnO, SnO_2 and Nb_2O_5, or chalcogenides such as CdSe, are studied for photoelectrochemical applications. Amongst them, TiO_2 is by far the most commonly used, because of its energetic properties and its stability, which allows attachment of dyes. Figure 6.8 shows the SEM image of a nanocrystalline TiO_2 film with a grain size in the range of 10–80 nm [37], which is a low cost material and is widely available. TiO_2 is used in its low-temperature stable form, anatase (pyramide-like crystals).

6.2.4.2 *Dye Sensitizer*

The dye is the photoactive element of the photovoltaic device, harvesting the incident light for the photon-to-electron conversion. The ideal characteristic of the dye should cover a wide range of the solar spectrum. More than 50% of the solar energy is emitted in the region from 400 to 800 nm. Thus ruthenium dyes that adsorb throughout the visible spectrum (panchromatic) capture most of the sunlight. The extinction coefficient of the dye should be high throughout the entire absorption spectrum, in order to absorb most of the light with

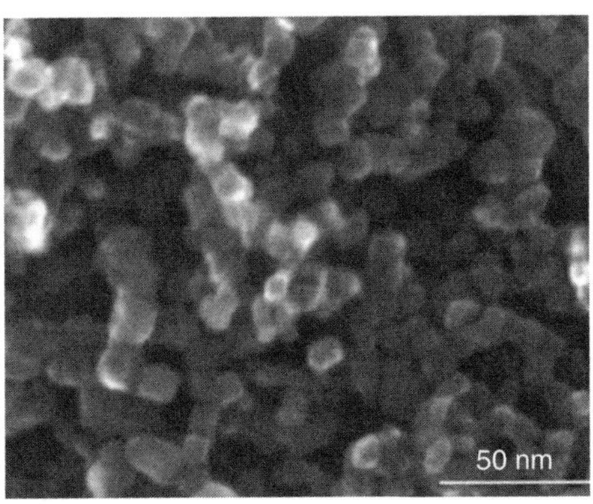

Figure 6.8 SEM image of nanocrystalline TiO_2 film.

a minimum quantity of dye material. Since the white light harvesting efficiency is limited by the absorption in the weaker absorbing regions, a high extinction coefficient at wavelengths different from the maximum absorption peak is particularly important. The dye excited state lifetime must be long enough for efficient electron injection. In general this requirement is not a limiting factor, as electron injection occurs on the sub-nanosecond time scale and thus can compete efficiently with other excited state decay processes.

The best photovoltaic performance has so far been achieved with polypyridyl complexes of ruthenium and osmium. Sensitizers having the general structure $ML_2(X)_2$, where L stands for 2,2_-bipyridyl-4,4_-dicarboxylic acid M is Ru or Os and X presents a halide, cyanide, thiocyanate, acetyl acetonate, thiacarbamate or water substituent, are particularly promising. The ruthenium complex cis-$RuL_2(NCS)_2$, as shown in Figure 6.9, known as N719 dye, has become widely used sensitizer for mesoporous solar cells. The fully protonated N719 has absorption maxima at 518 and 380 nm and the complex emits at 750 nm with the 60 ns lifetime.

6.2.4.3 *Electrolyte*

The electrolyte facilitates two important tasks in the DSSC process. It serves as the transport mechanism for the redox mediator from the TiO_2 electrode to the counter electrode where electron transfer will occur and the triiodide ions will be reduced to iodide as shown by the following reaction:

$$I_3^- + 2e^- \longrightarrow 3I^- \tag{6.1}$$

At the TiO_2 side, the electrolyte serves to regenerate the dye molecule, which has been oxidized following the electron injection into

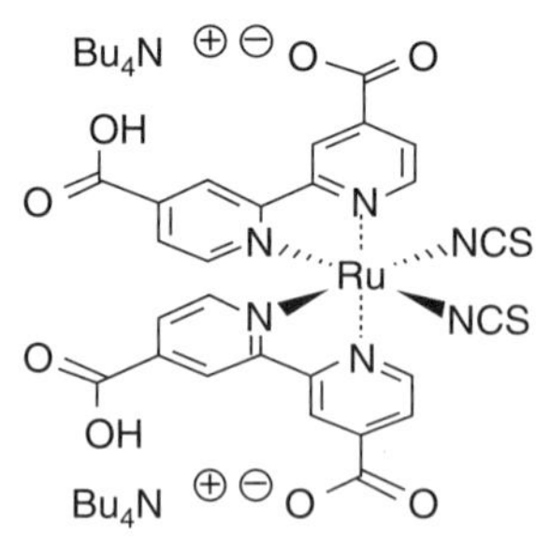

Figure 6.9 Chemical structure of ruthenium N719 dye.

the conduction band of the TiO_2. This regeneration is represented by the following reaction:

$$3I^- \longrightarrow I_3^- + 2e^- \qquad (6.2)$$

The dye regeneration is critical in preventing the dye molecules from being reduced via recombination of a TiO_2 electron. There are two methods of recombination of TiO_2 electrons. One is with oxidized dye molecules, which is unlikely because the regeneration from the iodide is very fast. The second likely recombination mechanism is to combine with acceptors in the electrolyte. The iodide/triodide redox couple exhibits longer lifetime (1–20 ms) compared to other redox couples tried earlier and so it has become the preferred redox couple for DSSCs. In order for the dye regeneration to occur, the difference between the oxidation potential of the dye and the redox potential of the electrolyte, given as (ΔG^0), has to be sufficient. For a typical ruthenium (Ru)-based dye and a standard I^-/I_3^- electrolyte, the driving force is given by $\Delta G^0 = 0.75$ eV. This driving force must be taken into account when choosing an electrolyte material, because if the redox potential is too high, the driving force is reduced, and if it falls much lower than 0.6 eV, the performance of the DSSC will be greatly reduced. The mediator should not have any significant light absorption in the visible range, to avoid internal filter effects.

The electrolyte should be able to transport a current of a few mA/cm^2 without diffusion limitation or significant Ohmic losses. Generally liquid electrolytes with adequate viscosity meet these requirements and so they are commonly used in DSSCs. Additionally, the electrolyte must have long-term stability, including chemical, thermal, optical, electrochemical, and interfacial stability, which does not cause desorption and degradation of the dye from the oxide surface. Liquid electrolytes have challenges that can cause evaporation and eventually degrade the cell performance over time.

6.2.4.4 *Platinum-Coated Counter Electrode*

The counter electrode is ideally a high conductivity material and exhibits an Ohmic contact to the hole conductor, which requires work function matching of hole conductor and counter electrode. A glass-coated transparent film deposited with a platinum layer constitutes the counter electrode. When a photon excites the dye molecule, it injects electrons into the semiconductor layer. After that, a redox reaction in the electrolyte ($3I^- \rightarrow I_3^- + 2e^-$) provides the electron back to the dye molecule. The triiodide in turn accepts electrons, and is reduced to iodine ($I_3^- + 2e^- \rightarrow 3I^-$) to regenerate the redox couple. The efficiency of a DSSC is based on different rate constants for iodine reduction between the working and counter electrodes. The iodine reduction at the counter electrode has to be several orders of magnitude faster than the recombination at the TiO2/electrolyte interface. The regeneration of the oxidized dye occurs in the nanosecond range, which is typically 100 times faster than any recombination reaction and about 10^8 times faster than the intrinsic lifetime of the oxidized dye.

6.2.4.5 *Equivalent Circuit of DSSC*

The equivalent circuit of conventional solar cells is well understood and employed to analyze cell devices and to improve cell performance. The same equivalent circuit worked well to analyze DSSC using the electrochemical impedance spectroscopy (EIS) method [38–40]. Figure 6.10(a) is an example of the Nyquist Plot, measured by the EIS method to interpret the internal resistance of DSSCs [41]. Three semicircles are observed in the measured frequency range from 100Hz to 1 MHz. The internal resistances of R_1, R_2, and R_3 describe the real parts of Z_1, Z_2 and Z_3, respectively. Because impedance over 1 MHz could not be measured due to instrument limitations, the resistance element in this frequency region is defined as R_h. Figure 6.10(b) depicts the equivalent circuit of DSSCs. R_{sh} is the shunt resistance inside DSSCs.

R_h, which locates high frequency range > 10^6 Hz area, represents the sheet resistance of the TCO substrate and is primarily influenced by the thickness of the conductive layer on the electrode. R_1 is related to the carrier transport resistance at the surface of Pt counter electrode

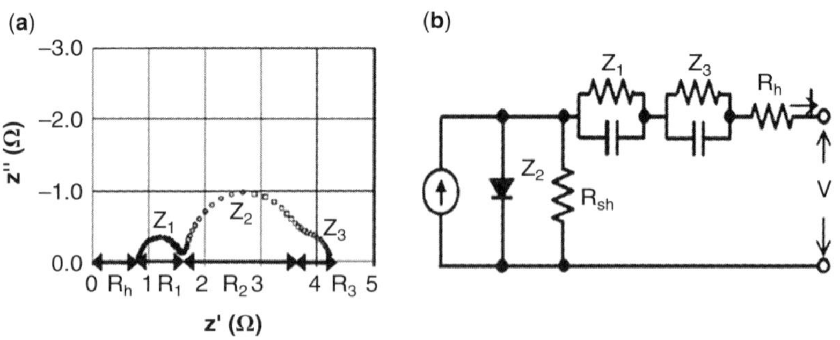

Figure 6.10 (a) Nyquist plot and (b) equivalent circuit of DSSC.

and can be varied changing the roughness factor of counter electrode. Changing the distance between TCO and Pt counter electrode can optimize R3, which represents the diffusion of iodide and triiodide within the electrolyte. R_2 shows the resistance of the diode element in DSSCs, and reflects the properties of the photoinjected electrons within the TiO_2. The series internal resistance, Rs, is defined as the sum of the resistance R_1, R_3, and R_h. The variation of each resistance pinpoints a different direction to optimize the efficiency of DSSCs.

6.3 DSSC Key Performance Parameters

A DSSC, which performs a similar function as a silicon solar cell, provides electrical power to a load under illumination. The standard characterization techniques of DSSC include the determination of the DC current-voltage characteristic and analysis of the photocurrent-voltage (I-V) curves.

The following device parameters can be extracted from the I-V curves:

a. **Short-circuit current (I_{SC}):** Cell current measured at an applied potential of zero volt. I_{SC} is a function of the illumination intensity.
b. **Open-circuit potential (V_{OC}):** Cell potential measured when current in the cell is zero, corresponding to almost flat valence and conduction bands.
c. **Maximum power output (P_{max}):** For a given bias voltage the power output of the cell is the product of the measured cell current and the voltage P(V)=I*V. I_{max} and V_{max} are the coordinates of the maximum in the P(V) curve (maximum power point). The maximum power corresponds visually to the area of the largest rectangle, which can fit inside the current voltage curve.
d. **The fill factor (*FF*)** is the ratio of the maximum power to the external short and open circuit values:

$$FF = \frac{I_{max}.V_{max}}{I_{sc}.V_{oc}} \tag{6.3}$$

This parameter indicates the deflection of the current-voltage characteristic from a square-like curve and is therefore

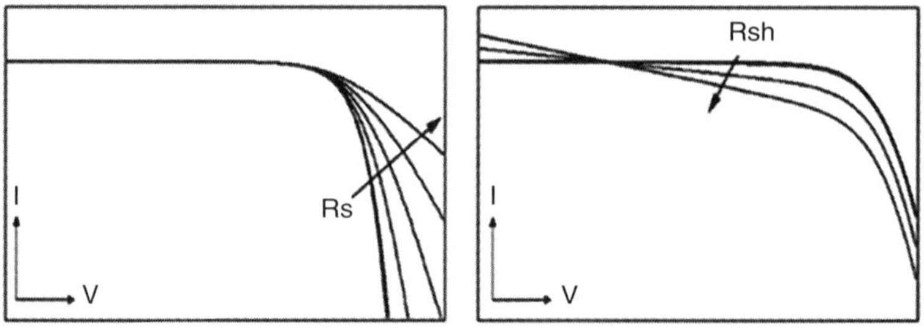

Figure 6.11 Illustration of the influence of increasing series resistance R_s and decreasing shunt resistance R_{sh} on the I-V curve of DSSC.

dependent on series and shunt resistance. Figure 6.11 illustrates the influence of increasing series resistance, R_s and decreasing shunt resistance, R_{sh} on the current-voltage characteristic. To obtain high fill factors, R_s has to be as small as possible, while R_{sh} needs to be as high as possible.

e. The **Efficiency (η)** describes the performance of DSSC and is defined as the ratio of the maximum electric power extracted to the radiation power incident on the solar cell surface:

$$\eta = \frac{P_{max}}{P_{in}} = \frac{I_{sc}.V_{oc}.FF}{P_{in}} \tag{6.4}$$

The efficiency is a function of the V_{OC}, the I_{SC} and the FF of the cell. Accordingly, improvement of the photovoltaic yield is the result of the optimization of these three parameters.

6.4 Device Improvements

The research efforts our group has focused on are the following: preparation of the working electrode, counter electrode, cell size and experimentation with different electrolytes. Optimization of these parameters is covered in this section. The standard I-V characterization techniques are performed to evaluate the device performance using solar simulator (Figure 6.12). The solar simulator is calibrated with a standard cell prior to taking the data.

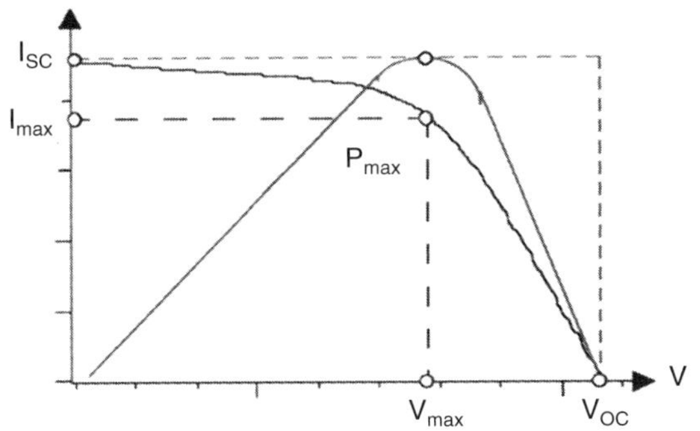

Figure 6.12 Illustration of maximum power output from I-V curve.

6.4.1 Experimental

6.4.1.1 Working Electrode Preparation

Prior to coating the electrode, the TiO_2 paste was sonicated first for about 5 minutes to eliminate the aggregation of TiO_2 particles. The desired thickness of TiO_2 layer was first determined using different layers of 3M Magic Tape (~100um) on the PVC mask while a 0.6 cm x 0.6 cm area on the tape was cut by the laser machine. The working electrode was attached underneath the tape and was placed on the metallic plate of the Coatema tool to carry out the TiO_2 coating. After the coating, TiO_2 coated FTO glass electrode was kept under the petri dish for 1 hour so that the paste could settle to reduce the irregularity of the surface before sintering. After working electrode and counter electrode were coated with TiO_2 and platinum respectively, both electrodes were then sintered in the furnace at 400°C for one hour. The thickness and quality of TiO_2 film was verified with SEM images. Both electrodes after sintering are shown in Figure 6.13.

After preparing the electrodes the next step is dye immersion. Ruthenium dye, $Ru(II)L_2(NCS)_2$, N719 from Soloronix is diluted with 40 ml of pure ethanol (>99%) and the dye solution is sonicated for uniform dye dispersion prior to immersing the electrode. After sufficient dye absorption by the TiO_2 film a change from white to brownish color is indicated as shown in Figure 6.14.

6.4.1.2 Cell Assembly

The critical operation of cell fabrication is assembly. A good sealing procedure is necessary to prevent electrolyte leakage and evaporation. The 25 μm thick sealant, (Meltonix, 1170–25, Solaronix) was

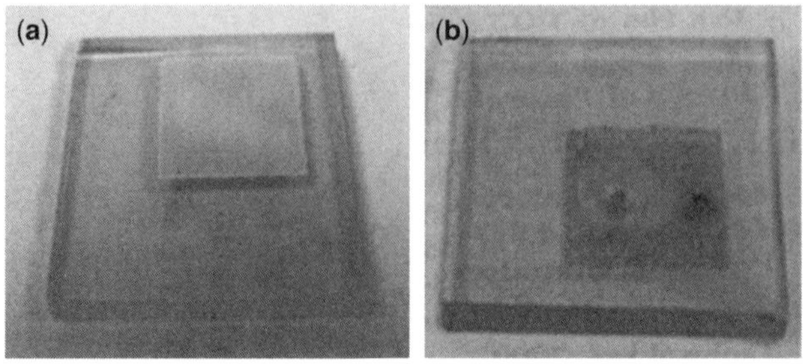

Figure 6.13 (a) Working electrode with sintered TiO_2 layer; (b) counter electrode with sintered Pt. layer.

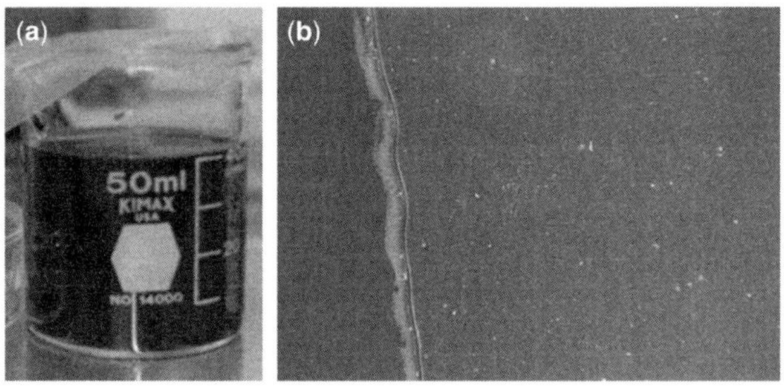

Figure 6.14 (a) Image of N719 dye solution; (b) TiO_2 layer after dye absorption.

used for sealing the cell. An electrode area of 0.5 cm x 0.5 cm was defined on the sealant by using the laser machine. The sealant was sandwiched between the two electrodes and secured the electrode assembly using the binder clips at each end applying adequate pressure for sealing. The electrodes were placed in the furnace at 135°C for 12 minutes to seal the electrodes.

6.4.1.3 Electrolyte Injection

The electrolyte, (Iodolyte R-50, Solaronix) was injected through the hole from counter electrode by using a syringe. Additional precautions were taken to minimize the trapped air bubbles. Finally the hole used for electrolyte injection was sealed to prevent electrolyte evaporation.

6.4.2 DSSC Performance Results

6.4.2.1 TiO_2 Film Thickness Optimization

TiO_2 paste containing 13 nm particles (Ti-Nanoxide D Series, Solaronix) is used for determining the optimized film thickness for the

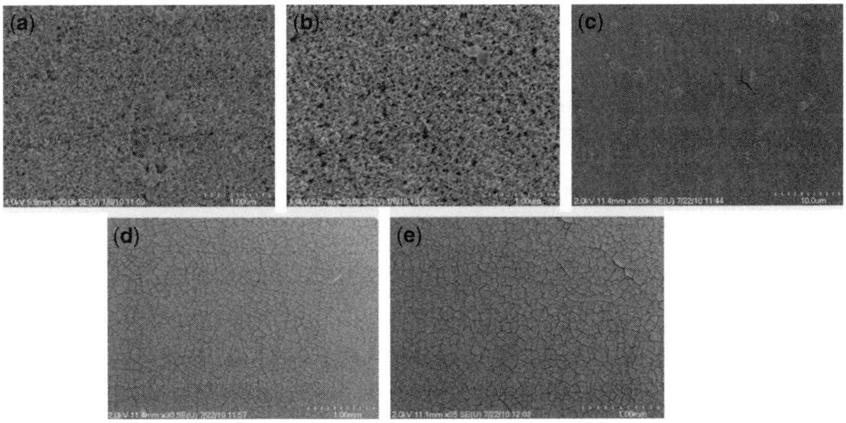

Figure 6.15 SEM images of (a) 2, (b) 3, (c) 4, (d) 5, and (e) 6 mask layers of TiO_2 film.

working electrode of DSSC. 3M Magic tape with different layers (2–6) are used for varying the film thickness on different working electrodes. The electrode coating with TiO_2 is done using screen printing technique. The quality and thicknesses of electrodes were studied using SEM images shown in Figure 6.15.

The working electrodes were fabricated only with 2–4 TiO_2 mask layers based on the information extracted from the SEM images. The TiO_2 layer thicknesses are verified under SEM and the values of the working electrodes are shown in Table 6.1.

Due to microcracks, working electrodes with five and six layers were discarded for DSSC assembly and the remaining electrodes were assembled for performance evaluation.

In order to optimize the TiO_2 film thickness for DSCC, several samples with thickness ranging from 20–60 μm were prepared and the assembled cells were characterized to generate I-V curves and electrochemical impedances (Nyquist Plot) using Parastat 2273 Advanced Electrochemical System; Newport Xe OF Arc Lamp, (Newport) was used for AM 1.5 solar simulator ($100mWcm^{-2}$) as light source. The results of this part of the experiment are shown in Figure 6.16.

It is clearly evident from Figure 6.16 that both open circuit voltage and short circuit current have indicated dependence on the TiO_2 layer thickness. With an increase of the semiconductor layer thickness the short-circuit current increased due to the higher amount of dye absorption that might have contributed to higher electron injection rate to provide higher short-circuit current; however, the open circuit voltage showed an opposite trend. To understand this

Table 6.1 Thickness of TiO_2 film corresponding to the mask layers.

No. of mask layers	TiO_2 Film Thickness
2 layer	~20 μm
3 layer	~40 μm
4 layer	~66 μm

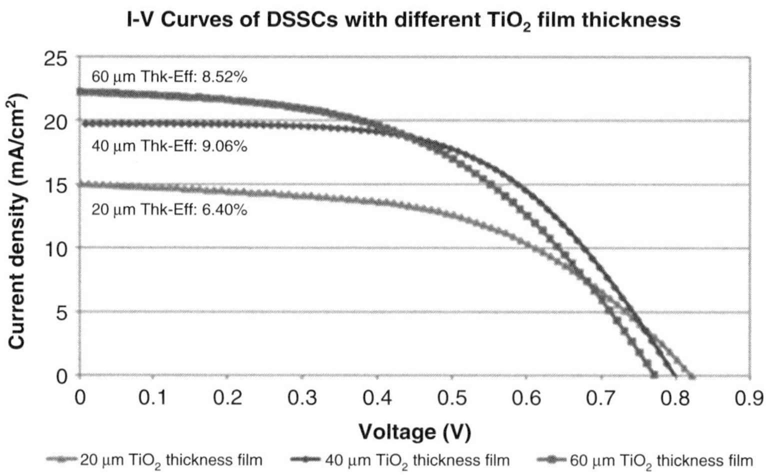

Figure 6.16 Optimization of TiO_2 film thickness.

effect completely further experimentation is required. However, dependence of the cell efficiency on the film thickness is not that well-defined. The sample with 40 μm film thickness showed higher efficiency compared to other samples. Based on the Nyquist plots it was noted that the internal resistance of the samples increased with higher film thickness, and this could have denied the benefit of higher absorption dye in thicker films.

6.4.2.2 Optimization of Nanoparticle Size in TiO_2

For this study the cell with 40 μm film thickness was selected because of its better efficiency obtained in the earlier study. The study of nanoparticle size for enhancing the photo conversion efficiency of the DSSC was carried out using the particle sizes of 13nm, 20nm and 37nm (Ti-Nanoxide D Series, Solaronix) in the TiO_2 layer. The I-V characterization results of this study are shown in Figure 6.17.

The performance data in Figure 6.17 clearly indicated cell with 20nm size particles performed better than the other two. The open circuit voltage and fill factor did not change but the short-circuit current showed variation with different particle sizes. Figure 6.18 shows the Nyquist plots to assess the cell internal resistance as a function of nanoparticle size.

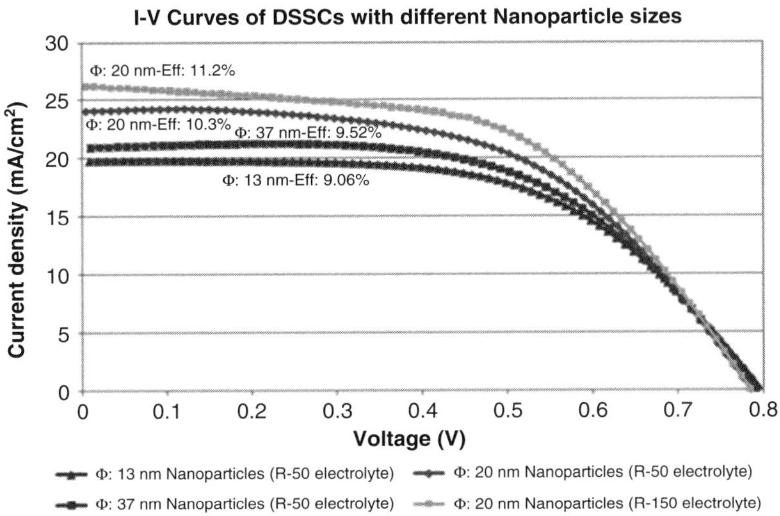

Figure 6.17 DSCC performance study with different nanoparticles size of TiO_2.

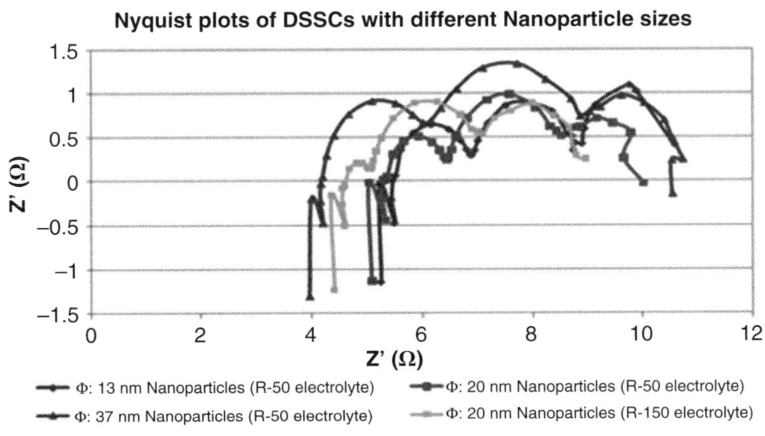

Figure 6.18 Nyquist plots of DSSCs with different nanoparticles sizes.

Extracting from the Nyquist plots the interface resistance, R_2 between TiO_2/dye/electrolyte interface somewhat correlated to the nanoparticle size. The interface resistance increased with bigger particle size. The sheet resistance of the cell with 20nm particles showed lowest compared to other cells. The Table 6.2 below gives the quantitative values of the internal resistance of the cell.

6.4.2.3 Scaling Down the DSS Cell Size

Several studies in the literature indicated improvement of DSSC efficiency provided the cell size is scaled down. Three different cell sizes were fabricated to study the impact of cell size on performance. Indeed, our results also confirmed the fact that cell size has greater influence on the performance. Table 6.3 below shows the results of our study.

6.5 DSSC Performance with Different Electrolytes

6.5.1 Liquid Electrolyte

One of the key components of the DSSC system is electrolyte. The desired features of the electrolyte for improving device performance

Table 6.2 DSSC performance characteristics with different nanoparticle sizes.

40 nm Thickness	**φ : 13 nm R-50 Electrolyte**	**φ : 20 nm R-50 Electrolyte**	**φ : 20 nm R-150 Electrolyte**	**φ : 37 nm R-50 Electrolyte**
Rh	5.24	5.08	4.4	3.96
R1	1.7	1.44	0.72	2.2
R2	**1.79**	**2**	**1.98**	**2.71**
R3	1.84	1.49	1.86	1.67
Rs	8.78	8.01	6.98	7.83
Efficiency (%)	**9.06%**	**10.30%**	**11.20%**	**9.52%**
Voc (V)	0.79	0.78	0.78	0.79
Isc (niA.cm^{-2})	20.57	24.1	26.21	20.92
FF	0.57	0.55	0.55	0.58

Table 6.3 DSSC cell size scaling.

Cell Scaling	**6 × 6 cm^2**	**2.5 × 2.5 cm^2**	**1.25 × 1.25 cm^2**
Sample cell			
Area	6 cm^2	1 cm^2	0.25 cm^2
Efficiency	< 2.0 %	3.8 %	8.4 %

were previously mentioned in this article. Both liquid and semi-solid clay-based electrolytes are experimented on in this study. The difficulty with liquid electrolyte is that it remains a challenge for both stability and longevity of the device. First several liquid electrolytes with higher concentration of iodolyte were used to study the influence of the electrolyte on device performance. The electrolytes used for this study were R-150, AN-50, PN-50 and MPN-100, commercially available from Solaronix [42]· The results of this work are depicted in Figure 6.19. Both electrolyte injection and cell sealing were found to be very critical. However, manual sealing was done to assemble the cells with different electrolytes in the laboratory environment. Based on the I-V curves data, R-150 electrolyte performed better compared to the standard electrolyte (Iodolyte R-50) and as well as other experimental electrolytes.

6.5.2 Quasi-Solid Electrolyte

An alternative to the liquid electrolyte is quasi-solid electrolyte, which is a gel-based material possessing redox couple similar to the liquid electrolyte. The advantage quasi-solid electrolyte offers is greater stability compared to liquid electrolytes. However, the higher viscosity associated with quasi-solid material causes degradation in cell efficiency due to restricted ionic mobility. For this study high molecular weight polymethymethacrylate (PMMA) and montmorillonite (MMT) clay were used to create a polymer nanocomposite gel electrolyte that increased the conductivity from 7.6×10^{4} S/cm^2 to 9.1×10^{-4} S/cm^2.

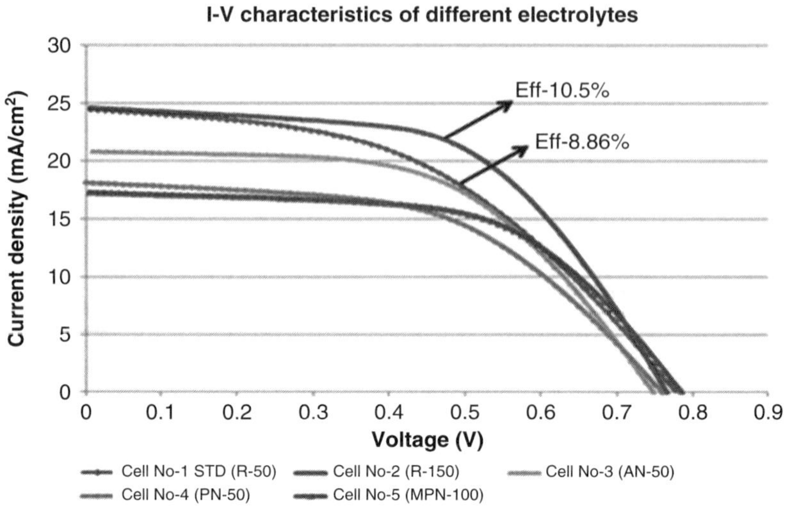

Figure 6.19 DSSC performance study with different liquid electrolytes.

Table 6.4 Performance data for the cells with differfent electrolytes.

Electrolyte Type	V_{OC} (V)	I_{SC} (mA/cm^2)	η (%)	Lifetime (days)
Liquid	0.78	12.88	5.19	8.15
Gel	0.82	14.92	5.96	19.96

The gel was prepared by adding 7 wt% of nanoclay Naomer@ (1.31PS, montmorillonite clay surface modified with 15–35% octadecylamine and 0.5–5 wt% aminopropyltriethoxysilane, procured from Aldrich) to the liquid electrolyte (Iodide/triodide) and stirred for approximately 24 hours using a magnetic stirrer to ensure complete dispersion of the nanoclay particles. The resulting gel electrolyte was more viscous but thin enough to allow injecting with a standarad syringe into the DSS cell assembly. The experimental data with quasi-solid electroyte with control group of liquid electrolyte is shown in Table 6.4.

The most staggering difference observed between liquid and nanoclay gel cells was that the gel cells lasted more than twice as long compared to their liquid counterparts. This data is very preliminary and further investigation is still in progress.

6.6 Conclusion and Outlook

Dye-sensitized solar cells are gaining importance as an alternate for silicon-based solar cells for indoor applications. Significant global research efforts are being undertaken to commercialize this technology. However, several challenges are still being pursued to achieve the maturity equivalent to the existing silicon-based technology. This article covers some background information related to the DSSC technology, the basic operation principle of DSSC, and challenges associated with improving the performance of the DSSC devices by optimizing several components of the DSSC system. The research results from our laboratory are included to address some of the challenges to improve the DSSC performance.

References

1. B. O'Regan and M. Grätzel, *Nature* Vol. 353, 737–740 (1991).
2. http://www.nrel.gov/ncpv/images/efficiency_chart.jpg, accessed in July 2013.
3. Y. Chiba, A. Islam, Y. Watanabe, R. Komiya, N. Koide and L. Han, Dye-sensitized solar cells with conversion efficiency of 11.1%, *Jpn. J. Appl. Phys.* Vol. 45, 638–640, (2006).
4. A.J. McEvoy and M. Grätzel, Sensitization in photochemistry and photovoltaics, *Solar Energy Mat. Solar Cells* Vol. 32, 221–227 (1994).
5. A. Hagfeldt and M. Grätzel, Molecular photovoltaics, *ACC. Chem. Res.* Vol. 33, 269–277 (2000).
6. E. Becquerel, *Comptes Rendus* Vol. 9, 561 (1839).
7. M. Grätzel, Review dye-sensitized solar cells, *J. Photoch. Photobio. C - Photoch. Rev.* Vol. 4, 145–153 (2003).
8. H. Gerischer and H. Tributch, *Ber. Bunsenges. Phys. Chem.* Vol. 72, 437 (1968).
9. D.H. Dung, N. Serpone and M. Grätzel, Integrated systems for water cleavage by visible-light sensitization of TiO_2" particles by surface derivatization with ruthenium complexes, *Helv. Chim. Acta* Vol. 67, 1012–1018 (1984).
10. J. Desivestro, M. Grätzel, L. Kavan, J. Moser and J. Augustynski, Highly efficient sensitization of titanium dioxide, *J. Am. Chem. Soc.* Vol. 107, 2988–2990 (1985).
11. M. Grätzel, *Prog. Photovolt. Res. Appl.* Vol. 8, 171 (2000).
12. J. Krüger, *Interface Engineering in Solid-State DSSC*, University of Heidelberg, (2003).
13. F. Cao, G. Oskam and P. Searson. *J. Phys. Chem.* Vol. 99, 17071 (1995).
14. A.F. Nogueira, M.A. De Paoli, I. Montanari, et al., *J. Phys. Chem. B* Vol. 105, 7517–7524 (2001).
15. W.A. Gazotti, A.F. Nogueira, E.M. Girotto, et al., *Synth. Met.* Vol. 108, 151–157 (2000).
16. W. Kubo, K. Muakoshi, T. Kitamura, et al., *Chem. Lett.* 1998, 1241.
17. N. Papageorgiou, Y. Athanassov, M. Armand, et al., *J. Electrochem. Soc.* Vol. 143, 3099–3108 (1996).

18. K. Tennakone, G. Kumara, I. Kottegoda, et al., *J. Phys. D-Appl. Phys.* Vol. 31, 1492–1496 (1998).
19. K. Tennakone, G.R.R.A. Kumara, A.R. Kumarasinghe, et al., *Semicond. Sci. Tech.* 1995, 10, 1689–1693.
20. K. Tennakone, A.R. Kumarasinghe and P.M. Sirimanne, *Semicond. Sci. Tech.* 1993, Vol 8, Iss 8, 1557–1560.
21. K. Tennakone, G.R.R.A. Kumara, K.G.U. Wijayantha, et al., *Semicond. Sci. Tech.* 1998, 13, 134–138.
22. B. O'Regan and D.T. Schwartz, *J. Appl. Phys.* 1996, 80, 4749.
23. B. O'Regan and D. Schwartz, *Chem. Mater.* 1998, 10, 1501–1509.
24. J. Hagen, W. Schaffrath, P. Otschik, et al., *Synth. Met.* 1997, 89, 215–220.
25. K. Murakoshi, G. Kano, Y. Wada, et al., *J Electronanal. Chem.* 1995, 396, 27.
26. K. Murakoshi, R. Kogure and S. Yanagida, *Chem. Lett.* 1997, 5, 471–472.
27. A.I. Kontosa, A.G. Kontosa, D.S. Tsouklerisa, M.C. Bernardc, N. Spyrellis, P. Falarasa, Nanostructured TiO2 films for DSSCS prepared by combining doctor-blade and sol–gel techniques, *J. Mater. Process. Tech.* 196 243–248 (2008).
28. L. Grinis, S. Dor, A. Ofir, A. Zaban, Electrophoretic deposition and compression of titania nanoparticle films for dye-sensitized solar cells, *J. Photoch. Photobio. A-Chemistry* 198 52–59 (2008).
29. W. Tana, X. Yina, X. Zhoua, J. Zhanga, X. Xiaoa, Y. Lina, Electrophoretic deposition of nanocrystalline TiO_2 films on Ti substrates for use in flexible dye-sensitized solar cells, *Electrochim. Acta* 54 4467–4472 (2009).
30. G.S. Kim, H.K. Seo, V.P. Godble, Y.S. Kim, O.B. Yang, H.S. Shin, Electrophoretic deposition of titanate nanotubes from commercial titania nanoparticles: Application to dye-sensitized solar cells, *Electrochem. Commun.* 8 961–966 (2006).
31. W. Jarernboon, S. Pimanpang, S. Maensiri, E. Swatsitang, V. Amornkitbamrung, Optimization of titanium dioxide film prepared by electrophoretic deposition for dye-sensitized solar cell application, *Thin Solid Films* Vol. 517, 4663–4667, 2009, Elsevier.
32. www.Solaronix.com, accessed in July 2013.
33. http://cnsi.ctrl.ucla.edu/nanoscience/pages/solarCells, accessed in July 2013.
34. B. O'Regan and M. Grätzel. A low-cost, high efficiency solar cell based on dye sensitized colloidal TiO2 films, *Nature* Vol. 353, 737–740 (1991
35. C.J. Barbé, et al., Nanocrystalline titanium oxide electrodes for photovoltaic applications, *J. Am. Ceram. Soc.* Vol. 80, 3157–3171 (1997).
36. B. O'Regan and D. Schwartz, *Chem. Mater.* 10, 1501–1509 (1998).
37. J. Hagen, W. Schaffrath, P. Otschik, et al., *Synth. Met.* 89, 215–220 (1997).
38. K. Murakoshi, G. Kano, Y. Wada, et al., *J. Electronanal. Chem.* Vol. 396, 27 (1995).
39. W.A. Gazotti, A.F. Nogueira, E.M. Girotto, et al., *Synth. Met.* Vol. 108, 151–157 (2000).
40. C.J. Barbé, et al., Nanocrystalline titanium oxide electrodes for photovoltaic applications, *J Am. Ceram. Soc.* Vol. 80, 3157–3171 (1997).
41. A. Hagfeldt, G. Boschloo, L. Sun, L. Kloo and H. Pettersson, Dye-sensitized solar cells, *Chem. Rev.* Vol. 110, pp. 6595–6663, 2010.
42. http://www.solaronix.com/products/iodideelectrolytes/iodolytez50/, accessed in July 2013.

7

Nanoimprint Lithography for Photovoltaic Applications

Benjamin Schumm[1] and Stefan Kaskel[1,2,*]

[1]*Department of Chemical Surface and Reaction Technology, Fraunhofer Institute for Material and Beam Technology, Dresden, Germany*
[2]*Inorganic Chemistry Department, Dresden University of Technology, Dresden, Germany*

Abstract

Nanoimprint lithography has recently attracted considerable interest as a low-cost and high-throughput tool for nanofabrication in photovoltaic cell processing. While nanoimprint lithography offers a cheap and simple alternative to conventional lithography techniques on the one hand, it can also be used for large-scale roll-to-roll processes. These aspects enable the technique to be applied in solar cell processing steps. Nanostructured antireflection layers on the substrate or wafer outside can be produced with photocurable polymers. Acting as etch masks layers patterned by nanoimprint lithography can be used to generate a large variety of three-dimensional surface reliefs. By patterning the inside of thin-film solar cell substrates light-trapping effects can be generated. Light-trapping can also be achieved by the preparation of plasmonically active metal layers. After an introduction of the nanoimprint lithography technique, different applications in solar cell manufacturing are presented in this chapter.

Keywords: NIL, nanoimprint, PDMS, roll-to-roll, antireflection, light-trapping, plasmonics, etching mask

**Corresponding author*: stefan.kaskel@chemie.tu-dresden.de

Atul Tiwari, Rabah Boukherroub, and Maheshwar Sharon (eds.) Solar Cell Nanotechnology, (185–202) 2014 © Scrivener Publishing LLC

7.1 Introduction

The reduction of process costs with increasing light conversion efficiencies is one of the most important R&D issues in the photovoltaic (PV) sector. This obstacle can be overcome by combining thin-film technologies with high-throughput approaches such as roll-to-roll-based processes and printing methods. While printing technologies allow fast surface functionalization within a broad range of applicable materials, roll-to-roll processes enable fast and automated substrate supply and packaging [1].

As a scalable high-throughput and low-cost multi-purpose tool, nanoimprint lithography (NIL) has been introduced in a broad range of academic research. Various applications of this relatively young patterning technique have been demonstrated. Recently, NIL processes were adapted to first industrial processes, showing the relevance of this technique and its potential for solar cell manufacturing [2]. The number of applications for NIL processes in the PV sector is as high as the one for cell technologies. However, for all the different approaches, the scalability, and thus the applicability, in low-cost roll-to-roll processes is given.

Summarizing the different approaches for photovoltaics, NIL and related techniques are mainly used for the preparation of antireflective films, for the preparation of protective masks applicable in etch processes, for molding three-dimensional surfaces, generating high interface areas between different materials and for the deposition of plasmonically active films.

These principles, all being relatively young techniques, will be presented and discussed in this chapter. Prior to this, the basic principles of the process and the used materials will be introduced.

7.2 Soft Lithography

7.2.1 Soft Lithography Methods

In the 1990s, Whitesides *et al.* developed a wide range of non-photolithographic techniques for patterned surface modification on the micro- and even nanoscale [3]. Since a stamp instead of "hard" radiation (UV/Vis, X-ray or e-beam) is used for patterning, these methods are called "soft lithography" techniques. Elastomers like polydimethylsiloxane (PDMS) are often used as flexible stamp material, giving another hint for this denomination. Besides pure stamping methods, masking and molding processes were developed. These techniques are fast, simple and cheap tools for micro- and nanostructuring of surfaces. Large and nonplanar (rough or curved) areas can be treated with a variety of "ink" materials [4]. Figure 7.1 summarizes frequently used techniques and principles.

The earliest developed soft-lithographic method, microcontact printing, represents a form of contact printing that uses a high-resolution elastomeric stamp soaked with a chemical ink capable of forming a self-assembled monolayer. This monolayer can guide material deposition on or removal from the substrate to yield patterns of other materials. It furthermore can act as bridging molecule for other substances or particles. Without expensive facilities, this technique offered the first way to form high-resolution patterns using standard laboratory equipment.

The advantages of microcontact printing have been adapted to replica molding as it has been used for the mass-production of a wide range of structured surfaces such as compact disks (CDs), diffraction gratings or holograms [5–7]. While these techniques are

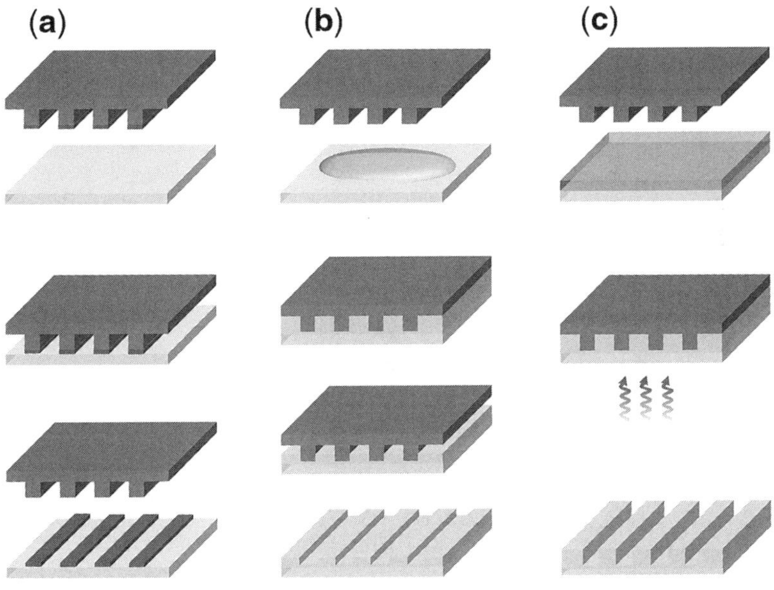

Figure 7.1 Illustration of frequently used soft lithography techniques: (a) Microcontact printing, (b) soft embossing NIL or SANIL, and (c) UV or thermal NIL.

based on molding against a rigid mold with an appropriate material (usually a thermoplastic polymer), the use of elastomers makes it easier to release small, fragile structures [8].

Originally nanoimprint lithography was presented by Chou as a process where a rigid stamp is pressed in a PMMA layer on a rigid silicon substrate under slight thermal treatment for obtaining a patterned PMMA layer after stamp removal [9]. Today, the term NIL is used in a broader view to describe the process of pressing a stamp (also flexible) in a prepolymer layer coated on a substrate, where either UV or thermal curing of the prepolymer leads to a patterned coating.

With the step-and-flash imprint lithography (S-FIL) the imprint process was extended by Wilson *et al.* to solutions of curable substances [10]. This allows the patterning directly after deposition of prepolymer droplets — a uniform prepolymer layer is not required. In further developments, embossing in polymer solutions or even sol-gel precursors was applied in the NIL process. These processes are called soft embossing NIL or solvent assisted NIL (SANIL). Several other approaches have been developed, which are not discussed in detail here. These include molding processes induced by capillary forces, microfluidic approaches, combined optical/soft lithography methods and many more [4].

7.2.2 Stamp Materials Used for Nanoimprint Lithography

As described above, the first stamp materials for NIL and related techniques utilized hard molds, e.g., made of silicon, silicon oxide or metals. With these materials feature resolutions below 10 nm have been realized [11]. If the restrictions in terms of resolution and aspect ratio are just moderate, PDMS is a suitable mold material for NIL as well. This allows large area and high-throughput processing by integration in efficient roll-to-roll processes [12].

The manufacturing process of the flexible PDMS stamp is shown in Figure 7.2. In the case of PDMS, a blend of silicone oligomers with a crosslinker and a catalyst are cast on a perfluorinated silicon "master"-mold. The master is fabricated using microlithographic methods. After thermal curing the crosslinked elastomeric stamp can be peeled off and used as it is or can be further functionalized [13]. Other patterned master materials such as polymers or metals can also be applied as mold.

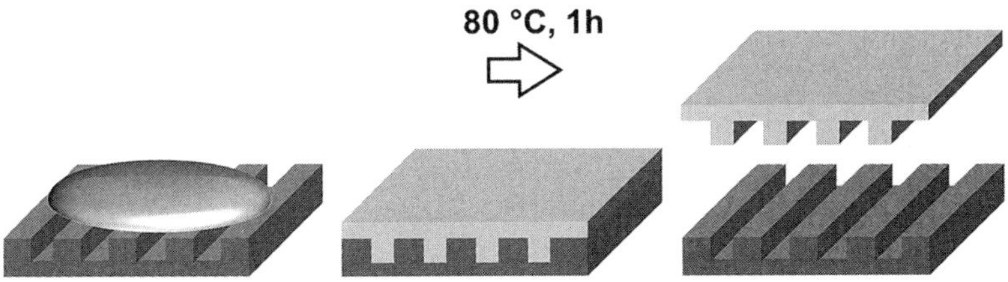

Figure 7.2 Scheme of the PDMS stamp preparation process. The silicone oligomer liquid is poured onto a patterned silicon wafer. After thermal crosslinking the flexible replica can be peeled off.

Polydimethylsiloxanes have a unique combination of properties resulting from the presence of a siloxane backbone and organic methyl groups attached to silicon [14]. In addition to its elasticity, the PDMS elastomer also has other properties making it extremely useful in soft lithography. Due to the low interfacial surface energy and good chemical stability, most molecules or polymers being patterned or molded do not adhere irreversibly to or react with the surface of PDMS. The hydrophobicity of PDMS avoids swelling of the material with humidity. Gasses and evaporated organic solvents pass the PDMS membrane easily. With its good thermal stability (up to ca. 180 °C in air) PDMS enables thermal molding or imprinting of various materials. UV-curing of prepolymers through the PDMS membrane is possible due to its optical transparency down to ca. 300 nm.

Oligomeric siloxanes are fluids at room temperature and can be readily converted into solid elastomers by crosslinking. For a broad range of applications the crosslinking of methylsiloxane oligomer with a vinyl terminated dimethylsiloxane oligomer in a catalytic hydrosilylation using a platinum-based catalyst is applied for the stamp preparation (Figure 7.3). Commercial kits for PDMS fabrication are available in large quantities from several companies (e.g. Dow Corning Sylgard 184).

Generally, PDMS is a satisfying mold material for many nanofabrication processes involving features on the order of 500 nm. However, there are some inherent disadvantages with this material that have to be overcome if a sub-100 nm resolution is required. The low modulus of 2.0 MPa limits the feature size and aspect ratio that can be replicated [15]. In addition, PDMS swells when exposed to some common organic solvents, and the fact that Sylgard 184 needs to be thermally cured limits the numbers of materials useable as master-mold [16].

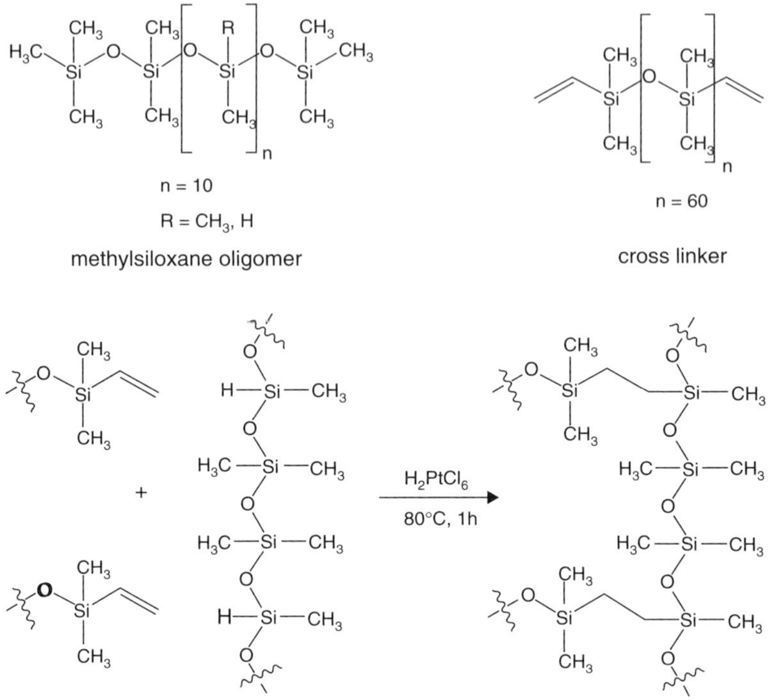

Figure 7.3 Chemical structure of methylsiloxane oligomer and crosslinker for the H_2PtCl_6-catalyzed hydrosilylation towards flexible PDMS.

To overcome these disadvantages, high-modulus and photocurable PDMS as well as thiol-ene (poly[(3-mercaptopropyl)-methylsiloxane] (PMMS)) acrylate-based molds have been developed [17, 18]. Higher resolutions and aspect ratios could be realized that could not be replicated in Sylgard 184.

DeSimone *et al.* have introduced perfluoropolyether-based (PFPE) elastomers that offer high modulus, low surface energy, solvent resistance, chemical stability and high visible light transparency. This flexible material allows a resolution down to 20 nm with high aspect ratios [19].

7.3 NIL-Based Techniques for PV

7.3.1 Antireflection Layers Prepared with NIL Methods

Reflection of incident light occurs on each interface of a PV device [20, 21]. As light losses due to reflection lower the cell efficiency,

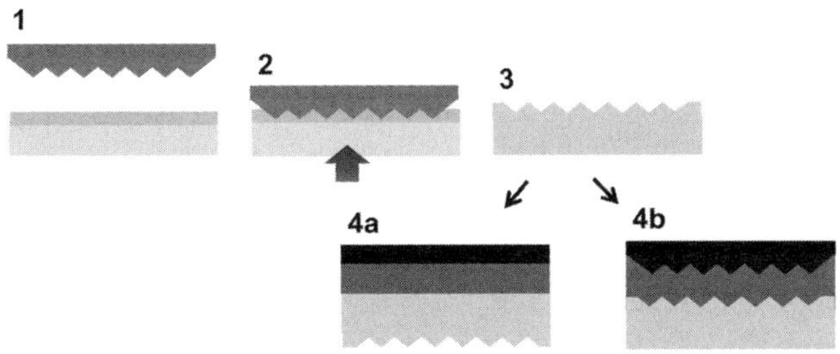

Figure 7.4 Scheme of the NIL-based preparation process for antireflection layers: (1) Prepolymer deposition, (2) NIL process to obtain a patterned substrate (3). The prepared rough surface can be used as antireflection coating on the substrate outside (4a) or as patterned superstrate (4b).

rough surfaces have been introduced to decrease these effects [22]. NIL offers lots of possibilities allowing the preparation of rough or patterned surfaces.

7.3.1.1 Structured Substrates — Outside

The first obvious interface where such a structure can be applied is the outside of a glass substrate or laminated foil. In this case the substrate is coated with a prepolymer, which can be structured by a thermal or UV-NIL process. After the processing, a rough surface results showing similar optical properties as the substrate. Usually, photocurable acrylate- or epoxy-based resins are utilized [23, 24]. Application of such a coating on the protection glass of a GaAs-based tandem solar cell by Shin and co-workers showed significantly reduced reflectance in the visible spectral range leading to a short-circuit current density improvement from 11.1 to 11.9 mA/cm^2 [25].

7.3.1.2 Structured Wafers

In a similar way, NIL structuring of the top side of cells affords reasonable antireflection effects [12]. As shown by Chen *et al.*, nanoimprinted layers on top of a crystalline silicon solar cell can lead to similar antireflective properties as achievable by SiN_x layers [26]. Poortmans *et al.* used these kinds of structures for antireflection coatings in very thin crystalline solar cells [27]. This improved the efficiency of 1 μm thick cells from 4.3 to 4.8%. Compared to state-of-the-art 200 μm thick wafers, the required amount of silicon is reduced to 1/200 while the efficiency is just reduced to 1/5.

Han *et al.* showed that the basic NIL process is easily transferrable to other cell concepts by imprinting a polymer on the top surface of a GaInP/Ga(In)As/Ge concentrator solar cell [28]. Especially in this case, where the surface is irradiated with high light intensity, decreased surface reflection losses have a strong influence on the device performance.

Barbé *et al.* extended the process for c-Si cells by SANIL imprinting in a TiO_2 precursor layer. The solvent evaporates during the imprint process and a patterned nanocrystalline titanium oxide layer results after stamp removal. Subsequently to a final sintering step, a pure TiO_2 layer without further organic impurities can be obtained. This layer combines two properties. On the one hand an antireflective effect results due to surface roughness, and on the other hand the passivation of the silicon can be achieved [29].

The above mentioned possible applications of NIL for PV applications point out that almost any kind of surface pattern is achievable. Thus, the realization of optimized structures as a result of theoretical simulations on real devices becomes possible. Theoretical and measured data of actual devices can be obtained in good agreement as shown, e.g., by Yamada *et al.* However, one ideal structure for any kind of cell cannot be predicted. Geometrical optimazations have to be done for each new cell layout.

7.3.1.3 Structured Substrates — Inside

Superstrate configurations, as they are often used for thin-film solar cell manufacturing, offer not only the possibility to pattern the interface air/cell, but also the interface superstrate/cell. This may induce some sensibilities during further cell processing, but besides this it is much more effective due to the occurance of photonic light-trapping effects. Similar to the above mentioned processes, the superstrate can be structured with NIL patterning of transparent polymer or oxidic sol-gel-derived materials. Subsequently, the deposition of the solar cell layers is carried out. Depending on the pattern geometry and the thickness of the following films, the structure can be transferred more or less conformal into the active layers. As shown, for example, by Bessonov *et al.* and Pei *et al.* a light-trapping effect can be generated by this method [32, 33].

In the case of polymer-based substrates, there is no need for the additional deposition of curable polymers. In fact, the plastic substrate itself can be structured by thermal embossing leading to enhanced cell efficiencies [34]. For this procedure especially hard stamps are suitable to obtain the desired geometries.

Especially for silicon thin-film solar cells, light-trapping surfaces are of particular interest as they have a strong impact on the cells' performance. Escarré *et al.* have developed a NIL process, allowing the transfer of pyramid structures into a UV-curable lacquer on glass substrates. As master they used the well-established LPCVD ZnO pyramid films (Figure 7.5 b, c). On the structured lacquer an indium oxide-based electrode is deposited prior to micromorph silicon solar cell deposition [35, 31]. This allows the decoupling of optical TCO properties from surface morphology [36]. In analogy, the deposition steps can be carried out in opposite order, where silver is deposited on a patterned polymer film creating a patterned back electrode [37].

7.3.2 NIL-Patterned Films as Etching Masks

The previous section focused on the preparation of three-dimensional surface structures by patterning a layer of a curable material. As a common way of thin-film or surface structuring, etching processes have been described. These methods are either based on anisotropic dissolving of a material in a solvent or on selectively etching of a film that is partly covered by a protective mask. The fabrication of the latter is usually done by photolithography methods or similar processes. Very often polymethylmethacrylate (PMMA) is used as protective layer. As shown by Mondin *et al.*, a structured PMMA surface and even freestanding PMMA membranes can be easily obtained by means of soft lithography methods [38]. Resolutions of 100 nm have been demonstrated and were reproduced on 1 cm^2 with high pattern quality. In this manner almost any kind of pattern can be generated and be used as etching mask. As shown by various groups, this principle can be applied for different PV applications [39, 40]. While Wang *et al.* used the technique for etching FTO substrates in order to apply them to dye-sensitized solar cells, Tommila *et al.* showed that the surface of AlInP/GaAs-based solar cells can be structured with this kind of process for anti-reflective properties.

The general purpose of the NIL-based etching mask is illustrated in Figure 7.6. In a SANIL-approach a solution of PMMA is applied to the surface. Subsequently, the PDMS-stamp is pressed into the PMMA layer. After solvent evaporation, a patterned PMMA layer remains on the substrate. Alternatively, a curable polymer film is

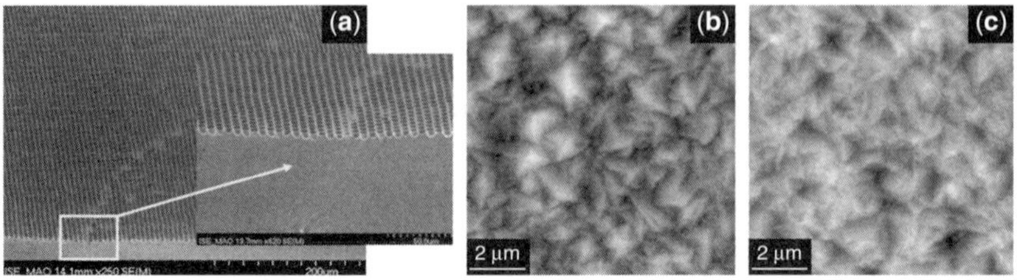

Figure 7.5 SEM cross section of a NIL- patterned c-Si wafer (a) [30]. © SPIE 2012. Reproduced with permission of SPIE; AFM images of a LPCVD ZnO master (b) and its inverse texture obtained by UV-NIL (c) [31]. © IOP Publishing. Reproduced with permission of IOP Publishing. All rights reserved.

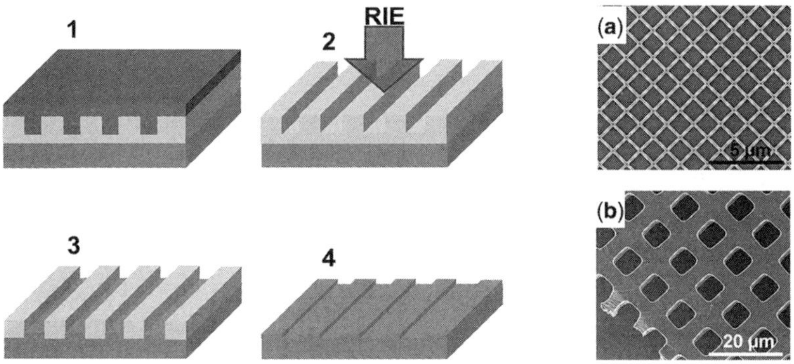

Figure 7.6 Scheme of a NIL-based process for etching masks (left): (1) SANIL patterning, (2) etching process, and (3) protective layer removal to obtain the patterned surface. SEM images of PMMA masks (right) on glass (a) and freestanding membrane (b).

coated on the surface and a NIL process is carried out for structuring. In both cases, an etching step follows — either liquid, gaseous or plasma-based. A common plasma-based etching method is reactive-ion etching (RIE), where reactive species are created by introducing gases or vapors in a low-pressure plasma chamber. Since by RIE even parts of the protective layer can be etched, a thin residual layer between the mask structures is no problem for the etching process. In the last step the remaining protective layer has to be removed, e.g., by dissolving in a suitable solvent.

7.3.3 NIL for Organic Solar Cell Processing

Remarkably high numbers on NIL-based applications for photovoltaics have been published in the field of organic solar cells throughout the last years [41, 42]. One reason for this is the suitability of NIL for the preparation of structured polymeric material, and thus this technique can be applied to organic devices in a straightforward manner. The process was used to pattern the substrate, the transparent polymer electrode or the active layers. The approach of substrate patterning, that has basically been discussed in Section 7.3.1, offers the possibility of roll-to-roll fabrication, especially for polymeric substrates [43]. As described above, either crosslinkable precursors can be deposited and cured during the imprint process or the polymeric substrate itself can be embossed by high-pressure processes [34]. In both cases, a surface relief pattern with light-trapping properties can be obtained inducing a pattern transfer into the following solar cell layers (Figure 7.7).

As described for inorganic solar cells, the patterning of the transparent front electrode by NIL methods is applicable to organic devices as well. In this case, the conductive polymer PEDOT:PSS can be patterned. Chou *et al.* achieved enhanced device performance by depositing the active layers on an embossed PEDOT:PSS layer [44].

Figure 7.7 illustrates the importance in patterning the active layers in organic devices. Especially for bulk heterojunction (BHJ) solar cells, a high internal interface has to be created resulting in effective charge carrier separation. Instead of using a donor-acceptor blend, one of the materials can be patterned in the sub-micron range prior to the deposition of the other material. Many groups demonstrated the beneficiary effect of a patterned organic donor layer resulting in an ordered nanopillar BHJ solar cell. Very often poly-3-hexylthiophane (P3HT) is NIL-patterned (usually by hot embossing or

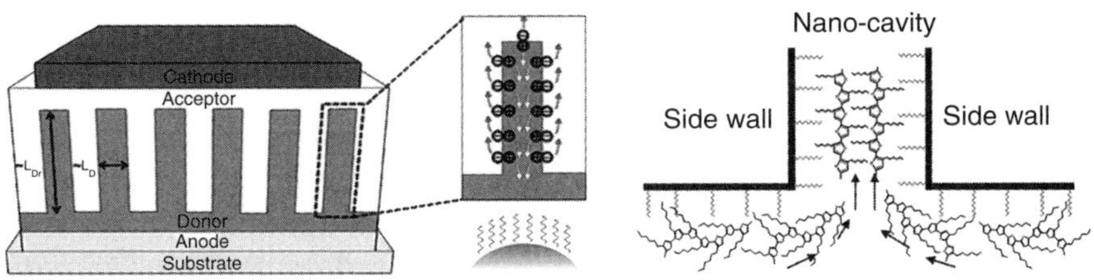

Figure 7.7 Illustration of an ordered nanopillar BHJ and carrier diffusion pathways (left). Reprinted with permission from [41]. © 2009 American Chemical Society. Effect of chain alignment by mold release during P3HT imprinting (right). Reprinted with permission from [51]. © 2009 American Chemical Society.

SANIL) and subsequently covered with phenyl-C_{61}-butyric acid methyl ester (PCBM) as acceptor material [45–50]. Hu *et al.* discussed the effect of chain alignment of P3HT induced by interactions with the stamp sidewall during the stamp release [51, 52]. In combination with beneficiary π-π-stacking effects the charge carrier mobility is enhanced and exciton travel length for charge dissociation is decreased. This leads to higher conversion efficiencies compared to devices with simple polymer blend absorbers [53].

The approach of ordered bulk-heterojunction organic solar cells can also be realized as an organic-inorganic hybrid. Respective devices are prepared by SANIL nanostructuring of sol-gel titania precursors. After thermal annealing the titania acceptor is covered with P3HT donor material forming the heterojunction [54, 55]. The reason for improved efficiency results from the efficient charge separation and reduction of effective series resistance by increasing TiO_2–P3HT interface area [56].

7.3.4 Plasmonic Films Prepared with NIL Methods

Metallic nanoparticles with localized surface plasmon resonances have recently been discussed as alternative light-trapping materials for silicon thin-film solar cells [57]. These nanostructures show strong interaction with visible light and can concentrate it and couple it into the semiconductor films of solar cells. An enhanced absorption in the case of geometry-optimized structures can be achieved. In general two concepts were developed: On the one hand the metal particles on the top of the cell scatter the light into the material with the higher refractive index — in this case the silicon absorber. On the other hand, red light can be coupled into the guided modes of the semiconductor by periodic metal structures on a metallic back reflector. The latter case was realized in the groups of Atwater and Polman by NIL methods [58, 59]. Periodic sol-gel-derived patterns of oxidic materials have been produced onto which silver was deposited by PVD methods. The resulting devices show significantly enhanced efficiencies.

McGehee *et al.* used the same effect for dye-sensitized solar cells by NIL-patterning a TiO_2-precursor deposited on a FTO substrate. The periodic grooves were filled with silver by PVD resulting in a metal back contact with periodic silver pillars [60] (Figure 7.8). The weak absorption of a ruthenium-based absorber in the 600–900 nm range could be significantly increased.

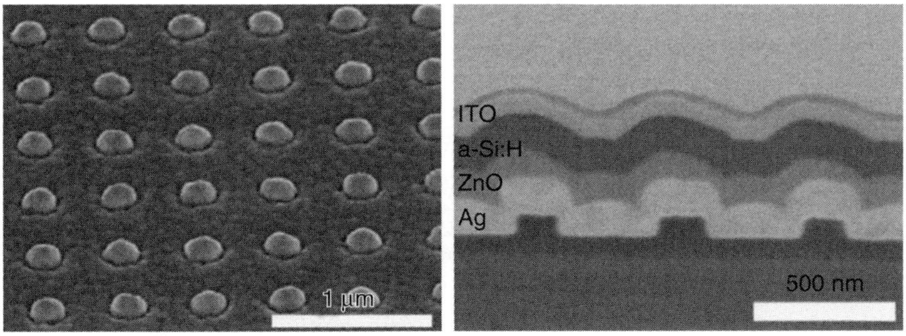

Figure 7.8 SEM image of a periodic array of silver-coated sol-gel particles (left). SEM image of an FIB cross section of a full nip type a-Si:H solar cell grown on the patterned back contact (right) [59]. © IOP Publishing. Reproduced with permission of IOP Publishing. All rights reserved.

7.3.5 Up-Scaling Potential of NIL Processes

NIL processes offer a great potential for up-scaling and high-throughput as they can be run under ambient conditions and the soft stamps are roll-to-roll compatible. By using fast curable photoresists, Guo *et al.* demonstrated a UV-NIL process on 10 cm wide PET foil and feed rates of 1 m/min [43]. The feed rate was found to have direct influence on the residual resist layer thickness. However, sub-micrometer gratings could be obtained. Only a few approaches of roll-to-roll NIL-processed layers for photovoltaic devices have been reported so far. Most of the publications concern the introduction of antireflective layers. As presented by Bläsi *et al.*, the patterning of UV-curing resist layers on brittle, stiff and opaque substrates (e.g. silicon) can be carried out in a continuous process flow [61]. A surface texturing of monocrystalline silicon wafers was applied by reactive-ion etching resulting in decreased reflection losses.

Beyond these approaches, Gonzalez Lazo *et al.* were able to establish a UV-NIL process for texturizing polymer foils to be used as substrates for amorphous silicon solar cell depositing [2]. They could show the general adaptability of existing light-trapping concepts to enhance conversion efficiencies of the devices.

These results illustrate that NIL-techniques are on the way to be used in commercial scales for further cost reduction. Simple replication and templating methods can be used instead of expensive vacuum technologies. Light trapping is the first concept that is applied in roll-to-roll NIL processes for PV devices, but it even

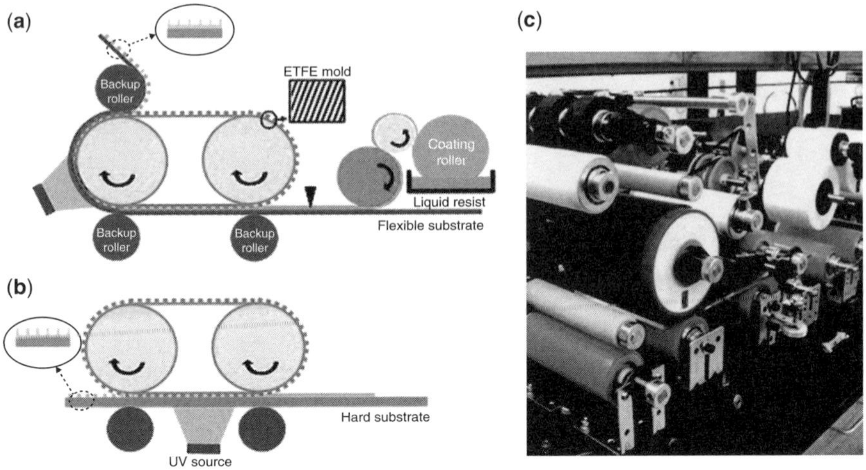

Figure 7.9 Schematics of (a) roll-to-roll- NIL and (b) roll-to-plate-NIL process. (c) Photograph of 6 in.-capable roll-to-roll apparatus. Adapted with permission from [43]. © 2009 American Chemical Society.

shows a great potential for cost reduction (Figure 7.9). However, it is very likely only a matter of time before other techniques presented in this chapter will be transferred from lab-scale to cheap roll-to-roll processes.

7.4 Conclusion and Outlook

As can be seen in the previous sections, nanoimprint lithography has developed from a scientific lab-scale patterning technique into a real alternative for conventional patterning processes throughout the last years. With NIL-tools the solar cell manufacturing procedures have been enhanced by another powerful nanotechnology method. Existing solar cell concepts with textured or patterned surfaces or interfaces could successfully be transferred to soft lithography techniques. The main applications are the preparation of antireflective or light-trapping layers and etching masks. Organic and plasmonic devices based on a NIL process have been presented as well.

It was shown that NIL-based antireflective structures on wafer surfaces or substrate outsides can be easily adapted to almost any cell technology. And as NIL structures can be prepared in a huge variety, an alignment to the respective cell technology is straightforward. Besides the improvements in photocurrent yielded by reduced reflection losses, the more challenging application of NIL for the preparation of effective light-trapping films inside the cell has been discussed. The nanoimprint technique can be used for both, either for TCO or for plasmonic light-trapping, clearly demonstrating the process flexibility. As it is a cheap, simple and fast patterning tool for high-quality structures with a resolution in the lower nanometer range, the great potential of NIL as a real nanotechnology tool for further cost reduction in photovoltaics becomes obvious. However, to use this potential, the presented methods have to be developed from a basic stage and need to be further evaluated on a production scale. In the same manner, the discussed scaling approaches towards an in-line roll-to-roll production of NIL-patterned cells have to be adapted to other cell concepts. Together with a continuous geometry optimization, this would allow the reduction of production costs with increasing cell efficiency at the same time.

References

1. M. Izu and T. Ellison, *Solar Energy Materials and Solar Cells*, Vol. 78, 613–626, 2003.
2. M.A. González Lazo, R. Teuscher, Y. Leterrier, J.-A.E. Månson, C. Calderone, A. Hessler-Wyser, P. Couty, Y. Ziegler and D. Fischer, *Solar Energy Materials and Solar Cells*, Vol. 103, 147–156, 2012.
3. A. Kumar and G.M. Whitesides, *Applied Physics Letters*, Vol. 63, 2002–2004, 1993.
4. Y. Xia and G.M. Whitesides, *Annual Review of Materials Science*, Vol. 28, 153–184, 1998.
5. B. Terris, H.J. Mamin and D.D. Rugar, *Applied Physics Letters*, Vol. 68, 141–143, 1996.
6. M.C. Hutley, *Techniques of Physics, London: Academic Press*, Vol. 1, 1982.
7. M. Nakano and N. Nishida, *Applied Optics*, Vol. 18, 3073–3080, 1979.
8. E. Kim, Y. Xia, and G.M. Whitesides, *Journal of the American Chemical Society*, Vol. 118, 5722–5731, 1996.
9. S.Y. Chou, P.R. Krauss and P.J. Renstrom, *Applied Physics Letters*, Vol. 67, 3114–3116, 1995.
10. L.J. Guo, *Advanced Matererials*, Vol. 19, 495–513, 2007.
11. L.J. Guo, *Journal of Physics D: Applied Physics*, Vol. 37, R123, 2004.
12. H. Hauser, P. Berger, B. Michl, C. Müller, S. Schwarzkopf, M. Hermle, B. Bläsi, H. Thienpont, P. van Daele, J. Mohr and H. Zappe, Nanoimprint lithography for solar cell texturisation, in: *Micro-Optics 2010*, SPIE, pp. 77160X1–77160X9, 2010.
13. J.A. Rogers and R.G. Nuzzo, *Materials Today*, Vol. 8, 50–56, 2005.
14. S.J. Clarson and J.A. Semlyen, *Siloxane Polymers*, Prentice Hall, Englewood Cliffs, NJ, 1993.

15. H. Schmid and B. Michel, *Macromolecules*, Vol. 33, 3042–3049, 2000.
16. J.N. Lee, C. Park and G.M. Whitesides, *Analytical Chemistry*, Vol. 75, 6544–6554, 2003.
17. K.M. Choi and J.A. Rogers, *Journal of the American Chemical Society*, Vol. 125, 4060–4061, 2003.
18. L.M. Campos, T.T. Truong, D.E. Shim, M.D. Dimitriou, D. Shir, I. Meinel, J.A. Gerbec, H.T. Hahn, J.A. Rogers and C.J. Hawker, *Chemistry of Materials*, Vol. 21, 5319–5326, 2009.
19. S.S. Williams, S. Retterer, R. Lopez, R. Ruiz, E.T. Samulski and J.M. DeSimone, *Nano Letters*, Vol. 10, 1421–1428, 2010.
20. J. Zhao and M.A. Green, *IEEE Transactions on Electron Devices*, Vol. 38, 1925–1934, 1991.
21. A. Shah, H. Schade, M. Vanecek, J. Meier, E. Vallat-Sauvain, N. Wyrsch, U. Kroll, C. Droz and J. Bailat, *Progress in Photovoltaics: Research and Aapplications*, Vol. 12, 113–142, 2004.
22. J. Müller, B. Rech, J. Springer and M. Vanecek, *Solar Energy*, Vol. 77, 917–930, 2004.
23. K.-S. Han, J.-H. Shin and H. Lee, *Solar Energy Materials and Solar Cells*, Vol. 94, 583–587, 2010.
24. V. Satzinger, V. Schmidt, L. Kuna, C. Palfinger, R. Inführ, R. Liska, J.R. Krenn, H. Thienpont, P. van Daele, J. Mohr and M.R. Taghizadeh, in: *Micro-Optics 2008*, SPIE, pp. 699217–699217–10, 2008.
25. J.-H. Shin, K.-S. Han and H. Lee, *Progress in Photovoltaics: Research and Aapplications*, Vol. 19, 339–344, 2011.
26. J. Chen, W.-L. Chang, C. Huang and K. Sun, *Optics Express*, Vol. 19, 14411–14419, 2011.
27. C. Trompoukis, O. El Daif, V. Depauw, I. Gordon and J. Poortmans, *Applied Physics Letters*, Vol. 101, 1039011–1039014, 2012.
28. K.-S. Han, J.-H. Shin, W.-Y. Yoon and H. Lee, *Solar Energy Materials and Solar Cells*, Vol. 95, 288–291, 2011.
29. J. Barbé, A.F. Thomson, E.-C. Wang, K. McIntosh and K. Catchpole, *Progress in Photovoltaics: Research and Aapplications*, Vol. 20, 143–148, 2012.
30. B. Bläsi, H. Hauser, C. Walk, B. Michl, A. Guttowski, A. Mellor, J. Benick, M. Peters, S. Jüchter, C. Wellens, V. Kübler, M. Hermle, A.J. Wolf, R. Wehrspohn and A. Gombert, Photon management structures for solar cells, in: *Photonics for Solar Energy Systems IV*, SPIE, pp. 84380Q, 2012.
31. J. Escarré, C. Battaglia, K. Söderström, C. Pahud, R. Biron, O. Cubero, F.-J. Haug and C. Ballif, *Journal of Optics*, Vol. 14, 24009, 2012.
32. A. Bessonov, Y. Cho, S.-J. Jung, E.-A. Park, E.-S. Hwang, J.-W. Lee,M. Shin and S. Lee, *Solar Energy Materials and Solar Cells*, Vol. 95, 2886–2892, 2011.
33. L. Pei, A. Balls, C. Tippets, J. Abbott, M.R. Linford, J. Hu, A. Madan, D.D. Allred, R.R. Vanfleet and R.C. Davis, *Journal of Vacuum Science and Technology A*, Vol. 29, 21017, 2011.
34. S.-Y. Chuang, C.-C. Yu, H.-L. Chen, W.-F. Su and C.-W. Chen, *Solar Energy Materials and Solar Cells*, Vol. 95, 2141–2150, 2011.
35. C. Battaglia, J. Escarré, K. Söderström, L. Erni, L. Ding, G. Bugnon, A. Billet, M. Boccard, L. Barraud, S. de Wolf, F.-J. Haug, M. Despeisse and C. Ballif, *Nano Letters*, Vol. 11, 661–665, 2011.
36. C. Battaglia, J. Escarré, K. Söderström, M. Charrière, M. Despeisse, F.-J. Haug, and C. Ballif, *Nature Photonics*, Vol. 5, 535–538, 2011.

37. W. Soppe, M. Dörenkämper, J.-B. Notta, P. Pex, W. Schipper and R. Wilde, *Physica Status Solidi A*, Vol. 210, 707–710, 2013.
38. G. Mondin, B. Schumm, J. Fritsch, R. Hensel, J. Grothe and S. Kaskel, *Materials Chemistry and Physics*, Vol. 137, 884–891, 2013.
39. J. Tommila, V. Polojärvi, A. Aho, A. Tukiainen, J. Viheriälä, J. Salmi, A. Schramm, J. Kontio, A. Turtiainen and T. Niemi, *Solar Energy Materials and Solar Cells*, Vol. 94, 1845–1848, 2010.
40. F. Wang, N.K. Subbaiyan, Q. Wang, C. Rochford, G. Xu, R. Lu, A. Elliot, F. D'Souza, R. Hui and J. Wu, *ACS Applied Materials and Interfaces*, Vol. 4, 1565–1572, 2012.
41. Y. Yang, K. Mielczarek, M. Aryal, A. Zakhidov and W. Hu, *ACS Nano*, Vol. 6, 2877–2892, 2012.
42. J.-T. Chen and C.-S. Hsu, *Polymer Chemistry*, Vol. 2, 2707, 2011.
43. S.H. Ahn and L.J. Guo, *ACS Nano*, Vol. 3, 2304–2310, 2009.
44. W.-Y. Chou, J. Chang, C.-T. Yen, F.-C. Tang, H.-L. Cheng, M.-H. Chang, S. Lien-Chung Hsu, J.-S. Chen and Y.-C. Lee, *Applied Physics Letters*, Vol. 99, 1831081–1831083, 2011.
45. M. Aryal, F. Buyukserin, K. Mielczarek, X.-M. Zhao, J. Gao, A. Zakhidov and W. Hu, *Journal of Vacuum Science and Technology B*, Vol. 26, 2562, 2008.
46. D. Cheyns, K. Vasseur, C. Rolin, J. Genoe, J. Poortmans and P. Heremans, *Nanotechnology*, Vol. 19, 424016, 2008.
47. M.-S. Kim, J.-S. Kim, J.C. Cho, M. Shtein, L.J. Guo and J. Kim, *Applied Physics Letters*, Vol. 90, 1231131–1231133, 2007.
48. X. He, F. Gao, G. Tu, D.G. Hasko, S. Hüttner, N.C. Greenham, U. Steiner, R.H. Friend and W.T.S. Huck, *Advanced Functional Materials*, Vol. 21, 139–146, 2011.
49. J.Y. Park, N.R. Hendricks and K.R. Carter, *Langmuir*, Vol. 27, 11251–11258, 2011.
50. J. Wang, G. Min, Z. Song, X. Ni, W. Zhou, J. Zhan, Y. Zhang, J. Zhang and L. Shi, *Journal of Materials Chemistry*, Vol. 22, 21154, 2012.
51. M. Aryal, K. Trivedi and W. Hu, *ACS Nano*, Vol. 3, 3085–3090, 2009.
52. H. Hlaing, X. Lu, T. Hofmann, K.G. Yager, C.T. Black and B.M. Ocko, *ACS Nano*, Vol. 5, 7532–7538, 2011.
53. M. Zhou, M. Aryal, K. Mielczarek, A. Zakhidov and W. Hu, *Journal of Vacuum Science and Technology B*, Vol. 28, C6M63, 2010.
54. S.S. Williams, M.J. Hampton, V. Gowrishankar, I.-K. Ding, J.L. Templeton, E.T. Samulski, J.M. DeSimone and M.D. McGehee, *Chemistry of Materials*, Vol. 20, 5229–5234, 2008.
55. H.-J. Her, J.-M. Kim, C. Kang and Y.-S. Kim, *Journal of Physics and Chemistry of Solids*, Vol. 69, 1301–1304, 2008.
56. W.-H. Baek, I. Seo, T.-S. Yoon, H.H. Lee, C.M. Yun and Y.-S. Kim, *Solar Energy Materials and Solar Cells*, Vol. 93, 1587–1591, 2009.
57. H.A. Atwater and A. Polman *Nature Materials*, Vol. 9, 205–213, 2010.
58. V.E. Ferry, M.A. Verschuuren, H.B. Li, R.E. Schropp, H.A. Atwater and A. Polman, *Applied Physics Letters*, Vol. 95, 1835031–1835033, 2009.
59. P. Spinelli, V.E. Ferry, J. van de Groep, M. van Lare, M.A. Verschuuren, R.E.I. Schropp, H.A. Atwater and A. Polman, *Journal of Optics*, Vol. 14, 24002, 2012.
60. I.-K. Ding, J. Zhu, W. Cai, S.-J. Moon, N. Cai, P. Wang, S.M. Zakeeruddin, M. Grätzel, M.L. Brongersma, Y. Cui and M.D. McGehee, *Advanced Energy Materials*, Vol. 1, 52–57, 2011.

8

Indoor Photovoltaics: Efficiencies, Measurements and Design

Monika Freunek (Müller)

Institute for Microsystem Technologies, Albert-Ludwigs-University of Freiburg, Freiburg, Germany

Abstract
Recent reductions of the energy demand of small electronic devices down to a few microwatts have changed the opportunities of indoor photovoltaics (IPV) from a niche market to a prospering segment of photovoltaic applications. Typically, the maximum allowable IPV module area is a few centimeters, the intensity is as low as 0.1–10 Wm^{-2}, and the spectrum consists of solar radiation transmitted through a window, diffuse radiation reflected from the surrounding indoor environment, and narrow-banded artificial light sources. This chapter gives an introduction to typical indoor spectra and the resulting maximum theoretical photovoltaic efficiencies, such as 50% for a fluorescent tube and a bandgap of E_g = 1.9 eV. Effects in real cells, both the technological aspects and their impact on the design of IPV devices, as well as strategies for further optimization, such as nanophotonics, are discussed in the following sections. Finally, practical aspects and current indoor efficiency records are provided.

Keywords: Indoor photovoltaic, photovoltaic efficiency, solar cell efficiency, indoor irradiance, low intensity, nanophotonic applications, fluorescent collectors

8.1 Introduction

The first wave of photovoltaically powered consumer products for indoor applications started in the late seventies, when the oil crisis

**Corresponding author*: monika.freunek@imtek.uni-freiburg.de

Atul Tiwari, Rabah Boukherroub, and Maheshwar Sharon (eds.) Solar Cell Nanotechnology, (203–222) 2014 © Scrivener Publishing LLC

encouraged the development of regenerative power supplies, and the technological progress in integrated circuits and the first microprocessors enabled new applications. Besides postage weighing scales [1], kitchen scales, and office table decorations, solar calculators [2] implying amorphous silicon converters were the most common products. However, low electrical energy output and the need of an active orientation to powerful indoor radiation sources, such as windows, led to low user acceptance and a more playful character of indoor photovoltaics (IPV).

A major change to this niche market has been introduced in recent years by the achieved reduction of the average electric power needs of small electronic devices, such as wireless sensor nodes, to the microwatt scale with demonstrated standby power in the picowatt range [3], and the introduction of efficient narrow banded lighting technologies, such as LED (light emitting diodes) and fluorescent tubes. Within the concept of "micro energy harvesting," i.e., to power electronic devices using ambient energy [4–6], optical radiation is one of the most promising forms due to high power densities and good predictability. Together with the increasing automatization of buildings, e.g., in order to save energy, and high conversion efficiencies, IPV can be expected to be a growing market within the next years.

The required power of consumer electronics ranges between 10 mW for the standby of a mobile phone and 10–100 W for a tablet PC [7]. For these applications, optimized IPV devices can be used for the enhancement of battery life.

The current available commercial products are mainly based on amorphous or crystalline silicon, while materials investigated in research also include CdTe, CIGS, CIS, $Cu(In,Ga)Se_2$, dye-sensitized cells, GaAs, GaInP, and GaAs/GaInP multijunction cells, and P3HT:PCBM [8–17]. Exemplary IPV devices consisting of amorphous and crystalline silicon, and GaInP are depicted in Figure 8.1.

Typically, the area allowed for an IPV module is a few centimeters, the incident intensity is as low as 0.1–10 Wm^{-2}, and the spectrum consists of solar radiation transmitted through a window, diffuse radiation reflected from the surrounding indoor environment, and narrow-banded artificial light sources, such as fluorescent tubes. Therefore, the maximum theoretical efficiency, the optimum material, and the design of IPV devices differ from other photovoltaic disciplines.

Figure 8.1 Indoor photovoltaic converters and products. Source: Burzlaff, IMTEK, 2013.

The characterization of IPV devices and available radiation has not yet been standardized, and many reported methods lack reproducibility, which increases the risk of design errors and product failure. The use of pyranometers, retraceable reference cells together with spectral mismatch calculations for measuring available radiation, and the use of common light sources, such as fluorescent tubes of daylight spectrum, are the current most reliable methods for measuring indoor performance [18].

This chapter is structured as follows: First, typical indoor spectra and the resulting maximum theoretical photovoltaic effiencies are evaluated. Then effects in real cells and their impact on the design of IPV devices are discussed. The following section provides aspects of the characterization of IPV devices and current measured indoor efficiencies. The chapter closes with an outlook on optimization strategies for IPV including nanophotonics and fluorescent collectors.

8.2 Indoor Radiation

8.2.1 Spectra and Intensities

The spectral photon flux $\phi_{p\lambda}$ per area A of a radiation source is defined by the number of photons N per second and per unit bandwith.

The energy of a photon is Eq. 8.1:

$$E=h\,c/\lambda, \tag{8.1}$$

with the vacuum speed of light $c = 2.998\ 10^8\ ms^{-1}$ and Planck's constant $h = 6.626\ 10^{-34}$ Js. The radiant power density or irradiance E_e is obtained by multiplication of the spectral photon flux with the photon energy and integration over the total incident spectrum. The electrical current density of a photovoltaic converter induced by a photon flux $\phi_{p,\lambda}$ is known as photocurrent density j_{ph} and calculated to be Eq. 8.2,

$$j_{ph} = q\int_0^\infty \phi_{p,\lambda}(\lambda)\, EQE(\lambda)\, d\lambda, \tag{8.2}$$

where $q = 1.602\ 10^{-19}$ A is the elementary charge. The external quantum efficiency EQE of the converter depends on the choice of materials, and the overall optical and electrical cell design. The knowledge of the radiation spectrum is thus essential for photovoltaic design. For basic indoor efficiency calculations, every incident photon is assumed to be absorbed, i.e., EQE=1, and the short-circuit current density j_{sc} is obtained by assuming $j_{sc} \approx j_{ph}$.

Under indoor conditions, $\phi_{p,\lambda}$ is a mixture between artificial lighting and solar radiation, which is usually transmitted through a window of a spectral transmittance $T(\lambda)$. Depending on the latitude and on the purpose of the building, $T(\lambda)$ will be optimized for sun or heat protection as well as for application specific requirements, such as noise or radio wave protection. Figure 8.2 depicts the terrestrial solar irradiance for an Air Mass Factor (AM) 1.5 (black line) and $T(\lambda)$ for different types of windows.

For lighting applications, the sensitivity of the human eye is mimicked by a spectral function $V(\lambda)$ with a maximum at 555 nm and a range from 380–780 nm. $V(\lambda)$ is also known as "visibility function," and evaluates a radiation for its usability for the human eye. Figure 8.3 depicts the spectral irradiance of typical indoor radiation sources, the terrestrial solar AM 1.5 reference spectrum, the spectral response of amorphous and crystalline silicon, and $V(\lambda)$.

The photo current density j_{ph} from direct radiation is Eq. 8.3.

$$\mathrm{j}_{ph} = q\int_0^\infty \mathrm{d}\lambda\, EQE(\lambda)\left[\phi_{p,\lambda}^{solar}(\lambda)T(\lambda) + \phi_{p,\lambda}^{artificial}(\lambda)\right] \tag{8.3}$$

Although for the position directly behind windows short-term maximum values can reach the 100 Wm^{-2} range, the typical indoor

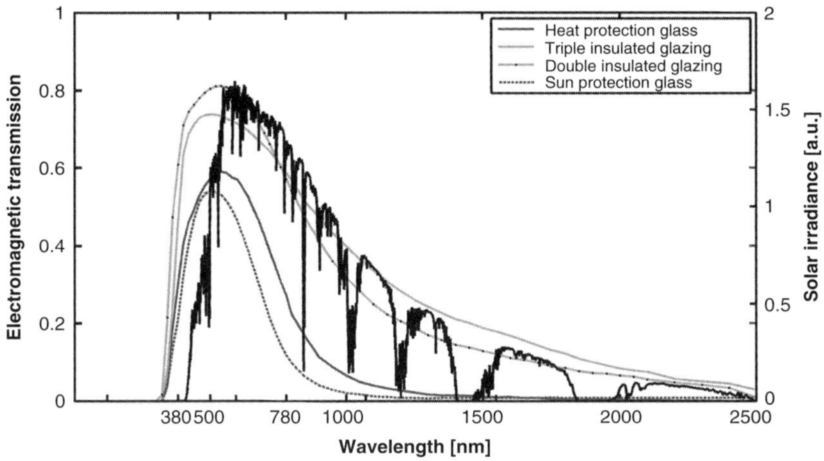

Figure 8.2 Terrestrial solar irradiance and transmission $T(\lambda)$ of different types of windows.

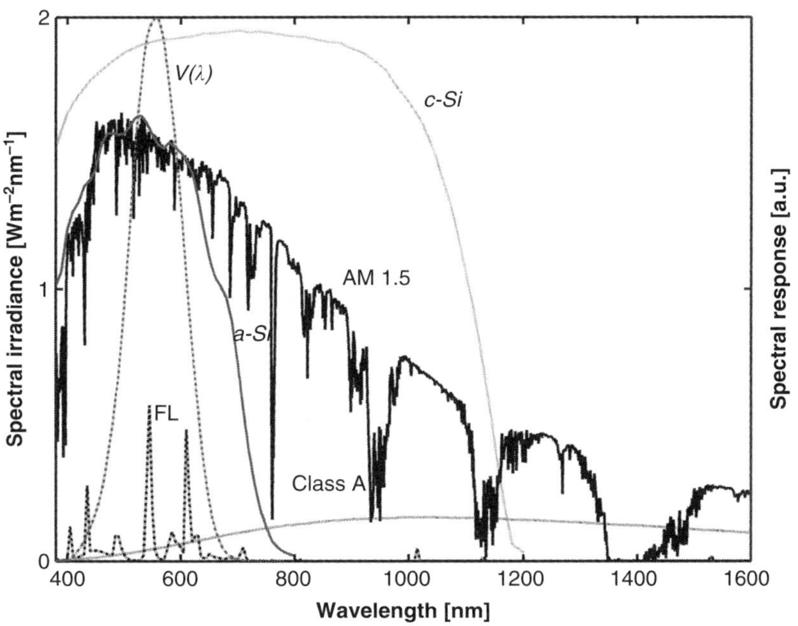

Figure 8.3 Spectral irradiance of indoor light (FL: fluorescent tube; Class A: incandescent bulb) and terrestrial solar radiation AM 1.5, spectral response of a- and c-Si, and the spectral sensitivity function of the human eye, $V(\lambda)$.

irradiance ranges between 0.1–10 Wm^{-2} [13, 19]. This is up to four orders of magnitude below outside conditions. Depending on the point of installation and the reflectivity $R(\lambda)$ of the major materials of the room, indirect radiation can contribute the main part of the incident power. In these cases, solid angles dependencies can no longer be omitted, which yields Eq. 8.4.

$$j_{ph} = q\int_0^{\infty}\int_0^{\Omega'} d\lambda d\Omega\, EQE(\lambda,\Omega)\left[\phi_{p,\lambda}^{solar}(\lambda,\Omega)T(\lambda,\Omega)R(\lambda,\Omega)+\phi_{p,\lambda}^{artificial}(\lambda,\Omega)R(\lambda,\Omega)\right] \tag{8.4}$$

8.3 Maximum Efficiencies

8.3.1 Maximum Indoor Efficiencies and Ideal Materials

The photovoltaic efficiency η for a given irradiance E_e is Eq. 8.5,

$$\eta = \frac{p_{mp}}{E_e} \tag{8.5}$$

with the maximum power density p_{mp} at the operation point (Eq. 8.6),

$$p_{mp} = j_{sc}V_{oc}\ FF. \tag{8.6}$$

The fill factor FF is a rating of the ideality of the cell, and V_{oc} denotes the open circuit voltage. Ideally, a higher bandgap enables both a higher fill factor and a higher open circuit voltage. Figure 8.4 shows the maximum photovoltaic efficiencies calculated from the detailed balance limit [20] for the spectral photon flux of a fluorescent tube with daylight spectrum (OSRAM Lumilux 840/18 W), a halogen tube, a phosphor and an RGB white LED, a sodium discharge lamp, an incandescent bulb, and, as reference, for the terrestrial standard spectrum ASTM-G173 AM1.5 global. For any Planckian radiator colder than the sun, such as halogen lamps and light bulbs, and for the typically employed materials, the photovoltaic efficiency limit will be below the limit for solar radiation. The ideal bandgap shifts to the range of 0.8 eV, which is close to the germanium bandgap of 0.67 V, with maximum efficiencies around 30%.

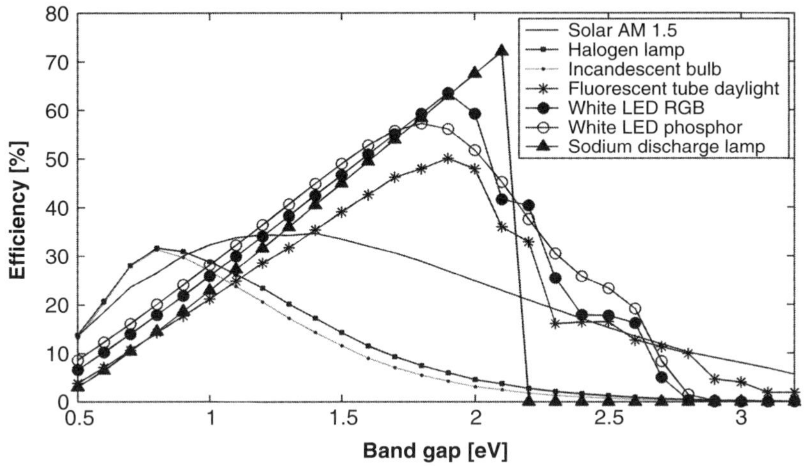

Figure 8.4 Energy gap E_g and limiting Shockley-Queisser efficiency of an ideal photovoltaic converter for the spectral photon flow $\phi_{p,\lambda}$ of a fluorescent tube with daylight spectrum (*OSRAM Lumilux* 840/18W), terrestrial solar light with standard spectrum AM1.5, a halogen lamp, an incandescent bulb, a RGB white LED, a phosphor white LED, and a sodium discharge lamp.

For narrow banded lighting sources optimized to the visibility function with a maximum at 555 nm, the ideal photovoltaic bandgap is around 1.9–2.0 eV with maximum efficiencies between 50% for a fluorescent tube with daylight spectrum and 64% for a RGB white LED. Similar results can be expected for energy saving lamps. For a sodium discharge lamp with monochromatic emission at λ = 589 nm, the calculated efficiency is $\eta_{sodium} = 72\%$.

Figure 8.5 depicts the calculated photon yield. Indoor lighting sources are optimized for an emission of photons in the visible range from 380 nm to 780 nm. As a consequence, IPV devices can exploit the advantage of a higher open circuit voltage V_{oc} when shifting to larger bandgaps, while simultaneously benefiting from a high photon yield. At a bandgap of 1.9 eV with a cutoff wavelength λ_c = 653 nm, 87% of the incident light is absorbed by an ideal photovoltaic converter under fluorescent irradiance. For terrestrial solar light (AM1.5), only 29% of the incident photons account for the short-circuit current.

8.3.2 Monochromatic Radiation

In some applications it will be possible to include the choice of artificial light source in the overall system design. This might be the

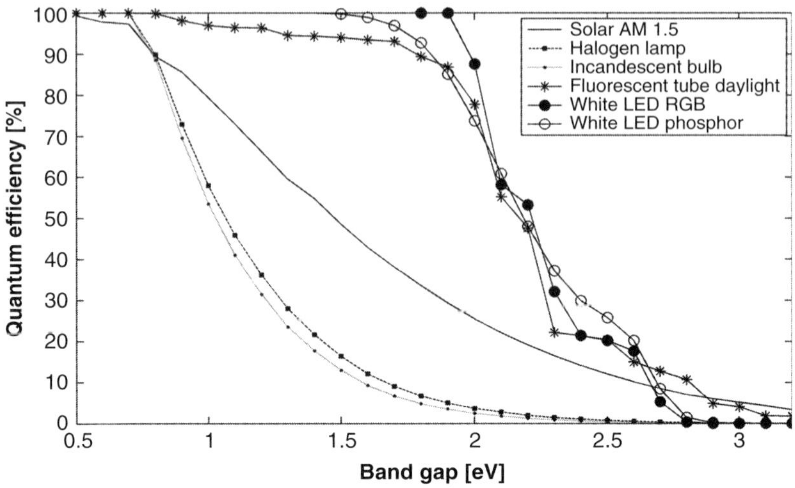

Figure 8.5 Photon yield in percent of an ideal photovoltaic converter as a function of the energy gap E_g for the spectral photon flux $\phi_{p\lambda}$ of various indoor radiation sources, and terrestrial solar light with standard spectrum AM1.5 (NREL 2013).

case for power beam applications using laser radiation in industrial settings or for the construction of an automatized building. In this case, the energy gap E_g of the photovoltaic converter would be matched to the indoor radiation source. For a monochromatic radiator of a wavelength λ, and of energy $E(\lambda) = E_g$, i.e., matched to the photovoltaic converter, the efficiency is Eq. 8.7.

$$\eta = \frac{E_g \int_{E_g}^{\infty} \phi_{p,\lambda} \mathrm{d}E}{\int_0^{\infty} E(\lambda) \phi_{p,\lambda} \mathrm{d}E} = 1. \tag{8.7}$$

As only photons of $E(\lambda) > E_g$ can deliver work, the ideal bandgap energy needs to be just below the photon energy. The ideal efficiency of such a matched system from this straightforward model is close to but less than unity.

The photon flow emitted from the matched radiation source is Eq. 8.8.

$$\phi_\lambda = \frac{P}{AE_g} \tag{8.8}$$

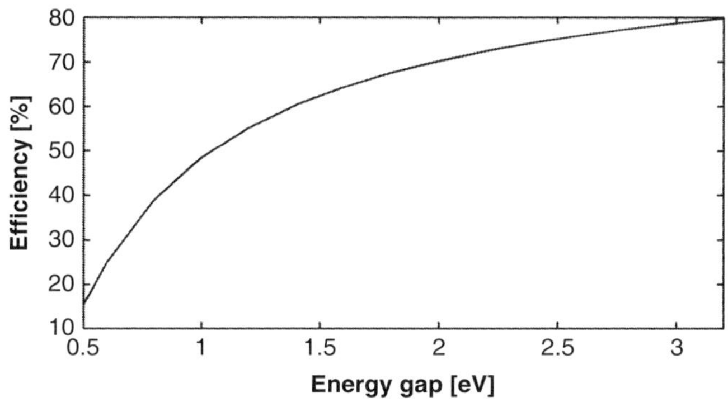

Figure 8.6 Maximum efficiency for a matched monochromatic radiation source, calculated following the detailed balance model.

Figure 8.6 shows the results calculated from the detailed balance limit, i.e., taking into account the radiative recombination losses within the photovoltaic cell. For a radiation power $P = 1$ W and $E_g = 2.3$ eV, matched to a monochromatic radiation source of 555 nm, η reaches 73%. The increase of efficiency with the bandgap can be explained by the corresponding decrease of radiative recombination current in this model.

8.3.3 Intensity Effects

The open circuit voltage V_{oc} is calculated from Eq. 8.9,

$$V_{oc} = \frac{k_B T}{q} \ln\left(\frac{I_{SC}}{I_0}\right), \tag{8.9}$$

where T is the temperature, $k_B = 8.617\ 10^{-5}$ eVK^{-1} is Boltzmann's constant, and $q = 1.602\ 10^{-19}$ As the elementary charge. The operation temperature for indoor applications is typically around room temperature, and temperature effects can be neglected. However, as shown above, typical indoor intensities and thus the short-circuit currents are up to four orders of magnitudes below outside conditions. This also leads to a slight reduction of V_{oc}.

The total saturation current I_0 is influenced by the choice of material, the cell manufacturing processes, and the operation temperature and intensity. I_0 consists of surface recombinations and recombination in defect states (Shockley-Read-Hall recombination), drift currents, radiative losses, and phonon-assisted recombination (Auger recombination). For indoor applications and poor design, I_0 can reach the range of I_{sc}. As can be seen from Eq. 8.9, this will affect the open circuit voltage. For indoor and low intensity applications, the reduction of dark current losses is therefore critical. Especially for indirect semiconductors, such as silicon, Shockley-Read-Hall losses need to be minimalized, i.e., via surface passivation, and low defect state densities. Direct semiconductors suffer less from Shockley-Read-Hall recombination, and have no phonon-based Auger recombination, but emit photons (radiative losses). As the radiative recombination current depends on the radiation intensity, this is mainly important for power beam applications.

The different Ohmic losses due to imperfections resulting from the manufacturing process and electrical contacts are often lumped together as a series resistance R_s and a shunt resistance R_{shunt} [21]. Their effect on the electrical characteristics of a photovoltaic converter is given by Eq. 8.10,

$$\text{In}\left(\frac{I+I_{ph}}{I_0}-\frac{V-IR_s}{I_0R_{shunt}}+1\right)=\frac{q}{k_BT}(V-IR_s). \qquad (8.10)$$

In general, besides for power beam applications, the series resistance losses can be neglected for indoor applications. The shunt resistance needs to be maximized for low irradiance applications, otherwise it will affect both open circuit voltage and fill factor [22]. The order of magnitude of the minimum required shunt resistance can be approximated by Eq. 8.11 [9].

$$R_{sh}^{\min}\approx\frac{V_{oc}}{I_{sc}} \qquad (8.11)$$

Hence, for an AM 1.5 global spectrum with an intensity of 1000 W/m^2 and an ideal short-circuit current of 45 mA for Si (17 mA for GaP), a shunt resistance in the range of a few ohms is sufficient, while for a 4 Wm^{-2} fluorescent spectrum with a short-circuit current of 186 μA (165 μA) the minimum resistance increases to the range of kΩ. For some materials the shunt resistance also varies with intensity. Figure 8.7 depicts the effect of the shunt resistance calculated using the one-diode model, a 4 W/m^2 fluorescent tube spectrum, and the solar 1000 W/m^2 AM 1.5 spectrum, respectively.

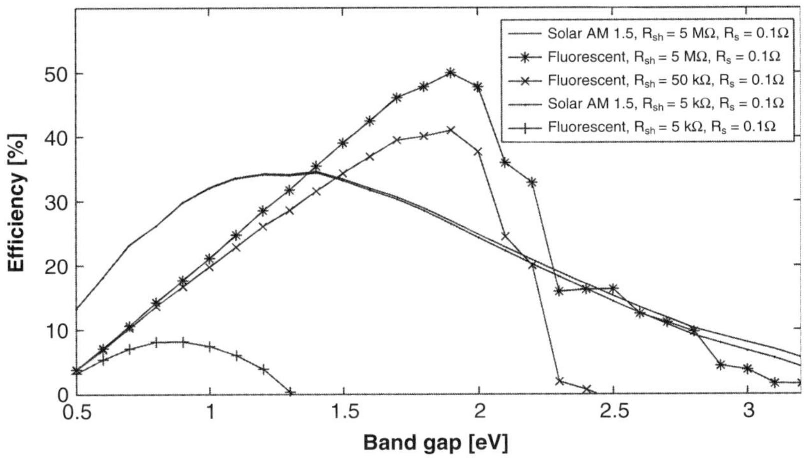

Figure 8.7 Effect of the shunt resistance calculated using the one-diode model, and a 4 W/m² fluorescent tube spectrum, and the solar 1000 W/m² AM 1.5 spectrum, respectively.

8.4 Optimization Strategies

The special conditions of IPV, i.e., low irradiance, diffuse radiation even as a major radiation source, and radiation from both artificial light and filtered solar spectra, make specific and new design approaches necessary. Especially of interest are the various approaches from nanophotonics and plasmonics, some of which are discussed in the following section.

Due to the low distance of the IPV converter to the indoor radiation sources, and due to the large contribution of incident diffuse radiation, angular dependencies and related losses are important design factors in IPV. There are currently two main strategies in order to improve the optical efficiency: nanophotonics using photonic crystals and fluorescent collectors.

Fluorescent collectors were intentionally designed as a low-cost approach to reduce spectral losses in terrestrial photovoltaics [23]. The principle is shown in Figure 8.8. A collector material of refractive index *n* is doped with fluorescent dyes and optimized for total internal refraction. Incident photons of energy E_1 interact with the dye molecules, and a photon of energy E_2 is emitted. Most of the photons are trapped, until they are collected at the edges, where the photovoltaic converter is mounted. As the energy level E_2 is defined by the choice of dye, the attached photovoltaic converter can be

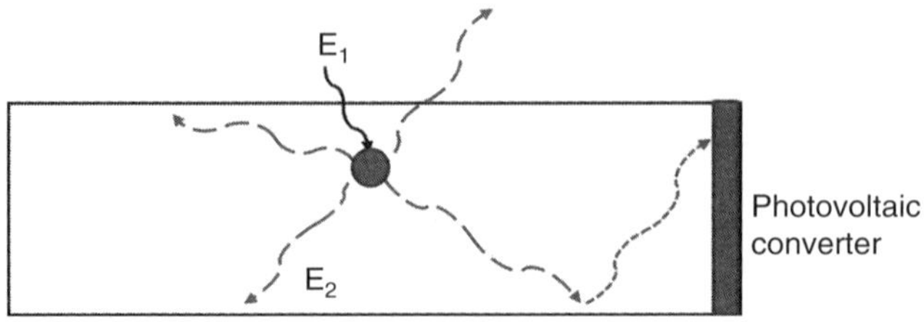

Figure 8.8 Principle of a fluorescent collector. A photon of energy E_1 is absorbed by a fluorescent dye molecule within the collector. The molecule emits a photon in a random distribution at energy level E_2. The photon is then emitted from the collector or totally reflected within the collector, until it is absorbed by the photovoltaic converter attached to one side.

matched to E_2, thus minimizing spectral losses. Furthermore, every angle of incidence is accepted, which renders this approach attractive to IPV applications with major contributions from diffuse radiation, such as wall integrated systems. The system proposed by Beattie *et al.* uses acrylic plastic with fluorescent dyes on a-Si:H [24]. Van Sark and coworkers also demonstrated the use of quantum dots [25].

In addition to downshift of radiation, which has been described above, there are up and down conversion of radiation, which involves the absorption (emission) of more than one photon [25]. Regarding narrow-banded artificial light sources, down conversion and shifting are of interest for IPV in order to improve the efficiency of crystalline and amorphous silicon photovoltaics, i.e., low cost IPV.

The ideal bandgap for IPV with currently employed narrow-banded artificial light sources is between 1.9–2.0 eV, depending on the light sources. Typical bandgaps of organic and GaInP converters are around 1.9 eV. Being a direct semiconductor, GaInP devices naturally have a large shunt resistance, and the highest indoor efficiency currently measured is 15.8% for a GaInP cell. The critical parameters of III-V cells are cost and the use of rare earth materials. In most applications, cost will be more important than size, i.e., efficiency or voltage stability. The main efforts in order to reduce cost focus on the replacement of rare earth substrates with low-cost semiconductors, such as Si. However, the lattice mismatch reduces the efficiency. The current record is an efficiency of 10% under solar AM 1.5 conditions for a GaAs nanowire cell grown on Si [26].

The current measured record of organic photovoltaic converters under indoor conditions is 3.5% compared to 3.1% under solar conditions for the same sample [9]. Dye-sensitized cells showed an increase from 3.0% to 6.8%. Costs for printable technologies consisting of organic materials are low. Major issues in organic photovoltaic cells are impurities, internal recombination, and long-term stability. Here, nanostructuring and confinement are employed in order to increase the electrical conductivity.

Chalcogenide materials, especially CdTe with its bandgap of 1.45 eV, are well suited for indoor conditions. The maximum theoretical efficiency is above 30% for both fluorescent and LED light. Gemmer calculated a maximum efficiency of 8% for a CIGS device with R_{shunt}=10 kΩ, R_s=0.5 Ω, and a fluorescent spectrum with an intensity of 10 Wm^{-2} [10]. Up to now, an IPV efficiency of 11% has already been demonstrated for a 9% CdTe cell (STC). GaAs with its direct bandgap of 1.44 eV suffers less from internal losses. Both materials are well suited for use in applications with mixed solar and artificial spectra, because the efficiency is similar for both lighting scenarios.

Crystalline silicon devices have often been proposed for indoor applications, because there are large production capacities and sufficient raw material SiO_2 at low cost. However, due to their bandgap of 1.12 eV the efficiency limit decreases to below 30%. In addition, being an indirect semiconductor, the efforts in order to reduce the shunt resistance losses are much higher than for GaInP as material of high purity is required. Thus, the cost of IPV Si devices will be above the ones for state-of-the-art commercial cells for outdoor applications. The current measured indoor record is 8.2% for a cell from Fraunhofer ISE. The limit calculated by Gemmer was 15% for R_{shunt}=100 kΩ, R_s=0.5 Ω, and a fluorescent spectrum with an intensity of 10 Wm^{-2} [10].

Amorphous Si is well established and cost effective for indoor and low irradiance applications, but suffers from internal recombination losses due to defect states. It was originally employed for use with light bulbs. The current indoor records are 7.1% for a commercial Schott Solar cell compared to 7.6% of a crystalline Si commercial device [9]. For real a-Si devices and R_{shunt}=10 MΩ, R_s=1 Ω, and a fluorescent spectrum with an intensity of 10 Wm^{-2}, Gemmer calculated a maximum efficiency of 14% [10]. For commercially available IPV devices, large spreads of measured efficiencies for different manufacturers using the same technology, and even amongst the same product series, have been reported [13, 27]. This renders cost-effective reliable design of IPV products difficult. A possible reason for these spreads might be found in the characterization methods also used by the manufacturer. The characterization of IPV devices is outlined in the following section.

8.5 Characterization and Measured Efficiencies

Outdoor solar cells are characterized under the conditions defined in IEC 60904–3, Ed. 2 (2008), sometimes referred to as Standard Test Conditions (STC) [28]. These conditions include a standard solar spectrum derived for an Air Mass Factor 1.5, and an intensity of 1000 W/m^2. As there are no characterization standards for indoor photovoltaic conditions, results obtained under STC are sometimes also found on datasheets for IPV products. As shown in the above sections, results from STC are of low predictive value for IPV. Another practice is to list efficiency parameters measured at different solar intensities. Provided that the intensity has been decreased to at least 10 Wm^{-2}, information on the voltage stability may be obtained. A practice including artificial spectra is the use of measurement results obtained under photometric conditions. Typically they employ a fluorescent tube of daylight spectrum, as they are the most common office lamps, and an illuminance of 200 Lux. Although this approach is close to the operation conditions of many IPV products, the method suffers from its low reproducibility. Provided a photovoltaic converter of a spectral response $SR(\lambda)$ and a short-circuit current proportional to the photon current, the spectral mismatch factor *MMF* for two different spectra E_{ref} and E_e can be calculated to be Eq. 8.12.

$$MMF = \frac{\int_0^\infty I_{sc,ref} E_{e,ref} \, \mathrm{d}\lambda}{\int_0^\infty I_{sc} E_e \, \mathrm{d}\lambda} \tag{8.12}$$

An illuminance E_v is obtained by evaluating irradiance with the human visibility function $V(\lambda)$ (Eq. 8.13),

$$E_V = K_m \int_{380nm}^{780nm} E_e V(\lambda) \, \mathrm{d}\lambda \tag{8.13}$$

where the luminous coefficient K_m is 683 lm/W. The spectrum of the light source used for the data and the spectral response of the luxmeter would therefore be required for reproducibility. Luxmeters are calibrated to Class A incandescent bulbs. The mismatch for other light sources is strongly depending on the applied filter of the luxmeter [19], therefore the type of luxmeter needs to be stated.

8.6 Irradiance Measurements

For outside irradiance measurements, the standard instruments are pyranometers and silicon sensors. Pyranometers are based on thermopiles, which are shielded with a glass dome. The typical spectral range is 280–4000 nm with a hemispheric field of view. These instruments are preferred due to their accuracy. Especially in operation monitoring of silicon plants, silicon sensors are preferred due to their comparable low cost. Both devices are available with temperature correction.

There are no standard instruments for characterization of indoor photovoltaic conditions. In principle, pyranometers can be used, provided a sufficient signal-noise-ratio is obtained. For applications, the most reliable method is the use of retraceable reference cells characterized under STC and spectral mismatch calculations following Eq. 8.12 (see section above).

8.7 Characterization

There are no characterization standards for IPV devices yet. Measurement results of photovoltaic devices need to be reproducible and retraceable to SI units, i.e., to the World Photovoltaic Scale [29]. Such a radiometric measurement campaign using two fluorescent tubes with daylight spectral distribution (OSRAM Lumilux 840/18 W daylight) and silicon reference cells for low intensity applications has been reported recently [9, 13]. The types of lamps had been chosen, because fluorescent tubes with daylight spectral distribution and energy saving lamps are the most common types of office lighting. High-efficiency monocrystalline silicon cells for low intensity applications were used as reference cells [11]. Their short-circuit current, I_{sc}, and their spectral response were determined at an intensity of 9.1 Wm^{-2}, the spectral distribution and further conditions following IEC 60904–3 Ed.2 (2008) [28]. The IPV measurement setup was then calibrated to the I_{sc} resulting from the spectral mismatch calculations following Eq. 8.12 and the fluorescent tube spectrum at 9.1 Wm^{-2}. The irradiance was regulated electrically, whereas the spectrum had been characterized with a spectroradiometer. The device temperature was kept at 25 ± 2°C, in order to be compliant with IEC 60904–3 Ed.2 (2008). The electrical characteristics of the IPV devices were measured with a Keithley 6514 Electrometer and a Keithley 2400 SourceMeter.

The efficiencies measured in this study are listed in Table 8.1. The relative change of efficiency compared to outdoor conditions matches the theoretical predictions. The highest efficiency measured was 16% for a GaInP cell. Organic photovoltaic samples consisting of P3HT:PCBM showed a maximum efficiency of 3.5%, which is in the range of some amorphous silicon devices. Silicon devices showed a relative decrease of efficiency of at least 50%,

Table 8.1 Photovoltaic efficiencies measured [9, 13] sorted by best indoor performance in descending order.

Material	Manufacturer	Type	Area [cm²]	η_{STC} [%]	η_{IPV} [%]
GaInP	ISE	prototype	16.00	14.5	15.8
CdTe	TU Darmstadt	prototype	1.56	9.5	10.9
Si	ISE	prototype	2.10	20.0	8.2
Si	IXYS	XOB 17–12x1	1.47[a]	13.0	7.6
a:Si:H	Schott Solar	ASI 2Oi07/ 037/022JJF	8.25	6.4	7.1
dye	ISE	prototype	74.27[b]	3.0	6.8
Si	EPS Soltec	prototype	9.00	10.0	5.3
a:Si:H	Sanyo	AM-1417 9 NSJM	4.55	3.2	4.4
Flexible	Flexcell	Gaia 3.6 A -0070	32.00	3.0	4.2
P3HT:PCBM	ISE/FMF	prototype	1.15	3.1	3.6
Flexible	Plastecs	SP 1.5–50–3	23.34	0.5	3.3

[a] active (total: 1.54)

[b] active (total: 99.97)

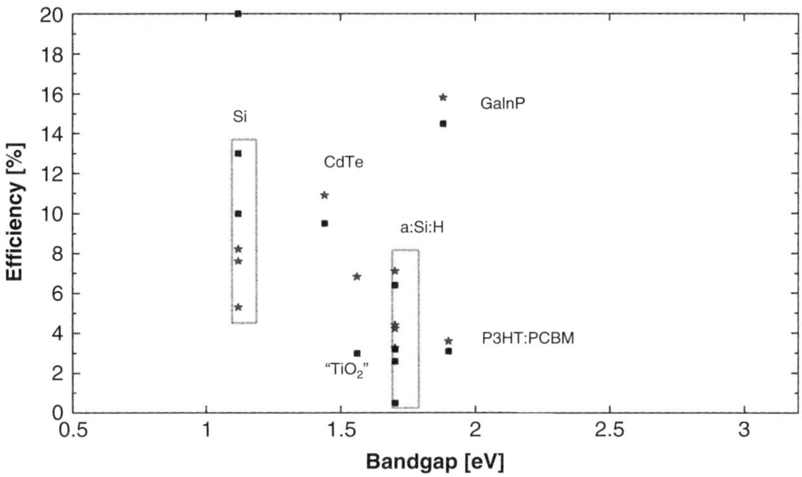

Figure 8.9 Typical bandgap and efficiencies measured [9,13] under fluorescent tube of daylight spectrum (OSRAM Lumilux 840/18 W, black squares), and under terrestrial solar radiation (AM 1.5, *IEC 60904-3*, grey stars).

while the dye-sensitized device more than doubled its performance. As IPV is a niche market, there are few commercial products in the centimeter scale available. For some technologies only low sample numbers consisting of research prototypes, some of which were designed for other purposes, such as Concentrated Photovoltaics, could be investigated in the study. Future photovoltaic devices of these technologies especially designed for IPV are therefore expected to outperform the results shown in Table 8.1. Also, for commercial products higher efficiencies seem feasible, i.e., matching the light source used for characterization during optimization processes. This is especially valid for some amorphous and crystalline silicon devices. Figure 8.9 shows the relative change of efficiency measured in the study in relation to the bandgaps. The energy gaps were assumed as follows: 1.12 eV for crystalline Silicon, 1.44 eV for CdTe, an effective bandgap of 1.62 eV for the dye-sensitized photovoltaic device, 1.7 eV for a: Si: H and flexible devices, 1.88 eV for GaInP, and 1.9 eV for P3HT: PCBM.

8.8 Conclusion and Outlook

Due to the development of narrow-banded indoor lighting technologies and energy-efficient electronic devices with average power consumption in the range of microwatts, indoor photovoltaics promises to become an exciting discipline in photovoltaics. The theoretical maximum efficiency of a photovoltaic converter for typical irradiance sources calculated in this chapter is close to 64% for E_g = 1.9 eV for a RGB LED, and 50% for E_g = 1.9 eV and a fluorescent tube.

The highest efficiencies measured under indoor conditions are 15.8% for a GaInP cell with a bandgap of 1.88 eV, and 10.9% for two CdTe cells. These prototypes had been originally designed for concentrated photovoltaic applications and standard outdoor applications, respectively. As IPV is a young photovoltaic field of research, future devices especially designed for IPV are expected to outperform these results. Depending on the choice of material, different optimization strategies have been presented in this chapter. Basic design strategies applying to all IPV applications include the matching of the shunt resistance to the irradiance level and of the bandgap to the spectrum mainly applied, respectively.

Currently, some commercial IPV devices show a large spread of efficiency even amongst the same product series. This can also be caused by characterization under conditions differing from typical IPV lighting, i.e., characterization under STC. Therefore, the field of IPV requires standard characterization methods retraceable to SI units. Typical indoor irradiance levels and a radiometric characterization procedure based on silicon photovoltaic cells for low irradiance applications and spectral mismatch calculations for fluorescent lighting have been presented in this chapter. The same method can be used for LED.

Traditionally, amorphous photovoltaic cells are applied for indoor applications. However, state-of-the-art P3HT:PCBM organic devices already reach or even exceed the demonstrated efficiencies of amorphous silicon in IPV. Depending on the required lifetime, organic photovoltaic can already be used in IPV. For long-term applications, Ga(In)P, CdTe, GaAs, and high quality Si are of interest.

The special opportunity inherent in IPV is the possibility to match both the radiation source and the corresponding photovoltaic converter, which minimizes spectral losses and enables a cost-effective system of high efficiency. Power beam applications can be an application example of a monochromatic emitter, where the efficiency calculated from the detailed balance model increases with bandgap. Due to the high intensities, the series resistance needs to be reduced in these systems, i.e., optimization of the front contact design.

In the second case, the lighting system of a building and the photovoltaic converter are designed at the same time, e.g., halogen lamps lighting a conference room. Depending on the required efficiency, the ideal photovoltaic material can be chosen. For applications with large contributions of diffuse radiation, e.g., hallway lighting consisting of LED covering a wall, fluorescent collectors based on fluorescent dyes or quantum dots with a matched photovoltaic converter will be of interest. For all these technologies, nanotechnology can be expected to contribute.

References

1. J.A. Conti, Electronic postage weighing scale, Pitney-Bowes, Inc., Apr. 11, 1978.
2. S.A. Roen, Solar powered portable calculator, Litton Business Systems, Inc., Apr. 12, 1977.
3. S. Hanson, M. Seok, Y.S. Lin, Z. Foo, D. Kim, Y. Lee, N. Liu, D. Sylvester, and D. Blaauw, *Journal of Solid-State Circuits*, Vol. 44, 1145–1155, 2009.
4. J.A. Paradiso, and T. Starner, *IEEE Pervasive Computing*, Vol. 4, 18–27, 2005.
5. S. Roundy, P.K. Wright, and J.M. Rabaey, *Energy Scavenging for Wireless Sensor Networks: With Special Focus on Vibrations*, Kluwer Academic Publishers, 2004.
6. W.S. Wang, T. O'Donnell, L. Ribetto, B. O'Flynn, M. Hayes, C. O'Mathuna, *Energy Harvesting Embedded Wireless Sensor System for Building Environment Applications*,IEEE,http://dx.doi.org/10.1109/WIRELESSVITAE.2009.5172418 New York, 2009.
7. N.W. S. Beeby, *Energy Harvesting for Autonomous Systems*, Artech House, London, 2010.
8. N.R.e. al., Weak Light performance and spectral response of different solar cell types, in: 20th European Photovoltaic Solar Energy Conf., Barcelona, Spain, pp. 2120–2123, 2005.
9. M. Freunek, M. Freunek, and L.M. Reindl, *Photovoltaics, IEEE Journal of*, Vol. 3, 59–64, 2013.
10. C.E.M. Gemmer, *Analytische und numerische Untersuchungen von Solarzellen unter wechselnden Beleuchtungsbedingungen*, Der Andere Verlag, Osnabrueck, Germany, 2003.
11. S.W. Glunz, J. Dicker, M. Esterle, M. Hermle, J. Isenberg, F.J. Kamerewerd, J. Knobloch, D. Kray, A. Leimenstoll, F. Lutz, D. Osswald, R. Preu, S. Rein, E. Schaffer, C. Schetter, H. Schmidhuber, H. Schmidt, M. Steuder, C. Vorgrimler, G. Willeke, *High-efficiency silicon solar cells for low-illumination applications*, New York, 2002.
12. M. Hermle, J. Dicker, W. Warta, S.W. Glunz, and G. Willeke, *Analysis of edge recombination for high-efficiency solar cells at low illumination densities*, Wcpec-3 Organizing Committee, Tokyo, 2003.
13. M. Müller, *Energieautarke Mikrosysteme am Beispiel von Photovoltaik in Gebäuden*, Der Andere Verlag, Osnabrueck, Germany, 2010.

14. J.F. Randall, and J. Jacot, *Renewable Energy*, Vol. 28, 1851–1864, 2003.
15. N.H. Reich, W. van Sark, E.A. Alsema, R.W. Lof, R.E.I. Schropp, W.C. Sinke, and W.C. Turkenburg, *Solar Energy Materials and Solar Cells*, Vol. 93, 1471–1481, 2009.
16. Photovoltaische Energieversorgung für Geräte im kleinen und mittleren Leistungsbereich, Fraunhofer Institut für Solare Energiesysteme. Freiburg, Breisgau, 1991.
17. A. Virtuani, E. Lotter, and M. Powalla, *Solar Energy Materials and Solar Cells*, Vol. 90, 2141–2149, 2006.
18. M. Muller, J. Wienold, W.D. Walker, and L.M. Reindl, Characterization of indoor photovoltaic devices and light, *34th IEEE Photovoltaic Specialists Conference, Vols 1–3*, New York, pp. 491–496, 2009.
19. M.H. Müller, H.; Walker, W.D.; Reindl, L.M. , Simulations and Measurements for Indoor Photovoltaic Devices, in: 24th European Photovoltaic Solar Energy Conference, Hamburg, Germany, 2009.
20. H.J.Q. W. Shockley, *J. Appl. Phys.* , Vol. 32, 1961.
21. M.B. Prince, *Journal of Applied Physics*, Vol. 26, 534–540, 1955.
22. H.J. Hovel, *Solar cells*, Academic Press, New York, 174, 1975.
23. A. Goetzberger, and W. Greubel, *Applied Physics*, Vol. 14, 123–139, 1977.
24. N. Beattie, T. Themistokleous, G. Zoppi, I. Forbes, and R. Miles, Vol., 2012.
25. W.G.J.H.M. van Sark, *Renewable Energy*, Vol. 49, 207–210, 2013.
26. J.V. Holm, H.I. Jørgensen, P. Krogstrup, J. Nygård, H. Liu, and M. Aagesen, *Nat Commun*, Vol. 4, 1498, 2013.
27. N. Reich, W. van Sark, E. Alsema, S. Kan, S. Silvester, A. van der Heide, R. Lof, and R. Schropp, *20th European Photovoltaic Solar Energy Conference and Exhibition*, Vol. 10, 2005.
28. IEC, Photovoltaic devices - Part 3: Measurement principles for terrestrial photovoltaic (PV) solar devices with reference spectral irradiance data, in, 2008.
29. C.R. Osterwald, S. Anevsky, K. Bücher, A.K. Barua, P. Chaudhuri, J. Dubard, K. Emery, B. Hansen, D. King, J. Metzdorf, F. Nagamine, R. Shimokawa, Y.X. Wang, T. Wittchen, W. Zaaiman, A. Zastrow, and J. Zhang, *Progress in Photovoltaics: Research and Applications*, Vol. 7, 287–297, 1999.

9

Photon Management in Rare Earth Doped Nanomaterials for Solar Cells

Jiajia Zhou[1], Jianrong Qiu[1,2,*]

[1]State Key Laboratory of Silicon Materials, Zhejiang University, Hangzhou, China
[2]State Key Laboratory of Luminescent Materials and Devices, South China University of Technology, Guangzhou, China

Abstract
In the last decade, innovations in photonic material design and large-area nanostructure fabrication technique have brought a new era of high-efficiency solar cells. Photon management approach containing up-conversion (UC) and down-conversion (DC) has received considerable attention as an efficient way to improve the photoelectric conversion efficiency of solar cells by breaking through the Shockley-Queisser efficiency limits. In this chapter, we introduce the basic aspects of the photon management approach, typical experimental results, and discuss the suitability and existing problems of UC and DC nanomaterials for solar cells.

Keywords: Up-conversion, down-conversion, rare earth, nanomaterials, solar cells

9.1 Introduction

Solar cells are devices that absorb energy from sunlight to produce electricity and are regarded as a key technology for a sustainable energy supply. Due to the increased demand for renewable energy sources, the manufacturing of solar cells has advanced considerably in recent years. However, it is well known that the low photoelectric conversion efficiency has limited the extensive use of solar cells

**Corresponding author*: qjr@zju.edu.cn

Atul Tiwari, Rabah Boukherroub, and Maheshwar Sharon (eds.) Solar Cell Nanotechnology, (223–240) 2014 © Scrivener Publishing LLC

up to now. There are many types of approaches to improve the conversion efficiency, in which the photon management strategy based on the manipulation of the energy distribution of solar irradiation shows various advantages such as ease of implementation in solar cells without structural change, environmental friendliness, and low cost, etc. Benefiting from the abundant energy levels of rare earth (RE) ions, it is convenient to fabricate RE-doped nanomaterials for enhancement of conversion efficiency of solar cells. In this chapter, we would like to introduce the basic aspects of the photon management approach for solar cells and discuss the suitability of the up-conversion (UC) and down-conversion (DC) nanomaterials for this goal.

9.2 Basic Aspects of Solar Cell

9.2.1 Mechanism of Efficiency Limitation

Shockley and Queisser applied the concept of detailed balancing to estimate the efficiency limits of solar cells for the first time [1]. The principle of detailed balance dictates that at thermal equilibrium, every photon absorption event must be countered by a photon emission event, with the balance holding at every frequency and solid angle. As the solar cells always operate far from thermal equilibrium, the principle of detailed balance cannot be directly applied to it. However, Shockley and Queisser recognized that the emission spectrum away from thermal equilibrium is different from the emission spectrum at equilibrium by only a scaling factor, and this recognition aided the understanding of fundamental efficiency limits of solar cells. Based on the calculations, the Shockley-Queisser maximum theoretical efficiency was found to be 30% for an energy gap of 1.1ev and a radiative recombination fraction (fc) of 1. The upper limit of the conversion efficiency of solar cell is caused by the intrinsic energy loss among the generation, trapping, recombination, and transport of electron-hole pairs throughout the semiconducting material and within the contact electrodes after photon incidence. A typical p-n junction band diagram with related energy loss mechanisms is shown in Figure 9.1 [2]. The two major loss mechanisms that need to be overcome to improve the photoelectric efficiency are lattice thermalization (process ①) and transparency to sub-band gap photons (process ②). Further loss mechanisms

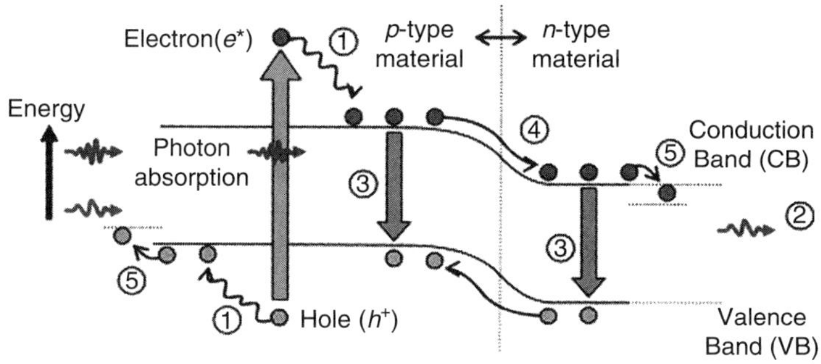

Figure 9.1 Loss processes in a single-junction solar cell: (1) lattice thermalization loss, (2) transparency, (3) recombination loss, (4) junction loss, and (5) contact voltage loss [2].

contain recombination of photoexcited e-h pairs (process ③), junction loss (process ④) and contact voltage loss (process ⑤).

9.2.2 EQEs of Solar Cells

All of the above mentioned energy loss processes constrain the response of each solar cell in a featured wavelength region in accordance with the bandgap of the absorbing material. The wavelength-dependent response of a device can be described by its external quantum efficiency (EQE), defined as the ratio of the number of electron-hole pairs generated to the number of photons incident on the front surface of the cell.

The normalized EQE spectra of typical single-junction cells are shown in Figure 9.2 and Figure 9.3 [3]. In Figure 9.2, the narrow bandgap solar cells with the absorbing materials like $CuInGaSe_2$ (CIGS), multicrystalline silicon (mc-Si), monocrystalline silicon (c-Si), etc., show their featured response to the wavelength range from 300 nm to 1200 nm. "Miasole CIGS" presents the CIGS module with 13.8% efficiency per m^2 fabricated by Miasole and measured by National Renewable Energy Laboratory (NREL) [4]. "Kyocera mc-Si" presents a multicrystalline silicon module with an efficiency record of 17.3% for 1.3 m^2 aperture area fabricated by Kyocera and measured by the Japanese National Institute of Advanced Industrial Science and Technology (AIST). "Mitsubishi mc-Si" also relates to multicrystalline silicon cell, where an efficiency increase to 19.3% has been confirmed by AIST for 218 cm^2 area fabricated by Mitsubishi Electric [5]. Another result of "ZSW

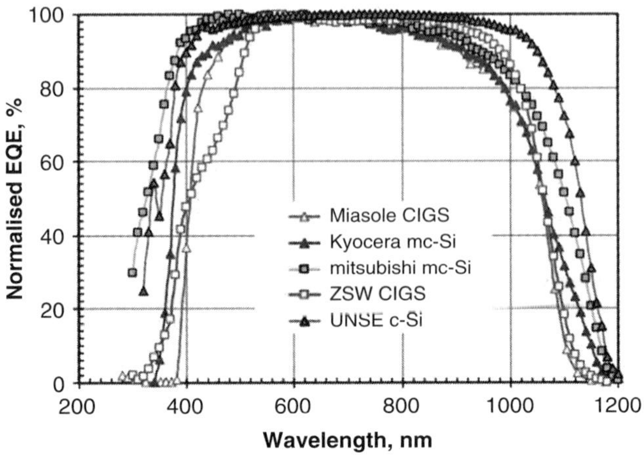

Figure 9.2 Normalized EQE spectra of five narrow bandgap solar cells, which were fabricated by different organizations [3].

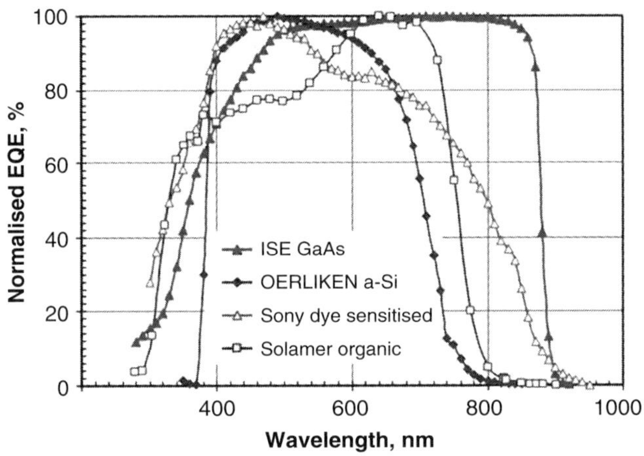

Figure 9.3 Normalized EQE spectra of several wide bandgap solar cells, which were fabricated by different organizations [3].

CIGS" is the improvement of a small area (0.50 cm^2) CIGS cell fabricated by Zentrum für Sonnenenergie-und Wasserstoff-Forschung (ZSW), Stuttgart to 20.1% efficiency as measured by Fraunhofer Institute for Solar Energy Systems (FhG-ISE). "UNSW c-Si" presents the monocrystalline cell with an efficiency approaching 25% as fabricated and measured by the University of New South Wales (UNSW) [6]. The wide bandgap solar cells, including GaAs, amorphous Si, dye-sensitized and organic solar cells, also can work well in the whole visible region, but their absorption is generally limited to a maximum wavelength of ~900 nm (Figure 9.3).

9.2.3 Photon Management Approaches to Enhance the Efficiency of Solar Cell

The absorption materials of solar cells limit their usable EQE in a certain wavelength region. In the face of such a problem, photon management approaches containing up-conversion, down-conversion and down-shifting have been proposed for use in wide and efficient solar energy harvesting [2, 7, 8]. Figure 9.4 plots the AM1.5G spectrum, the fraction of this light that can be absorbed by a bulk Si device, and the fractions available for up-conversion and down-conversion. Silicon cells with a bandgap of 1.12 eV own a response which approaches the ideal in the intermediate wavelengths from 400 nm to 1100 nm, as shown in Figure 9.5. At short wavelengths below 400 nm, the covered glass absorbs most of the light and the cell response is very low. Hence, DC and DS, which offer the wavelength shifting ability from high frequency to low, could be used to overcome poor solar cell performance to UV/blue light. However, in general, benefiting from the DC principle that two-photon emission occurs for a single-photon absorption and the efficiency approaches 200%, the maximum cut-off wavelength of DC for silicon solar cells is always set to 550 nm according to double energy of Si bandgap, as labeled with shadow in visible

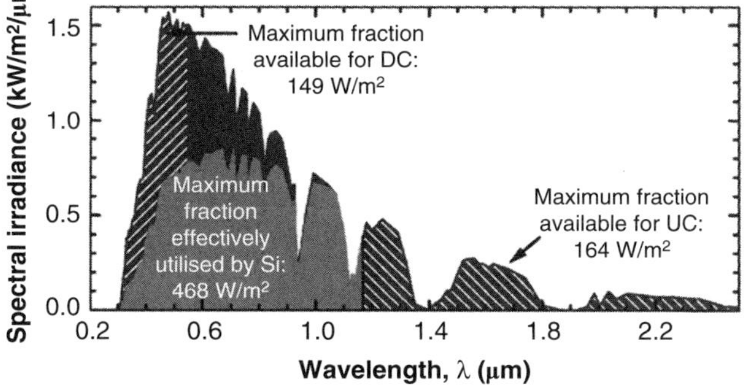

Figure 9.4 AM1.5G spectrum showing the fraction that is currently absorbed by a thick silicon device and the additional regions of the spectrum that can contribute towards up-and-down conversion [2].

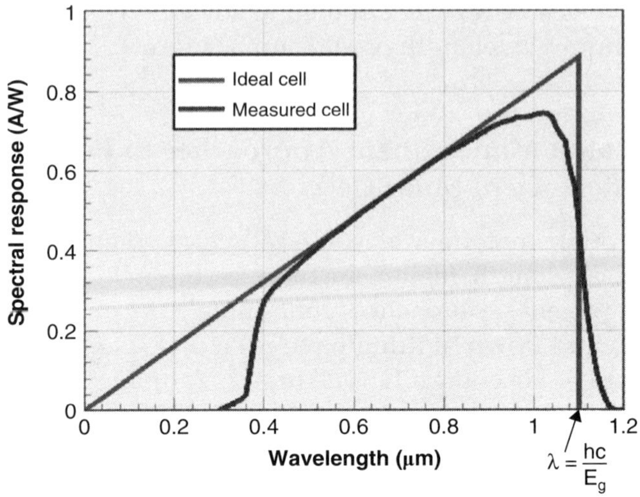

Figure 9.5 The ideal and measured spectral response of a silicon solar cell under glass [9].

region in Figure 9.4. At long wavelengths, the responses of Si cell fall back to zero, which means the optical transparent property of silicon in the wavelength region longer than 1200 nm. To overcome this limitation, UC is available through absorbing the light above 1100 nm and converting it into emission around 1000 nm. It should be noted that compared to conventional c-Si type of narrow bandgap solar cells, wide bandgap solar cells would benefit much more from incorporation of an up-converting layer due to the dominant transmission losses. Specifically, up to now, most of the up-converting layers were based on Yb^{3+} absorbing (around 1000 nm) and were followed with emission in wavelength region shorter than 900 nm, which is more appropriate for wide bandgap cells according to their EQE spectra (Figure 9.3).

9.3 Up-Conversion Nanomaterials for Solar Cell Application

9.3.1 Principles of Photon Up-Conversion

Up-conversion is known as an anti-Stokes process coupled with simultaneous two-photon absorption (TPA) and second-harmonic generation (SHG) [10]. In general, UC could be defined as a process

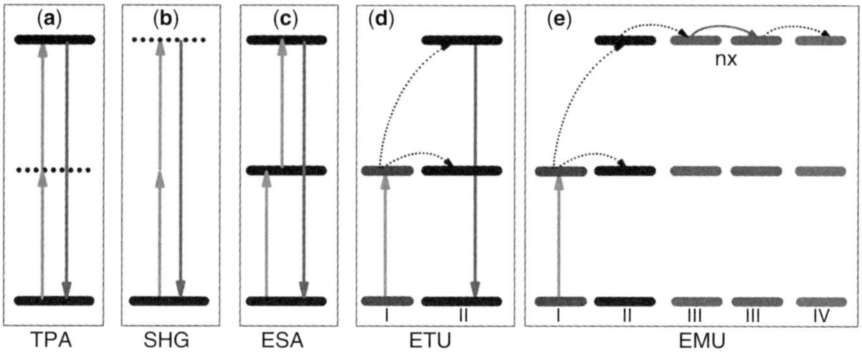

Figure 9.6 Simplified energy level diagrams depicting the anti-Stokes processes.

that refers to a system absorbing two or more photons, combining their energies, and emitting one higher energy photon. Typical classification of UC includes excited state absorption (ESA) and energy transfer up-conversion (ETU) through the use of physically existing intermediary energy states of RE ions [10, 11]. Recently, a new UC process named energy migration up-conversion (EMU) was proposed by Wang *et al.* which involves the use of four types of RE ions and a core-shell design [12]. All of the above mentioned anti-Stokes processes have been simplified and described in Figure 9.6 [12].

9.3.2 Spectroscopy Analysis and Application Demonstration

Trivalent erbium (Er^{3+}) is an ideal candidate for single wavelength NIR up-conversion due to its ladder of nearly equally spaced energy levels that are multiples of the $^4I_{15/2} \rightarrow {}^4I_{13/2}$ (~1540 nm) transition, as shown in Figure 9.7(a) [13, 14]. $Gd_2(MoO_4)_3$: Er^{3+} nanophosphors, Figure 9.7(b) exhibits its SEM image, and have been proposed as potential luminescent materials to enhance silicon solar-cell NIR response by Zhang *et al.* [15]. Upon excitation with low-energy near-infrared photons provided by a tunable laser ranging 1510–1565 nm, intense up-converted emissions at 545, 665, 800, and 980 nm have been achieved with conversion efficiencies of 0.12%, 0.05%, 0.83%, and 1.35%, respectively. In Figure 9.7(c), colloidal $LiYF_4$: Er^{3+} nanocrystals (NCs) with bright green emission upon 1490 nm excitation have been reported by Prasad *et al.* [16]. The total up-conversion quantum yield (UCQY) has been measured to be as high as $1.2 \pm 0.1\%$, which is almost 4 times higher than the UCQY for the most efficient UC NCs

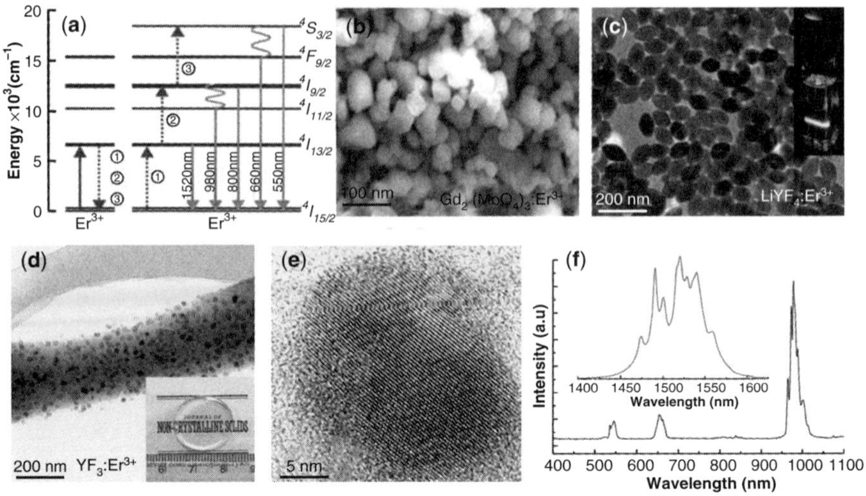

Figure 9.7 (a) Three-step UC process between two Er^{3+} ions. (b) SEM image of $Gd_2(MoO_4)_3:Er^{3+}$ nanophosphors. (c) TEM image of $LiYF_4$:10% Er^{3+} NCs; inset shows the digital photograph of UC in 1 wt% colloidal $LiYF_4$:10% Er^{3+} NCs under unfocused laser excitation at 1490 nm of 4 W/cm^2. (d) TEM and (e) HRTEM images of Er^{3+}: YF_3 NCs embedded transparent glass; inset of (d) shows the digital photograph of the bulk sample. (f) UC emission spectrum of Er^{3+}: YF_3 NCs under 1530 nm laser excitation; inset shows the absorption spectrum of NCs in the wavelength range of 1400–1650 nm.

reported to date. Transparent glass containing YF_3 NCs, which could be clearly observed by transmission electron microscopy (TEM) and high resolution TEM as shown as Figure 9.7(c,d), also has been investigated as Er^{3+} host to generate up-converted emissions [17]. The typical emission spectrum of $YF_3:Er^{3+}$ NCs under 1530 nm laser excitation and the absorption spectrum of NCs in the wavelength range of 1400–1650 nm are shown in Figure 9.7(f) [17].

The large spectral overlap between Yb^{3+}/Er^{3+} and the efficient ETU for this lanthanide couple, makes Yb^{3+}/Er^{3+} a widely investigated up-conversion couple in a variety of host materials. The Yb^{3+} ion absorbs around 980 nm and transfers the energy from the $^2F_{5/2}$ level to the $^4I_{11/2}$ level of Er^{3+}. Subsequent energy transfer from a second excited Yb^{3+} ion to Er^{3+} ($^4I_{11/2}$), excites Er^{3+} ion to the $^4F_{7/2}$ excited state. After multiphonon relaxation to the lower lying $^4S_{3/2}$ and $^4F_{9/2}$ states, green and red emissions are observed. In 2010, Shan and Demopoulos reported for the first time the application of an Yb^{3+}/Er^{3+} co-doped LaF_3-TiO_2 layer to fabricate a "triple layer" working electrode on fluorine-doped tinoxide (FTO) glass for a

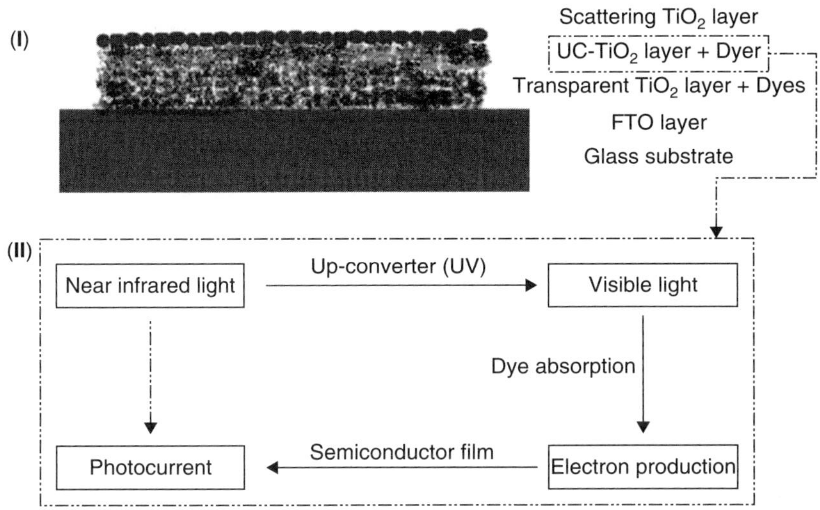

Figure 9.8 The structure (II) and mechanism of UC-TiO_2-based work electrode [18].

dye-sensitized solar cell (DSSC), as shown in Figure 9.8. In their investigation, 0.40 V of Voc and 0.036 mA of Isc were obtained under illumination of a 980 nm fiber laser (fiber core diameter: 105 μ m) with 2.5 W power supply [18]. In 2013, Liang *et al.* for the first time employed the highly uniform bifunctional core/double-shell-structured b-$NaYF_4$: Er^{3+},Yb^{3+}@SiO_2@TiO_2 hexagonal sub-microprisms for high-performance dye-sensitized solar cells [19]. Due to the scattering effect of the sub-micrometer dimension and the surface protecting effect of the amorphous SiO_2 insulating layer, at a certain mixing mass ratio of 10%, an efficiency of 8.65% was obtained, which is 120% higher than that of a device based on bare $NaYF_4$:Er^{3+}, Yb^{3+} crystals, and an enhancement of about 10.9% was achieved when compared to the reference cell. Furthermore, to enhance the NIR light harvesting, the first demonstration of photoelectrochemical electrodes with up-conversion nanoparticles embedded in porous photonic crystals, which render the nanoparticles spatially positioned in close proximity to the CdSe quantum dots, was reported by Su *et al.* [20]. In their report, the hetero-nanostructured photoanode exhibits a photocurrent of 0.02 mA upon near-infrared laser excitation.

Recently, for more efficient improvement on solar cell efficiency by the up-conversion approach, Zou *et al.* reported that using an organic near-infrared dye as an antenna for the b-$NaYF_4$:Yb, Er nanoparticles could dramatically enhance (by a factor of ~3,300) the up-conversion as a result of increased absorptivity and overall broadening of the absorption spectrum of the up-converter [21]. Consequently, a tendency named "up-conversion goes broadband," which promotes the nanofabrication of rare earth ions doped structure, was proposed by Liu *et al.* [22]. Er^{3+}:$NaGdF_4$@Ho^{3+}:$NaGdF_4$@$NaGdF_4$ active-core/active-shell/inert-shell NCs, which can extend the NIR wavelength excitable range for UC emissions, were successfully fabricated for the first time by Chen *et al.* [23]. However, for the above mentioned methods for broadening of the absorption spectrum, intrinsic shortages exist. For example, the poor photostability of dye and the unavoidable detrimental energy transfer between co-doped rare earth ions. Thus, a new approach named multi-color excitation based on the GSA, ESA, and phonon-coupled absorption of RE ions was proposed by our group, which not only broadened the absorption spectrum of up-converter but also greatly enhanced the up-conversion due to the most efficient ETU mechanism.

9.4 Down-Conversion Nanomaterials for Solar Cell Application

9.4.1 Principles of Photon Down-Conversion

Down-conversion, which is also named as quantum cutting, is well known through the photon conversion in Eu^{3+}-doped $LiGdF_4$, where two visible photons can be emitted by Eu^{3+} through an efficient two-step energy transfer from Gd^{3+} to Eu^{3+} with a quantum efficiency that approaches 200% upon excitation of Gd^{3+} with a high-energy photon [24]. In principle, DC is a process related to the phosphor that could generate two or more low-energy photons for each incident high-energy photon absorbed. For solar cell, the absorption of the two or more low-energy photons emitted by the DC converter, which is usually located on the front surface of a conventional single-junction solar cell, leads to the generation of more than one electron-hole pair in the solar cell per incident high-energy photon [8].

The concept of DC is illustrated (Figure 9.9) with two types of ions, I and II, with hypothetical energy level schemes [24]. Type I is an ion for which emission from a high energy level can occur. Type II is an ion to which energy transfer takes place. For NIR quantum cutting applied for solar cells, the type II ion usually is focused on Yb^{3+} ion, and the type I ion as a donor should be with a level that matches with Yb^{3+}' level, i.e., an energy level with twice the energy difference

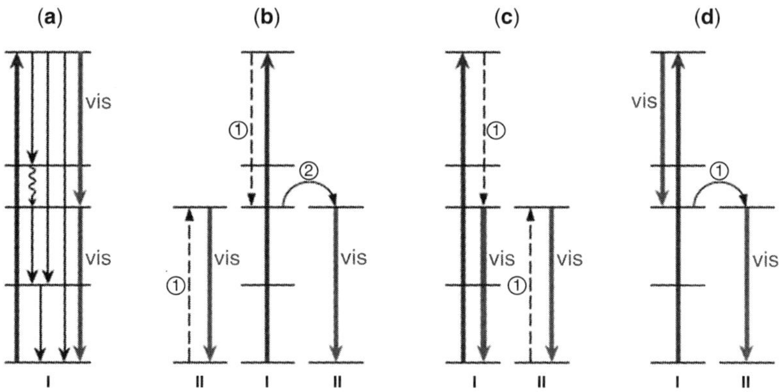

Figure 9.9 Energy level diagrams for two (hypothetical) types of lanthanide ions (I and II) showing the concept of down-conversion. Type I is an ion for which emission from a high energy level can occur. Type II is an ion to which energy transfer takes place. (a) Quantum cutting on a single ion I by the sequential emission of two visible photons. (b) The possibility of quantum cutting by a two-step energy transfer. In the first step (indicated by ①), a part of the excitation energy is transferred from ion I to ion II by cross-relaxation. Ion II returns to the ground state by emitting one photon of visible light. Ion I is still in an excited state and can transfer the remaining energy to a second ion of type II (indicated by ②), which also emits a photon in the visible spectral region, giving a quantum efficiency of 200%. (c) and (d)The remaining two possibilities involve only one energy transfer step from ion I to ion II. This is sufficient to obtain visible quantum cutting if one of the two visible photons can be emitted by ion I [24].

to $^2F_{5/2} \rightarrow {}^2F_{7/2}$ transition of Yb^{3+}. There are two types of donor including those with and without an intermediate level at approximately 10000 cm^{-1}, which is in accordance with the first-order and second-order energy transfer, respectively. Evaluation of the Dieke diagram reveals that potential donors are Pr^{3+}, Nd^{3+}, Ho^{3+}, Er^{3+}, Tb^{3+}, and Tm^{3+} ions, in which the first four have the intermediate levels. Also, Eu^{2+}, Ce^{3+}, Yb^{2+} ions are new kinds of donors with f-d transition nature, which exhibit broad absorption bands in the UV/VIS region and ligand controlled bandgap [25–29]. The energy transfer between Eu^{2+}, Ce^{3+}, and Yb^{2+} ions and Yb^{3+} usually takes place through the cooperative down-conversion mechanism due to the lack of intermediate level form NIR to UV of these donors.

9.4.2 Experimental and Spectroscopy Analysis

NIR down-conversion with prospects for increasing the energy efficiency of crystalline Si solar cell was first reported by Vergeer *et al.* in $Yb_xY_{1-x}PO_4:Tb^{3+}$, where luminescence decay curves of the $^5D_4$ emission from Tb^{3+} show an excellent agreement with simulations based on cooperative energy transfer via dipole-dipole interaction [30]. In 2008, Ye *et al.* demonstrated the NIR quantum cutting with the maximum quantum efficiency approaching 155% in Tb^{3+}, Yb^{3+} co-doped transparent glass ceramics containing CaF_2 nanocrystals (inset of Figure 9.10(a)) [31]. As shown in Figure 9.10(a), the

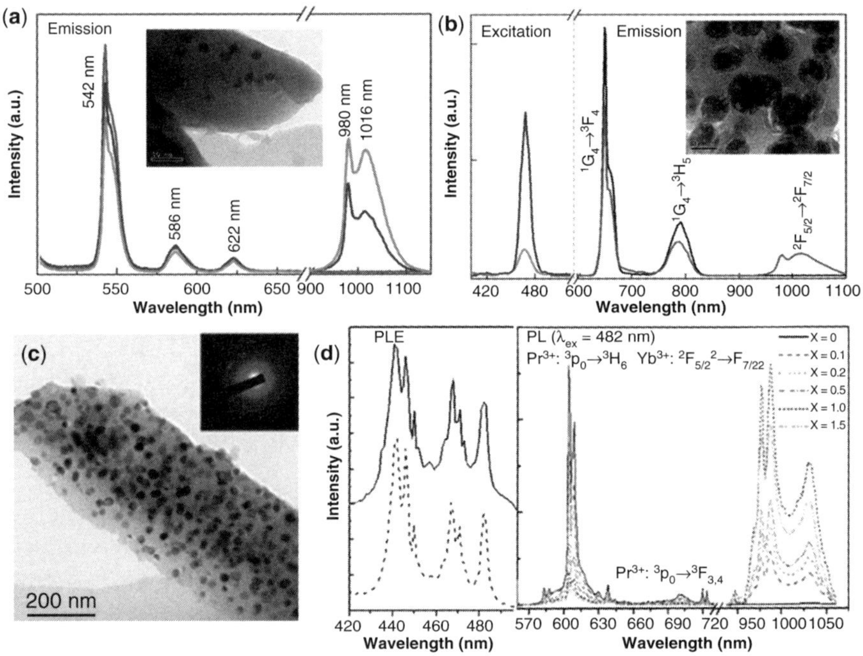

Figure 9.10 (a) Emission spectra of glass ceramics with different Yb^{3+} concentrations under excitation at 484 nm (solid line in red: 0 mol % Yb^{3+}, blue: 4 mol % Yb^{3+}, green: 6 mol % Yb^{3+}). Inset shows the TEM image of glass ceramic containing CaF_2 nanocrystals. (b) Left: Excitation spectra of Tm^{3+} 651 nm emission monitored in $0.5Tm^{3+}$ single doped glass ceramic (blue line) and of Yb^{3+} 1016 nm emission monitored in $0.5Tm^{3+}$-$8Yb^{3+}$ co-doped glass ceramic (green line). Right: Emission spectra of $0.5Tm^{3+}$ single doped (black linne) and $0.5Tm^{3+}$-$8Yb^{3+}$ co-doped (red line) glass ceramics under the excitation of 468 nm. Inset shows the TEM image of glass ceramic containing LaF_3 nanocrystals. (c) TEM image and the corresponding selected area electron diffraction pattern of the glass ceramic containing β-YF_3 nanocrystals. (d) Photoluminescence excitation (PLE) spectra of the Pr^{3+}: $^3P_0 \rightarrow {}^3H_6$ emission (605 nm, dotted) and the Yb^{3+}: $^2F_{5/2} \rightarrow {}^2F_{7/2}$ emission (976 nm, solid) in the glass ceramic; Visible-NIR PL spectra of the glass ceramics with different Yb^{3+} concentrations (x%) on excitation at 482 nm.

emission bands of Tb^{3+}: $^5D_4 \rightarrow {}^7F_J$ transitions decreased with increasing Yb^{3+} concentrations from 0, 4, to 6 mol % under 484 nm excitation, while a sharp peak at 980 nm and a broader peak at 1016 nm of Yb^{3+}: $^2F_{5/2} \rightarrow {}^2F_{7/2}$ transition increased. A similar cooperative DC process with a maximum quantum efficiency value of 162% was also observed in Tm^{3+}-Yb^{3+} codoped glass ceramics containing LaF_3 nanocrystals (inset of Figure 9.10(b)) [32]. Figure 9.10(b) depicts the typical excitation and emission spectra of Tm^{3+}-Yb^{3+} ion pair with energy transfer. More efficient DC with optimal quantum efficiency close to 200% involving the emission of two near-infrared photons for each blue photon absorbed was realized by Chen *et al.* in transparent glass ceramics with embedded Pr^{3+}/Yb^{3+}: b-YF_3 nanocrystals (inset of Figure 9.10) [33]. In the PLE spectra shown in Figure 9.10(d), intense excitation bands centered at 482, 467, and 441 nm corresponding to Pr^{3+}: $^3H_4 \rightarrow {}^1I_6$, ^3P_j(j=0,1,2) transitions were measured by monitoring both the Pr^{3+}: $^3P_0 \rightarrow {}^3H_6$ transition at 605 nm and the Yb^{3+}: $^2F_{5/2} \rightarrow {}^2F_{7/2}$ transition at 976 nm, verifying the existence of energy transfer from Pr^{3+} to Yb^{3+}. PL spectra show the Yb^{3+} concentration dependent emission bands of Pr^{3+} in visible region and Yb^{3+} in NIR region.

9.4.3 Evaluation

EQE evaluation of down-converter covered solar cell wafers was first demonstrated by our group in Pr^{3+}-Yb^{3+} ion pair [34]; $0.4Pr^{3+}/1Yb^{3+}$ (mol %) doped oxyfluoride glass ($60SiO_2$–$20Al_2O_3$–$20CaF_2$) and host glass with size of 2 cm×2 cm×2 mm are covered on Si solar cell wafers. The EQE values of the cell covered by the doping glass are lower than that of cell covered by the host glass in the visible and NIR regions, as shown in Figure 9.11(a). From the difference of the two curves shown in Figure 9.11(b) it was easy to see that absorption of the light due to the doping ions is dominant rather than the expected improvement of EQE in the absorption wavelength region, due to DC of rare earth ions. Quantitative analysis of EQE values, absorbance and solar irradiation flux in various wavelength regions (Figure 9.11(c)) implied that the DC process may contribute to the increase in the EQE, though the increasing effect is not high enough to compete with the absorption effect. Unfortunately, there are no more positive results that report down-conversion of rare earth ion pair greatly improve the solar cell EQE at this time.

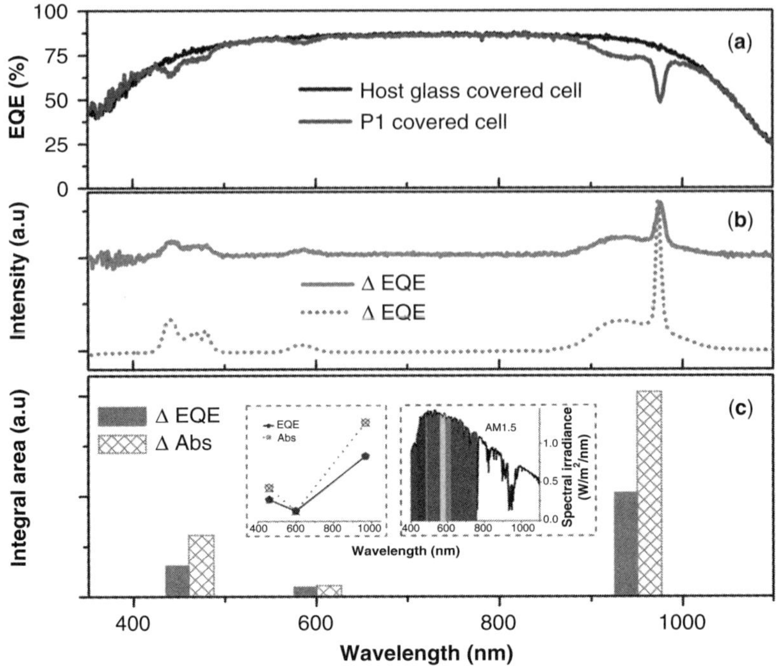

Figure 9.11 (a) EQE values of silicon solar cell covered with the host glass and P1(0.4Pr^{3+}/1Yb^{3+} doped glass). (b) The difference of the EQE values of silicon solar cell covered with the host glass and P1, and the absorbance difference between the host glass and P1. (c) The integral areas of the EQE difference and the absorbance difference in three different regions: 410–510 nm, 550–650 nm and 850–1100 nm. The insets give the quantitative integral areas normalized to the middle region and the solar spectrum (AM1.5) [34].

9.5 Conclusion and Outlook

In this chapter, fundamentals of the photon management approaches relying on up-conversion and down-conversion mechanisms for solar cells have been presented. Rare earth ions doped nanomaterials, including nanocrystals and NCs embedded glass ceramics, have been used as spectral converters owing to their advantages like easy structuralization with different matrixes and weak particle scattering. Spectroscopy analysis of up-conversion implied that it is efficient to convert the NIR energy to visible energy with the help of Yb^{3+} sensitization. As a result, in their studies researchers favor using Yb^{3+}-Er^{3+} doped up-converter to enhance the efficiency of wide bandgap solar cells, especially for dye-sensitized solar cells. One of the major challenges in using up-converter for solar cell is how to change the excitation source from intense laser light to natural solar light. Down-conversion, which is regarded as an efficient approach for using the above high-energy bandgap, has been widely investigated in spectroscopy from narrow-band sensitization with f-f transition RE ions to broadband sensitization with f-d transition RE ions. However, little effort has been given to the practical demonstration of a down-converter embedded in a solar cell. In an attempt to achieve the practical application purpose of photon management in rare earth ions doped nanomaterials, material form and efficiency of the spectral conversion process is important. Thus, we propose that the following issues be considered in future work.

9.5.1 Solution-Processable Nano-Coating for Broadband Up-Converter or Down-Converter

The thickness of the converter, which affects the energy loss in the optical transmission process, plays an important role in the efficiency of the spectral conversion. Solution-processable nano-coating is considered to be a possible way to control the thickness of the converter. For broadband UC or DC, dyes or quantum dots could be introduced as the sensitizer for rare earth ions. In 2010, Yan *et al.* successfully synthesized the CdSe/$NaYF_4$:Yb, Er nanoheterostructures using a seeded-growth method [35]. Benefiting from the strong coupling between CdSe and $NaYF_4$ in the nanoheterostructures, as shown in the TEM images in Figure 9.12, sufficient

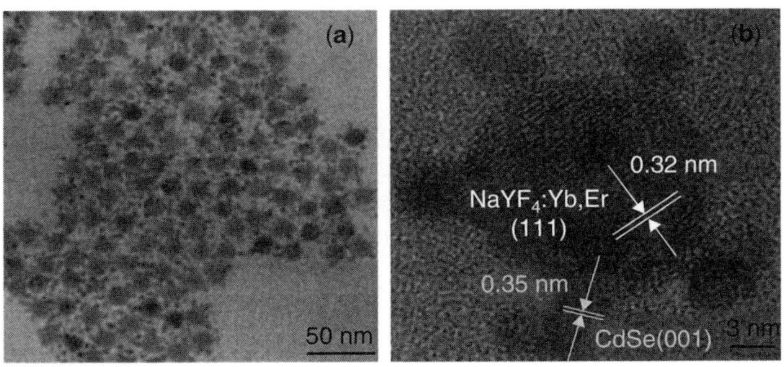

Figure 9.12 (a) TEM and (b) HRTEM images of CdSe/$NaYF_4$:Yb, Er nanoheterostructures [35].

energy transfer from rare earth ions to quantum dots was observed. This implies that such nanoheterostructures may also prove useful in the realization of broadband down-converter owing to the strong absorbance of CdSe in UV/Visible region and the energy transfer possibility between quantum dots and REs. Besides, these nano-sized structures obtained in wet chemical methods are easy to be surface modified and finally dispersed in various solutions.

9.5.2 Efficient Photon Management Using Nanoplasmonic Effect

It is well known that metal nanostructures (i.e. plasmonics substrates) can efficiently collect light and enhance the light intensity in their vicitniy due to surface plasmon resonances, and these effects have been widely used for enhancing various optical processes, such as the Raman scattering, down-conversion luminescence, and up-conversion luminescence [36–45]. For instance, in 2012, Zhang *et al.* reported a 310-fold UC enhancement uniformly over a large area, and an 8-fold reduction in the luminescence decay time were observed in $NaYF_4:Yb^{3+}/Er^{3+}$ co-doped nanocrystals using a 3D plasmonic nanoantenna architecture: disk-coupled dots-on-pillar antenna array [45]. Similarly, Li *et al.* have also investigated the UC nanomaterials with Au nanostructures applied to flexible amorphous silicon solar cells, where 16-fold to 72-fold improvement of the short-circuit current under 980 nm illumination compared to a cell without up-converters were obtained [46]. Thus, it is possible to improve the luminescence efficiency of the spectral converters using nanoplasmonic effect, which can ultimately lead to an effective photon management process.

References

1. W. Shockley and H.J. Queisser, *Journal of Applied Physics*, Vol. 32, no. 3, p. 510, 1961.
2. B.S. Richards, *Solar Energy Materials And Solar Cells*, Vol. 90, no. 15, p. 2329, 2006.
3. M.A. Green, K. Emery, Y. Hishikawa and W. Warta, *Progress In Photovoltaics*, Vol. 18, no. 5, p. 346–352, 2010.
4. Miasole. http://www.miasole.com, 2010.
5. D. Niinobe, K. Nishimura, S. Matsuno, H. Fujioka, T. Katsura, T. Okamoto, T. Ishihara, H. Morikawa and S. Arimoto, *23rd European Photovoltaic Solar Energy Conference and Exhibition, Valencia, Spain. Session: 2CV.5.74*, p. 1824, 2008.

6. J. Zhao, A. Wang, M.A. Green and F. Ferrazza, *Applied Physics Letters*, Vol. 73, p. 1991, 1998.
7. T. Trupke, M.A. Green and P. Wurfel, *Journal of Applied Physics*, Vol. 92, no. 7, p. 4117, 2002.
8. T. Trupke, M.A. Green and P. Wurfel, *Journal of Applied Physics*, Vol. 92, no. 3, p. 1668, 2002.
9. PVCDROM, http://pvcdrom.pveducation.org/CELLOPER/spectral.htm.
10. F. Auzel, *Chemical Reviews*, Vol. 104, no. 1, p. 139, 2004.
11. J.F. Suyver, A. Aebischer, D. Biner, P. Gerner, J. Grimm, S. Heer, K.W. Kramer, C. Reinhard and H.U. Gudel, *Optical Materials*, Vol. 27, no. 6, p. 1111, 2005.
12. F. Wang, R.R. Deng, J. Wang, Q.X. Wang, Y. Han, H.M. Zhu, X.Y. Chen and X.G. Liu, *Nature Materials*, Vol. 10, no. 12, p. 968, 2011.
13. A. Shalav, B.S. Richards, T. Trupke, K.W. Kramer and H.U. Gudel, *Applied Physics Letters*, Vol. 86, no. 1, 2005.
14. A. Shalav, B.S. Richards, K. Kramer and H. Gudel, *Conference Record of the Thirty-First IEEE Photovoltaic Specialists Conference - 2005*, p. 114, 2005.
15. X. Liang, X. Huang and Q. Zhang, *Journal of fluorescence*, Vol. 19, no. 2, p. 285, 2009.
16. G. Chen, T.Y. Ohulchanskyy, A. Kachynski, H. Ågren and P.N. Prasad, *ACS nano*, Vol. 5, no. 6, p. 4981, 2011.
17. D. Chen, Y. Wang and M. Hong, *Nano Energy*, Vol. 1, no. 1, p. 73, 2012.
18. G.B. Shan and G.P. Demopoulos, *Advanced Materials*, Vol. 22, no. 39, p. 4373, 2010.
19. L. Liang, Y. Liu, C. Bu, K. Guo, W. Sun, N. Huang, T. Peng, B. Sebo, M. Pan, and W. Liu, *Advanced Materials*, Vol. 25, no. 15, p. 2174, 2013.
20. L.T. Su, S.K. Karuturi, J. Luo, L. Liu, X. Liu, J. Guo, T.C. Sum, R. Deng, H.J. Fan and X. Liu, *Advanced Materials*, Vol. 25, no. 11, p. 1603, 2013.
21. W. Zou, C. Visser, J.A. Maduro, M.S. Pshenichnikov and J.C. Hummelen, *Nature Photonics*, Vol. 6, no. 8, p. 560, 2012.
22. X. Xie and X. Liu, *Nature Materials*, Vol. 11, no. 10, p. 842, 2012.
23. D. Chen, L. Lei, A. Yang, Z. Wang and Y. Wang, *Chemical Communications*, Vol. 48, no. 47, p. 5898, 2012.
24. R.T. Wegh, H. Donker, K.D. Oskam and A. Meijerink, *Science*, Vol. 283, no. 5402, p. 663, 1999.
25. J. Zhou, Y. Zhuang, S. Ye, Y. Teng, G. Lin, B. Zhu, J. Xie and J. Qiu, *Applied Physics Letters*, Vol. 95, no. 14, p. 141101, 2009.
26. X. Liu, Y. Teng, Y. Zhuang, J. Xie, Y. Qiao, G. Dong, D. Chen and J. Qiu, *Optics Letters*, vol. 34, no. 22, p. 3565, 2009.
27. Y. Teng, J. Zhou, X. Liu, S. Ye, and J. Qiu, *Optics Express*, Vol. 18, no. 9, p. 9671, 2010.
28. J. Zhou, Y. Teng, G. Lin, X. Xu, Z. Ma and J. Qiu, *Journal of The Electrochemical Society*, Vol. 157, no. 8, p. B1146, 2010.
29. Y. Teng, J. Zhou, S. Ye and J. Qiu, *Journal of The Electrochemical Society*, Vol. 157, no. 10, p. A1073, 2010.
30. P. Vergeer, T. Vlugt, M. Kox, M. Den Hertog, J. Van der Eerden and A. Meijerink, *Physical Review B*, Vol. 71, no. 1, p. 014119, 2005.
31. S. Ye, B. Zhu, J. Chen, J. Luo and J.R. Qiu, *Applied Physics Letters*, Vol. 92, p. 141112, 2008.

32. S. Ye, B. Zhu, J. Luo, J. Chen, G. Lakshminarayana and J. Qiu, *Optics Express*, Vol. 16, no. 12, p. 8989, 2008.
33. D.Q. Chen, Y.S. Wang, Y.L. Yu, P. Huang and F.Y. Weng, *Optics Letters*, Vol. 33, no. 16, p. 1884, 2008.
34. J. Zhou, Y. Teng, S. Ye, G. Lin and J. Qiu, *Optical Materials*, Vol. 34, no. 5, p. 901, 2012.
35. C. Yan, A. Dadvand, F. Rosei and D.F. Perepichka, *Journal of The American Chemical Society*, Vol. 132, no. 26, p. 8868, 2010.
36. S. Baluschev, F. Yu, T. Miteva, S. Ahl, A. Yasuda, G. Nelles, W. Knoll and G. Wegner, *Nano Letters*, Vol. 5, no. 12, p. 2482, 2005.
37. E. Verhagen, L. Kuipers and A. Polman, *Nano Letters*, Vol. 7, no. 2, p. 334, 2007.
38. V.K. Rai, L. de S. Menezes, C.B. de Araújo, L.R. Kassab, D.M. da Silva and R.A. Kobayashi, *Journal of Applied Physics*, Vol. 103, p. 093526, 2008.
39. S. Schietinger, T. Aichele, H.Q. Wang, T. Nann and O. Benson, *Nano Letters*, Vol. 10, no. 1, p. 134, 2010.
40. Y. Lu and X. Chen, *Applied Physics Letters*, Vol. 94, p. 193110, 2009.
41. E. Verhagen, L. Kuipers and A. Polman, *Optics Express*, Vol. 17, no. 17, p. 14586, 2009.
42. A. Wokaun, H.P. Lutz, A. King, U. Wild and R. Ernst, *The Journal of Chemical Physics*, Vol. 79, p. 509, 1983.
43. M. Moskovits, *Reviews of Modern Physics*, Vol. 57, no. 3, p. 783, 1985.
44. H. Zhang, D. Xu, Y. Huang and X. Duan, *Chemical Communications*, Vol. 47, no. 3, p. 979, 2011.
45. W. Zhang, F. Ding and S.Y. Chou, *Advanced Materials*, Vol. 24, no. 35, p. OP236, 2012.
46. Z.Q. Li, X.D. Li, Q.Q. Liu, X.H. Chen, Z. Sun, C. Liu, X.J. Ye and S.M. Huang, *Nanotechnology*, Vol. 23, no. 2, 2012.

Part 3

DEVELOPMENTS IN PROSPECTIVE

10

Advances in Plasmonic Light Trapping in Thin-Film Solar Photovoltaic Devices

J. Gwamuri[1], D. Ö. Güney[2], and J. M. Pearce[1,2,*]

[1]*Department of Materials Science & Engineering, Michigan Technological University, Houghton, Michigan, USA*
[2]*Department of Electrical & Computer Engineering, Michigan Technological University, Houghton, Michigan, USA*

Abstract

This chapter reviews the recent promising advances in the use of plasmonic nanostructures forming metamaterials to improve absorption of light in thin-film solar photovoltaic (PV) devices. Sophisticated light management in thin-film PV has become increasingly important to ensure absorption of the entire solar spectrum while reducing semiconductor absorber layer thicknesses, which reduces deposition time, material use, embodied energy, greenhouse gas emissions, and economic costs. Metal nanostructures have a strong interaction with light, which enables unprecedented control over the propagation and the trapping of light in the absorber layer of thin-film PV. The literature is reviewed for both theoretical and experimental work on multiple nanoscale geometries of both plasmonic absorbers and PV materials. Finally, the use of nanostructures to improve light trapping in PV is outlined to guide development.

Keywords: Plasmonic nanostructures, light trapping, optical enhancement, metamaterials, perfect absorbers, solar photovoltaic, solar energy, photovoltaic, plasmonic solar cells, plasmonics

**Corresponding author*: professor.pearce@gmail.com

Atul Tiwari, Rabah Boukherroub, and Maheshwar Sharon (eds.) Solar Cell Nanotechnology, (243–270) 2014 © Scrivener Publishing LLC

10.1 Introduction

The growing environmental concerns for the combustion of fossil fuels are driving a renewed interest in developing solar photovoltaic (PV) devices, which convert sunlight directly into electricity. PV is an established, technically-viable and sustainable solution to society's energy needs [1]. However, in order to out-compete highly-subsidized fossil fuels to reach mass deployment at the terawatt scale, further decreases in the levelized cost of electricity from solar are needed [2]. Conventional (1st generation polycrystalline and single crystalline silicon [c-Si]) PV have extremely limited abilities to improve performance and/or reduce costs [3]. Second generation thin-film PV technology does have this ability. Unfortunately, as these thin-film technologies evolve, their costs become progressively dominated by those of the constituent materials such as the top cover sheet and encapsulation [4]. Thus, the most promising approach to further drive down costs for clean solar electricity is to copiously increase the conversion efficiency of thin-film PV [3, 5]. Practically this means that more sunlight must be absorbed by the PV and converted into electricity. Various methods of optical enhancement have been developed such as the two classical methods of (i) scattering from a top roughened anti-reflection coating and (ii) using a detached back reflector [6, 7]. More sophisticated light management in thin-film PV is important to ensure absorption of the entire solar spectrum, while reducing semiconductor absorber layer thicknesses, which reduces deposition time, material use, embodied energy and greenhouse gas emissions, and economic costs. Recent advances in optics, particularly in plasmonics and nanophotonics provide new methods to improve the optical enhancement in thin-film PV. Theoretical work on the use of plasmonic nanostructures on PV devices has indicated that absorption enhancement of up to 100% is possible for thin-film PV devices using nanostructured metamaterials [6, 8]. This chapter reviews the recent promising advances in the use of plasmonic nanostructures to improve absorption of light in thin-film solar PV devices. Metal nanostructures have a strong interaction with light, which enables unprecedented control over the propagation and the trapping of light in the absorber layer of thin-film PV. Both the theoretical and experimental literature is reviewed on multiple nanoscale geometries of metamaterial absorbers and PV materials. Finally, the future of the use of novel nanostructures to improve light trapping architectures and enhance photocurrent generation in PV devices is outlined.

10.1.1 Plasmonics Basics

Plasmonics represent a rapidly growing field for the application of surface/interface plasmons towards realization of a variety of surface-plasmon-based devices including: biosensors, nano-imaging, waveguides, data storage, PV, light-emitting devices, and others [9, 10]. Plasmonics mainly involve two types of plasmons: 1) Surface Plasmon Polaritons (SPP), and 2) Localized Surface Plasmons (LSP).

In general, surface plasmons (SPs) can be described as the non-propagating collective vibrations of the electron plasma near the metal surface [11, 12]. SPs are collective oscillations of surface electrons whereas SPP describes a coupled state between an SP and a photon. On the other hand, LSP are surface excitations in bounded geometries such as metallic nanoparticles or voids of various topologies [13]. The fluctuations of surface charge results in highly localized and significantly enhanced electromagnetic fields in the vicinity of metallic surfaces. A detailed treatment of theory and fundamentals of plasmonics is given in [11–14]. In PV applications, light impinging on a metal surface produces surface waves along the metal-dielectric interface when they interact with the collective oscillations of free electrons in the metal. These surface waves (SPP), have shown potential for making plasmonic metamaterial "perfect absorbers" to enhance the efficiency of PV devices [15]. Perfect metamaterial absorbers can be designed with broadband, are polarization-independent, and have wide-angle optical absorption [16]. These critical features, lacking in most optical enhancement schemes for solar cell designs, are ideally required to maximize the efficiency of PV devices. Wide-angle reception, for example, is particularly important to increase solar energy conversion efficiency for maximized temporal and spatial response of the modules to solar radiation. This then eliminates the need for a physical tracking mechanism, which is beneficial as tracking schemes are known to increase both PV system's initial costs and operating costs.

Various configurations used to excite SPPs for different applications are illustrated in Figure 10.1 [13]. The attenuated total reflectance (ATR) configuration is often used for sensing applications. The geometry of the metallic structures can be manipulated to achieve

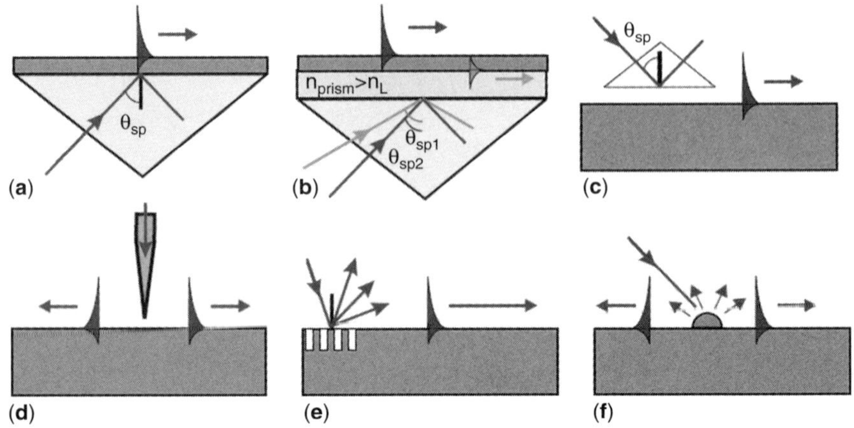

Figure 10.1 SPP excitation configurations: (a) Kretschmann geometry/ATR, (b) two-layer Kretschmann geometry/ATR, (c) Otto geometry, (d) excitation with a SNOM probe, (e) diffraction on a grating, and (f) diffraction on surface features. Figure reproduced with permission from A.V. Zayats [13].

desired surface plasmon resonances and propagating properties. Typically, surface plasmon resonances exhibit a strong relationship to the size, shape and the dielectric properties of the surrounding medium. The resonances of noble metals are mostly in the visible or infrared region of the electromagnetic spectrum, which is the range of interest for photovoltaic applications [17].

10.1.2 Metamaterials

Metamaterials are rationally designed geometries of artificial optical materials [14, 18], with electromagnetic response in nearly arbitrary frequencies [19]. Usually metamaterials are in the form of periodic structures with sub-wavelength features. The uniqueness of these novel materials is that they exhibit optical properties not observed in their constituent materials, i.e., they derive properties mainly from structure rather than from composition and they have enabled unprecedented flexibility in manipulating light waves and producing new functionalities [20]. Some of the proposed applications for metamaterials include superlenses that allow sub-wavelength resolution beyond the diffraction limit, and the electromagnetic cloak, which promises the ultimate optical illusion — invisibility [18].

Ultrathin metamaterial nanostructures exhibiting broadband and wide-angle resonant absorption have been proposed for PV devices and demonstrated both theoretically and practically, through device simulations and fabrication [17, 19, 21, 22]. Different cell geometries have been investigated in diverse configurations, with the metal-insulator-metal (MIM) being the most common.

Recently, more complex structures have been developed. The main difference between plasmonic metamaterial absorbers and other plasmonic approaches for enhancing solar power conversion efficiency is that the metamaterial approach provides more degrees of freedom in terms of impedance matching to minimize reflection and more flexible management of light in terms of polarization independence and wide angular reception. Therefore, plasmonic metamaterial absorbers can potentially provide better management of light. However, metamaterial absorbers have not been studied systematically in the literature for solar cell applications. The "super absorber" proposed by Aydin *et al.* [21], for example, uses a lossless spacer where almost all the power is absorbed by metal, hence resulting in significant Ohmic loss (i.e. heating). Nevertheless, there is potential to transfer this energy lost in the metal to the semiconductor if a semiconductor spacer replaces the dielectric spacer in the perfect (or super) absorbers, which can then be used in thin-film PV.

10.2 Theoretical Approaches to Plasmonic Light Trapping Mechanisms in Thin-film PV

The ability of metallic nanostructures to sustain coherent electron oscillations SPPs and LSPs leading to electromagnetic fields confinement and enhancement has attracted substantial attention in the PV scientific community [18, 23, 24]. Plasmonic solar cells have been investigated around the following main configurations: i) metal nanoparticles, ii) nanowires, and iii) metallic nanostructures (MIM/IMI), which can also be regarded as simple metamaterial absorbers.

Designs have been proposed for two-dimensional, ultra-thin, wide-angle perfect absorber structures for infrared light with a theoretical limit of nearly 100% absorption at the tunable resonant wavelength [24]. In 2011, another group reported an average measured absorption of 0.71 over the entire visible spectrum (400–700 nm) from their ultrathin (260 nm), broadband and polarization independent plasmonic super absorber consisting of a MIM stack [21]. They further demonstrated through simulation that absorption levels in the order of 0.85 are possible. By using suitably engineered metal nanostructures integrated in thin-film PV designs, effective light trapping mechanisms and geometries can be achieved. Plasmonic light-trapping geometries for thin-film PV such as those proposed by Atwater and Polman [6] have the potential of enhanced light absorption throughout the entire AM 1.5 solar spectrum. Using these same geometries proposed in [6], researchers experimentally demonstrated superior cell efficiencies of ultra-thin n-i-p hydrogenated amorphous silicon (a-Si:H) solar cells with plasmonic back reflectors compared to randomly textured cells [26]. These absorber layer thickness savings are particularly important for a-Si:H-based PV as this would reduce the light-induced degradation as well as save on manufacturing costs [27]. Recently, another research group designed and fabricated ultra-thin (100 nm) film silicon on insulator (SOI) Schottky photo-detectors for photocurrent enhancement in the spectral range 600–950 nm and achieved a spectral integrated 31% enhancement [28].

The advances recorded so far in the field of plasmonic PV devices can be attributed to the availability of electromagnetic (EM) field modeling tools such as COMSOL Multiphysics and Lumerical FDTD, which have provided important insights into both device design and optimization [29, 30]. Simulation studies have been the principal guiding tool in the area of plasmonics by providing a holistic picture of the interactions between EM fields and nanostructures at sub-wavelength scale. Simulation information such as the scattering or resonant absorption by nanoparticles behaving like dipoles have not only provided deep insights into the various processes such as the near field and coupling effects, but also served as accurate optimization tools [31, 32].

10.2.1 Optimal Cell Geometry Modeling

Optimization of plasmonic light trapping in a solar cell is a complex balancing act in which several physical parameters must be taken into account. However, the problem is simplified by the use of recently developed tools for nanoscale fabrication and nanophotonics characterization, as well as the emergence of powerful electromagnetic simulation methods [6]. Cell geometry simulations have been based on theoretical parameters from other reported models for optimization by continually changing model parameters [33].

The flexibility in this type of a model usually results in impressive theoretical PV performance parameters being reported, such as a FF > 0.73 and J_{sc} > 15 mA/cm^2. The other common approach is to use parameters from a physical model, which are experimentally determined using tools such as variable angle spectroscopic ellipsometry for determining optical properties. An example of such a model device and metal top surface structure are shown in Figure 10.2.

In this kind of model, usually the physical parameters such as thickness and dielectric properties of layers are the initial inputs for optimizing plasmonic meta-structures. To ensure consistency between the prepared cell samples and the simulated models, both the layer thicknesses and optical parameters are kept constant as the period, height and width of the nanostructures are varied. The metallic parts (back contact and front nanostructure) are made of silver. Optical constants of silver (Ag), a-Si:H, indium tin oxide

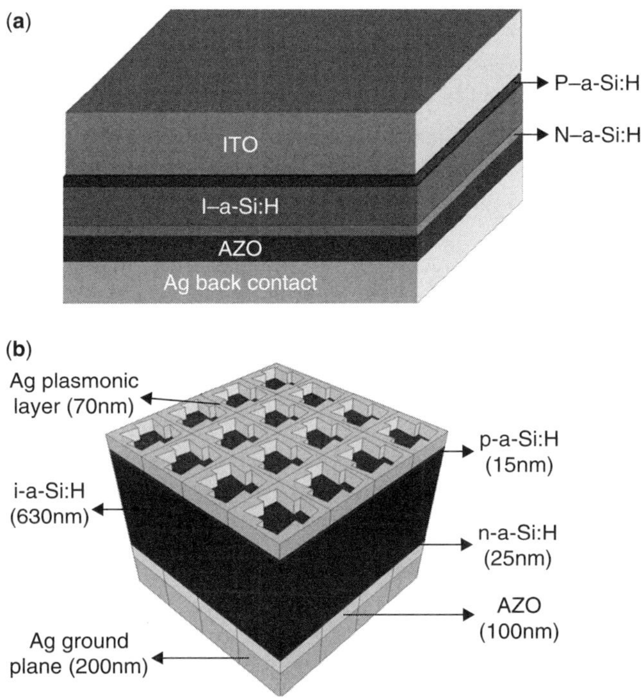

Figure 10.2 Typical schematic device design for (a) the standard single junction of a-Si:H solar cell, (b) the plasmonic enhanced n-i-p a-Si:H solar cell with top surface Ag metallic structure, which takes the place of the standard TCO (ITO or ZnO) contact (not to scale).

(ITO), aluminum-doped zinc oxide (ZnO:Al) and silicon nitride (Si_3N_4) are available [34, 35]. However, if the material specifications of the samples differ from that reported in literature, the parameters of interest can be physically determined using spectroscopic ellipsometry [36].

In designing optimal cell geometry for plasmonic PV cells, it is paramount to maximize the absorption within the semiconductor region while minimizing it within the metallic regions over the entire solar spectrum for a wide range of incidence angles. The radius of curvature technique (RCT) [37] can be employed to minimize the metallic losses, so that optical absorption is directed toward the semiconductor layers. This technique requires identification of resonant modes of the metamaterial absorber that contributes to optical absorption, followed by a geometric tailoring of the nanostructure based on underlying resonant currents. It is also necessary to develop and optimize the metamaterial absorbers to satisfy the requirements of wide-angle reception [38] and polarization-independent operation. To achieve this task different nanostructure geometries such as a square-grid structure (i.e., cross structures formed by perpendicularly placed metallic stripes) or any isotropic design found in the literature [25, 33, 39–41] can be used. Broadband response can be achieved by employing the nanoantenna concept [41, 42] or the "super absorber" design demonstrated in Ref. [21].

The model output helps to inform in selecting the nanostructure parameters giving the optimum absorption (see Table 10.1) for the actual integration into experimental cells. At these parameters, the nanostructures start to exhibit both features (i.e., polarization independence and broadband response) within a single design and the potential to operate over the entire visible spectrum (400–700 nm) and wide angular ranges.

10.2.2 Optical Properties Simulations

The SPP coupling must be stronger in the semiconductor than in the metal for efficient light absorption to occur. Hence metal nanostructure design must be such as to result in increased propagation length in the desired solar-cell geometry. For high efficiency, the PV must be optically thick, but electronically thin — having minority carrier diffusion lengths several times the material thickness if all the photogenerated carriers are to be successfully collected [6].

Table 10.1 Selected plasmonic solar cells parameters (adapted from Gu *et al.*, 2012).

Mechanism	Solar cell type	Plasmonic material and location	Fabrication method	Enhancement (Relative)	Enhancement (Absolute)
Metallic nanoparticles (on top of the solar cell) dipole scattering – *far field effect.* 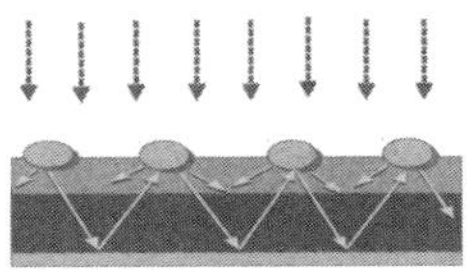	a-Si thin-film, substrate-configured [58]	Au NPs, front	Synthesis + coating	J_{SC} = 8.1 %, η = 8.3 %	J_{SC} from 6.66 to 7.2 mA/cm^2, η from 2.77 % to 3 %
	multi-Si wafer, commercial [58]	Au NPs, front	Chemical synthesis + dip coating	η = 35.2 %	η - 15.2 %, J_{SC} - 32.3 mA/cm^2, FF -76 %, V_{oc} – 0.62 V (Best cell results)
	a-Si/c-Si heterojunction [45]	Au NPs, front	Sputtering on heated substrate	J_{SC} 20 %, η = 25 %	J_{SC} from 19 to 22.9 mA/cm^2, η from 7.5 % to 9.4 %
	p-n Si:μcSi [59]	Ag NPs, front and Al back contact	Deposition (e-beam) + annealing	Highest values for λ *less than 550 nm*	η - 9.06%, V_{oc} = 0.57V, J_{SC} - 21.49 mA/cm^2, FF - 73.84%
	GaAs [60]	Ag NPs, front	Evaporation through anodic Al template	J_{SC} - 8 %, η - 25 % J_{SC} from 11 to 11.9 mA/cm^2	η - from 4.7 % to 5.9 %
	Organic, P3HT:JSCBM [44]	Ag NPs, front	Synthesis + spin coating	η from 1.3 % to 2.2 %	J_{SC} - 58 %, η - 69 % J_{SC} from 4.6 to 7.3 mA/cm^2
	Organic, P3HT:JSCBM [61]	Au NPs, front	Electrostatic assembling	J_{SC} - 36 %, η - 20 %	J_{SC} from 10.74 to 11.13mA/cm^2, η from 3.04 % to 3.65 %

(Continued)

Table 10.1 (*Cont.*)

Mechanism	Solar cell type	Plasmonic material and location	Fabrication method	Enhancement (Relative)	Enhancement (Absolute)
Metallic embedded NPs– *near field effect.*	InGaN[62]	Ag NPs, embedded	(Numerical simulation)	η - 27 % (calculated)	η from 10.59 % to 13.53 % (calculated)
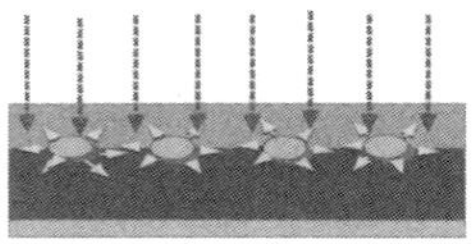	Organic, CuJsc-PTCBI Tandem [11]	Ag clusters, at interface of two sub-cells	Deposition	J_{SC} - 29 % , η - 26 % (calculated)	J_{SC} from 3.65 to 4.72 mA/ cm^2 , η from 1.9 % to 2.4 % (calculated)
	poly-Si thin-film, superstrate-configured [46]	Ag NPs, rear	Deposition + annealing	J_{SC} - 44 %	J_{SC} from14.85 to 21.42 mA/ cm^2
Metallic nanostructures (back contact / semiconductor interface) *SPP/ photonic mode coupling.*	a-Si thin-film, superstrate-configured [63]	Ag NPs, rear	Synthesis + coating	J_{SC} - 14.3 % , η - 23 %	Highest η - 8.1 %
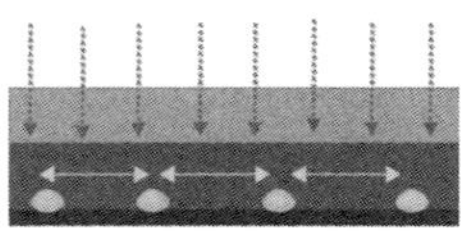	a-Si thin-film, substrate-configured [26]	Ag nanostructure, rear	Nanoimprint lithography	J_{SC} - 47 % , η - 15 %	J_{SC} from 11.52 to 16.94 mA/ cm^2 , η from 6.32 % to 9.6 %

Table 10.1 (*Cont.*)

Mechanism	Solar cell type	Plasmonic material and location	Fabrication method	Enhancement (Relative)	Enhancement (Absolute)
Metallic nano-structures (back contact/semicon-ductor interface) *SPP/photonic mode coupling.* 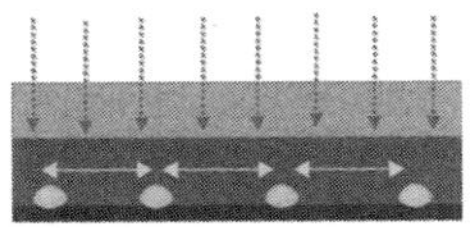	*n-i-p* a-Si:H[56]	Ag nanostructure, rear and front grating.	Nanoimprint Lithography (sol-gel imprint and FIB milling).	–	η - 6.16% Voc 0.81V, J_{SC} 12.5 mA/cm^2 FF 61%, λ from 600 to 800 nm
	n-i-p a-Si:H[64]	Ag nanostructure, rear	Deposition (Sputter/evaporation) + annealing	EQE enhancement factor of 2.5	Highest values; Voc 0.82 V, J_{SC} from 12.12 mA/cm^2, FF - 50% , λ from 700 to 800 nm
	n-i-p a-Si:H[54]	Ag nanostructure, rear	Nanoimprint Lithography (sol-gel imprint)	–	Highest values; η - 6.6%, Voc 0.89V, J_{SC} 13.4 mA/cm^2 , FF 66%, λ from 550 – 650 nm

Designing PV cells optimized for enhanced optical absorption is a sophisticated task that often involves a balancing of the physical parameters of the device. For example, small particles can be used to create a very intense forward scattering anisotropy with albedos greater than 90%, however they exhibit equally high Ohmic losses. Balancing between nanostructure size and Ohmic losses is a critical design task. Increasing the effective scattering cross section can be achieved by spacing the nanostructure further away from the active semiconductor substrate at the expense of the near-field coupling, hence another balancing act. EM modeling tools thus play a crucial role in helping researchers find an optimal balance of design paramaters by revealing insights into light interactions with different possible modes within the substrate material at subwavelength scale.

It is difficult to separate optical properties simulation from cell geometry modeling as device models must maintain broadband, wide angle and polarization-independent absorption. All these properties fundamentally determine the final device's optical characteristics in general, particularly the quantum efficiency (QE). The arbitrary absorption, A, is one parameter that is usually simulated for both the transverse electric (TE) mode and the transverse magnetic (TM) mode with the ultimate objective being making both as close to unity as possible. It is important to note that absorption in this case is determined from:

$$A = 1 - T - R \tag{10.1}$$

where T and R represent the transmitted and the reflected radiation respectively. However, as no radiation is transmitted through the PV devices with silvered back-contacts, Eq. 10.1 can therefore be reduced to:

$$A = 1 - R \tag{10.2}$$

As seen in Eq. 10.2, the absorptivity of a cell is determined by its reflectivity; hence the goal of optical device simulation is aimed at reducing R as much as possible. Conventionally, reflectivity in solar cells are reduced by incorporating an anti-reflecting coating (ARC) on the front surface of the cell, and ITO is commonly used although it is not ideal. For example, the use of ITO as a front contact layer results in poor a-Si:H absorption at short wavelengths and increased sheet resistance, which is undesirable in high-efficiency solar cells [33]. In plasmonic PV devices, ITO is replaced with metallic nanostructures (metamaterial perfect absorbers), which are often embedded in a thin layer of transparent and non-conducting Si_3N_4, resulting in enhanced energy conversion efficiency over the entire visible spectrum [12]. Reflection losses are not the only optimization as it is important to also maximize the spectral absorption A(λ) in the active absorber as that determines cell performance under solar illumination. Optimizing the device for fabrication involves changing nanostructure, size, shape and their periodic spacing until a significant enhancement in comparison to the reference device is achieved. To obtain a realistic device performance, the AM 1.5G spectrum [43] is integrated into the simulation resulting in A(λ) over the entire spectrum. The incident power, P (λ), and output power, $P_{out}(\lambda)$ can be monitored, and using the relationship below, A(λ) can be determined for different situations.

$$A(\lambda) = \frac{P_{in}(\lambda) - P_{out}(\lambda)}{P_{incident}(\lambda)} \tag{10.3}$$

10.2.3 Electrical Properties Simulations

When simulating for electrical properties, two simplifying assumptions are usually made: 1) all absorbed photons will excite electron-hole (e-h) pairs, and 2) all photogenerated e-h pairs are successfully collected. Furthermore, these assumptions help to establish a link between simulated optical properties and electrical properties hence making it possible to use Eq. 10.4 to calculate photocurrent density resulting from integrating A(λ) and the normalized AM1.5G spectral photon flux density, $\phi_{1.5}(\lambda)$ over the range of wavelengths:

$$J_{sc} = q \int \frac{A(\lambda)\, \phi_{1,5}(\lambda)\, \lambda}{hc}\, d\lambda \tag{10.4}$$

The most important cell performance parameters are its short-circuit current, I_{sc}, maximum power current, I_{mp}, open circuit voltage, V_{oc} and the maximum power voltage, V_{mp}. However, the other performance parameter, the fill factor, FF, is commonly used to collectively describe the degree to which V_{mp} matches V_{oc} and I_{mp} matches I_{sc}.

Photocurrent enhancements have been reported in various studies involving plasmonic solar cells with different enhancement schemes. Relative photocurrent enhancement of 47% (11.52–16.94 mA/cm^2) in a-Si thin film using Ag rear nanostructures [26], and 58% (4.6–7.3 mA/cm^2) in organic P3HTPCBM using front Ag nanoparticles [44] with efficiency enhancements of 15% and 69% respectively, have been achieved using the scattering (far field) mechanism from either the front or back of the cell. Near field enhancement has achieved slightly lower photocurrent enhancement compared to the far-field enhancement: 20% (19–22.9 mA/cm^2) for a-Si/c-Si hetero-junction using front Au nanoparticles [45], 44% (14.85–21.42 mA/cm^2) for poly-Si thin film with rear Ag nanoparticles [46] and 60% for a-Si thin film with Ag front nanowire array was achieved through numerical simulation [47]. Efficiencies reported for metamaterials geometries related to solar cells have been mainly based on numerical simulation and mostly on MIM geometries instead of real PV devices. Although most models have achieved wide-angle broadband absorption, some have failed to quantify the photocurrent enhancement that could be achieved from their proposed cell geometries. Notable exceptions are Wang *et al.*, who used numerical simulation to calculate a maximum J_{sc} value of 19.7 mA/cm^2 for their a-Si:H-based Ag metamaterials effective film cell [48]. They also reported a FF of 0.7 at V_{oc} of 0.87 V and J_{sc} of 4.6 mA/cm^2 [48]. In addition, Wu *et al.* reported a 41% efficiency improvement in their solar thermo-photovoltaic (STPV) cell also from numerical simulation [49].

10.3 Plasmonics for Improved Photovoltaic Cells Optical Properties

10.3.1 Light Trapping in Bulk Si Solar Cells

For the last 30 years it has been assumed that surface texturing in optical sheets can provide an intensity enhancement factor of $4n^2(x)$ for indirect bandgap semiconductors such as crystalline silicon [50]. Conventional thick c-Si-based solar cells have incorporated a pyramidal surface texture as a standard design for commercial devices [15]. In bulk c-Si solar cells, material costs account for 40% of the final module price [15]. This fact has been the main driver behind the growth of thin-film PV technologies as the reduction of material volume directly provides a reduction in materials costs.

Unfortunately, similar light enhancement through surface texturing is unsuitable for thin-film PV due to:

i. Geometrical limitations: Pyramidal texture features (microns in size) are usually much larger than the actual thin-film cell thickness [15, 17].
ii. Technical reasons: Although texturing of submicron features can be achieved through plasma etch techniques, this can damage the absorber and degrade the cell resulting in greatly reduced device efficiency [43]. Furthermore, texturing will result in increased surface area, which promotes greater minority carrier recombination.

New mechanisms for enhancing light trapping in thin-film PV have been proposed and investigated, and have shown that light coupling/enhancement into the substrate without increased recombination losses can be achieved through use of plasmonics.

10.3.2 Plasmonic Light-Trapping Mechanisms for Thin-Film PV Devices

Plasmonic light-trapping techniques may provide considerable reduction in the PV material layer thickness by a factor ranging from 10 to 100 [6], while enhancing absorption through folding and concentrating [15] the incident light energy into the thin-film semiconductor layer. Three different techniques are used as shown schematically in Figure 10.3.

In the mechanism described in Figure 10.3(a), metallic nanoparticles act as subwavelength scattering elements to preferentially scatter light into the high-index semiconductor substrate resulting in the anti-reflection effect over a broad spectrum [6, 15, 17] at the surface of the cell. Also, the angular redistribution of the scattered light folds light into thin absorber layers and further increases its optical path length. In addition, at the right nanostructure sizes and specific dielectric material properties, intense localized fields due to SP resonances may result and the energy from the surface plasmons may be emitted as light and scattered in the direction of the semiconductor through a dipole effect [17, 51]. Incident sunlight is collected by subwavelength-scale metal particles with their large extinction cross section and reradiated into semiconductor in

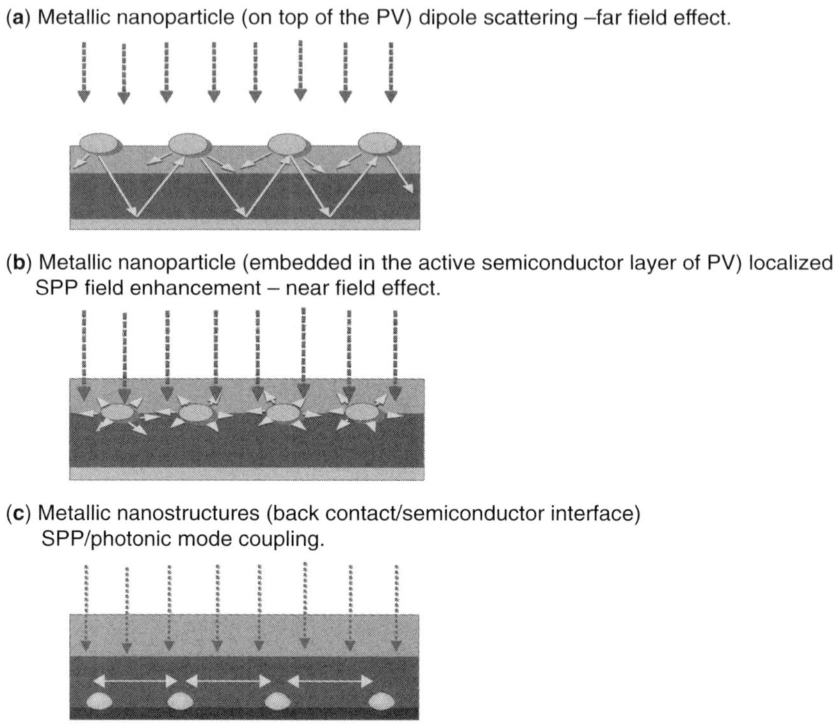

Figure 10.3 Schematic representation of three different plasmonic light-trapping schemes for thin-film PV devices.

multiple angles to increase optical path length in thin-film photovoltaic layers.

The second mechanism, shown in Figure 10.3(b), takes advantage of nanoparticles being used as subwavelength antennas with the ability to directly couple the plasmonic near-field into the semiconductor thin-film layer through the near-field dipole effect. The plasmonic field near an emitting dipole is intense and localized to a few tens of nanometers around the nanoparticle, hence decreasing the spacing and particle size will further enhance this effect [17]. Furthermore, placing nanoparticles directly inside the semiconductor have been observed to directly excite electron-hole pairs leading to enhanced photogenerated charge collection facilitated by the presence of an enhanced electric field. Optical steering effects have also been suggested to contribute to overall enhancement [52]. The optical steering of light from subwavelength scale structures, which can be achieved by modifying the nano-antenna radiation pattern through changing the LSP resonances results in a form of wavelength dispersion at the nanoscale where light of different frequencies is scattered into different directions. This mechanism has shown potential in organic and direct bandgap semiconductors where carrier diffusion lengths are small and electron-hole pairs are usually generated close to the collection junction [6, 17].

The third mechanism, shown in Figure 10.3(c), involves the direct patterning of the metal back contact into some form of a back diffraction grating to trap the incident light at long wavelengths into guided SPP modes, in such a way to increase photon lifetime in the active medium [15, 53]. Generally, a solar cell designed for these types of plasmonics absorb the blue side of the spectrum effectively leaving the longer red wavelengths, which are then coupled into the waveguide or/and SPP modes propagating in the plane of the semiconductor layer [54]. However, for each photonic mode, part of the light will be absorbed into the semiconductor layer and part will be lost. The advantage of this mechanism, though, is that the incident solar flux is effectively rotated through an angle of $\pi/2$ radians therefore providing for light to be absorbed along the lateral dimensions of the solar cell [6] in which the cell dimensions are considerably larger than the optical absorption path length. Incident sunlight is coupled into surface plasmons propagating at the semiconductor-metal interface via subwavelength-size grooves to increase the optical path by switching the light direction from normal to the PV absorber layer to lateral to the PV absorber layer [55]. Since metallic back contacts are part of the standard design for most solar cells, nanotexturing of the back contact can be integrated into current fabrication techniques, although a substrate effect can hamper optimal semiconductor growth.

Considering design challenges associated with each of the proposed mechanisms above, currently, having nanoparticles/nanostructures on the front surface or back contact have been the most preferred options by many researchers especially for inorganic thin-film PV technologies. The advantage of these approaches is that they enable semi-independent optimization of plasmonic properties of the cell as the optical properties are isolated from the electrical properties of the cell. However, having nanostructure within the semiconductor layer is still an active research area for groups working with organic and quantum dots based PV technologies.

10.3.3 Experimental Results

The most recent experimental enhancements and efficiencies of plasmonic PV devices are summarized in Table 10.1.

10.4 Fabrication Techniques and Economics

Current research in fabrication of plasmonic structures for PV is focused on reducing the fabrication cost in order to make plasmonic-based light-trapping technologies economically viable. Plasmonic coupling mechanisms described in this chapter require stringent control of process parameters for integration of dense arrays of metal nanostructures with dimension tolerances at the nanometer scale. The majority of fabrication techniques used in most of the reviewed literature is based on cleanroom techniques such as electron-beam lithography (EBL) or focused ion-beam lithography (FIB).

These techniques allow precise control of design parameters of nanostructures at sub-wavelength scale; however, they are too expensive and not suitable for large-scale production. Large-scale, inexpensive and scalable fabrication techniques must be developed to enable giga-watt production of plasmonic-based PV modules.

10.4.1 Lithography Nanofabrication Techniques

Some of the most common fabrication techniques are summarized in Table 10.2. Despite having well-established lithographic technologies for PV devices fabrication, challenges still exist regarding the cost of implementing the processes. The challenges are further compounded by the fact that most of the well-established processes are not only expensive, but they also have low throughput rendering them unattractive for large-scale industrial fabrication regardless of their capability to produce highly uniform sub 10 nm critical dimensions (CDs). To achieve large-scale fabrication and bring down the price of solar cells, fabrication processes such as EBL, FIB and other next generation lithographic (NGL) techniques must be adapted for high throughput so as to reduce the production costs through the economics of scale. The lithographic technique that has shown great potential for large-scale adoption is nanoimprint/soft-imprint lithography (NIL) [6, 15, 17, 26, 54, 56–57]. However, the stamp creation process is still considerably tedious and expensive.

Table 10.2 Common lithographic nanofabrication techniques for PV device fabrication.

Lithographic Technique	Types	Application in PV fabrication	Feature sizes	Comments
Electron beam lithography (EBL)	Direct write EBL	- Top surface nano-patterning - Back contact texturing	- Approximately 10 nm (single line) - 20-40 nm (3D line)	- Well established technology - Good reproducibility - High spatial resolution - High precision - Fast turn-around - Low throughput and expensive (ideal for R&D but not suitable for large industries)
	Electron Projection Lithography (EPL)	- Top surface nano-patterning	- Approximately. 130 nm or less	- High precision - Critical dimensions still a concern - Incapable of fabricating complex nanostructures - Low throughput
Ion-beam lithography	Focused ion-beam (FIB)	- Top surface nanopatterning	- 4 to 8 nm using Ga ion source	- High precision - Well established technology - Slow and expensive
	Deep-ion beam lithography (DIBL)	- Back contact texturing	- Less than 200 nm	- Expensive

(Continued)

Table 10.2 (*Cont.*)

Lithographic Technique	Types	Application in PV fabrication	Feature sizes	Comments
Next generation lithography	Nano-imprint lithography (NIL)	- Top surface nano-patterning - Back contact texturing - Embossing silica sol-gel on glass substrate.	- 5 nm CDs are possible	- Most promising technology - Potentially economic - Creation of mask/stamp still time consuming and expensive
	Substrate conformal imprint lithography (SCIL)			
	Colloidal lithography	- Nanowires - Nanocones - Nanoshells - Nanowells	- sub 10 nm feature sizes	- Good precision - Not suitable yet for large scale production - Expensive
Others	X-ray, Extreme UV, Holographic, Dip pen lithography	- Si solar cell back and front contact	- uses 13.4 nm λ to achieve less than 30 nm features for extreme UV - Approx. 0.3 μm or less for x –ray.	- 26 nm line achieved for EUV. - low throughput and expensive

10.4.2 Physical/Chemical Processing Techniques

Physical and chemical techniques can be described as the backbone of the semiconductor industry for both very traditional large-scale integration (VLSI) fabrication and the novel giga-watt PV industry. Deposition techniques such as atomic layer deposition (ALD) and molecular beam epitaxy (MBE) have presented researchers the flexibility to control precise device parameters such as doping and layer thickness through *in-situ* control of crystal growth at the atomic level. This has resulted in excellent film quality resulting in high-quality devices. However, these two techniques have low throughput and are currently expensive. On the contrary, chemical deposition processes such as plasma assisted/enhanced chemical vapor deposition (PACVD/PECVD), atmospheric pressure chemical vapor deposition (APCVD) and metal-organic chemical vapor deposition (MOCVD) are well established large-scale fabrication techniques. PECVD is currently the standard fabrication technique for thin-film a-Si:H PV devices with film growth rates of 1 A°/s at 13.56 MHz giving optimum quality device performance [54, 65, 66]. Increasing this rate would be beneficial for large-scale fabrication and higher rates have been achieved with PECVD. Nanostructure and thin-film fabrication techniques are summarized in Table 10.3.

10.5 Conclusion and Outlook

The potential of thin-film plasmonic-enhanced solar cells is evident from simulations reviewed in this chapter. There is a reason to be optimistic about the future of high-efficiency solar cells as demand and the cost of land escalates [2, 6, 17]. As fabrication techniques evolve, it will be possible to fabricate less expensive and higher efficiency thin-film solar cells based on plasmonic quantum dots, optical nano-antenna and other novel technologies.

Quantum dot solar cells are being investigated and the results reported are promising. Quantum dots (or semiconductor nanocrystals) exhibit optical properties that are size dependent. Their absorption spectrum can be tuned, which has advantages for solar cells, where the nanocrystal size can be selected to optimize the absorption of nearly the entire AM 1.5 solar spectrum resulting in increased efficiency. QDs have demonstrated the ability to extract energy from "hot electrons," which is typically lost in conventional

Table 10.3 Physical and chemical processes commonly used for PV device fabrication.

Physical/Chemical processing	Types	Application in PV fabrication	Advantages	Disadvantages
Physical Deposition	Atomic – Layer Deposition (ALD),	thin films isolating layers	- high aspect ratio and uniformity for nanomorphology fabrication - pin-hole and particle free deposition is achieved - excellent film quality and step coverage	- high vacuum process - low throughput - expensive process
	Molecular Beam Epitaxy (MBE)	thin films	- precise control of compositions and morphology - in situ control of crystal growth at atomic level	- low throughput - expensive process
Physical Vapor Deposition (PVD)	Evaporation	thin films nanoparticles/ nanostructures	- large scale fabrication - less costly	- less surface uniformity
	Sputtering	thin films etching nanostructures	- high film uniformity - high etch rates	- low rates - can introduce defects
Chemical depositions	Chemical Vapor Deposition (PECVD, PACVD, APCVD, MOCVD, HWCVD)	thin film (semiconductor and organic) Si nanowires Single walled nanotubes ARC/ITO layers	- high throughput for large scale fabrication. - Well established technologies. - low cost	- high vacuum equipment - highly toxic gases (requiring post process conditioning)

Table 10.3 (*Cont.*)

Physical/Chemical processing	Types	Application in PV fabrication	Advantages	Disadvantages
Nanoparticles fabrication processes	Wet chemistry (Sol-gel)	nano-layers	- high purity from raw materials - good homogeneity from raw materials - high shape uniformity - low temperature process	- high cost of raw materials - long processing times - health hazards of organic solutions - cracking and shrinkage
	Gas phase processing (pyrogenic and controlled detonation synthesis)	nanoparticles	- precise control of features sizes	- complex and expensive processes
	Mechanical milling	nanopatterning	- precise control of features sizes around 10 nm scale	- complex and expensive processes
Others	High velocity deformation, mechanochemical processes, plasma and flame spray	coatings metal nanoparticles	- high film uniformity	- less established

solar cells [67-69]. Some of the QDs-based solar cells demonstrated so far include a 20-nm-thick layer of CdSe semiconductor quantum dots deposited on a Ag film [70], and 11.6% (Au nanoparticles) or 10.9% (Ag nanoparticles) [49].

Future solar cells must provide low cost at high efficiencies closer to the thermodynamic limit of 93% [3]. Using tandem multi-junction cells with a series of varying bandgap materials stacked in decreasing order of bandgap is a promising option to circumvent the Shockley-Queisser limit of single bandgap devices [17, 55]. A balance limit calculation estimated that a 36-junction cell ideally would reach 72% efficiency at a concentration of 1,000 suns relative to the 37% for a 1-junction device [72]. Plasmonic tandem solar cells can be designed by stacking layers of QDs of increasing size from the top to the bottom of the cell, or nanoparticles imbedded in semiconductor material in a similar manner. By variation of particle or dots size, it is possible to absorb from the blue (at the top of the cell) to the red (at the bottom) wavelengths of the solar spectrum using minimal material resources. An alternative structure would be to have semiconductors of different bandgaps stacked on top of each other separated by plasmonic structured metal films, each optimized for different optical spectral bands.

Acknowledgements

Authors would like to acknowledge helpful discussions with W. Knudsen and support from Fulbright (Science and Technology), and the National Science Foundation (CBET-1235750).

References

1. J.M. Pearce, *Futures* Vol. 34, p. 663, 2002.
2. K. Branker, M.J.M. Pathak, J.M. Pearce, *Renew. Sustain. Energy Rev.* Vol. 15, p. 4470, 2011.
3. M.A. Green, *Third Generation Photovoltaics: Advanced Solar Energy Conversion*, Springer-Verlag, Berlin, 2003.
4. J.M. Woodcock, H. Schade, H. Maurus, B. Dimmler, J. Springer, A. Ricaud, *Conf. Rec. 14th Euro. Photo. Solar Energy Conf.* pp. 857-860, 1997.
5. M.A. Green, *Third Generation Photovoltaics*, Springer, 2005.
6. H.A. Atwater and A. Polman. *Nature Materials*, Vol. 9, no. 3, p. 205, 2010.
7. E.L. Hu, M. Brongersma, A. Baca, *Nanotechnology Research Directions for Societal Needs in 2020*, Springer Netherlands, pp. 417–444, 2011.

8. *F. Zhu, P.J. Jennings, J.C.L Cornish, G. Hefter, C.P. Lund, 35th Ann. Conf. of the Australian and New Zealand Solar Energy Soc., December, Canberra, 1997.*
9. S.A. Maier, *Plasmonics: Fundamentals and Applications*. Springer, 2007.
10. N.J. Halas, *Nano Lett.*, Vol. 10, no. 10, p. 3816, 2010.
11. *B.P. Rand, P. Peumans, S.R. Forrest, J. Appl. Phys.* Vol. *96, 7519 (2004).*
12. *D.M. Schaadt, B. Feng, E.T. Yu, Appl. Phys. Lett.* Vol. *86, 063106 (2005).*
13. V. Zayats, I.I. Smolyaninov, A. A. Maradudin, *Phys. Reports*. Vol. 408, no. 3–4, pp. 131–314, 2005.
14. J.M. Pitarke, V.M. Silkin, E.V. Chulkov, P.M. Echenique, *Rep. on Prog. in Phys.* Vol. 70, no. 1, pp. 1–87, 2007.
15. P. Spinelli, V.E. Ferry, J. van de Groep, M. van Lare, M.A. Verschuuren, R.E.I. Schropp, H.A. Atwater, A. Polman, *J. Opt.*, Vol. 14, no. 2, p. 024002, 2012.
16. S. Hayashi, T. Okamoto, *J. Phys. D: App. Phys.*, Vol. 45, no. 43, p. 433001, 2012.
17. S. Pillai, M.A. Green, *Sol. Energ. Mat. and Sol. Cells*, Vol. 94, no. 9, p. 1481, 2010.
18. W. Cai, V. M. Šalaev, *Optical Metamaterials: Fundamentals and Applications*. Springer, 2009.
19. S.A. Maier, H.A. Atwater, *J. Appl. Phys.* Vol. 98, p. 011101, 2005.
20. A. Sihvola, *Metamaterials*, Vol. 1, no. 1, p. 2, 2007.
21. K. Aydin, V.E. Ferry, R.M. Briggs, H.A. Atwater, *Nat. Comm.*, Vol. 2, p. 517, 2011.
22. S.Y. Chou, W. Ding, *Opt. Express*, Vol. 21, p. A60, 2013.
23. S. Lal, S. Link, N.J. Halas, *Nat. Photon.*, Vol. 1, no. 11, p. 641, 2007.
24. N. Verellen, Y. Sonnefraud, H. Sobhani, F. Hao, V.V. Moshchalkov, P.V. Dorpe, P. Nordlander, S.A. Maier, *Nano Lett.*, Vol. 9, no. 4, p. 1663, 2009.
25. C. Wu, Y. Avitzour, G. Shvets, *Proc. of SPIE* Vol. 7029, p. 70290W, 2008.
26. V.E. Ferry, M.A. Verschuuren, C. van Lare, E.I. Ruud, H.A. Atwater, A. Polman, *Nano Lett.* Vol. 11, p.4239, 2011.
27. C.R. Wronski, J.M. Pearce, R.J. Koval, X. Niu, A.S. Ferlauto, J. Koh, R.W. Collins. *Mat. Res. Soc. Symp. Proc.* Vol. 715, p. A13.4, 2002.
28. J. Trevino, C. Forestiere, G. Di Martino, S. Yerci, F. Priolo, L. Dal Negro, *Optics Express*, Vol. 20, no. S3, p. A418, 2012.
29. A. Aubry, D.Y. Lei, A.I. Fernández-Domínguez, Y. Sonnefraud, S.A. Maier, J.B. Pendry. *Nano Letters*, Vol. 10, no. 7, p. 2574, 2010.
30. F.J. Beck, S. Mokkapati, A. Polman, K.R. Catchpole. *App. Phys. Lett.*, Vol. 96, no. 3, pp. 033113, 2010.
31. J.R. Cole, N.J. Halas. *App. Phys. Lett.*, Vol. 89, no. 15, p. 153120, 2006.
32. M.A. Green, S. Pillai, *Nat. Photon.*, Vol. 6, no. 3, p. 130, 2012.
33. I. Massiot, C. Colin, N. Pere-Laperne, P. Roca i Cabarrocas, C. Sauvan, P. Lalanne, J.-L. Pelouard, S. Collin. *Appl. Phys. Lett.*, Vol. 101, no. 16, p. 163901, 2012.
34. P.B. Johnson, R.W. Christy. *Phys. Rev. B*, Vol. 6, no. 12, p. 4370, 1972.
35. E.D. Palik, *Handbook of optical constants of solids*. Orlando: Academic Press, 1985.
36. H. Fujiwara, *Spectroscopic ellipsometry: principles and applications*. Wiley, 2007.
37. *D.Ö. Güney, T. Koschny, C. M. Soukoulis, Phys. Rev. B,* Vol. *80, p. 125129, 2009.*
38. M. Diem, Th. Koschny, C. M. Soukoulis, *Phys. Rev. B*, Vol. 79, p. 033101, 2009.
39. X., Liu, T., Starr, A.F., Starr, W.J., Padilla, *Phys. Rev. Lett.*, Vol. 104, p. 207403, 2010.

40. K.B. Alici, A.B. Turhan, C.M. Soukoulis, E. Ozbay, *Opt. Express,* Vol. 15, p. 14260, 2011.
41. C. Wu, G. Shvets, *Opt. Lett.,* Vol. 37, p. 308, 2012.
42. Y. Cui, J. Xu, K.H. Fung, Y. Jin, A. Kumar, N.X. Fang, *Appl. Phys. Lett.* Vol. 99, p. 253101, 2011.
43. S. Pillai, K.R. Catchpole, T. Trupke, M.A. Green, *J. Appl. Phys.,* Vol. 101, no. 9, p. 093105, 2007.
44. A.J. Morfa, K.L. Rowlen, T.H. Reilly, M.J. Romero, J. van de Lagemaat, *Appl. Phys. Lett.* Vol. 92, p. 013504, 2008.
45. M. Losurdo, M.M. Giangregorio, G.V. Bianco, A. Sacchetti, P. Capezzuto, G. Bruno, *Solar Energy Mat. and Sol. Cells,* Vol. 93, no. 10, p. 1749, 2009.
46. Z. Ouyang, X., Zhao, S., Varlamov, Y., Tao, J., Wong, Pillai, Supriya, *Prog. Photovoltaics Res. Appl.* Vol. 19, p. 917, 2011.
47. C. Rockstuhl, S., Fahr, F., Lederer, *J. Appl. Phys.* Vol. 104, p. 123102, 2008.
48. Y. Wang, T., Sun, T., Paudel, Y., Zhang, Z., Ren, K., Kempa, *Nano Lett.* Vol. 12, p. 440, 2011.
49. C. Wu, B. N. Iii, J. John, A. Milder, B. Zollars, S. Savoy, G. Shvets, *J. Opt.,* Vol. 14, p. 024005, 2012.
50. E. Yablonovitch, G. D. Cody. *IEEE Trans. Elec. Dev.* Vol. 29, no. 2, p. 300, 1982.
51. B. S. Luk'yanchuk, M. I. Tribel'ski, V. V. Ternovski, *J. Opt. Technol.,* Vol. 73, no. 6, p. 371, 2006.
52. A., Djalalian-Assl, D.E., Gómez, A., Roberts, T.J., Davis, *Opt. Lett.* Vol. 37, p. 4206, 2012.
53. M. Yang, Z. Fu, F. Lin, X. Zhu, *Opt. Express,* Vol. 19, no. S4, p. A763, 2011.
54. V. E. Ferry, M. A. Verschuuren, H. B. T. Li, E. Verhagen, R. J. Walters, R. E. I. Schropp, H. A. Atwater, A. Polman. *Proc. 25th EU-PVSEC Conference,* Valencia, 2010.
55. K. Tanabe, *Energies,* Vol. 2, no. 3, p. 504, 2009.
56. M. J. Madou. *Fundamentals of Microfabrication and Nanotechnology: Manufacturing techniques for microfabrication and nanotechnology.* CRC Press, 2011.
57. *D. Derkacs, S. H. Lim, P. Matheu, W. Mar, E. T. Yu. Appl. Phys. Lett.* Vol. *89, p. 093103, 2006.*
58. N. F. Fahim, B. Jia, Z. Shi, M. Gu, and S. Kim, *Opt. Express* Vol. *20, p. A694–A705,* 2012.
59. T. L. Temple, G. D. K. Mahanama, H. S. Reehal, D. M. Bagnall, *Solar Energy Mat. and Solar Cells,* Vol. 93, no. 11, p. 1978, 2009.
60. K. Nakayama, K. Tanabe, H.A. Atwater, *Appl. Phys. Lett.* Vol. 93, p. 121904, 2008.
61. J. H. Lee, J. H. Park, J. S. Kim, D. Y. Lee, K. Cho,. *Organic Electronics,* Vol. 10, no. 3, p. 416, 2009.
62. J.-Y. Wang, F.-J. Tsai, J.-J. Huang, C.-Y. Chen, N. Li, Y.-W. Kiang, and C. C. Yang, *Opt. Express,* Vol. 18, no. 3, p. 2682, 2010.
63. *X. Chen, B. Jia, J. K. Saha, B. Cai, N. Stokes, Q. Qiao, Y. Wang, Z. Shi, M. Gu, Nano Lett.,* Vol. *12, no. 5, p. 2187, 2012.*
64. C. Eminian, F.-J. Haug, O. Cubero, X. Niquille, and C. Ballif, *Prog. Photovolt.: Research and Appl.,* Vol. 19, no. 3, p. 260, 2011.
65. R. E. I. Schropp, K. F. Feenstra, E. C. Molenbroek, H. Meiling, J. K. Rath, *Phil. Mag. Part B,* Vol. 76, no. 3, p. 309, 1997.

66. R.W. Collins, A.S. Ferlauto, G.M. Ferreira, C. Chen, J. Koh, R.J. Koval, Y. Lee, J.M. Pearce, C. R. Wronski, *Solar Energy Mat. and Solar Cells*, Vol. 78, p. 143, 2003.
67. A.J. Nozik, *Phys. E Low-Dimens. Syst. Nanostructures* Vol. 14, p. 115, 2002.
68. A., Luque, A. Martí, A.J. Nozik, *MRS Bull.* Vol. 32, p. 236, 2007.
69. A.J. Nozik, M.C. Beard, J.M. Luther, M. Law, R.J. Ellingson, J.C. Johnson, *Chem. Rev.* Vol. 110, p. 6873, 2010.
70. D. Pacifici, H. J. Lezec, H. A. Atwater. *Nat Photon*, Vol. 1, no. 7, p. 402, 2007.

11

Recent Research and Development of Luminescent Solar Concentrators

Yun Seng Lim*, Shin Yiing Kee, and Chin Kim Lo

Universiti Tunku Abdul Rahman, Department of Electrical and Electronic Engineering, Faculty of Engineering and Science, Kuala Lumpur, Malaysia

Abstract

Luminescent solar concentrator (LSC) is a transparent plate containing luminescent materials with photovoltaic (PV) cells attached to its edges. Sunlight entering the LSC is absorbed by the luminescent materials which in turn emits light at a different wavelength. The emitted light propagates across the plate and arrives at the PV cells through the total internal reflection. The ratio of the area of the relatively cheap polymer plate to that of the expensive PV cells is high, thus reducing the cost of power generated by solar (US$/kWh). To improve their emission performance, simulation modeling of LSCs becomes essential. Ray-tracing modeling and thermodynamic methods are the two tools commonly used for simulating luminescent solar concentrator. Recently, a hybrid model of the thermodynamic and ray-tracing models has been proposed to provide a fast, powerful and cost-effective solution for the design of LSC. Apart from that, various luminescent materials and acrylic materials have been used in LSC to investigate its performance. The objective of this chapter is to provide the latest findings on the research and development of LSC.

Keywords: Luminescent solar concentrator, thermodynamic, backward ray-tracing.

**Corresponding author*: yslim@mail.utar.edu.my

Atul Tiwari, Rabah Boukherroub, and Maheshwar Sharon (eds.) Solar Cell Nanotechnology, (271–292) 2014 © Scrivener Publishing LLC

11.1 Introduction

Producing highly efficient solar cells at a reduced cost has been the key motivation in the research and development of photovoltaics (PV) for many decades. At present, the prices of PV modules are still high, creating a barrier to the widespread installation of photovoltaic technologies. This barrier can be minimized if financial support schemes such as investment subsidies or feed-in-tariffs are introduced. Alternatively, an innovative and cost-effective means of concentrating a large area of solar irradiation to a small area on which solar cells are attached is being pursued. With the use of the solar concentrator, the amount of the solar cell materials is reduced, hence making the cost of solar electricity ($/kWh) drop. One example of the solar concentrators is the luminescent solar concentrator.

A luminescent solar concentrator (LSC) consists of a polymer plate doped with luminescent materials, with solar cells attached to the plate edges. Figure 11.1 shows a diagram of a simple luminescent solar concentrator. The luminescent materials capture the incident solar light and then emit another light at a different wavelength. Figure 11.2 is a photograph of a LSC with quantum dots being used as luminescent materials. The solar cells receive the majority of the emitted light through total internal reflection [1]. The direct and diffused sunlight can be concentrated by a factor of 5 to 10 by using LSC without any tracking systems [2]. The cost of the transparent acrylic glass and luminescent materials is much lower than the solar cells. The net effect is that the cost of solar electricity can be reduced. The extended surface area of LSCs enables effective heat dissipation, allowing solar cells coupled at the edges to operate at the optimum temperature and hence avoiding any cooling systems. In addition, LSCs can be used as building facades such as roofs, windows, walls or pavements. The electricity generated

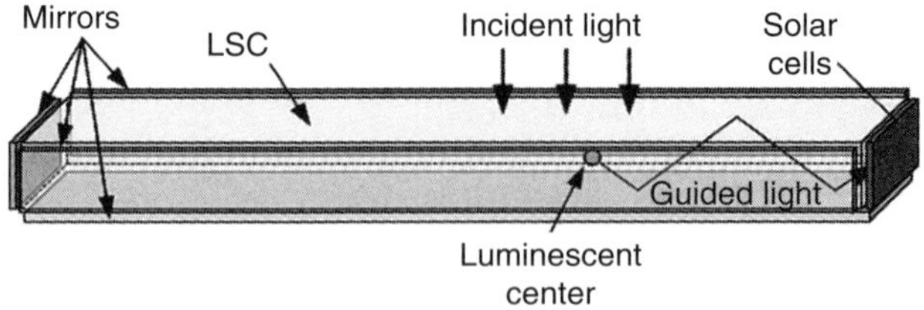

Figure 11.1 Luminescent solar concentrator.

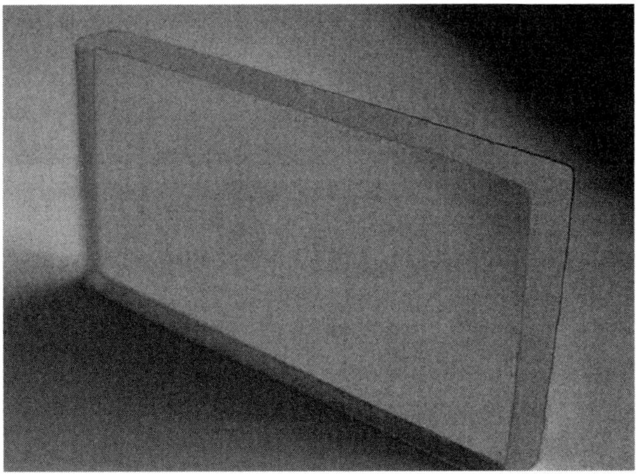

Figure 11.2 Photograph of a transparent plate containing CdSe core and multishell quantum dots [11].

by LSC systems is consumed directly by electricity customers. Transmission of the solar electricity across any electrical distribution systems can be avoided. As a result, power transmission losses can be reduced and technical issues caused by the integration of renewable energy sources with the distribution networks can be avoided [3–5].

Higher efficiency can be achieved by matching the peak quantum efficiency of the solar cells to the peak emission of a LSC. Several types of luminescent materials can be used in the LSC, such as laser dyes or organic dyes, semiconductor quantum dots [6], rare earth materials [7], and semiconducting polymers [8]. To further improve the efficiency of the LSC, materials such as photonic layers [9] and liquid crystals [10] have also been used to reduce the losses in the LSC.

To improve the radiation transfer of luminescent solar concentrators (LSCs), computer modeling of LSCs becomes very essential as it can be used to determine the optimum design parameters of LSCs. Those parameters are quantum efficiency of luminescent materials, concentration of luminescent materials, refractive index of the host, red-shift of the luminescent materials, thickness, geometries, diffuse or direct sunlight and surface contour of the LSCs. Computer simulation of the LSCs is required to fine-tune each of the parameters in order to achieve the maximum emission at the edges of the LSCs. This chapter reviews the recent findings on the development and research of LSCs, encompassing the modeling and experimental works.

11.2 Mechanisms of Power Losses in Luminescent Solar Concentrator

The optical quantum efficiency of a luminescent solar concentrator (LSC) is defined as the ratio between the incoming radiant flux (watts) and the outgoing power (watt). It is the fraction of the incoming energy that the LSC can deliver to the photovoltaic cells coupled at its edges. The concentrator factor is the ratio between the incoming irradiance and the emitted irradiance (w/m^2). If the incoming irradiance to the surface of the LSC is 1 kW/m^2, and the emitted irradiance received by the photovoltaic cells at the edges is 10 kW/m^2, then the concentration factor is 10. The concentration factor is a useful indication of how effective the LSC can concentrate the incoming irradiance. Solar cells without any LSC convert the incoming irradiance within a specified range of frequencies into useful output power. The output power is usually low because the solar cells are only sensitive to a narrow frequency bandwidth and the input irradiance is low and highly diffuse. The contribution of the LSC is to collect the incoming irradiance over a large area and produce emission at high intensity within the required range of the frequencies for the solar cells coupled at the edges of the LSC.

The majority of the emitted light is transported to photovoltaic (PV) devices coupled at the edge of the LSC sheet by total internal reflection. However, a minority of the emission can escape from the plate, hence reducing the quantum efficiency as well as the concentration factor of the LSC. There are two main power loss mechanisms in the LSC, namely reabsorption and escape cone losses as shown in Figure 11.3

Reabsorption loss is the process where the emitted photons are absorbed by the dye while traveling across the LSC. The amount of photons to be absorbed depends on the overlapping area between the absorption and emission spectra of the luminescent materials as indicated in Figure 11.4. If the overlapping area is large, then the amount of emitted light to be absorbed during the propagation is high. As a result, the total amount of emission reaching the edge of the LSC is reduced, hence lowering the luminescent quantum efficiency of the LSC. To minimize the reabsorption loss,

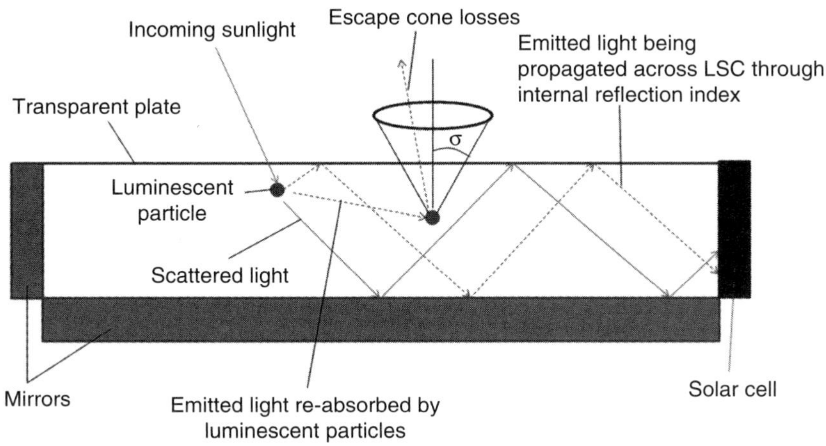

Figure 11.3 Cross-sectional diagram of a LSC with luminescent particles where σ is the critical angle of the escape cone.

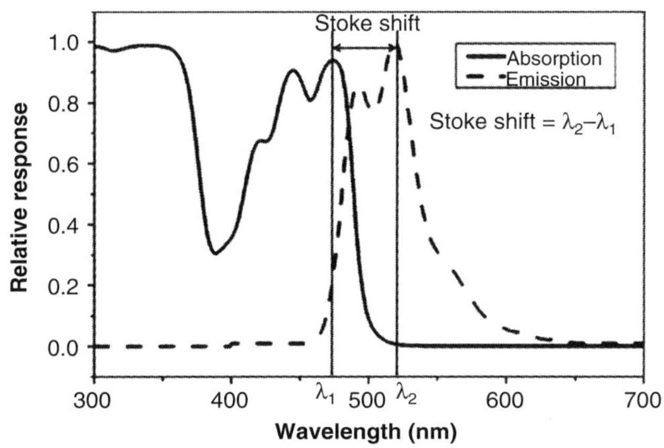

Figure 11.4 The Stoke's shift referred to as the distance between the emission and absorption peak for the luminescent material [12].

the luminescent particles to be used in the LSC should have a large Stoke's shift, the distance between the peaks of the absorption and emission spectra. Alternatively, the red-shift of the emission spectrum from a particular luminescent particle should be large. The concentration of the luminescent particles in the LSC is another factor affecting the performance of the LSC because the high concentration of dyes results in the high reabsorption loss. Therefore, the concentration of the dye must be optimum in order to maximize the collection of emission at the edge of the LSC. The geometric of the LSC is another factor influencing its luminescent quantum efficiency. This is because the emission at the edges of the LSC is the total contribution of emitted lights from luminescent particles that occupy a great area of the LSC. However, if the area of the LSC is too large, then the reabsorption loss can be significant to reduce the total emission. Therefore, the dimension of the LSC must be appropriate in order to achieve the maximum emission at its edges.

A small amount of the emitted light can escape from the transparent plate through a boundary known as the escape cone, which is bounded by the critical angle (σ) as shown in Figure 11.3. The critical angle (σ) of the escape cone depends on the refractive index difference between plate and air, where the plate must have the higher index. A large refractive index difference increases the amount of emission trapped inside the plate, hence reducing the loss of emission through the escape cone.

If the refractive index of the transparent plate is about 1.5, then the quantum efficiency of the LSC is about 75 to 80%, theoretically. Under ideal conditions, the efficiency of the LSC encompassing the photovoltaic cell is about 20%. Such a setup of the LSC with a high concentration factor and efficiency can offer an impressive cost reduction in the photovoltaic cells. However, the practical efficiency of the LSC is lower than its full potential. The reabsorption loss is the major cause of the low efficiency. Numerous research efforts have been carried out in order to increase the efficiency and concentration factor of the LSC.

11.3 Modeling

Currently, there are four approaches of modeling luminescent solar concentrators (LSCs), namely, thermodynamic modeling [13], ray-tracing modeling [14], hybrid model of thermodynamic and ray-tracing methods [15] and Monte Carlo method [16]. The details of the simulation methods are given below.

11.3.1 Thermodynamic Modeling

Thermodynamic modeling can be easily used to simulate a simple rectangular shape of LSCs under direct sunlight. The simulation results agree with the experimental results on test devices [6, 8].

The thermodynamic approach uses detailed balance arguments to relate absorbed and emitted light under equilibrium condition. The differential equations in the model were derived from the Chandrasekhar's general radiative transfer equation [17], simplified using two-stream approximation [18] with Schwarzschild-Milne [19] type of sampling. The equations were then integrated over the volume of the concentrator and applied with appropriate reflection boundary conditions. The thermodynamic model can predict the Stokes shift and calculate the total flux escaping each LSC surface [20].

Theoretical maximum concentration factor achieved by the LSC has been calculated by the thermodynamic model. An idealized LSC being doped with CdSe/ZnS quantum dots has been modeled, with mirrors attached at the three sides as well as the bottom of the LSC. GaInP solar cells are attached at the other side of the LSC. The quantum dots emission is tuned to match with the solar cell spectral response. The concentration factor of the idealized LSC is found to be 4.18 [2]. However, the concentration factor can be increased up to 12.98 when a wavelength-selective mirror is used on top of the LSC that reflects the dye emission back to the LSC and transmits the rest [20].

11.3.2 Ray Tracing Modeling

In standard ray-tracing modeling, a ray that represents a photon with a certain wavelength moving at a certain direction is traced from the light source until it leaves the LSC. The 3D ray-tracing model for the LSC extends the standard ray-tracing model to handle absorption and emission by luminescent dye particles in the LSC [21]. Instead of modeling individual luminescent dye particles, the model applies statistical averaging of the absorption to reduce computation time. Simulation results from this model agree with the experimental results of LSC reflection and transmission measurements, as well as its photoresponse.

A few case studies have been performed using the ray-tracing model to find the theoretically achievable LSC efficiency. The efficiency is calculated to be 2.45% for the ordinary setup of LSC doped with 2 different dyes with a monocrystalline silicon solar cell on one edge and mirrors on the three remaining edges as well as the bottom [22]. It is also noted that, by replacing the side mirrors on the edges by air-gap mirrors (mirror with air-gap between the mirror and LSC) and the bottom mirror by reflective Lambertian air-gap mirror, the efficiency of the LSC can be increased up to 2.94%. Reducing the reabsorption loss and increasing the refractive index of the transparent polymer matrix can increase the efficiency further to 3.8% [23]. By replacing the monocrystalline silicon solar cell with InGaP solar cells, the efficiency can be boosted up to 9.1% due to their larger open-circuit voltage [24]. Alternatively, adding a separate LSC plate with infra-red emitting luminescent dye at the bottom of the ordinary LSC can also increase the overall efficiency to 4.5% [25].

A set of numerical experiments have been performed by using the ray-tracing approach to evaluate the efficiency of LSC when the thickness and material properties of the outer layer are varied [26]. The rectangular and the cylindrical LSC are the two different shapes being studied under various conditions. The results show that the thickness of LSC and the attenuation coefficient are the two dominant factors that influence the performance of the cylindrical LSC as compared to the rectangular one.

11.3.3 Hybrid of Thermodynamic and Ray-Tracing Method

In the hybrid simulation approach, the two-flux thermodynamic model as developed by the authors in [13] is modified and used in conjunction with the open source ray-tracing program named Radiance [27]. The input parameters required by the modified two-flux thermodynamic model are LSC dimension, reflective index, luminescent dye concentration, absorption cross section, luminescent quantum efficiency and temperature. The input data required by the ray-tracing program are light source radiation spectrum and profile, LSC dimension, reflective index and absorption coefficient. The ray-tracing model simulates the average irradiation spectrum received by the photovoltaic cells without considering the contribution from the dye emission. It also simulates the average irradiation spectrum as seen by the dye particles at the top surface of LSC. The thermodynamic model simulates the average irradiation spectrum contributed by the dye emission. The average irradiation spectra from the two models are then combined to provide the overall irradiation output spectrum. This spectrum is considered as the incident light to the photovoltaic cells at the edge of LSC. The details of this hybrid simulation model are described in [15]. This hybrid simulation model has advantages over the existing thermodynamic and ray-tracing models in that it does not require a large amount of input data as compared to the thermodynamic model, and the simulation time is much less than the ray-tracing approach. The difference between the simulation and experimental results are within the range of 5 to 10% as reported in [15].

11.3.4 Monte Carlo Simulations

Monte Carlo simulation has been used to study the performance of a wedge-shaped LSC with respect to the conventional planar LSC with the same exposed area [16]. The results show that the average efficiency in which the planar LSC performs is about 6.3%, better than the efficiency of the wedge LSC which is about 3.5% when the sun is high in the sky. However, the efficiency of the wedge LSC increases up to 32.8%, much greater than that of the planar LSC which is only 7.6% when the sun is low in the sky. This makes the wedge LSC produce relatively higher electrical energy per square meter of PV cells than the planar LSC.

11.4 Polymer Materials

Poly(methyl methacrylate) (PMMA) is the standard host matrix for the luminescent solar concentrator (LSC). PMMA has relatively low cost, good optical clarity, high refractive index, and satisfactory photostability. However, there are a number of drawbacks of using PMMA. The photostability of luminescent particles in PMMA deteriorates very rapidly because the required additives react negatively with luminescent species or by the presence of monomer residues due to incomplete polymerization. Further research has been carried out to improve the stability of luminescent particles in transparent materials. Bomm *et al.* [11] synthesized the luminescent quantum dots with multiple inorganic shells. The multiple shells of quantum dots can be mixed well with the monomer of lauryl methacrylate (LMA) without any chemical reaction before poly(lauryl methacrylate) (PLMA) is formed. This is because multiple inorganic shells surrounding the quantum dots are able to protect the quantum dots from being attacked by radicals or additives, hence minimizing the luminescence quenching. Also, the agglomeration of the nanocrystals can be avoided by using PLMA.

Unsaturated polyester (UP) is a possible alternative host material for LSCs. Photodegradation of organic dyes during the curing process can be minimized because only thermal treatment is required to cure the mixture instead of ultraviolet curing, without needing to use any additives to initiate the polymerization. UP has double covalent bonds in its structure which will form a three-dimensional structure during the polymerization process, hence making its mechanical strength high. As a result, LSCs can be used as roofs, windows, walls or floors in order to increase sunlight captivity for electricity generation. To enhance the optical properties of LSCs, UP resin can be mixed with methylmethacrylate (MMA) to modify the chemical structure during polymerization. This modified structure has stable optical clarity and refractive index over a long period of thermal and ultraviolet treatments [28].

11.5 Luminescent Materials for Luminescent Solar Concentrator

11.5.1 Organic Dyes in LSC

Several types of organic dyes such as naphtalimide, perylene, coumarin and xanthene dyes have been used in luminescent solar concentrators (LSCs). The advantages of using organic dyes are their availability in a wide range of colors and also their extremely high luminescence [29]. The price of organic dyes is cheaper than inorganic dyes [28]. However, organic dyes have relatively large self-absorption of the dyes emission and short lifetime. The organic dyes in the LSCs will degrade rapidly after exposing to sunlight for several weeks. The degradation rate of organic dyes in polymer matrix is also influenced by the minor compounds, which may be present in the LSCs such as remaining monomer, stabilizers, initiator and other additives [29].

The PMMA luminescent concentrator has been doped with coumarin dyes under the commercial name of Makrolex fluorescent red G dye. Organic solar cells fabricated using zinc-phthalocyanine (ZnPc) and fullerene C_{60} have been attached to the concentrator as described in [30]. These solar cells have their main absorption in the wavelength range between 600 and 800 nm. The emissions spectrum of coumarin dyes in the LSCs is towards the red wavelength. If the second solar cells harvesting the red light are attached to the edges of the concentrator, the photocurrent density is increased compared to a single solar cell of equal active area from about 8.5 up to 10 mA/cm^2.

The performance and stability of luminescent flat plate concentrator (LFPC) in combination with monocrystalline silicon cells have been studied [31]. Different types of dyes being used in the study are listed in Table 11.1. These dyes have been embedded in four different types of matrices. The first matrix is P(MMA/HEMA) matrix by mixing MMA (methylmethacrylate monomer) with HEMA ((2-hydroxyethyl)methacrylate) monomer for synthesis. The second matrix is Plexit 55 Plate. The third is to mix PMMA and dye together before it is coated onto glass plates. The final one is to mix Paraloid B72 with the dyes before it is coated onto glass plates. The results show that the electrical current of silicon solar cell attached to the luminescent plate contacting Makrolex fluorescent red G is

Table 11.1 Overview of the organic dyes with the corresponding absorption, emission peak wavelengths and luminescent yields.

Commercial Product	Organic Dye	Absorption λ_{max} (nm)	Emission λ_{max} (nm)	Luminescence quantum yield (%)
Makrolex fluorescent red G	Coumarin	520	600	87
Lumogen F Blue 650	Naphthalimide	377	411	>80
Lumogen F Violet 570	Naphthalimide	378	413	94
Lumogen F Yellow 083	Perylene	476	490	91
Lumogen F Yellow 170	Perylene	505	528	>90
Lumogen F Orange 240	Perylene	524	539	99
S13	Perylene	526	534	100
Lumogen F Red 305	Perylene	578	613	98

improved by a factor of 1.5. In addition, screening of the stability of several other dyes indicates that the stability is strongly dependent on the types of dyes and the polymer matrix such as additives or the monomer residues. The most stable dyes tested are Lumogen F570, 650 and F300.

The surface and edge emissions from dye-filled and dye-topped polycarbonate and polymethyl methacrylate luminescent solar concentrators have been measured as described in [32]. Lumogen F Red 305 is used as fluorescent dye in this study. The surface energy losses from luminescent solar concentrator waveguide are more than 40%, translating into a photon loss of 50% to 70%. The losses from the top surface area are greater than the losses from the bottom for more heavily doped waveguides.

An organic solar concentrator with its surface being coated by a mixture of two dyes, namely 4-(dicyanomethylene)-2-t-butyl-6-(1,1,7,7-tetramethyljulolidyl-9-enyl)-4H-pyran (DCJTB) and platinum tetraphenyltetrabenzoporphyrin [Pt(TPBP)], has been constructed as described in [33]. Inorganic solar cells are attached at the edges of the concentrator. The quantum efficiencies of such a solar concentrator can be more than 50%, with the power conversion efficiency of approximately 6.8%.

The effect of adding white scattering layers to the bottom side of the luminescent solar concentrator waveguides are evaluated as illustrated in [34]. A polycarbonate sheet waveguide containing 35 ppm of Lumogen red 305 is used to study the effects of the scatterer. It is found that a rear scatterer separated from the waveguide by an air gap results in 37%–50% increase of energy output from the waveguides.

A method to evaluate the parameters defining the efficiency of luminescent solar concentrators is presented in [35]. Lumogen F Red 305 is dispersed in a PMMA layer on top of a transparent glass. Silicone grease is used as an index matching material between LSC and the solar cell. The light harvesting and self–absorption properties of the thin-film LSCs on glass substrates are determined by using optical spectroscopy. The maximum measured photon flux gain is 1.96.

The design, synthesis and study of a new organic dye, perylene, with a high Stokes shift of 300 meV are described in [36]. The LSC doped with the new organic dye has the fluorescent quantum efficiency of 70% with high chemical and photochemical stability. The quantum yield in the LSC is very high due to the completely separated absorption and emission peaks.

A study on the photodegradation mechanism of a perylene-based thin-film organic luminescent solar concentrator (OLSC) is discussed in [37]. By exposing the OLSC to the sunlight with the presence of oxygen, substantial changes to the molecular structure of the organic dye are noticed. Such changes are identified by means of photoluminescence, UV–vis and FTIR spectroscopy. The potential degradation mechanism in the dye molecule could therefore be proposed. These findings provide an important insight into the photodegradation mechanism of perylene-based OLSC devices. As a result, various stabilizing ideas to prolong the lifespan of OLSC could be developed. The use of radical scavenging molecules is one of the options for maintaining the photostability of the dyes. Hindered amine light stabilizers (HALS) can be used to scavenge radicals generated during the sunlight exposure. In addition, protective films with refractive index similar to that of the polymer matrix could be used to minimize the molecular damage.

The simulation and experimental studies that have been carried out to investigate the effects of dye concentration on the efficiency of a luminescent concentrator based on perylene diimide (PDI) derivatives in poly(methyl methacrylate) films coated on transparent plates are described in [38]. By increasing the PDI concentration, the dye can be aggregated, hence causing a large Stokes shift between emission and absorption spectra. This effect can reduce luminescence reabsorption and improve the efficiency of the concentrator. However, the aggregation of PDI can lower its fluorescence quantum yield and limit its overall absorption of light for luminescence yield.

11.5.2 Quantum Dots

Quantum dots have been doped in PMMA to produce LSCs. The possible advantages of using quantum dots over organic dyes are their ability to sustain its emission more than organic dyes over the exposure to ultraviolet light and the capability of collecting specific wavelengths in the solar spectrum by choosing the appropriate sizes of the quantum dots. However, the optical efficiency and the concentration factor of the quantum dot solar concentrators are restricted by the low luminescent quantum yields of the quantum dots and the large reabsorption loss. The luminescent quantum efficiency was found to be about 80%, lower than the 100% luminescent quantum efficiency of organic dyes [39]. Moreover, the preparation of good quality acrylic plates containing the quantum dots by polymerization is more complicated than those using organic dyes because the quantum dots are passivated by hydrophobic ligands, which causes the quantum dots to form turbid dispersions in the hydrophilic monomer such as MMA. The luminescence of hydrophobic quantum dots in hydrophilic media is quenched as a consequence of the formation of agglomerates [2].

Cadmium sulphide (CdS) is a luminescent quantum dot that has the potential to be applied in biology, solar cells and gas sensors. In 2008, CdS embedded in thin film by sol-gel spin coating on silica matrix was developed in order to study and control the optical properties of the quantum dot solar concentrator [40]. The CdS crystallite structure in the silica matrix is hexagonal and the size of the CdS crystallites is 3.5–6.5 nm. The CdS crystallite sizes increase with rising annealing temperature. A red-shift in the absorption and emission spectra is observed in the nanocrystalline CdS-doped silica by increasing the annealing temperature from 373 to 673 K in order to increase the particle size. The photostability of CdS-based solar concentrators is examined after 4 weeks of exposure to sunlight. The results show that the absorption loss increases after the sample is exposed to sunlight. The quantum dots are sensitive to oxygen and sunlight. Oxidation of the quantum dots occurs under sunlight. It is also shown that the sample annealed at lower temperature increases the emission at the edges of the luminescent solar concentrator.

LSCs made of the combined quantum dots, namely CdSe cores and ZnS shells, have been developed as described in [41]. The emission intensity and photostability of the combined quantum dot solar concentrators are assessed and compared with that of the LSCs containing the organic dye, Lumogen F Red 300 (LR). The measured fluorescence quantum yields of the combined quantum dots are lower than that of LR. However, the photodegradation rate of the quantum dot LSC is approximately five times slower than the LR LSC under the same sunlight exposure. The photodegradation of the quantum dot LSC's absorption completely recovers during a prolonged dark cycle.

The optical efficiency and concentration factor of a single-plate quantum dot solar concentrator using near infra-red (NIR) emitting quantum dots has been studied as described in [42]. The results show that the reabsorption losses can account for 58% and 57%

of incident photons absorbed in LSC if it contains the commercially available green and orange visible-emitting quantum dots. However, if the near infra-red emitting quantum dot is used in the LSC, the reabsorption loss can be reduced to 43% of incident photons. The optical efficiency is higher than that of the green and orange quantum dot solar concentrator.

Quantum dot LSC based on CdSe core and multishell quantum dots in a polymer matrix, namely poly(lauryl methacrylate-co-ethylene glycol dimethacrylate), is synthesized and characterized as discussed in [11]. The inorganic shells of the quantum dots consist of 2 monolayers of CdS, 3 monolayers of $Cd_{0.5}Zn_{0.5}S$ and 2 layers of ZnS. The multishells are able to avoid agglomeration of the nanocrystals. The results show that the inorganic fluorophores are stable under intense illumination over long periods of time. The LSC has a final quantum yield of 45% and an overall power conversion efficiency of 2.8%.

The simulation of the loss mechanisms in a LSC has been carried out in order to study the effects of the self-absorption by using type-II CdTe/CdSe heteronanocrystals with other luminophores [43]. The simulation results show that the self-absorption is very much reduced after type-II heteronanocrystals are used in the LSC. Therefore, the type-II semiconductor heteronanocrystals can be the best quantum dots for the LSC as far as efficiency is concerned.

Plasmonic excitation enhanced fluorescence of CdSe/ZnS core-shell quantum dots (QDs) in the presence of Au nanoparticles (NPs) has been studied for the application in the quantum dot solar concentrator [44]. In the optimal concentration of Au NPs, a maximum quantum efficiency of 53% is achieved for the particular QD/Au NP composite.

11.5.3 Rare Earth

Rare earth elements such as Europium (III) complex, Ytterbium (III), Neodymium (III) and Erbium (III) chelates have been used as luminescence materials in PMMA luminescent solar concentrator (LSC). Rare earth complexes are attractive because of their large stoke shifts, narrow emission bandwidths and long emission lifetime [45].

LSCs using neodymium and chromium doubly doped on the specially made lithium aluminium phosphate glass are described in [46]. The dependence of the light concentration coefficient versus dopant concentration and attenuation of the guided light in the LSC have been studied. The best results show that a 0.25m^2 triangular plate LSC receives 8-fold concentrated power at the edge, equivalent to 30W. The light concentration efficiency is estimated to be 3% for the 1 m^2 rectangular solar concentrator containing chromium and neodymiun.

A uniform and transparent LSC has been prepared by dip coating the organically modified silicates (Ormosils) solutions together with europium phenanthroline complex, $[Eu(phen)_2]CI_3$ into quartz plates as presented in [47]. The Ormosil solutions are derived from tetraethoxysilane (TEOS) and diethoxydiphenylsilane (DEDPS). The characteristics of the luminescent solar concentrators coupled with crystalline solar cells are studied. The results show that the photovoltaic outputs are increased 10–15% as compared with the values of uncoated LSC. This is because the Ormosil solution mixed with $[Eu(phen)_2]CI_3$ converts ultraviolent radiation to the high red emission to be utilized by the photovoltaic cells for power generation. Recently, a solar concentrator with minimum self-absorption loss and maximum geometric gain by doping a rare earth complex, $Eu(TTA)_3Phen$, into a polymer optic fiber has been achieved as presented in [48].

11.5.4 Semiconducting Polymer

The efficiency of the LSCs consisting of semiconducting polymers in liquid encased in glass has been studied and compared with that of organic dyes and quantum dots as discussed in [8]. The results suggest that commercially available quantum dots may not be viable LSC dyes because of their large absorption/emission band overlaps and relatively low quantum yields. Materials such as red polyfluorene (Red F) demonstrate that semiconducting polymers with high quantum yield and small absorption and emission band overlap are good candidates for LSCs.

11.6 New Designs of Luminescent Solar Concentrator

Recently, there have been several new designs of LSCs for achieving high-efficiency and concentrator factors. For example, LSC windows with front-facing photovoltaic (PV) cells have been developed as presented in [49]. In the conventional LSCs, a photovoltaic (PV) cell is attached at the edge. In this new design, the solar cells are placed facing the front together with the LSC. A piece of acrylic glass is used to cover the LSCs and the solar cells. In this case, the solar cells are able to capture both direct sunlight and waveguided luminescence emitted from the dye, Lumogen Red 305, embedded in LSC. The results show that, for the cell coverage of 5% out of the total area of the LSC, the power output of the solar cells can be increased by 2.2 times as compared to the solar cells alone. The design can achieve the lowest cost per watt if the power efficiency is made at 3.8% and the gain in the concentration factor is about 1.6 times.

Polysiloxane rubber has been used as the host material and doped with Lumogen Red 305 as reported in [50]. Polysiloxane rubber has the advantage of physical flexibility in that it can be easily bent or curved. Samples with different dye concentrations have been produced and their optical properties studied by absorption and fluorescence spectroscopy measurements. GaAs photovoltaic cells and a silicon photodiode have been coupled to the polysiloxane waveguides. Results show that polysiloxane rubber can be effectively employed as the host material of LSC for the collection, conversion and transmission of sunlight. The results also show that Lumogen Red can be dispersed in polysiloxane rubbers at concentrations up to 0.01 wt% for the maximum luminescence yield. If the concentration of the dye is further increased, then quenching effects become significant and hence the luminescence at the edges is reduced.

11.7 Conclusion and Outlook

The integration of luminescent solar concentrators (LSCs) with solar cells is a potential option for reducing the cost of solar electricity without any tracking and cooling systems. The LSCs can be used as building facades such as roofs, windows, walls and pavements. The electricity generated by the LSC system can be consumed on site. The circuit for the interconnection between the utility grid and the photovoltaic systems can be avoided. If the total efficiency of the LSC coupled with photovoltaic cells is improved to 20%, then the cost of the solar electricity can be significantly reduced and become very competitive with the cost of the fossil-fuel generated electricity.

In the past 20 years, numerous research efforts have been directed towards the improvement of the LSCs. Several simulation tools have been developed to study the effects of adjusting several parameters of the LSCs for reducing energy losses across LSC. The energy losses can be reduced by minimizing the overlapping area between the absorption and emission spectral and increasing the refractive index of the host matrices of LSCs. The most recent findings show that wedge LSC can perform much better than planar LSC when the sun is low in the sky. The maximum efficiency that can be achieved by the wedge LSC is about 32%.

Poly(methyl methacrylate) (PMMA) often seems to be the standard host matrices for LSCs. This is because it is cheap, has excellent optical clarity and high refractive index, and is stable under a prolonged intense illumination. However, the photostability of the luminescent particles can degrade very rapidly in PMMA because the required additives can react with the luminescent particles, hence causing the luminescence quenching of the particles. Multiple shells of quantum dots are then proposed to be used in PMMA because the shells can protect the quantum dots from being attacked by additives. Recently, unsaturated polyester (UP) has been found to be another possible host material for the LSCs. The photogradation of the dyes can be reduced because additives are not required to form the transparent glass from UP.

There are a several types of luminescent materials being used in LSCs. Organic dyes are popular materials because they have high luminescent yield, and low self-absorption loss. Furthermore they are a cheap material. However, their luminescence degrades in the presence of the additives used in the synthesis of the concentrator. Therefore, quantum dots with multishells become a possible alternative luminescent material for the LSCs. The multishells are able to protect the quantum dots from agglomeration, making the quantum dots stable under intense illumination over a period of time. The incorporation of Au nanoparticles (NPs) into CdSe/ZnS core-shell quantum dots has made LSCs with an improved efficiency of 53%. Rare earth elements and semiconducting polymers are other possible luminescent materials for the LSCs because of their large Stoke's shifts, narrow emission bandwidths and long emission lifespan. A few new designs of LSCs have been proposed and proved to be an effective means of concentrating the incoming sunlight for producing solar electricity at a reduced cost. It is strongly believed that the continuous efforts of research and development could lead to the production of an effective luminescent solar concentrator so that the affordable cost of solar electricity can be realized one day.

References

1. B.C. Rowan, L.R. Wilson and B.S. Richards, *Selected Topics in Quantum Electronics, IEEE Journal of*, Vol. 14, p. 1312–1322, 2008.
2. W.G. Van Sark, K.W. Barnham, L.H. Slooff, A.J. Chatten, A. Büchtemann, A. Meyer, S.J. Mc.Cormack, R. Koole, D.J. Farrell, R. Bose, E.E. Bende, A.R. Burgers, T. Budel, J. Quilitz, M. Kennedy, T. Meyer, S.H. Wadman, G.P. van Klink, G. van Koten, A. Meijerink and D. Vanmaekelbergh, *Optics Express*, Vol. 16, p. 21773–21792, December 22, 2008.
3. P. Trichakis, P. Taylor, P. Lyons and R. Hair, *Renewable Power Generation, IET*, Vol. 2, p. 249–262, 2008.
4. L. Cipcigan and P. Taylor, *Renewable Power Generation, IET*, Vol. 1, p. 160–166, 2007.
5. N. Stannard, J. Bumby, P. Taylor and L. Cipcigan, *Renewable Power Generation, IET*, Vol. 1, p. 123–130, 2007.
6. K. Barnham, J.L. Marques, J. Hassard and P. O'Brien, *Applied Physics Letters*, Vol. 76, p. 1197, 2000.
7. M.H.V. Werts, J.W. Hofstraat, F.A.J. Geurts and J.W. Verhoeven, *Chemical physics letters*, Vol. 276, p. 196–201, 1997.
8. V. Sholin, J. Olson and S. Carter, *Journal of applied physics*, Vol. 101, p. 123114–123114, 2007.
9. U. Rau, F. Einsele and G. Glaeser, *Appl. Phys. Lett.*, Vol. 87, p. 171101–171103, 2005.
10. M. Debije, D. Broer and C. Bastiaansen, in: *Proc. 22nd Eur. Photovolt. Solar Energy Conf. Exhib.*, 2007, pp. 87–89.
11. J. Bomm, A. Büchtemann, A.J. Chatten, R. Bose, D.J. Farrell, N.L.A. Chan, Y. Xiao, L.H. Slooff, T. Meyer, A. Meyer, W.G.J.H.M. van Sark and R. Koole, *Solar Energy Materials and Solar Cells*, Vol. 95, p. 2087–2094, August 2011.
12. A. Earp, G. Smith, P. Swift and J. Franklin, *Solar Energy*, Vol. 76, p. 655–667, 2004.
13. A. Chatten, D. Farrell, B. Buxton, A. Büchtemann and K. Barnham, in: *Proc. 21st Eur. Photovolt. Solar Energy Conf.*, 2006, pp. 315–319.
14. S. Gallagher, P. Eames and B. Norton, *International Journal of Ambient Energy*, Vol. 25, p. 47–56, 2004.
15. C.K. Lo, Y.S. Lim, S.G. Tan and F.A. Rahman, *Energies*, Vol. 3, p. 1831–1860, 2010.
16. M.D. Hughes, C. Maher, D.-A. Borca-Tasciuc, D. Polanco and D. Kaminski, *Renewable Energy*, Vol. 52, p. 266–272, 2013.
17. S. Chandrasekhar, *Radiative Transfer*, Dover Publications, 1960.
18. R. Siegel and J. Howell, *Thermal Radiation Heat Transfer*, New York: McGraw-Hill, 1972.
19. E.A. Milne, *Monthly Notices of the Royal Astronomical Society*, Vol. 81, p. 361–375, 1921.

20. A.J. Chatten, D.J. Farrell, R. Bose, M.G. Debije, A. Büchtemann and K.W.J. Barnham, in: *Proceedings of Twenty Second European Photovoltaic Solar Energy Conference*, 2007, pp. 349–353.
21. A.R. Burgers, L.H. Slooff, R. Kinderman and J. Van Roosmalen, in: *Proc. 20th Eur. Photovolt. Solar Energy Conf.*, 2005, pp. 394–397.
22. L.H. Slooff, R. Kindermana, A.R. Burgersa, A. Büchtemannb, R. Danzb, T.B. Meyerc, A.J. Chattend, D. Farrelld, K.W.J. Barnhamd and J.A.M. van Roosmalena, in: *Proceedings of SPIE*, 2006, pp. 61970k1–8.
23. P. Polishuk, *Communications Magazine, IEEE*, Vol. 44, p. 140–148, 2006.
24. A. Zastrow, Physikalische Analyse Der Energievelustmechanismen Im Fluoreszenzkollektor, Ph.D. Thesis, 1981.
25. B.S. Richards and K.R. McIntosh, in: *Proceedings of the 21st European Photovoltaic Solar Energy Conference*, 2006, pp. 185–188.
26. G. Colantuono, A. Buckley and R. Erdelyi, *Journal of Lightwave Technology*, Vol. 31, p. 1033–1044, 2013.
27. G. Larson and R. Shakespeare, *Rendering with Radiance*, San Francisco, Calif: Morgan Kaufmann, 1998.
28. Y.S. Lim, C.K. Lo and G.B. Teh, *Renewable Energy*, Vol. 45, p. 156–162, September 2012.
29. W.G. Van Sark, K.W. Barnham, L.H. Slooff, A.J. Chatten, A. Büchtemann, A. Meyer, S.J. Mc.Cormack, R. Koole, D.J. Farrell, R. Bose, E.E. Bende, A.R. Burgers, T. Budel, J. Quilitz, M. Kennedy, T. Meyer, S.H. Wadman, G.P. van Klink, G. van Koten, A. Meijerink and D. Vanmaekelbergh, *Optics Express*, Vol. 16, p. 21773–21792, December 22, 2008.
30. R. Koeppe, N. Sariciftci and A. Buchtemann, *Applied Physics Letters*, Vol. 90, p. 181126–181126, 2007.
31. R. Kinderman, L. Slooff, A. Burgers, N. Bakker, A. Büchtemann, R. Danz and J. van Roosmalen, *Journal of solar energy engineering*, Vol. 129, p. 277–282, 2007.
32. M.G. Debije, P.P.C. Verbunt, B.C. Rowan, B.S. Richards and T.L. Hoeks, *Applied Optics*, Vol. 47, p. 6763–6768, 2008.
33. M.J. Currie, J.K. Mapel, T.D. Heidel, S. Goffri and M.A. Baldo, *Science*, Vol. 321, p. 226–228, 2008.
34. M.G. Debije, J.P. Teunissen, M.J. Kastelijn, P.P.C. Verbunt and C.W.M. Bastiaansen, *Solar Energy Materials and Solar Cells*, Vol. 93, p. 1345–1350, 2009.
35. T. Dienel, C. Bauer, I. Dolamic and D. Brühwiler, *Solar Energy*, Vol. 84, p. 1366–1369, 2010.
36. A. Sanguineti, M. Sassi, R. Turrisi, R. Ruffo, G. Vaccaro, F. Meinardi and L. Beverina, *Chemical Communications*, Vol. 49, p. 1618–1620, 2013.
37. G. Griffini, L. Brambilla, M. Levi, M. Del Zoppo and S. Turri, *Solar Energy Materials and Solar Cells*, Vol. 111, p. 41–48, April 2013.
38. C. Haines, M. Chen and K.P. Ghiggino, *Solar Energy Materials and Solar Cells*, Vol. 105, p. 287–292, October 2012.
39. X. Peng, M.C. Schlamp, A.V. Kadavanich and A. Alivisatos, *Journal of the American Chemical Society*, Vol. 119, p. 7019–7029, 1997.
40. S. Reda, *Acta Materialia*, Vol. 56, p. 259–264, 2008.
41. M.G. Hyldahl, S.T. Bailey and B.P. Wittmershaus, *Solar Energy*, Vol. 83, p. 566–573, 2009.

42. M. Kennedy, S. McCormack, J. Doran, and B. Norton, *Solar Energy*, Vol. 83, p. 978–981, 2009.
43. Z. Krumer, S.J. Pera, R. van Dijk-Moes, Y. Zhao, S. de Brouwer, E. Groenveld, W. van Sark, R. Schropp and C. de Mello Donega, in: *Renewable Energy and the Environment Optics and Photonics Congress*, Optical Society Of America, November 11, 2012, p. PW2B.3.
44. S. Chandra, J. Doran, S.J. McCormack, M. Kennedy and A.J. Chatten, *Solar Energy Materials and Solar Cells*, Vol. 98, p. 385–390, March 2012.
45. H. Jiu, J. Ding, Y. Sun, J. Bao, C. Gao and Q. Zhang, *Journal of Non-Crystalline Solids*, Vol. 352, p. 197–202, 2006.
46. B. Jeżowska-Trzebiatowska, E. Łukowiak, W. Stręk, A. Buczkowski, S. Patela, J. Radojewski and J. Sarzyński, *Solar Energy Materials*, Vol. 13, p. 267–277, 1986.
47. K. Machida, H. Li, D. Ueda, S. Inoue and G. Adachi, *Journal of Luminescence*, Vol. 87, p. 1257–1259, 2000.
48. W. Wu, T. Wang, X. Wang, S. Wu, Y. Luo, X. Tian and Q. Zhang, *Solar Energy*, Vol. 84, p. 2140–2145, 2010.
49. C. Corrado, S.W. Leow, M. Osborn, E. Chan, B. Balaban and S.A. Carter, *Solar Energy Materials and Solar Cells*, Vol. 111, p. 74–81, 2013.
50. M. Buffa, S. Carturan, M.G. Debije, A. Quaranta and G. Maggioni, *Solar Energy Materials and Solar Cells*, Vol. 103, p. 114–118, August 2012.

12

Luminescent Solar Concentrators – State of the Art and Future Perspectives

M. Tonezzer[1,2,*], D. Gutierrez[2], and D. Vincenzi[3]

[1]*University of Trento, Department of Materials Engineering and Industrial Technologies, Trento, Italy*
[2]*Leitat Technological Center, Barcelona, Spain*
[3]*University of Ferrara, Department of Physics and Earth Science, Ferrara, Italy*

Abstract

In this chapter the historical development and current status of luminescent solar concentrators (LSCs) are reviewed. LSCs are novel, low-cost photovoltaic devices basically constituted by highly luminescent slab coupled with small, high-efficiency solar cells. LSC devices convert both direct and diffuse sunlight, are very flexible covering a variety of possible shapes and colors, and are predicted to offer potentially much lower cost per power unit compared to conventional photovoltaic systems. These features give them the capacity to find extended use on the market, especially as building-integrated photovoltaic (BIPV) systems. LSC devices, their working principle, and the mechanisms which influence their efficiencies are discussed in this chapter. In addition, various designs, material aspects, and recent experimental results are presented. Finally, some possible pathways for future developments are offered, with additional research paths that could result in the affirmation of LSC devices in the market within a few years.

Keywords: Photovoltaic, concentrating photovoltaic (CPV) systems, building-integrated photovoltaic (BIPV), luminescence solar concentrators, organic photovoltaic

**Corresponding author*: Michele.Tonezzer@lnl.infn.it

Atul Tiwari, Rabah Boukherroub, and Maheshwar Sharon (eds.) Solar Cell Nanotechnology, (293–316) 2014 © Scrivener Publishing LLC

12.1 Introduction to the Third Generation of Photovoltaic Systems

Global energy demand is projected to double by mid-century. Procuring adequate energy supplies without large carbon dioxide emissions is one of society's most pressing objectives. Solar power is the most promising energetic source for exceeding this challenge. In fact, it is plentiful, widely distributed, available to all, and blind to geographical and geological luck. One of the major technologies for transducing sunlight to electrical power is the photovoltaic (PV) one. Unfortunately, the prices of PV modules are not actually competitive with fossil energy sources principally due to the high price of semiconductor materials. In order to become competitive, photovoltaic electricity prices need to drop below 1 €/Wp.

Several routes are being pursued to reach lower cost per installed capacity (€/W) within "Third Generation PV," all directed towards an improvement of the system's efficiency and the reduction of the semiconductor quantity within the system. The most promising solution is represented by concentrating photovoltaic (CPV) systems, in which optical concentrators allow for a reduction in semiconductor quantity and an increase in system efficiency by using high-efficiency small solar cells (i.e. III-V semiconductors). There are two main typologies of CPV systems: the "imaging" concentrators which accurately track the sun across the sky and are unable to convert diffuse solar radiation, and "non-imaging" concentrators. Among this last category, one of the most promising technologies is represented by the luminescent solar concentrators (LSCs).

12.2 Luminescence Solar Concentrators (LSCs)

12.2.1 Description of LSC Devices

Luminescent solar concentrators (LSCs) are static concentrators basically constituted by slabs of transparent materials with added luminescent dyes and equipped with arrays of small size photovoltaic cells (PV) (see Figure 12.1). Other forseen LSC designs are luminophores deposited as thin film on the top of the waveguide or contained as liquids between the two glass plates [1–3]. However this chapter will focus on bulk LSC devices.

LSC devices work as follows: i) the luminescent dyes absorb the incident solar radiation re-emitting photons at longer wavelength;

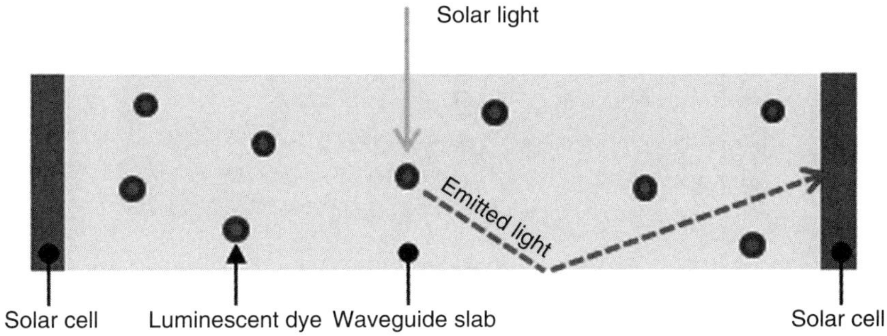

Figure 12.1 Working principle of LSCs. Incoming sunrays (yellow arrow) enter the waveguide and are absorbed by the luminophores which re-emit the light at longer wavelengths (red arrow). The light is trapped by the slab through total internal reflection (TIR) and brought to the solar cells which convert the light into electricity.

ii) the slab, by a process of total internal reflection (TIR), directs the emitted light on the PV cells arranged on its edges; iii) the solar cells convert the incident photons into electrical energy.

LSC devices show several advantages over CPV ones. Firstly they are able to use the diffuse component of solar radiation [4, 5], which is very important for countries with frequent cloud coverage or areas of persistent shady conditions (i.e. northern part of US and central Europe) and more generally in building environments (i.e. cities). Secondly LSCs are architecturally highly integrable: in fact, they can be made in a variety of colors, shapes, and transparencies, Moreover they are flexible, do not require expensive and voluminous solar tracking, and weigh less than Si panels, conferring them the viability for mounting to the side of a building [6]. Furthermore LSCs can be optimized for matching the luminescence energy and the solar cell spectral response: this way spectral and thermalization losses, which represent severe drawbacks in CPV technology, can be strongly reduced. Finally LSC devices promise very low production costs [7]: in particular, it has been estimated that LSC devices could produce electricity at around 30–50% of the cost of conventional PV [3].

12.2.2 The Efficiency and Losses Mechanism in LSC Devices

A scheme of a LSC device is shown in Figure 12.2.

LSC conversion efficiency depends on several mechanisms: one of the main objectives of the field consists in optimizing their

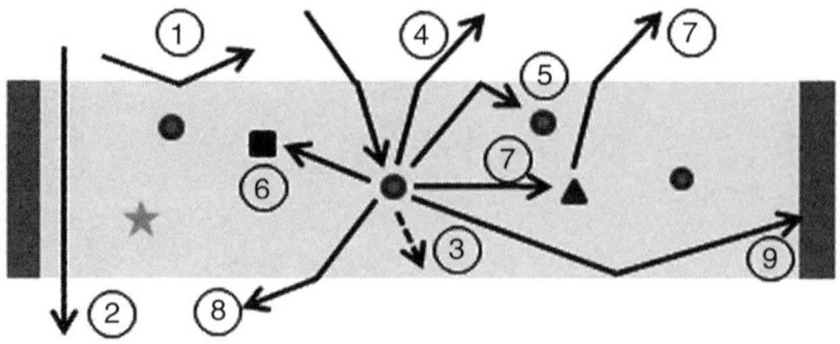

Figure 12.2 Loss mechanisms in LSC devices: (1) solar light reflected from the waveguide surface; (2) solar light is not absorbed by the luminophore; (3) internal quantum efficiency of the luminophore is less than one; (4) light emitted outside of "capture cone"; (5) reabsorption of emitted light by another luminophore; (6) absorption of emitted light by the waveguide; (7) internal waveguide scattering; (8) surface scattering; and (9) solar cell losses.

efficiencies. Loss mechanisms occurring in LSC devices can be distinguished in four main groups.

The first loss regards the solar light which does not reach the luminescent dye due to the reflection from the top slab surface (Figure 12.2(1)).

The second group of losses is imputable to the luminophores:

i. as the luminophores have specific absorption window, a part of the incident sunlight is lost through the bottom surface (Figure 12.2(2));
ii. some of the absorbed photons are not re-emitted by the luminophores but lost as heat due to emission quantum yields (QY < 1) (Figure 12.2(3));
iii. light emitted by the luminophore is refracted out of the waveguide through an «escape cone» rather than being reflected internally (Figure 12.2(4))
iv. re-absorption of the emitted photons by other dyes present in the slab due to the partial overlap of the emission and absorption bands of the luminophores (Figure 12.2(5)).

The third group of losses is due to the hosting slab and comprises:

i. slab material never is completely transparent over all the wavelength range of the emitted light giving a loss caused by parasitic absorption (Figure 12.2(6));
ii. imperfections in the «waveguide» bulk can scatter the emitted photons directed to the PV cells (Figure 12.2(7));

iii. the imperfections present on the surface of the «waveguide» slab can generate the loss of some photons (Figure 12.2(8)).

The final loss can be imputed to the photovoltaic cells: PV cells, in fact, have a nonuniform spectral response with a fraction of incident photons being lost due to the finite conversion efficiency of the PV cells (Figure 12.2(9)).

The overall LSC optical efficiency of an LSC (η_{opt}) indicates the fraction of incident solar power which reaches the edges and can be expressed as the product of the efficiencies of the different processes present inside the LSC [8],

$$\eta_{opt} = (1 - R)\, P_{TIR}\eta_{abs}\eta_{QY}\eta_{Strokes}\eta_{host}\eta_{TIR}\eta_{self} \tag{12.1}$$

where R is the surface reflection coefficient, P_{IR} is the total internal reflection efficiency, η_{abs} the absorption efficiency (the fraction of sunlight absorbed by the luminescent species), η_{QY} is the luminescent quantum yield of the luminescent species, η_{Stokes} is the Stokes efficiency, which is the energy loss between absorption and emission, η_{host} is the transport efficiency of the emitted light through the waveguide, η_{TIR} is the reflection the efficiency of the waveguide due to the quality of the waveguide surface, and η_{self} is the transport efficiency of the slab related to self-absorption of the luminescent species.

LSC conversion efficiency can be expressed as the product of the LSC optical efficiency (η_{opt}) and the efficiency of the solar cell (η_{PV}) (Eq. 12.2).

$$\eta_{LSC} = \eta_{opt}\eta_{PV} \tag{12.2}$$

The significance and magnitude of each term reported in Eq. 12.2 are discussed below.

Fraction of incident light transmitted into sheet (1-R), There is a small loss of incident light from the front surface of the LSC sheet due to Fresnel reflection (Eq. 12.3),

$$R = \frac{(n-1)^2}{(n+1)^2} \tag{12.3}$$

where n is the refractive index.

The reflectivity is approximately 4% for a slab with n = 1.5.

Probability of total internal reflection (P_{TIR}), Fluorescence light is trapped inside the LSC when it strikes the internal surface of the sheet at an angle greater than the critical angle Θ_C, where,

$$\Theta_c = \sin^{-1}\left(\frac{1}{n}\right) \tag{12. 4}$$

The probability that fluorescence emission is trapped inside the sheet, P_{TIR}, is given by Eq. 12.5.

$$P_{TIR} = \frac{\sqrt{n^2 - 1}}{n} \tag{12.5}$$

For a sheet with n = 1.5, the probability that a fluorescence photon reaches the edge of the sheet is approximately 75% [1].

Absorption efficiency (η_{abs}) is related to the fact that LSC, as common solar cells, LSC sheets absorb optical bands narrower than solar spectrum. For maximizing the power output, need to absorb the largest spectral window: for example, an ideal Si-based LSC which completely absorbs sunlight with $\lambda < 950$ nm can reach $\eta_{abs} = 0.71$ [9].

Quantum yield efficiency (η_{QY}) determines the probability that an excited fluorophore molecule will decay by the emission of a fluorescence photon (Eq. 12.6).

$$\eta_{QY} = \frac{Emitted\_Photons}{Excited\_Molecules} \tag{12. 6}$$

High quantum yields ($\eta_{QY} > 0.95$) are necessary for good LSC performance, especially for fluorophores with reabsorption losses: in fact, since the same QY rules all reabsorption events, the effects of η_{QY} are magnified by multiple reabsorptions.

"Stokes efficiency"(η_{Stokes}). Emitted photons are always characterized by longer wavelength than absorbed ones. This results in an energy loss, described by η_{Stokes} which generally is approximately 0.75.

"Host efficiency" (η_{host}) represents the fraction of emitted light transmitted by the host: generally it features high values at visible wavelengths [9] (i.e. visible η_{host} of PMMA = 0.95–0.98) but can be

Table 12.1 Experimental loss mechanism efficiencies measured for a 400mm * 400mm * 3mm LSC.

Quantity	Efficiency (%)
1–R	96
P_{TIR}	75
η_{abs}	20 – 30
η_{QY}	95 – 100
η_{Stokes}	75
$\eta_{host} + \eta_{TIR}$	90 – 95
η_{self}	75

lower for polymer-based LSC devices operating at near-infrared ($\lambda > 700$ nm) due to C-H and C-O bond stretches [10].

Efficiency of total internal reflection **(η_{TIR}).** Theoretically η_{TIR} should be 100%. However, the presence of defects on the sheet surface (i.e. dust, moisture droplets, physical damage) can cause light to be scattered out of the sheet. Nevertheless optimized and carefully cleaned PMMA surfaces generally feature high values ($\eta_{TIR} = 0.9998$ [11]).

Self-absorption (η_{self}) is related to the fraction of fluorescence photons lost due to the reabsorption process which occurs when there is a partial overlapping between absorption and emission spectra of luminophores. In fact, although subsequent emission can occur, the new photon will be emitted in a random direction increasing the probability of its leaving the sheet via the escape cones [12].

Experimentally determined loss mechanism efficiencies for a large size LSC are summarized in Table 12.1.

12.3 Components of LSC Devices

The elements which constitute LSC devices can be broadly subdivided into three main categories: i) "waveguide" slabs; ii) fluorophore dyes; and iii) solar cells. Descriptions of the materials commonly used are reported below.

12.3.1 Waveguide Slab

The host material of the LSC slab serves both as a medium into which the fluorophores are doped and as "waveguide." The most common host material is poly(methyl methacrylate) (PMMA) [13–16] thanks to its multiple advantages: i) low cost; ii) high optical transparency in the visible region of the spectrum [9, 17, 18]; iii) ease of doping with fluorophores, which can simply be dissolved in the monomer prior to polymerization; iv) good photostability [19]. PMMA shows absorption peaks above 700 nm [10] which can affect the LSC efficiency with NIR emitting dyes. Moreover the C-H and C-O vibrations can lead to an increased probability of non-radiative relaxation of the fluorophore, leading to a reduction in quantum yield [20]. Several attempts have been proposed in order to reduce these problems [21]: for example, deuterium and fluorine atoms have been added to the polymer in order to red-shift the absorption peaks thus reducing host absorption.

Two-component clear epoxy resins, characterized by around 30% lower absorption in the visible region than PMMA [10], have also been proposed as they can set at room temperature: however, epoxy resins show a lower photostability than PMMA.

Glass has also been used as a host material [16]. It has a high transparency at NIR wavelengths (>700 nm) and is not subject to photodegradation. Its major drawback, as the glass must be prepared from a melt at high temperatures (usually > 600°C) [22], is that the choice of fluorophores is restricted to those compounds which can withstand the required temperatures. Organic dyes, rare-earth complexes and quantum dots cannot be incorporated into the glass: nevertheless this compound can be deposited as thin films on glass slabs. Glass can also be prepared from a sol-gel: different fluorophores including organic dyes and rare-earth or inorganic ions have been inserted in sol-gel glasses [23].

Liquids contained between two glass plates to form a large, thin cuvette have also been used as hosting material [16] in order to prevent the limited photostability of the used organic dyes by replacing them once they had photodegraded. In fact, this solution allows for replacement of the dye once it had photodegraded. However, the development of stable dyes renders this an unnecessary method.

12.3.2 Fluorophore

Fluorophores are the driving force for concentration in LSCs and, as seen in Eq. 12.2, several potential loss mechanisms depends on them: i) the fraction of solar light absorbed, ii) the photoluminescent quantum yield; iii) the energy lost due to the heat generation during the absorption and emission events; iv) the transport efficiency of the waveguided photons related to reabsorption of the emitted photons; v) the spectrum of the photons reaching the PV cell. An effective luminophore must meet different requirements: i) broad spectral absorption; ii) high absorption efficiency over the whole absorption spectrum; iii) large Stokes shift; iv) high QY; v) good matching between the dye emitted spectrum and the external quantum efficiency (EQE) of the PV-cell; vi) high radiation hardness.

Different luminophores have been proposed and studied which can be subdivided into three categories: i) organic dyes; ii) quantum dots; iii) rare-earth materials.

12.3.2.1 *Organic Fluorescent Dyes*

Organic dyes are organic molecules in which all atoms of the conjugated chain lying in a common plane are linked by π -bonds. The π-electrons form a cloud around this plane resulting in absorption bands due to their promotion from a ground energy state to an excited higher energy state [24]. Organic dyes fluorescent processes can be basically described as follows: i) upon absorption of a photon an electron is excited from the ground electronic state (S_0) to one of the vibrational levels of the first excited singlet state (S^*); ii) the excited electron decays non-radiatively by internal conversion to a lower energetic triplet state (T^*); iii) the excited electron decays to one of the vibrational levels in S_0 by emission of a fluorescence photon.

Organic dyes have been widely investigated as luminophores in LSC devices as they feature high QY and can be easily incorporated in LSC slabs by dissolving them in a range of organic polymers. The main dyes proposed for LSCs applications are rhodamines [25, 26], coumarins [17, 27], dicyano methylenes [7, 17, 28], naphtalimides [29], and perylenebisimides [30, 31] derivates (see Figure 12.4).

Rhodamines and coumarins have been proposed in the earliest stages of LSC research, while the perylenes and perylenebisimides are mostly mentioned in the more recent papers. Rhodamines are known for their high QY and high molar extinction coefficients,

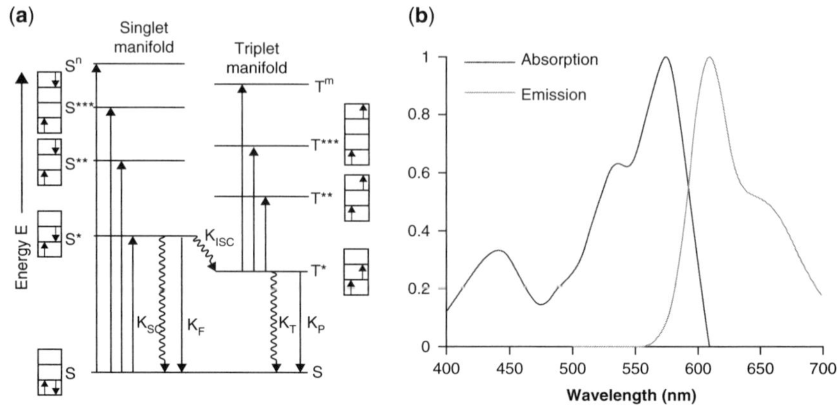

Figure 12.3 (a) Simplified Jablonski diagram depicting energy levels and pathways of activation and deactivation in organic molecules containing two electrons in their outer orbita. (b) Absorption and emission spectra of Lumogen F RED 305.

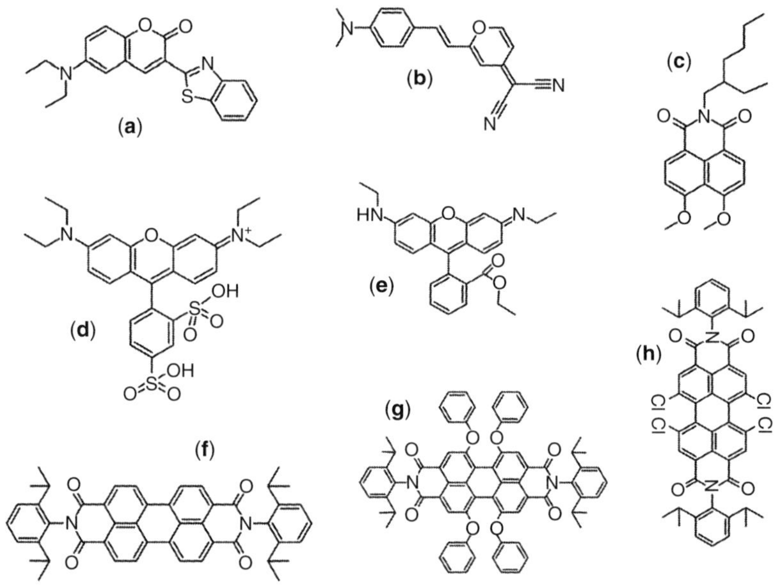

Figure 12.4 Structural formula of the main dyes used in LSCs: a) coumarin 6, b) DCM (4-dicyanomethylene-2 methyl-6-p-dimethylaminostyryl-4H-pyran), c) a naphtalimide derivate, d) sulfoRhodamine B, e) Rhodamine 6G, f) perylene-3,4,9,10 tetracarboxylic acid-bis-(2′-6′diisopropylanilide) (Perylenebisimide), g) perylene 1,7,8,12-tetraphenoxy-3,4,9,10 tetracarboxylic acid-bis-(2′-6′diisopropylanilide) (Perylenebisimide), h) perylene-1,7,8,12-tetrachloro-3,4,9,10 tetracarboxylic acid-bis-(2′-6′diisopropylanilide) (Perylenebisimide).

but their small Stokes shifts values lead to much reabsorption and consequently to increased LSC losses. Coumarins compounds also show high QY ranging from 87% to 98%. Coumarins show a photostability higher than rhodamines, but lower than perylene-based dyes. Perylene dyes and their derivates are characterized by intense fluorescence and good photostability: moreover by changing their side groups or the length of their core it is possible to alter their structural and optical properties, allowing for engineering of more suitable and robust molecules.

Two different types of dicyano methylenes have also been used in LSCs (see Figure 12.4): DCM and DCJTB [20]. DCM features a broad absorption spectrum, a large Stokes shift, and a reasonable quantum yield (≈ 80% in PMMA): nevertheless it suffers from a limited photostability. DCJTB used in combination with a platinum–porphyrin derivate allowed for production of a GaAs-coupled LSC with a 6.8% power conversion [20].

Other dyes have also been investigated in LSCs such as bipyridines [4], phtalocyanines [32], hematoporphyrin, and phycobilisomes [33].

Besides their very good properties, organic dyes are affected by some disadvantages such as low Stokes shift, narrow absorption bands, and low photostability. To solve these drawbacks several solutions are actually under investigation (see Section 12.4).

12.3.2.2 Quantum Dots

Quantum dots (QDs) promise several advantages in LSC applications over organic dyes such as broad absorption spectra and high absorption coefficients. Moreover the QDs absorption threshold can be tuned both by altering the diameter of the quantum dot [34, 35] and by varying QD materials. Also if QDs generally show a quite large absorption/emission spectra overlap, QDs with larger Stokes shift have been produced [36, 37].

For these reasons several authors have proposed LSC devices based on QDs [38, 39], among which the most used are cadmium selenide/zinc sulphide (CdSe/ZnS), lead sulphide (PbS), and lead selenide (PbSe). In particular, PbS QDs absorb all wavelengths below 900 nm and show a large Stokes shift (122 nm) [40]. Kennedy *et al.* showed that the optical absorption efficiency for NIR, orange, and green-emitting Cd-based QDs was, respectively, 23.1%, 21.7%, and 11.6%. PbS QDs based LSC show an optical efficiency of 12.6% [41] (Figure 12.5).

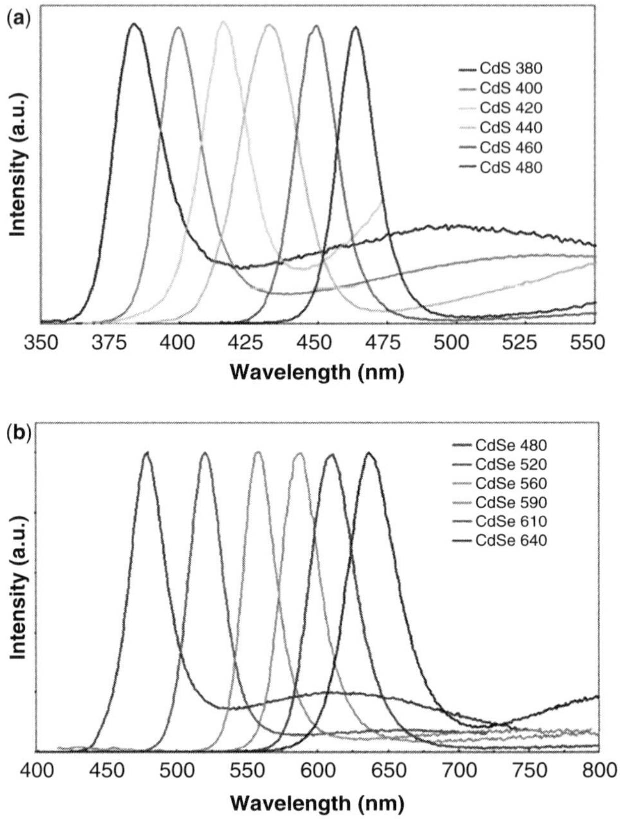

Figure 12.5 a) Cadmium sulfide (CdS) quantum dots in toluene with a typical nanocrystal size range from 1.6 nm (violet, at the left) to 7.3 nm (brown, at the right); b) Cadmium selenide quantum dots in toluene with a typical nanocrystal size range from 2.5 nm (blue, at the left) to 6.5 nm (brown, at the rigth).

The crystalline structure confers to Qds an intrinsic high optical stability: nevertheless their stability is quite sensitive to oxygen and light [42–45] becoming a challenge for module engineering. Different works have claimed that the incorporation of QDs into a solid matrix can lead to a blue-shift in both their absorption and emission spectra, caused by surface oxidation of the QD during the manufacturing process [46, 47]. Moreover, the hosting matrix can affect the emission intensity of the Qds owing to the scattering of particle clustering and matrix absorption of emitted light. Another drawback is related to the low QY of QDs [48], although important developments are actually in progress and QY of up to 80% have already been achieved for laboratory-produced Qds [49].

The other two drawbacks consist of the cost and potential toxicity of QDs: in fact, the cost of QDs [50, 51] is not competitive with organic dyes [52] and should be strongly reduced, and the potentially toxic nature of many QDs should be clarified [53, 54]. Several researches have actually been focused on studying and reducing the potential harmful effects of QDs [55–57].

12.3.2.3 Rare-Earth Materials

Rare-earth ions exhibit high QY, large Stokes shift, and excellent photostability [58, 59]. The properties of a range of rare-earth ions have been studied in several host materials such as glass, glass-ceramics, and solgels [60, 61] demonstrating QY in the range 90–100% [62, 62].

A drawback which can affect the efficiency of the emission process consists in intrinsic reabsorption of some rare-earth compounds: in fact in some ions, such as Yb^{3+} or Er^{3+}, the terminal state of the emission transitions is the ground state and absorption can occur at the same wavelength as the emission [64].

Taking into account the difficulty of casting bulk samples of glass or ceramic doped with lanthanide ions, which involves high-temperature melting/casting, several researches have been focused on developing nanoparticles lanthanide compounds which can be dispersed in polymer hosts similarly to organic dyes [20, 44]. Unfortunately, the mass absorption coefficients of the rare-earth nanoparticles are 10^5 times lower than conventional organic dyes: thus high concentrations of nanoparticles (up to 50% by weight) are required for absorbing a large fraction of incident sunlight resulting in expensive device.

A promising solution to the high required concentration and fluorescence reabsorption is the use of a complex or chelate of an organic dye molecule (or molecules) and a rare-earth ion [65, 66]. In these complexes, the ligands absorb energy and the electrons go from their S_0 to their S_1 state. From the single S_1 state, energy is transferred to the triplet state of the ligand (T_1) via intersystem crossing (ISC). From this triplet state, energy is transferred to the rare-earth ion. Due to all the energy transfers, these complexes feature a very high Stokes shift (> 200 nm). Different complexes have been synthetized [67]: the more interesting results were achieved with Eu^{3+} complexes [68] in which the decay of the ion to the $^7F_{0',1,2,3,4}$ levels gives rise to five narrow emission peaks in the range 570–700 nm (see Figure 12.6).

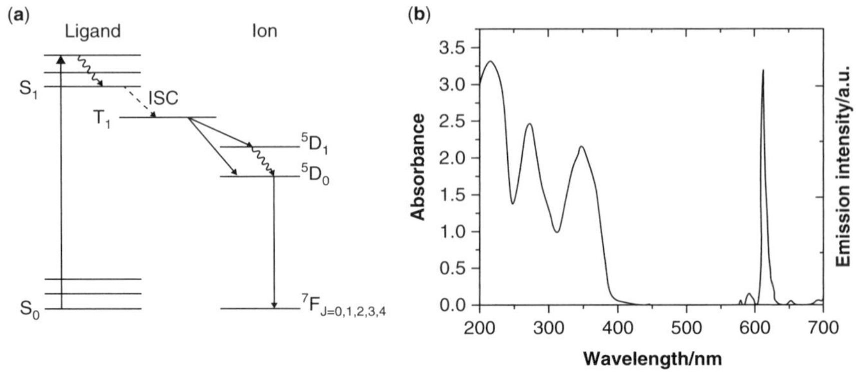

Figure 12.6 a) Jablonski energy-level diagram for an Eu^{3+} complex; b) Absorption (left) and emission (right) spectra of a $Eu(tta)_3phen$ film sublimated at 178°C on a quartz substrate (from ref. [69]).

Different Eu^{3+} complexes have been proposed for LSC applications [68], all characterized by a ligand absorption range at near UV region (200–400 nm) (i.e. β-diketonates, phenanthroline derivatives, bipyridine derivatives) and by high QY (up to 86%) [70]. LSC-type downconverters based on Eu^{3+} complexes demonstrated an increase in the efficiency of multicrystalline-silicon solar cells up by 17%.

12.3.3 Solar Cells

The emission spectrum irradiated by LSC edges is typically in the range 50–200 nm, much narrower than the solar spectrum, as showed by Figure 12.6a. Thus solar cells in LSC devices remain relatively cool under standard operating conditions, as they can convert the large part of the incident light. This allows the preservation of their high performance, as generally photovoltaics perform less well when heated [71]. Moreover, by matching the spectral response of the solar cell to the output spectra of the waveguides it is possible to obtain enhanced performance. Finally it is possible to optimize the cells over a smaller range of wavelengths thus increasing the cells efficiency: for example, by reducing their reflectance or better optimizing the doping concentration and junction depth.

Down-conversion process has been widely studied using luminophores attached directly to the surface of the PV cells [72] or within the encapsulation layers of the PVs [73] with the general purpose of shifting the incoming "blue" portion of the spectrum

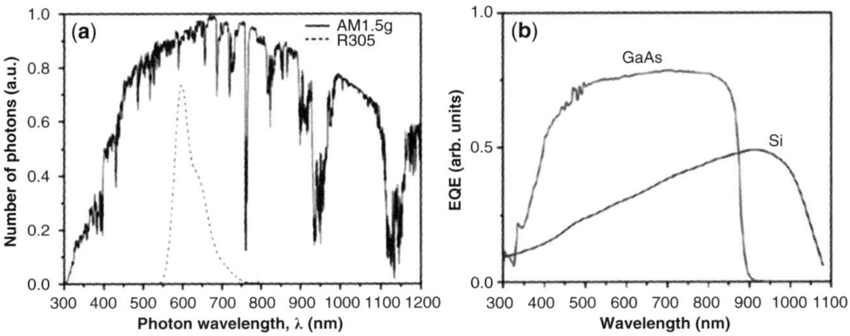

Figure 12.7 a) Comparison of AM1.5 solar spectrum and Rot 305 fluorescence emission spectrum (not to scale); b) External quantum efficiency (EQE) of a GaAs and a Si PV cell as a function of wavelength.

to longer wavelengths better suited to the PV. Although Si cells have been widely used in LSC research, other cells such as gallium arsenide (GaAs) and gallium indium phosphide (GaInP) can result in a higher efficiency. For example, GaAs-LSC showed efficiency approximately three times higher than similar Si-LSC [74]. Another very promising approach in LSC technology is the use of organic-based PV cells, which often have a "sweet spot" in the spectral range where the LSC emits [75].

12.3.4 Experimental Results

The first LSC devices were proposed in the late 1970s [16]. Extensive studies on LSC have been performed throughout the 1980s [16, 20]: Wittwer *et al.* [76] obtained an efficiency of $\eta_{LSC} = 4.0\%$ for a large-area LSC coupled to GaAs solar cell; Friedman used a mixed-dye thin-film module which allows achieving $\eta_{LSC} = 4.5\%$ with GaAs cells. Recent efforts have lead to several groups surpassing the 4% limit, as previously reviewed [38, 77]. Goldschmidt *et al.* [78] showed a conversion efficiency of 6.7% for a multiple stack coupled with GaInP solar cells, where the conversion efficiency was limited by dye spectral range (450–600 nm): the authors established that an efficiency of 13.5% could be reached enlarging the absorption spectrum to 650–1050 nm, and that the use of a photonic structure on top of the plate allows an increase in the device efficiency by 20%. Currie *et al.* [7] obtained conversion efficiencies as high as 6.8% for a tandem LSC based on thin film coupled with GaAs cells and calculated that, by using CdTe or $Cu(In,Ga)Se_2$ solar cells, conversion efficiency projections of 11.9% and 14.5% were respectively calculated. Slooff *et al.* obtained an efficiency of 7.1% with LSC based on PMMA plate functionalized with CRS040/Red305 dyes and coupled with GaAs cells [79].

Although the efficiency of an LSC is actually lower than an equivalent area of silicon panel, the reduced cost and tremendous flexibility in design could make them a viable alternative for the urban area in the near future.

12.4 Pathways for Improving LSC Efficiency

One of the main objectives in LSC development consists in achieving higher LSC efficiencies and consequently in reducing the main mechanism losses, which are: i) escape-cone losses; ii) absorption losses; iii) self-absorption losses.

12.4.1 Escape-Cone losses (P_{TIR})

Several attempts have been made for reducing P_{TIR} [80]. The most promising approaches are described below.

Front mirrors. To reduce the front cone losses, advanced wavelength-dependent mirrors have been proposed which are transparent to the large part of the solar spectrum and are selectively reflective at the wavelengths where the fluorophore emits. As a result, the front escape-cone losses are reflected back into the sheet [81]. Several wavelength selective mirrors, made from chiral nematic (cholesteric) liquid crystals and inorganics, have been used for LSC applications. Up to 30% of the light that had previously escaped the surface was turned into edge emission, translating into a 12%–20% LSC output improvement using inorganic reflectors and photonic structures, respectively. Similar results have been obtained with organic reflectors which are cast from solution resulting in a simpler and less expensive process than application of multilayer inorganic Bragg reflectors.

Dye alignment. Each molecule has a preferred emission direction thanks to the existence of the emission dipoles. Thus organic luminophores are often dichroic in absorption and transmission [82], opening new possibilities in controlling the spatial distribution of emitted light by ordering the dyes configuration. In particular, it is possible to align the dye molecules perpendicular to the waveguide surface so their emission direction is preferentially towards the edges: this approach has been demonstrated to decrease the surface losses to less than 10% [83]. LSCs based on aligned luminophores show an output power 15% higher than isotropically emitting ones [84].

12.4.2 Absorption Losses

Several researchers are involved in increasing absorption efficiency (η_{abs}). The main proposed solutions are discussed below

Back reflectors. The simplest way to increase the absorption efficiency consists of using a mirror to reflect the escape-cone losses back into the sheet: in fact, back reflectors reflect rear escape-cone photons back through the LSC sheet giving them a chance of being reabsorbed by the dye. This allows doubling the thickness of dye material: thus, the dye concentration and the related reabsorption losses can be halved. Different solutions have been proposed [85], among which are specular and diffuse mirrors [38, 86]. Diffuse reflectors, in particular, allow the reflection of solar photons directly onto the edge-mounted solar cells, mainly resulting in high efficiencies measured in small LSC modules [87].

Multiple dyes. Organic dyes typically show narrow absorption bands. In order to increase the overall absorption range multiple dyes with different absorption ranges can be incorporated into the same sheet [67–69]. Short-wavelength absorbed photons cascade through the different fluorescent dyes until they are finally emitted by the longest-wavelength dye in the mixture. Up to 70% of the solar spectrum can be absorbed by a sheet using a mixture of dyes. Unfortunately up to now NIR-emitting dyes are generally characterized by a low QY.

Multiple sheets. Another option to enhance the response of photovoltaic cells is to use individual dye-filled LSC waveguides, where each waveguide is separately attached to a solar cell specific for that emission wavelength. This configuration allows overcoming the drawback of the low QY of NIR-emitting dyes thus obtaining higher efficiencies. Wittwer *et al.* [88] obtained optical efficiencies of 15.8% for a three-sheet stacked system and Goetzberger predicted an efficiency of 23.7% for a four-sheet stacked LSC device. It is also possible to couple the LSC device directly on top of a second photovoltaic: in this configuration, the light that passes through the LSC due to its limited absorption range will be collected by the underlying cell [89].

12.4.3 Self Absorption Losses

The third main losses category which affects LSC efficiency is represented by self-absorption losses. Following the main solutions proposed to increase it are reported.

Novel luminophore systems. Most organic fluorophores are characterized by low Stokes shifts and consequently by large absorption/emission peak overlaps [90]. The first solution for solving this drawback consisted of designing new luminophores with larger Stokes shifts [4, 91], as reported in Section 12.3. Another approach consists of adding to the host matrix molecules with high mobility (i.e. dimethylsulfoxide (DMSO) [66]): these molecules stabilize the excited state, lowering its energy and resulting in a red-shifted emission spectrum maintaining unaffected the absorption spectrum. It has been found that concentrations of 5–10% of DMSO allow achieving red-shifts of 50–100 nm [66]. Another solution consists in using more than one luminophore employing Förster resonance energy transfer (FRET), which is the direct energy transfer from an excited molecule to another nearby molecule without the emission of a photon. By mixing different luminophores at high concentrations it is possible to transfer short-wavelength input light into long-wavelength output light at relatively high efficiency [92]. It is also possible to promote FRET process by developing ordered dye-nanochannel antennae (i.e. zeolites) [93]. Another possibility is to insert emitting dyes in active matrices: the matrix absorbs the solar radiation and transfers the energy to the dye which emits the light at higher wavelength. It was recently demonstrated that parylene- $Eu(TTA)_3$phen-based LSCs (in which parylene was the active matrix) showed an efficiency more than double that of parylene free ones [94].

Thin-film. Instead of using a bulk-doped LSC sheet, thin-film-based LSC devices can be produced in which luminescent films are coated on transparent substrates [95–97]. This allows confining the reabsorption processes into the thin dye-doped film [27, 78] and the photons transmission almost completely inside the un-doped substrates. This configuration has several advantages. Firstly it reduces NIR host absorption losses allowing the depositing of films of deuterated/fluorinated polymers onto conventional glasses. Secondly it minimizes the amount of polymer reducing the embodied energy, as the embodied energy for glass is typically 8–10 times lower than that for polymers [29, 81]. Furthermore, the thin film approach solves the inflammability problems of many polymers (i.e. PMMA) and improves the production feasibility. In fact, it is possible to envisage the large-area, inexpensive application of thin layers on either glass or polymer hosts using a variety of techniques, among which are casting, spraying, and printing [38, 98, 99].

12.5 Conclusion and Outlook

In this chapter a comprehensive overview of results obtained in the past 40 years on the work performed on luminescent solar concentrators was presented. Renewed interest in the LSCs over the last years has led to several developments in the different topics of the field: i) several models for describing LSCs performances, such as thermodynamic and ray-trace models, were developed; ii) new promising dyes, quantum dots, and lanthanide complexes have been synthetized; iii) novel material matrices for realizing LSC slabs for hosting different kinds of dyes have been produced; iv) several typologies of photovoltaic solar cells have been proposed for LSC applications. Both experimental data and theoretical models have demonstrated the technological and economic viability of LSC technology: for this purpose a variety of studies are actually being run by research groups all around the world with the final objective of affirming LSC technology in the market. Nevertheless some non-technological barriers have to be exceeded for the complete affirmation of LSC devices.

Firstly, a coordinated effort combining the many research teams has to be coordinated towards making LSCs a viable option for the marketplace. In fact, there are large numbers of researchers who are actually working independently on LSC technology around the globe. This fragmentation is probably an obstacle to more rapid performance improvement of LSC devices.

Secondly, LSC technology's place in global solar energy generation needs to be clearly defined by its functionality. In fact, LSC should not be considered as another type of organic photovoltaic (OPV), or as a solar cell, but as a multi-functional, highly architectural, integrated solar-energy generating system. The last obstacle for LSC affirmation resides in the misrepresentation of their capabilities: LSC devices are not competitors of conventional PV panels but complementary. In fact, LSC devices are low-cost devices with an aesthetical advantage designed to be brought directly into the public eye, given its flexibility in shape and color, rather than unaesthetic black silicon panels. There are huge areas which are exploitable by LSC devices and there are numerous examples of luminescent objects being incorporated in the environment already. It is thus necessary to discuss with architectural and building industries the potential development and use of LSC devices as the tremendous design freedom that they afford could be surely well-exploited in the future.

Acknowledgments

The authors thank Eng. Paolo Decarli and Dott. Federica Pasquini for their contribution. M.T. wants to thank the European Community's Seventh Framework Programme FP7/2007-2013 under grant agreement *Marie Curie 7th Framework Program - PCOFUND-GA-2008-226070, acronomy "Progetto Trentino"* within PHOTOFUTURE Project which first approached him about LSC technology, and Prof. Maso and Dott. Filippino for the helpful discussions.

References

1. A. Goetzberger, W. Greubel, *Appl. Phys.*, Vol. 14, p. 123, 1977.
2. R. Kondepudi, S. Srinivasan, *Sol. Energ. Mater.*, Vol. 20, p. 257, 1990
3. F. Vollmer, W. Rettig, *J. Photochem. Photobiol. A*, Vol. 95, p.143, 1996.
4. A. Goetzberger, *Appl. Phys.*, Vol. 16, p. 399, 1978.
5. M. Carrascosa, S. Unamuno, F. Agullo-Lopez, *Appl. Optics*, Vol. 22, p. 3236, 1983.
6. D. Chemisana, *Renew. Sust. Energ. Rev.*, Vol. 15, p. 603, 2011.
7. M.J. Currie, J.K. Mapel, T.D. Heidel, S. Goffri, M.A. Baldo, *Science*, Vol. 321, p. 226, 2008.
8. A. Goetzberger and V. Wittwer. *Solar Cells*, Vol 4, p. 23, 1981.
9. N.A. Bakr, A.F. Mansour and M. Hammam, *J. Appl. Polym. Sci.*, Vol. 74(14), p. 3316, 1999.
10. J. Ballato, S.H Foulger and D.W. Smith Jr., *J. Opt. Soc. Am. B*, Vol. 21(5), p. 958, 2004.
11. W.R.L. Thomas, J.M. Drake and M.L. Lesiecki, *Appl. Opt.*, Vol. 22(21), p. 3440, 1983.
12. J.R. Lakowicz, *Principles of Fluorescence Spectroscopy*, Plenum Press, 1983.
13. W.H. Weber and J. Lambe, *Appl. Opt*, Vol. 15 (10), p. 2299, 1976.
14. A. Goetzberger and O. Shirmer, *Appl. Phys.*, Vol. 19, p. 53, 1979.
15. J.S. Batchelder, A.H. Zewail and T. Cole, *Appl. Opt.*, Vol. 20(21), p. 3733, 1981.

16. J.M. Drake, M.L. Lesiecki, J. Sansregret and W.R.L. Thomas, *Appl. Opt.*, Vol. 21(6), p. 2945, 1982.
17. A.F. Mansour, H.M.A. Killa, S.A. El-Wanees and M.Y. El-Sayed, *Polym. Test.*, Vol. 24, p. 519, 2005.
18. R. Kinderman, L.H. Sloo, A.R. Burgers, N.J. Bakker, A. Buchtemann, R. Danz and J.A.M. van Roosmalen, J. *Sol. Energy Eng.*, Vol. 129, p. 277, 2007.
19. A.M. Hermann, *Solar Energy*, Vol. 29(4), p. 323, 1982.
20. Y.Hasegawa, Y. Wada and S. Yanagida, *J. Photochem. Photobiol. C: Photochem. Rev.*, Vol. 5(3), p. 183, 2004.
21. J. Gordon, J. Ballato, D.W. Smith Jr. and J. Jin., *J. Opt. Soc. Am. B*, Vol. 22(8), p. 1654, 2005.
22. W.G. Quirino, M.J.V. Bell, S.L. Oliveira, L.A.O, *J. Non-Cryst. Solid.*, Vol. 351, p. 2042, 2005.
23. R. Reisfeld, *Opt. Mat.*, Vol. 16, p. 1, 2001.
24. M.A. El-Shahawy, A.F. Mansour, *J. Mater. Sci. Mater. Elect.*, Vol. 7, p. 171, 1996.
25. J. S. Batchelder, A. H. Zewail, T. Cole, *Appl. Optics*, Vol. 20, p. 3733, 1981.
26. I. Baumberg, O. Berezin, A. Drabkin, B. Gorelik, L. Kogan, M. Voskobojnik, M. Zaidman, *Polym. Degrad. Stabil.* Vol. 73, p. 403, 2001.
27. W.G.J.H.M. van Sark, K.W.J. Barnham, L.H. Slooff, A.J. Chatten, A. Büchtemann, A. Meyer, S.J. McCormack, R. Koole, D.J. Farrell, R. Bose, E.E. Bende, A.R. Burgers, T. Budel, J. Quilitz, M. Kennedy, T. Meyer, C. d. M. Donega, A. Meijerink, D. Vanmaekelbergh, *Opt. Express*, Vol. 16, p. 21773, 2008.
28. J. Yoon, L. Li, A.V. Semichaevsky, J.H. Ryu, H.T. Johnson, R.G. Nuzzo, J.A. Rogers, *Nature Comm.*, Vol. 2, p. 343, 2011.
29. L.R. Wilson, B.S. Richards, *Appl. Optics*, Vol. 48, p. 212, 2009.
30. M.G. Debije, P. P. C. Verbunt, P. J. Nadkarni, S. Velate, K. Bhaumik, S. Nedumbamana, B. C. Rowan, B. S. Richards, T. Hoeks, *Appl. Optics*, Vol. *50*, p. 163, 2011.
31. A.F. Mansour, M.G. El-Shaarawy, S.M. El-Bashir, M.K. El-Mansy, M. Hammam, *Polym. Int*, Vol. 51, p. 393, 2002.
32. S.M. Reda, *Sol. Energ.*, Vol. 81, p. 755, 2007.
33. C.L. Mulder, L. Theogarajan, M. Currie, J.K. Mapel, M.A. Baldo, M. Vaughn, P. Willard, B.D. Bruce, M.W. Moss, C.E. McLain, J.P. Morseman, *Adv. Mater.*, Vol. 21, p. 3181, 2009.
34. H. Du, C. Chen, R. Krishnan, T.D. Krauss, J.M. Harbold, F.W. Wise, M.G. Thomas and J. Silcox., *Nano Lett.*, Vol. 2(11), p. 1321, 2002.
35. O.I. Mificific, H.M. Cheong, H. Fu, A. Zunger, J.R. Sprague, A. Mascarenhas and A.J. Nozik., *J. Phys. Chem. B*, Vol. 101, p. 4904, 1997.
36. O.I. Micic, C.J. Curtis, K.M. Jones, J.R. Sprague, A.J. Nozik, *J. Phys. Chem.*, Vol. 98, p. 4966, 1994
37. D.J. Norris, M.G. Bawendi, *Phys. Rev. B*, Vol. 53, p. 16338,1996.
38. A.J. Chatten, K.W.J. Barnham, B.F. Buxton, N.J. Ekins-Daukes, M.A. Malik, *Sol. Energ. Mater. Sol. C.*, Vol. 75, p. 363, 2003.
39. K. Barnham, J.L. Marques, J. Hassard, *Appl. Phys. Lett.*, Vol.76, p. 1197, 2000.
40. G.V. Shcherbatyuk, R.H. Inman, C. Wang, R. Winstron, S. Ghosh, *Appl. Phys. Lett.*, Vol. 96, p. 191901, 2010.
41. Y. Nosaka, *J. Phys. Chem*, Vol. 95, p. 5054, 1991.
42. S.J. Gallagher, P.C. Eames, B. Norton, *Int. J. Ambient Energy*, Vol. 25, p. 47, 2004.

43. C. de Mello Donega, S.G. Hickey, S.F. Wuister, D. Vanmaekelbergh, A. Meijerink, *J. Phys. Chem. B*, Vol. 107, p. 489, 2002.
44. M. Lomascolo, A. Cretì, G. Leo, L. Vasanelli, L. Manna, *Appl. Phys.Lett*, Vol. 82, p. 418, 2003.
45. X. Wang, J. Zhang, A. Nazzal, M. Xiao, *Appl. Phys. Lett.*, Vol. 83, p. 162, 2003.
46. S.J. Gallagher, B. Norton, P.C. Eames, *Sol. Energy*, Vol. 81, p.813, 2007.
47. S.M. Reda, *Acta Mater.*, Vol. 56, p. 259, 2008.
48. V. Sholin, J. D. Olson, S. A. Carter, *J. Appl. Phys*, Vol. 101, p. 123114, 2007.
49. K.R. Catchpole, A. Polman, *Opt. Express*, Vol. 16, p. 21793, 2008
50. N. Neuroth, R. Haspel, *Sol. Energ. Mater.*, Vol. 16, p. 230, 1987.
51. R. Reisfeld, Y. Kalisky, *Nature* , Vol. 283, p. 281, 1980.
52. A. Buch, M. Ish-Shalom, R. Reisfeld, A. Kisilev, E. Greenberg, *Proc. Int. Symp. Eng. Ceremics*, Vol. 71, p. 383, 1985.
53. R. Hardman, *Environ. Health Perspect*, Vol. 114, 2006.
54. J.L. Pelley, A.S. Daar, M.A. Saner, *Toxicol. Sci*, Vol. 112, p. 276, 2009.
55. S.J. Byrne, Y. Williams, A. Davies, S.A. Corr, A. Rakovich, Y.K. Gun'ko, Y.P. Rakovich, J.F. Donegan, Y. Volkov, *Small*, Vol. 3, p. 1152, 2007.
56. F. Erogbogbo, K.-T. Yong, I. Roy, G. Xu, P.N. Prasad, M.T. Swihart, *ACS Nano* Vol. 2, p. 873, 2008.
57. K. Fujioka, M. Hiruoka, K. Sato, N. Manabe, R. Miyasaka, S. Hanada, A. Hoshino, R.D. Tilley, Y. Manome, K. Hirakuri, K. Yamamoto, *Nanotechnology*, Vol. 19, p. 415102, 2008.
58. P.S. Friedman, *Opt. Eng.*, Vol. 20(6), p. 887, 1981.
59. R. Reisfeld, Y. Kalisky, *Chem. Phys. Lett.*, Vol. 80(1), p. 178–183, 1981.
60. T. Jin, S. Tsutsumi, Y. Deguchi, K.fichi Machida and G.-ya Adachi, *J. Electrochem. Soc.*, Vol. 143(10), p. 3333, 1996.
61. J. Lin and C.W. Brown, *Trends Anal. Chem.*, Vol. 16(4), p. 200, 1997.
62. S.M. Lima, A.A. Andrade, R. Lebullenger, A.C. Hernandes, T. Catunda and M.L. Baesso, *Appl. Phys. Lett.*, Vol. 78(21), p. 3220, 2001.
63. P.F. Liao and H.P. Weber. *J. Appl. Phys.*, Vol. 45, p. 2931, 1974.
64. G.A. Hebbink, L. Grave, L.A. Woldering, D.N. Reinhoudt and F.C.J.M. van Veggel, *J. Phys. Chem. A*, Vol. 107, p. 2483, 2003.
65. A.A. Earp, G.B. Smith, P.D. Swift and J. Franklin, *Solar Energy*, Vol. 76, p. 655, 2004.
66. J.C. Goldschmidt, M. Peters, A. Bosch, H. Helmers, F. Dimroth, S.W. Glunz and G. Willeke, *Solar Energy Materials & Solar Cells*, Vol. 93, p. 176, 2009.
67. M.J.V. Werts, J.W. Hofstraat, F.A.J. Geurts and J.W. Verhoeven, *Chem. Phys. Lett.*, Vol. 276, p. 196, 1997.
68. M.G. Debije, P.P.C. Verbunt, B.C. Rowan, B.S. Richards, T.L. Hoeks, *Appl. Opt.*, Vol. 47(36), p. 6763, 2008.
69. G. Maggioni, A. Campagnaro, M. Tonezzer, S. Carturan, A. Quaranta, *ChemPhysChem*, DOI: 10.1002/cphc.201300151, 2013
70. M. Lomascolo, A. Creti, G. Leo, L. Vasanelli, L. Manna, *Appl. Phys. Lett.*, Vol. 82(3), p. 418, 2003.
71. A. Royne, C.J. Dey, D.R. Mills, *Sol. Energ. Mater. Sol. C*, Vol. 86, p. 451, 2005.
72. N. Yamada, L.N. Anh, T. Kambayashi, *Sol. Energ. Mater. Sol. C*, Vol. 94, p. 413, 2010.
73. E. Klampaftis, B.S. Richards, *Prog. Photovolt: Res. Appl.*, Vol. 19, p. 345, 2011.
74. L.H. Slooff, E.E. Bende, A.R. Burgers, T. Budel, M. Pravettoni, R.P. Kenny, E.D. Dunlop, A. Büchtemann, *Phys. Status Solidi—R*, Vol. 2, p. 257, 2008.

75. A.W. Haines, Z. Liang, M.A. Woohouse, B.A. Gregg, *Chem. Rev.*, Vol. 110, p. 6689, 2010.
76. V. Wittwer, W. Stahl and A. Goetzberger, *Sol. Energy Mater.*, Vol. 11(3), p. 187, 1984.
77. M.G. Debije and P.P.C. Verbunt, *Adv. Energy Mater.*, Vol. 2, p.12, 2012.
78. X. Peng, M.C. Schlamp, A.V. Kadavanich, A.P. Alivisatos, *J. Am. Chem. Soc.*, Vol. 119, p. 7019, 1997.
79. T. Trupke, M.A. Green and P. Würfel, *J. Appl. Phy*, Vol. 92, p. 1668, 2002.
80. A. Luque, A. Marti, A. Bett, V.M. Andreev, C. Jaussaud, J.A.M. van Roosmalen, J. Alonso, A. Rauber, G. Strobl, W. Stolz, C. Algora, B. Bitnar, A. Gombert, C. Stanley, P. Wahnon, J.C. Conesa, W.G.J.H.M. van Sark, A. Meijerink, G.P.M. van Klink, K. Barnham, R. Danz, T. Meyer, I. Luque-Heredia, R. Kenny, C. Christodes, G. Sala and P. Benitez, *Solar Energy Materials & Solar Cells*, Vol. 87, p. 467, 2005.
81. M.G. Debije, J.-P. Teunissen, M.J. Kastelijn, P.P.C. Verbunt, C.W.M. Bastiaansen, *Solar Energy Materials and Solar Cells*, Vol. 93, p.1345, 2009.
82. M. van Gurp and Y.K. Levine, *The Journal of Chemical Physics*, Vol. 90, p. 4095, 1989.
83. S. McDowall, et al., *Journal of Applied Physics*, Vol. 108, p. 53508, 2010.
84. M.G. Debije, D.J. Broer and C.W.M. Bastiaansen, *Proceedings of the 22nd EUPVSEC 2007*, Milan.
85. J.C. Goldschmidt, M. Peters, L. Pronneke, L. Steidl, R. Zentel, B. Blasi, A. Gombert, S. Glunz, G. Willeke and U. Rau, *Phys. Stat. Sol. A*, Vol. 205(12), p. 2811, 2008.
86. L.H. Sloof, E.E. Bende, A.R. Burgers, T. Budel, M. Pravettoni, R.P. Kenny, E.D. Dunlop and A. Buchtemann, *Phys. Stat. Sol. (RRL)*, Vol. 2(6), p. 257, 2008.
87. J.C. Goldschmidt, M. Peters, A. Bosch, H. Helmers, F. Dimroth, S.W. Glunz and G. Willeke, *Solar Energy Materials & Solar Cells*, Vol. 93, p. 176, 2009.
88. A.M. Herman, *Solar Energy*, Vol. 29(4), p. 323, 1982.
89. R. Koeppe, *Appl. Phys. Lett.*, Vol. 90, p. 181126, 2007.
90. L.R. Wilson, B.C. Rowan, N. Robertson, O. Moudam, A.C. Jones, B.S. Richards, *Appl. Optics*, 49, 1651, 2010.
91. K. Barnham, J.L. Marques, J. Hassard, *Appl. Phys. Lett.*, 76, 1197, 2000.
92. S.T. Bailey, G.E. Lokey, M.S. Hanes, J.D.M. Shearer, J.B. McLafferty, G.T. Beaumont, T.T. Baseler, J.M. Layhue, D.R. Broussard, Y.-Z. Zhang, B.P. Wittmershaus, *Sol. Energ. Mater. Sol. C.*, 91, 67, 2007.
93. G. Calzaferri, R. Meallet-Renault, D. Brühwiler, R. Pansu, I. Dolamic, T. Dienel, P. Adler, H. Li, A. Kunzmann, *ChemPhysChem*, 12, 580, 2011.
94. M. Tonezzer, G. Maggioni, A. Campagnaro, S. Carturan, A. Quaranta, M. della Pirriera, D. Gutierrez Tauste, *Progress in Photovoltaics*, to be published.
95. M. Buffa, S. Carturan, G. Maggioni, M. Tonezzer, W. Raniero, A. Quaranta, G. Della Mea, *Proceedings of 25th EU PVSEC*, Valencia (Spain), p. 763–766, 2010.
96. G. Maggioni, M. Tonezzer, S. Carturan, A. Quaranta, M. Buffa, A. Antonaci, G. Della Mea, *Proceeding of 24th EU PVSEC*, Frankfurt (Germany), 2009.
97. M. Tonezzer, G. Maggioni, S. Carturan, A. Quaranta, G. Della Mea, *Proceedings of the 26th EU PVSEC*, Valencia (Spain), 2011.
98. M. Tonezzer, Metallo-organic solid films: Growing processes, physical features, and optical properties, in: *Handbook of Porphyrins: Chemistry, Properties and Applications*, p. 76–99, 2012.
99. M. Tonezzer, G. Maggioni, Physical vapour deposition techniques for producing advanced organic chemical sensors, in: *Advances in Chemical Sensors*, 2011.

13

Organic Fluorophores for Luminescent Solar Concentrators

Luca Beverina* and Alessandro Sanguineti

Department of Materials Science, University of Milan-Bicocca, Milan, Italy

Abstract

The luminescent solar concentrators (LSCs) technology is an established but still largely improvable approach to reduce the costs associated with common large-area, silicon-based photovoltaics. Organic LSCs, based upon a slab of polymeric transparent host material embedded with a suitable fluorescent material, capture sunlight (also diffuse radiation) and emit to longer wavelengths. Due to total internal reflection, the slab behaves as a planar waveguide for the down-converted radiation, eventually collected at the edges by small-area, low-cost, silicon solar cells. The luminescent material is the key component in LSCs. In this chapter we will report a survey of the most significant advancements in the development of fluorophores for LSCs pertaining to classes of organic dyes, lanthanide chelates and quantum dots. The chapter is structured through three main sections: structures and working principles of LSCs, classes of available fluorophores, focus on the best performing derivatives so far reported.

Keywords: Low-cost photovoltaic, organic solar cell, solar concentrator, luminescent solar collector, LSC, fluorophore, organic luminophore, lanthanide chelate, luminescent dye, chromophore, fluorescence, emission

**Corresponding author*: luca.beverina@mater.unimib.it

Atul Tiwari, Rabah Boukherroub, and Maheshwar Sharon (eds.) Solar Cell Nanotechnology, (317–356) 2014 © Scrivener Publishing LLC

13.1 Introduction

Solar radiation is probably the most attractive and sought after amongst all the other renewable energy sources (wind power, wave power, geothermal energy, etc.), because its clean, safe, reliable and practically inexhaustible [1]. The direct conversion of solar photons into electricity is in fact a well-established concept requiring low maintenance, and it has reduced environmental impact [2]. However, the full exploitation of photovoltaic energy is still limited by various factors, primarily the high costs associated with this technology and the employed materials (i.e. conventional crystalline silicon) as well as relatively low conversion yields [3]. As such, the quest for higher conversion efficiencies and lower costs represents the fundamental target in photovoltaic research. To reach lower cost per installed capacity ($/W), several routes are being pursued, all more or less directed toward a better use of the complete solar spectrum [4]: a) introduction of innovative semiconducting materials, b) research for new device architectures, and c) novel solar cell concepts and process methodologies. Major research efforts are currently devoted to both all plastic and hybrid organic/inorganic technologies (the so-called next or third generation photovoltaics) to increase the final device performance and meet cost-reduction perspectives [5–8]. Considering the evolution made so far in the field of photovoltaics, according to van Sark, enormous progress has been made in the past decades leading to present module cost between 1 and 2 $/Wp, depending on the specific technology. This will bring electricity cost close to the price of ~2 $/kWh that consumers pay today, but still far off from the large-scale electricity generation cost of 0.02–0.04 $/kWh for conventional, fossil-fuel-based plants [9]. Aside from such considerations, simply based on the costs of silicon photovoltaics, nowadays there is also an increasing interest in the so-called "building-integrated photovoltaics." That is, the integration of aesthetically pleasing photovoltaic solutions in the construction of energy efficient buildings. The use of commercially available silicon modulus is hardly an ideal solution in this respect. The various available concentrating solutions provide viable approaches to both cut the costs and improve aesthetical appeal.

Such technology basically operates collecting the solar radiation from large-area, low-cost devices onto smaller (and possibly more efficient) active PV cells, therefore allowing the generation of the same amount of power whilst reducing the amount of high-cost semiconductors. In this context the most diffuse and commonly employed category of concentrating devices is that of geometrical optical concentrators using lenses and reflectors (i.e. mirrors) for focusing incoming sunlight onto the PV cell [10]. These systems, while helping in minimizing the surface of active photovoltaic materials, suffer a few drawbacks that still negatively balance achievable cost reductions and enhanced efficiencies. The basic problem consists in the capacity of working only for a narrow angular range of impinging solar radiation, which severely limits the concentration factors below 5 suns for static mounted devices [11]. The standard solution to face this limitation consists in the introduction of complex and expensive tracking systems to continuously trace the sun during its apparent daytime path, and permanently align the pointing axis of the supported concentrator with the local sun vector [12]. Moreover, concentration leads to serious overheating parasitic phenomena, requiring the installation of specific high-cost cooling systems. Indeed these devices generally require large areas of installation and lack a straightforward building integration, being their main configurations are restricted to parabolic reflectors or to flatter panels suitable for rooftop installation [1].

Luminescent solar collectors (LSCs), the focus of this chapter, are a promising alternative not requiring tracking. In its most common embodiment a LSC consists of a slab of high-refractive index host matrix (glasses, polymers) containing an efficient luminescent species capable of first absorbing the best possible fraction of the incoming radiation, and thus downconverting it by either fluorescence or phosphorescence. By means of total internal reflection the luminescence radiation is guided through the slab toward the edges where small area, standard PV cells are located. This approach enables: higher collection of solar radiation (also exploiting diffuse radiation), larger concentration ratios, good heat dissipation and finally optimal spectral matching and tunability with respect to the maximum sensitivity of the employed solar cells. Particularly appealing among all these factors is the capability of collecting not only direct but also diffuse solar irradiation, which results in fact in the most abundant solar source in several specific environmental conditions (such as cloudy or shaded weather), and definitely constitutes the prevailing solar radiation in many regions of high latitudes (e.g. central Europe, where the total incoming radiation averaged over one year has been proven to consist of 60% diffuse radiation

[13]). Hermann [14], Rapp and Boiling used the term "Luminescent Solar Concentrator" to describe a broad range of luminescent collectors, whilst Lerner pioneered the assembly of solar collectors based upon a solution of dye laser between two glass sheets (1973 proposal to NSF). The very first publication on LSCs in the open literature traces back to 1976 when a Ford Scientific Laboratory Group [15] presented the realization of two different collectors of planar geometry using neodymium-doped glass and organic laser dyes. Since then, research literature on LSCs has boomed, leading to a plethora of publications and industrial patents throughout the mid 80s when, mainly due to the drop in the cost of oil, the researches in the field slowed down to a near abandonment. The past few years witnessed a renovated interest in LSCs (Figure 13.1).

This phenomenon is easily understood if one considers on the one hand the modern dwindling fossil-fuel availability (and consequently the urgency for economically attractive replacements to traditional high environmental impact sources), and on the other hand, the progress in the design of performing luminescent materials. Other peculiar features that shine a light on LSC technology are the low cost of the employed materials, the potential low cost of production/fabrication, the possibility of building integration

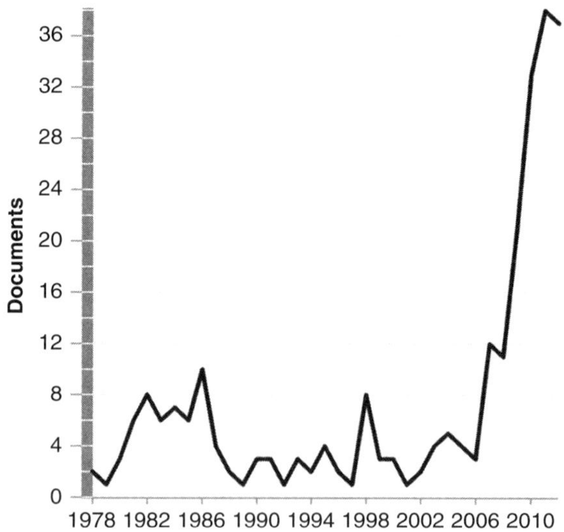

Figure 13.1 Analysis of the total number of documents for the topic "Luminescent Solar Concentrators" by year as obtained from [16].

matching architectural and design requirements (e.g. size, shape, color, lightness, transparency, flexibility), the adaptability to different solar cell technologies and materials (including state-of-the-art, all-organic devices). In the following section, the working principle of LSC, its constituting elements, the principal parameters to take into account and the main figures of merit for the efficiency evaluation are discussed.

13.2 LSCs: Device Operation and Main Features

The simplest luminescent solar concentrator consists of an emitting material (luminophore) dispersed in/deposited onto a transparent matrix (usually a high refractive index polymer or glass) which behaves substantially as a waveguide: the incident radiation is absorbed by the luminescent species and re-emitted at longer wavelength inside the matrix, where it is trapped by total internal reflection (TIR), and subsequently guided (i.e. concentrated) toward the PV cells located at the panel edges (Figure 13.2) [17–20]. A fundamental aspect to be considered in order to evaluate the efficiency of this technology is the maximum amount of re-emitted radiation that can be trapped inside the collector: this is limited

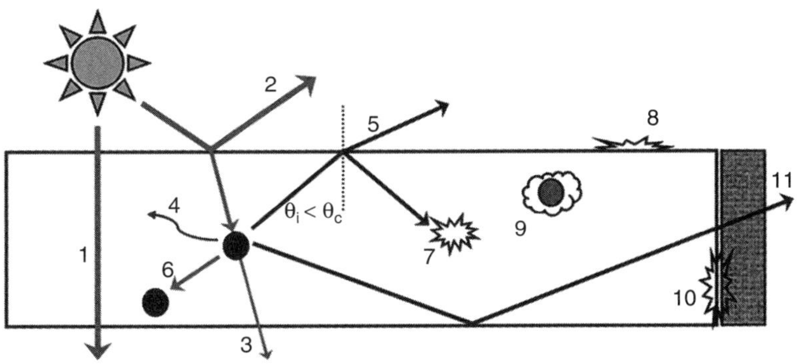

Figure 13.2 Cross-sectional diagram of an LSC panel with loss mechanisms as reported in the text: 1) Transparency to incoming radiation, 2) reflection at the waveguide surface, 3) limited absorption capacity of luminophore, 4) non-unity luminescence quantum efficiency of luminophore, 5) escape cone loss, 6) luminophore self-absorption of emitted radiation, 7) waveguide parasitic absorption, 8) waveguide surface scattering, 9) luminophore instability, 10) scattering at the panel-PV cell interface, 11) solar cell loss.

by the refractive index (n) of the employed matrix. According to Snell's law, the radiation impinging on the interface between two different materials (e.g. the panel matrix and air) at angles higher than a certain critical angle (θ_c) results in fact totally reflected, the critical angle being determined by the relation (Eq. 13.1):

$$\theta_c = \sin^{-1}\left(\frac{1}{n}\right) \tag{13.1}$$

The choice of a proper matrix material determines the light-entrapping and "light-pipe" behavior of the panel: high refractive index allows in fact to efficiently guide the emitted radiation but on the other hand it causes an increase of the material reflectivity, thus limiting the fraction of incoming light able to travel through the panel. A careful trade-off is therefore necessary. Focusing on polymer-based LSCs (the cost effective solution) reported in academic literatures and patents, poly(methyl methacrylate) (PMMA) is one of the most common polymeric materials due to its excellent transparency, well-established processability and favorable optical features: with a refractive index value of 1.49, the critical angle for PMMA is around 42.2°, and its trapping efficiency is 0.741 [9, 21–23]. Beside the aforementioned light trapping efficiency (η_{trap}), other specific mechanisms contribute to determine the overall optical efficiency of an LSC panel (η_{op}), according to the formalism introduced in the literature [9] the latter can be expressed as in Eq. 13.2,

$$\eta_{opt} = (1-R)\,\eta_{abs}\eta_{LQY}\eta_{Stokes}\eta_{trap}\eta_{mat}\left(1-\eta_{self}\right)\eta_{TIR} \tag{13.2}$$

where R is the surface reflectance, η_{abs} and η_{LQY} are the luminophore absorption efficiency and quantum emission efficiency respectively, η_{Stokes} is the Stokes efficiency (the ratio between the average energy of the emitted photons and the average energy of the absorbed photons), η_{mat} is the efficiency of the transport of light in the matrix (it takes into account the parasitic losses due to matrix absorption and scattering), η_{self} is the efficiency of self-absorption process for the luminescent species and η_{TIR} represents the efficiency of the total internal reflection mechanism. Accurate estimates of the final LSC performances and of the maximum achievable efficiencies are somewhat difficult to predict, as evidenced by the various involved processes just reported, therefore an accurate modeling of the LSC behavior is required. A detailed evaluation of the best predictive models is beyond the scope of this chapter which focuses on material issues. The interested reader is referred to the literature for an in depth discussion of this topic (hints can be found in reference [4]). Currently, the maximum efficiency of the LSC theoretically predicted using the Shockley-Queisser principle of thermodynamic detailed balance model sets around 33%, which results, however, about one order of magnitude higher than what can be experimentally obtained [24]. In fact the highest overall power efficiency reported for an LSC panel connected to photovoltaic cells (i.e., the product of the LSC optical efficiency time, the specific solar cell efficiency) has been obtained by Sloof *et al.* [25] with a 5.2% value for a 5 x 5 x 0.5 cm^3 when employing a multicrystalline silicon PV cell and a record value of 7.1% for the GaAs-based PV cell counterpart. In order to improve these still modest efficiencies, so far preventing large-scale exploitation of LSC technology, a detailed analysis of the main loss mechanisms is required.

Loss mechanisms pertain to three categories: losses related to the matrix material, losses due to the PV cells and finally processes directly connected with the luminophores [26]. Starting from the matrix materials and the slab structure, loss mechanisms mainly derive from reflections of the incoming light, scattering processes and parasitic absorption (occurring both at the waveguide surfaces and/or within the bulk of the host material). The so called "escape cone loss," directly related to the total internal reflection process, is also normally included in this class. Escape losses occur when the radiation from the luminescent species is emitted at an angle smaller then the critical angle, so that it is refracted out of the panel and lost through this "escape cone." Considering the PV cells, the nature of the active semiconductor (which is able to operate only for well-defined range of spectral frequencies) imposes a further restriction on the selection of the other LSC components. This issue dictates necessary trade-offs between PV semiconductors, matrix hosts and luminescent materials. Dealing with PV cells losses, we can also have an incomplete exploitation of the impinging radiation (due to the intrinsic finite conversion efficiencies of the cell) and scattering losses of the emitted light at the LSC Panel – PV cell interface. The main losses in the LSC panel are, however, due to the nature of the luminescent species and to its fundamental absorption/emission processes. In detail: a) absorption bands not completely covering solar emission spectrum (i.e. transparency to incoming radiation), b) limited absorption coefficient, c) non-unity luminescence quantum efficiency, d) self-absorption of the emitted light, e) limited thermal/photochemical stability.

In the following section we will detail all issues connected with the emitting materials, highlighting strength and weaknesses of the various classes of luminophores so far proposed. A particular emphasis will be given to the class of organic fluorophores most represented in the devoted scientific and patent literature: perylene diimides.

13.3 Luminophores in LSCs

Luminescent species, i.e., species capable of absorbing a certain fraction of the incoming radiation and down-converting it to lower wavelength by means of fluorescence emission, are the core element of the luminescent solar collector. The ideal luminophore must fulfill a number of specific requirements in order to maximize LSC efficiencies while keeping at bay costs and environmental impact. It is worthwhile to note that so far no luminophore simultaneously fulfills all such requirements and prototypes normally settle for trade-offs. Modern research is still focused on the quest for the "perfect" luminophore possessing all of the following characteristics at the same time [1]:

- Broad absorption range
- High absorption efficiency over the whole spectral range
- High (near unity) emission quantum efficiency
- Large energy difference between absorption and emission maxima (Stokes shift)
- Limited or null overlap between absorption and emission spectra
- High stability (thermal, photochemical, environmental, process)
- Good solubility in the host matrix
- Correct matching with the working frequencies of employed PV cells
- Low or null toxicity
- Low cost

The luminescent materials proposed during the years as possible luminophores for LSCs can be grouped in three main categories:

luminescent colloidal quantum dots (QDs), lanthanides chelates, and organic dyes.

13.3.1 Colloidal Quantum Dots (QDs)

Quantum dots (QDs) are semiconducting nanocrystals or nanoparticles, produced by a variety of methods including either colloidal or single-molecule precursor growth, whose peculiar electrical as well as optical properties are ultimately determined by the reduction of their dimensions to the nanoscale. The size of these semiconducting nanostructures is usually in the range between 10 and 100 nm [1], comparable or even smaller than the characteristic electron-hole pair natural radius (the so-called exciton Bohr radius). Thanks to the effects of such strong quantum confinement, the electronic and optical spectra of semiconductor QDs can be efficiently tuned simply through an accurate selection of their dimensions. Indeed it is possible to find a simple relation between the size of the nanocrystals and the energy of the electronic inter-band transitions occurring between electron and hole states (and therefore the frequencies of the absorbed and emitted radiation): the smaller the size, the greater the confinement energy, and consequently the shorter the absorption and emission wavelengths. Many innovative applications including biolabeling, sensing, imaging, drug delivery, nanoelectronics, photonics, quantum computing, light harvesting, display panel technology and solar energy conversion aim to exploit these peculiar features [27–29]. The use of semiconducting nanostructures as valuable luminophores for novel QD-LSCs was originally proposed in 2000 by Barnham *et al.* [30] who quantitatively predicted, through a thermodynamic model, how the separation between the absorption and luminescence peaks could be related to the spread of quantum dots size, and thus showing the possibility to optimize the solar collector performances by an efficient control of the reabsorption losses [31, 32]. Along with the size of the nanocrystals, their chemical composition can be varied leading to different nanostructures able to match the specific features required for high performance luminescent species. From this point of view, besides single composition QDs it is important to highlight the introduction of the so-called heteronanocrystals realized through the colloidal encapsulation of a "core" into a coating "shell" of different semiconducting species, to give a sort of "onion-like" structure, containing at least two semiconductors [33]. The

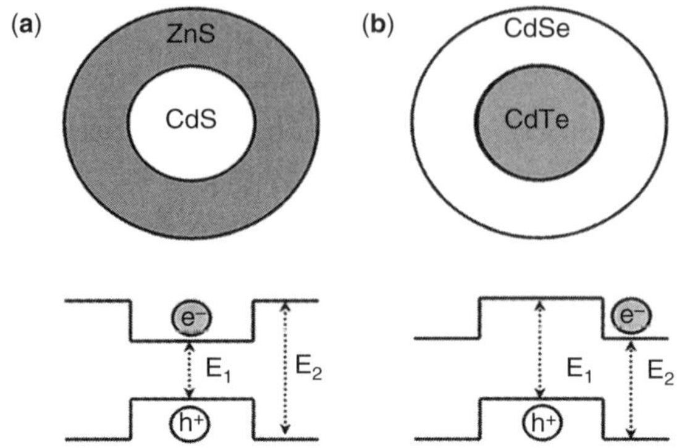

Figure 13.3 Schematic representation of type-I (a) and type-II (b) core-shell quantum dots and their band structures. Direct bandgap energies (E_1, E_2) are also reported.

specific combination of different core and shell materials enables the realization of type-I or type-II QDs in which the larger band-gap semiconductor is located in the shell or in the core, respectively (Figure 13.3) [34]. The introduction of such new nanostructures gives an enormous flexibility in QDs design, allowing enhancing performances such as: stability toward oxygen exposure (one of the main issues for uncapped QDs suffering from photoinduced oxidation), reduction of self-absorption losses, efficient tuning of the absorption spectra, increased luminescence quantum efficiency and enhanced matrix compatibility. Typical core-shell nanostructures comprise either II-VI or III-V semiconducting materials, usually type-I QDs involve CdSe/CdS, CdS/ZnS and ZnSe/ZnS semiconductor couples, whilst most diffuse type-II counterparts are based upon CdSe/CdTe and ZnSe/ZnTe [35].

QDs are clearly promising systems for LSCs, however they still do not replace more traditional emitting materials like laser dyes [36–38]. A few more aspects are in fact to be considered which still prevent a full exploitation of such technology: first of all the stability issue, which still is a challenge for a full employment in common environmental conditions and during manufacturing process due to photodegradation in presence of oxygen. Moreover a change (normally a blue-shift) in both absorption and emission spectra is frequently observed when the nanocrystals are dispersed into the matrix medium. The latter is associated with a reduction or quenching of the luminescence efficiency, particles clustering and surface oxidation. Besides, no QD so far matched the emission efficiency of the best performing organic dyes as their maximum reported emission efficiencies average around 80% [39]. Another fundamental issue to be considered is the difficulty encountered when trying to realize polymeric LSCs, because of the different chemical nature of the host and guest materials as well as the polymerization conditions employed (presence of very reactive radicals, high temperature and/or UV irradiation). Another final important limit is the toxicity of the elements constituting efficient QDs (Cd, Pb, Hg, As).

13.3.2 Luminescent Lanthanides Chelates

Exploiting organic lanthanide complexes as luminescent species for LSCs can be an efficient alternative to traditional luminophores, thanks to the possibility of achieving a useful combination of the peculiar optical properties of these ions (sharp and intense emission, high stability) and the absorption characteristics of organic chromophores (broad and strong absorptions in the UV-VIS range). Considering the electronic configurations of rare earth elements (i.e. $[Xe]4f^n$), the gradual electronic filling of the highly localized and shielded 4*f* orbitals, and consequently the high number of possible energy levels, are fundamentally the origin of their interesting properties. The most stable oxidation state for lanthanides is +3 and all the trivalent cations of the series are characterized by intense photoluminescence (except for Lutetium and Lanthanum), in particular Europium and Terbium are the most efficient ones, as testified by their use in imaging applications [40, 41]. The emission spectra for lanthanide ions originate from intraconfigurational *f-f* transitions inside the 4*f* shell, the latter being practically unperturbed by the chemical environment due to the high degree of shielding originated from the outer orbital states; this implies very sharp easily recognizable transitions. It is to be underlined that, if on the one hand well-defined, narrow-band emissions with long lifetimes of the excited states are originated, on the other hand very weak and spectral limited absorption bands are registered as a result of the *f* states involvement in the electronic transitions, forbidden according to Laporte's (or parity) selection rule [42, 43]. As a consequence, even though rare-earth ions have been investigated as luminophores for LSCs (above all Yb^{3+} and Nd^{3+}), usually high concentrations of lanthanides and co-doping [44, 45] are necessary to achieve sufficient light absorption (both in spectral range and intensity). The problem of the limited and weak light absorption can be overcome by exploiting the so-called *antenna effect*, profiting from the favorable absorption and excitation transfer capabilities of suitable molecular species (usually an organic chromophore). The mechanism involved was originally proposed by Crosby and Whan [46]: the solar radiation absorbed by the molecular ligand promotes the excitation of the latter with an electronic transition from its ground state level (S_0) to the singlet first excited state (S_1); favored by spin-orbit interaction (heavy atom effect of the lanthanide), the energy is thus transferred to the ligand triplet state (T_1) by intersystem crossing mechanism (ISC), and in cascade it is transferred (ET) to the coordinated lanthanide ion which emits it red-shifted [47]. Therefore by means of this process, it is possible to collect a broader spectral range of radiation frequencies (usually from UV to VIS) with high molar absorptivity with respect to the free ion alone, whilst retaining the same intense lanthanide emission at longer wavelengths, with practically null overlap between the absorption and emission spectra of the complex. It appears fundamental in this regard that the accurate choice of proper molecular ligands, which not only should possess the best possible absorption characteristics (i.e. strong and broad absorption behavior) but also meet specific requirements in term of coordinating sites and molecular energy levels, are required to match those of the rare earth ion. In order to populate the lanthanide emissive energy level the lowest triple state of the ligand needs to be located at a near equal energy or slightly higher energy with respect to the excited levels of the lanthanide, never below it (Figure 13.4) [43, 48].

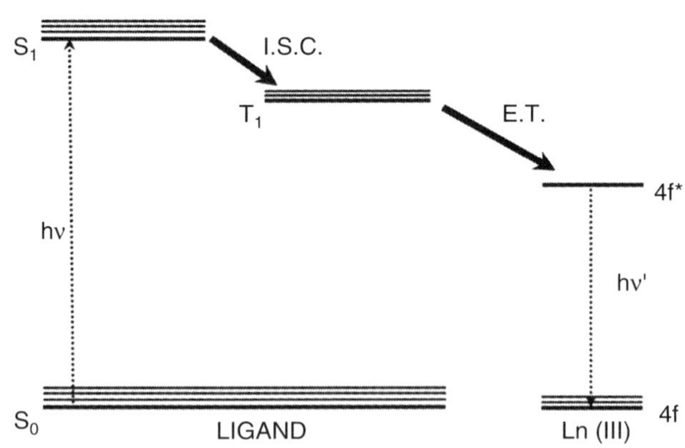

Figure 13.4 Schematic simplified representation of the photophysical processes involved in the sensitization of lanthanide ion via coordinating ligands.

The chelates employed for applications in LSCs are in most cases based upon Tb (III) and Eu (III) lanthanide ions, due to their high luminescence quantum efficiency (quantum efficiency in the range between 70–85% for β-diketonates complexes of Europium (III) with excited state lifetimes in the order of 0.80 ms have been reported [49–51], whilst for Terbium (III) fluorescence quantum yields between 80–96% and lifetimes between 1.70–2.60 ms have been registered [52]). Amongst the various available organic species capable of acting as chelating agents for lanthanides (Figure 13.5), the most commonly employed are b-diketonates (lanthanide preferentially bind to

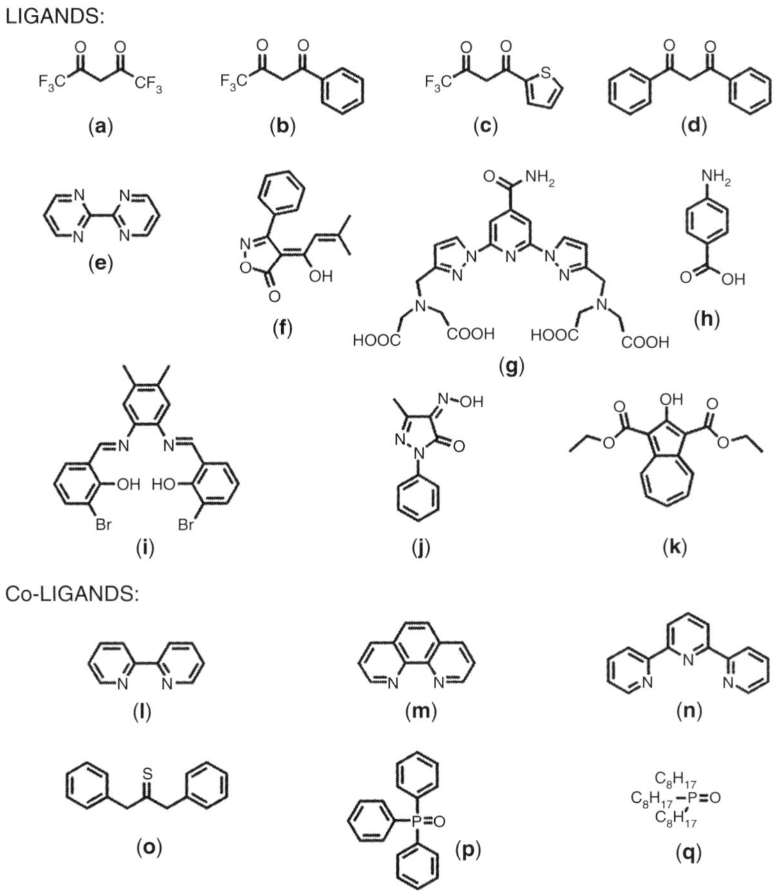

Figure 13.5 Examples of organic ligands (including β-diketonates: a) Hhfac, b) Hbtfac, c) Htta, and d) Hdbm usually employed for rare earth ion complexation, and commonly associated co-ligands (including: n) 2,2′-bipyridine, o) 1,10-phenanthroline and s) tri-n-octylphosphine oxide).

oxygen atoms). These complexes are actively investigated thanks also to the commercial availability of various diketonate ligands and to the straightforward access to the corresponding complexes.

The first coordination sphere of the trivalent rare-earth ions usually consists of eight or nine coordinating bonds; it therefore results unsaturated in these β-diketonate six-coordinate complexes. What is usually observed is thus an expansion of the lanthanide coordination sphere by different methods, namely: oligomers formation (bridging β-diketonates ligands), water molecules insertion, arrangement of four β-diketonate ligands around a single lanthanide trivalent cation with formation of a tetrakis complex [53]. Another approach, specifically studied in order to saturate the coordination sphere, consists in the insertion of different agents which behave as co-ligands (usually Lewis bases such as 1,10-phenanthroline, 2,2′-bipyridine or tri-*n*-octylphosphine oxide) through the formation of stable adducts (Figure 13.5) [54]. Despite the favorable characteristics of Terbium and Europium ions (high emission quantum efficiencies, exceptionally high Stokes shift in organic complexes), which in few cases have been effectively exploited for the realization of "zero self-absorption" luminescent solar collecting coatings and panels [55, 56], the possible application of these lanthanides in LSCs is limited by one major issue. The problem is related to the intrinsic emission wavelengths characteristic of such rare earth ions, which are located in the Vis region of the spectrum (around 550 nm for Tb^{3+} and around 620 nm for Eu^{3+}). This optical feature in fact restricts the possible choice of organic antennas to be used: as a consequence of the high Stokes shift obtained by the imposed energy level matching requirement previously reported, only those molecules showing absorption bands limited to the UV frequencies of the solar spectra can be exploited. A large range of the solar spectrum is therefore lost, limiting the final LSC efficiency. Different paths have been followed to bypass this problem: an interesting attempt based on pushing the absorption characteristics of the organic ligands toward the limits imposed by the physic of the energy transfer process, has been recently reported [57]. In this example, through a fine design of the employed chelating agents (Figure 13.6) it has been possible to induce a direct energy transfer to the lanthanide ion from the singlet excited state (S_1) of the organic ligand rather than from its triplet state (T_1) as commonly happens, thus giving the possibility to extend the absorption band of the ligands also into the visible part of the spectrum, where a

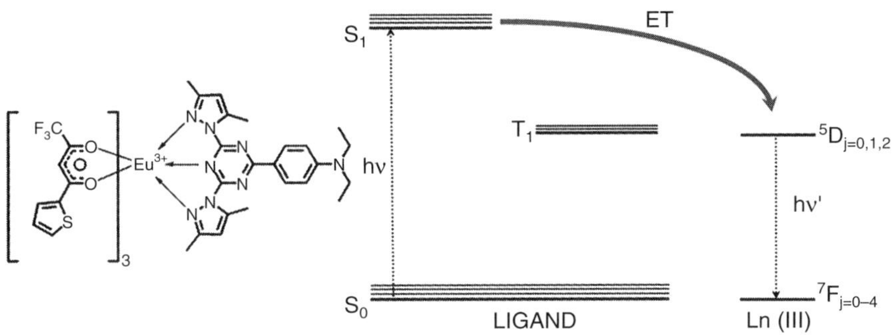

Figure 13.6 Molecular structure and luminescent mechanism for the Eu (III) complex reported in ref. [57].

consistent part of the solar spectrum is located. Such Europium chelate has been incorporated into a polyvinyl-butyral (PVB) matrix at a concentration of 1.6 wt% and then deposited onto the face of a K9 glass 3 mm thick; a crystalline silicon cell was then coupled to one edge of the LSC. Irradiation under a solar simulator (AM1.5G) afforded power conversion efficiency of 4.99%, a record value for lanthanide-based LSC. It is even more surprising considering the limited luminescence quantum yield of the complex (44%).

Another approach in the search for higher efficient lanthanide chelates consists in the replacement of Vis-emitting Terbium and Europium lanthanides with NIR-emitting ones, mostly Ytterbium(III). The energy of Yb^{3+} emitting state enables light harvesting at both UV and VIS frequencies. Also, Ytterbium emission energy corresponds to crystalline silicon energy gap (1.14 eV). Finally, when employing UV absorbing ligands, it is also possible to prepare colorless luminescent solar collectors (whose absorption and emission spectra are both outside of the Vis region) capable of replacing building windows without any chromatic distortion of the transmitted light. In this context a recent paper [58] demonstrates the successful design, synthesis, photophysical characterization and incorporation in a polymeric slab of a novel colorless NIR-emitting Ytterbium chelate (Figure 13.7) based upon nitrosopyrazolone ligands and ancillary phenanthroline co-ligand. The emission spectra registered in CCl_4 and methylmethacrylate (MMA) solution are identical, thus demonstrating the chelate stability even in strongly coordinating solvents. The luminescence quantum yield for solution sample reaches the value of 2.4% (CCl_4), in polymeric PMMA matrix the value drops to 1%. Focusing on the

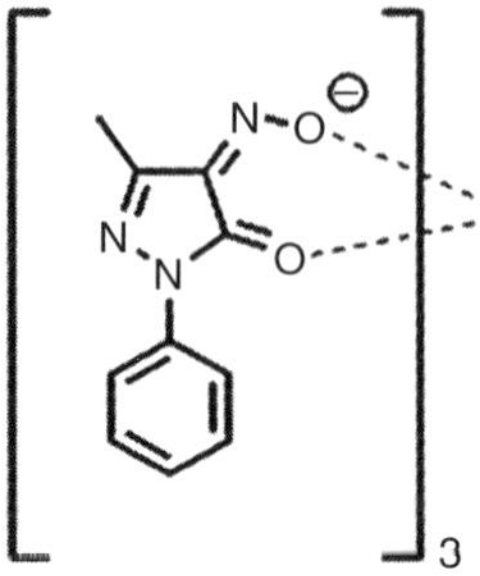

Figure 13.7 Molecular structures for the new colorless NIR-emitting Ytterbium chelate reported in ref. [58].

measured lifetimes, shorter values have been registered compared to the Ytterbium radiative one measured in inorganic matrices. The main reason for the latter outcomes is ascribed to the quenching of excitation by the overtones of high-energy vibrational centers (*vide infra*). It should be noted that the complex is fully compatible with radical initiated bulk polymerization.

The major challenge for research in the field is nowadays focused on the maximization of the complex emission efficiency. The most important parasitic phenomenon responsible for the drastic reduction of the Ytterbium emission is the vibrational quenching due to the energy transfer from the lanthanide ion to the overtones of C-H, N-H, and O-H stretching vibrations. Such parasitic phenomenon proceeds via a FRET (Förster Resonance Energy Transfer) mechanism, with a distance dependence between donor and acceptor inversely related to the distance between the involved species $(R_{DA})^{-6}$ [42]. Therefore it is of fundamental importance to exclude the presence in close proximity to the lanthanide ion of any C-H, N-H, and O-H species [59].

In fact, since the strength of the absorption tends to decrease by approximately an order of magnitude between each harmonic order, higher harmonics are generally weak enough not be of concern. Clearly, the highest energy vibrations will be those possessing high spring constant (stiffer bonds, such as double bonds) and/or small reduced mass. The smallest reduced mass occurs when one of the atoms is hydrogen, and the C-H aliphatic bond is typically used as the benchmark for infrared absorptions. Considering the positions and intensities of various vibrational overtone absorptions of interest, both C-H and O-H overtones are seen to be highly absorptive in the interested frequency window whereas C-F and C-D

Table 13.1 Wavelengths and intensities of some important vibrational overtones.

Bond	Overtone Order	Wavelength (nm)	Intensity (relative)
C–H	1	3390	1
C–H	2	1729	7.2×10^{-2}
C–H	3	1176	6.8×10^{-3}
C–F	5	1626	6.4×10^{-6}
C–F	6	1361	1.9×10^{-7}
C–F	7	1171	6.4×10^{-9}
C=O	3	1836	1.2×10^{-2}
C=O	4	1382	4.3×10^{-4}
C=O	5	1113	1.8×10^{-5}
O–H	2	1438	7.2×10^{-2}

overtones show extremely low absorption throughout the range of interest (Table 13.1) [60, 61].

The solution to this issue implies a complete exclusion of water molecules from the inner coordination sphere, for example by saturating the lanthanide coordination with co-ligand molecules such as those previously reported, and the replacement of the ligands with fully deuterated or perfluorinated analogues (Figure 13.8). The validity of this approach has been demonstrated in the literature [62, 63], with important contributions on Erbium chelates (a lanthanide suffering from the very same vibrational quenching phenomena) from our research group [64] and from Mancino *et al.* [65].

13.3.3 Organic Dyes

Organic dyes, the most widely investigated luminescent species over the last 30 years for application in LSCs, remain at the forefront of the research for innovative luminophores. The reason for their enduring success are mainly due to: a) their spectral characteristics (i.e. high fluorescence quantum yield, large absorptivity, broad absorption spectra), b) the intrinsic tunability of their optical and

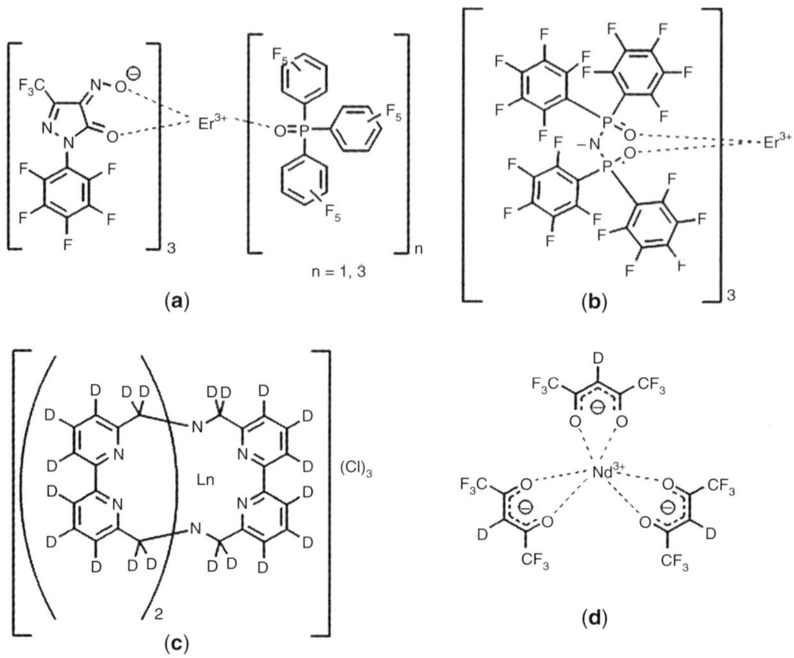

Figure 13.8 Molecular structures for perfluorinated lanthanides complexes; (a) from ref. [64], (b) from ref. [65], (c) fully deuterated,from ref. [63], and (d) mixed fluorinated-deuterated, from ref. [62].

chemical properties (through accurate molecular design), c) their availability by conventional synthetic procedures, d) their compatibility with polymeric matrixes (for full organic LSCs), e) the low costs associated with materials and manufacturing. Since the introduction of the LSC concept a wide variety of organic luminophores have been presented which belong to different classes of organic compounds. When designing the structure of suitable organic dyes, it is of fundamental importance to firstly consider the energies involved in the light-matter interaction. In other words, the intramolecular electronic transitions related to the light absorption process must set at energies comparable to those of the incoming solar radiation. In this regard, optical excitations for organic dyes usually concern molecular π-electrons and these compounds are therefore commonly based upon π-conjugated structures where the π-system form a cloud of delocalized and mobile electrons above and below the molecular backbone constituted of more energetic -bonds. Considering this electronic configuration, the absorption phenomenon is thus related to electronic transitions (transition moment oriented parallel to the main molecular axis) from the molecular ground state to the nearest excited state (usually π π^*), whilst the fluorescence emission process consists of a radiative de-excitation mechanism from the lowest excited state to the fundamental molecular level (short lifetime, singlet to singlet transition) [66]. A successful design strategy for organic luminophores, as a consequence, starts from finely designed π-conjugate polarizable molecules where the electron mobility can be greatly enhanced by introducing proper donor and acceptor substituents in specific positions on the molecular backbone, in order to affect the π-electron displacement though inductive and/or resonance effects. These factors rule the position of the HOMO and LUMO frontier orbital and consequently the optical properties. When designing organic fluorophores, beyond these basic electronic features, one must consider other fundamental elements which radically contribute in determining the molecular emission response (and thus the Stokes shift value) [67]: molecular planarity and symmetry, presence of intermolecular interaction and solvent effect.

A vast library of molecular structures with enhanced luminescent properties is reported in the literature for applications in LSCs (Figure 13.9), comprising: rhodamines [2, 21, 22, 68, 66, 69], coumarins [21, 22, 70, 71], perylene imides [21, 72–76], perylene bisimidazoles [77], (iso)violanthrones and mesabenzanthrones [78, 79], dicyanomethylenes (DCM) [17, 22, 80], boradipyrrin dyes (BODIPY) [81], benzothiadiazoles [82, 83], porphyrines and phthalocyanines [17, 84, 85], bipyridines [86–88]. Among all these heterogeneous classes, only a very limited number have so far resulted in being really suitable for LSCs, due to the strict requirements imposed in term of optical, chemical and physical properties, as well as overall stability [89]. Following, therefore, only the most prominent, extensively employed and promising organic luminophores (i.e. rhodamines, coumarins, rylene imides) will be presented in detail, focusing on innovative perspectives concerning rylene-imide derivatives.

Rhodamines and coumarins were the first examples of organic luminophores introduced in the pioneering publications on LSC in the early 80s [21, 22, 68]. Rhodamine dyes are a class of fluorophores belonging to the xanthenes family (together with fluorescein and eosin) characterized by excellent photophysical properties (high absorption coefficient, efficient emission in the Vis, phostability) and vast synthetic flexibility, which enable a widespread diffusion in

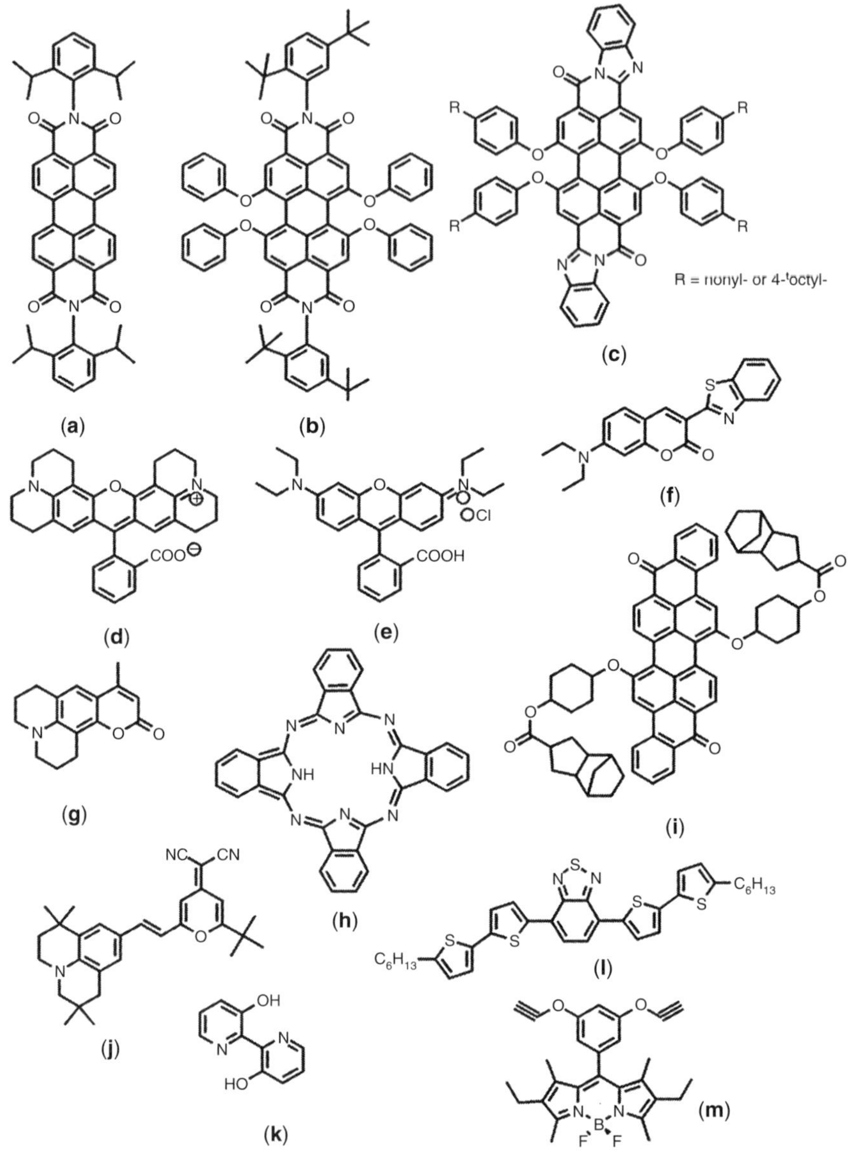

Figure 13.9 Molecular structures for organic fluorophores commonly encountered in LSCs: a) b) perylene diimides, c) perylenebisimidazole, d) e) rhodamines, f) g) coumarins, h) phtalocyanine, i) isoviolanthrone, l) DCM, m) bipyroidine, n) benzothiadiazole, and o) BODIPY.

the literature as: a) dye laser, b) molecular switches, c) fluorescence standards, d) fluorescence probes (above all for living cell imaging and biomolecular labeling), e) chemosensors (both *in vitro* and *in vivo*) for specific metal ions [90]. The introduction of multiple molecular modifications in precise structure positions guarantees an extreme flexibility in the molecular design of rhodamines, enabling a fine tuning of their optical as well as chemical characteristics to match the peculiar requests of each considered application. More specifically, three main types of elaboration mostly influence the rhodamine properties: a) modification of the amino groups on the xanthene moiety (positions 3, 6), b) modification of the carboxyphenyl ring (positions 4′, 5′), c) modification of the carboxylic functionality (position 2′) [91]. Commercially available and inexpensive rhodamine dyes usually employed in LSCs are rhodamine B, rhodamine 6G, rhodamine 101 and rhodamine 110 (Figure 13.10) (the cost for the first two rhodamines reported is below 2 \$/g).

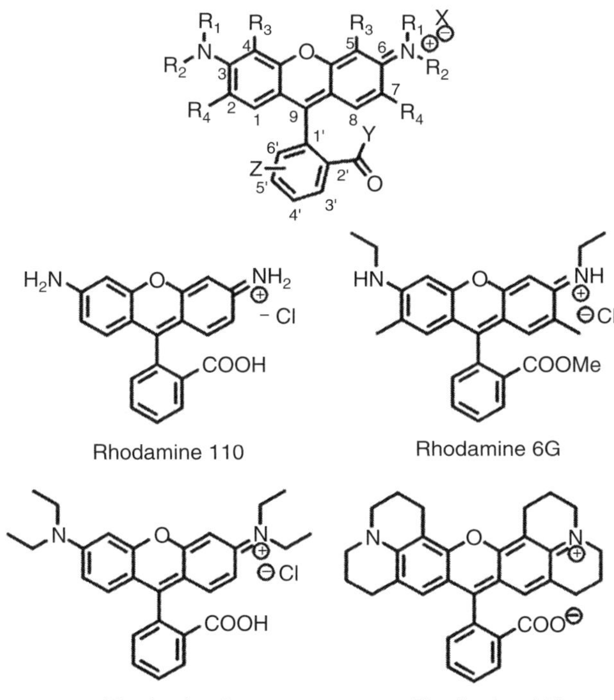

Figure 13.10 Generic molecular structure of rhodamine dyes and most representative exponents employed in LSC.

Focusing on the photophysical properties of rhodamines, they are firstly determined by their symmetrical π-electron system extending over the whole planar structure. Although rhodamines are generally highly fluorescent (due to their rigid molecular structure), depending on number, nature and position of the substituents (R_1, R_2, R_3, R_4, Z) they show major differences in their optical behavior (both in emission and absorption). Their emission efficiencies show in fact peculiar dependence on the substitution patterns: taking into account the functionalization of the amino groups of xanthene (positions 3, 6) with two alkyl substituents (R_1, R_2), the occurrence of non-radiative deactivation pathways due to internal conversion are observed. It is in this context possible to consider two distinct deactivating processes: the first one involves energy dissipation due to well-known energy coupling with vibration modes of C-H and N-H bonds, whilst the second one is associated with a non-fluorescent twisted intramolecular charge-transfer (TICT) state due to a torsion between the amino groups and xanthene ring [92]. It is possible to observe the direct effect of this substitution at the nitrogen atoms experimentally considering the rhodamine 6G molecule, presenting only one ethyl substituent on each nitrogen atom and a fluorescence quantum yield close to unity (0.95), the latter nearly independent of solvent polarity and temperature. On the other hand, for rhodamine B, where both the amino groups are fully alkylated, the fluorescence efficiency results were strongly depressed (0.50) and dependent on both solvent and temperature. A proof of the importance of molecular planarity for such systems is encountered if examining the optimal fluorescence response when inducing a planarization of the structure by the incorporation of the amino groups into six-member rings as for rhodamine 101. By means of this structural modification, a prevention of torsion or rotation of the amino group with respect to the molecular plan is achieved in the excited state and an experimental quantum yield of 0.96 is registered [93, 94]. It is to be also noted that whenever substitutions on the amino groups lead to unfavorable steric interactions with the molecular core, an enhancement of the nucleophilicity of the phenolic oxygen is produced which leads to rapid lactonization of the structure with consequent disruption of the conjugated system and quenching of the emission properties [91]. Positions 4′ and 5′ of the carboxyphenyl ring are effectively prone to functionalization; however, a mixture of isomers is usually obtained by simple functionalization of the commercially available unsubstituted precursor. In order to obtain isomeric pure compound a synthesis involving the whole molecular structure is therefore required with results that are obviously not economically advantageous for large-scale employment of the fluorophore. Commercially available 4′/5′ substituted products are indeed too expensive for practical applications. Moreover, if a modification of the carboxylic functionality is chosen, usually any synthetic pathway implies the use of both expensive coupling agents and severe reaction conditions, which automatically excludes this functionalization route for obtaining economically interesting fluorophores for LSCs.

The main drawback associated with the use of rhodamine dyes in LSCs is related to the pronounced overlap between their emission spectra and the low energy tail of the absorption band, which ultimately causes a detrimental reabsorption phenomenon (Figure 13.11). Also the stability in the polymeric matrix is still limited for most of this compounds [95].

The reabsorption issue and the limited photostability are somewhat less severe in the other big class of historically employed organic fluorophores for LSCs: coumarins. These systems result

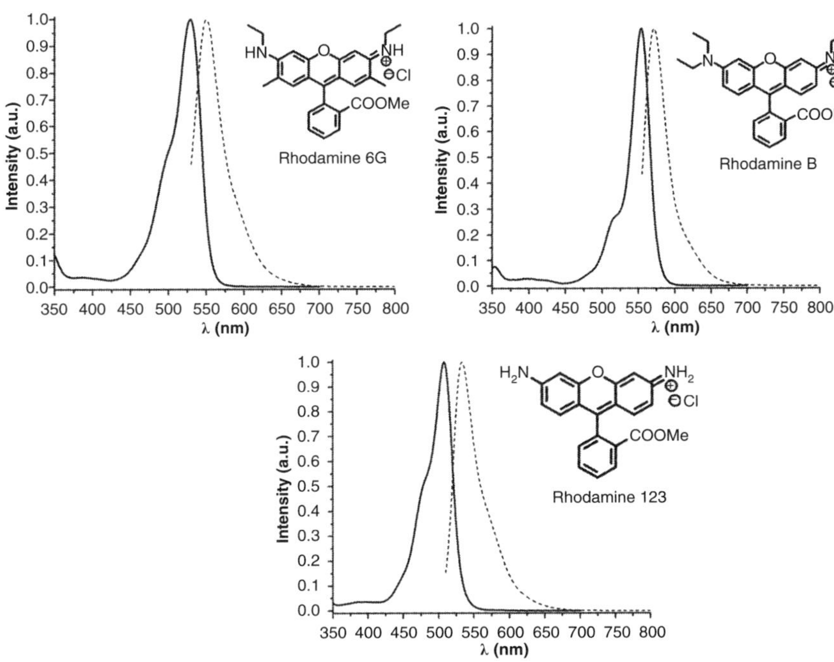

Figure 13.11 Absorption and emission spectra for most representative rhodamines dyes employed in LSC.

in fact more stable than most rhodamine dyes and higher Stokes shift are usually registered (for example, the overlap factor of the absorption and emission spectra for coumarin 504A is 0.12 whilst that of rhodamine 6G is 0.48 [1]) (Figure 13.12).

Considering the photochemical properties of coumarins and focusing on the fluorescence emission, even though near unity emission quantum yields are usually encountered [4], in a few cases a drastic lowering of the emission or a complete quenching is observed (Figure 13.13). An exhaustive structure-property

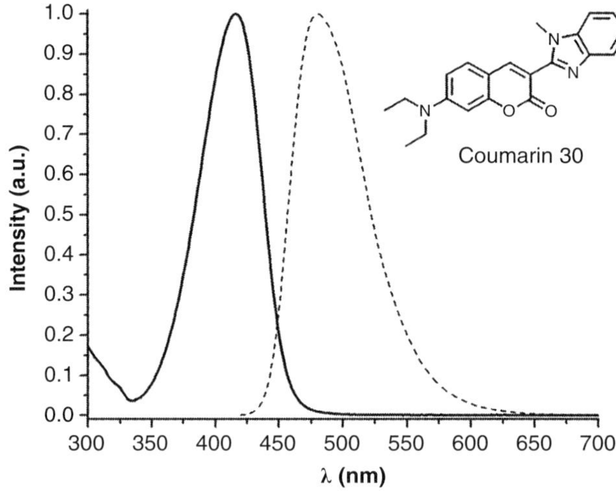

Figure 13.12 Absorption and emission spectra for representative coumarin 30.

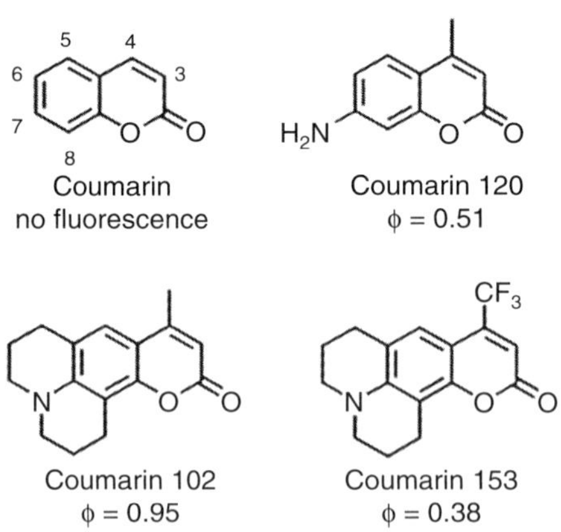

Figure 13.13 Structures of assorted coumarin dyes.

correlation is thus required in order to explain the optical response of these fluorophores and preliminarily select the best possible candidates for LSCs.

Unsubstituted coumarin does not fluoresce at all, mainly due to the nature and energy of its singlet lowest excited state involved in the emission process ($n\pi^*$ type with an energy of 3.4 eV). In fact, according to Kuznetsova *et al.* [70] the luminescence properties of coumarins are mainly connected with the relative positions of their $n\pi^*$ and $\pi\pi^*$ excited states: for structures with close lowest $n\pi^*$ and $\pi\pi^*$ excited states a sharp increase of non-radiative deactivation is registered, as a consequence of vibrational quenching leading to a displacement of these energy levels. The fluorescence responses of coumarins are affected by a number of non-radiative pathways: intersystem crossing to triplet states, non-radiative internal conversion to the ground states due to TICT state, vibrational quenching due to interactions with the solvent/matrix. Common functionalizations in coumarins involve positions 3, 4, 6 and 7 of the molecular structure (Figure 13.13), mainly throughout the introduction of electron-donating (EA) and electro-accepting (ED) groups. In the most common fluorescent members of the coumarin family, the 7-position results are generally occupied by an amino group: the introduction of an ED moiety (such as the amino group) in this position, while introducing only a slight variation in the energy of the singlet and triplet excited $n\pi^*$ states, indeed induces a significant decrease in the single and triplet excited $\pi\pi^*$ states which thus become the lowest excited states. This phenomenon is mainly due to an increase in the conjugation with the core aromatic system, especially with the pyrone carbonyl group. Along with its influence on the emission intensity, the introduction of the amino moiety in the 7 (or 6) position of the structure also induces a bathochromic shift (in both emission and absorption). The entity of this shift is directly related to the strength of the ED group: alkylated amino-groups usually intensify the effect that is, however, maximized when the nitrogen atom is introduced into a 6-membered ring, enforcing molecular planarization. In this regard, for coumarin 120 (non-alkylated amino group) the maximum of absorption and emission are experimentally found at 354 nm and 435 nm respectively, whilst for coumarin 102 (fully alkylated and planar amino substituent) they shift to 387 nm and 473 nm. The red-shift is enhanced if the position 4 of the coumarin structure is functionalized with an EA group, as for coumarin 153 where the methyl moiety is replaced by a trifluoromethyl group (absorption maximum at 421 nm, emission peak at 531 nm). An opposite effect is instead obtained when introducing in the same position an ED functionality. The role of a full alkylation and planarization of the amino group in the 7 position is also fundamental in order to obtain near unity fluorescence efficiency, as a result of the complete suppression of the reported parasitic deactivating mechanisms. When considering the application of luminophores for LSCs, besides the intrinsic fluorescence quantum efficiency, the molecular absorptivity, panchromaticity and the Stokes shift value are also crucial. By a rapid consideration of the optical features of coumarin dyes, it appears clear how these molecules do not possess all requisites, being capable of absorbing only the UV wavelengths and emitting the radiation downshifted in the blue-green part of the VIS spectrum. Only a limited fraction of available range of incident wavelength is thus collected, while the emission radiation results are scarcely adaptable to common crystalline silicon PV cell. However, these problems can be overcome by introducing a multiple-dye approach [2, 17, 96–99] and alternative semiconductors for the cell [25, 100, 101]. Different dyes possessing absorption bands in different spectral ranges can thus be employed together, either in a configuration consisting of stacked LSCs containing one specific dye or in a single multi-dye LSC architecture, in order to collect the maximum possible range of wavelengths. In the first geometry the solar spectrum is segmented by each single-dye concentrator coupled to a specifically designed bandgap-matched PV cell, in the second approach the LSC slab can intercept more of the solar radiation using a "donor" dye and transfer the energy to an "acceptor" dye whose emission can finally be matched to the proper employed active semiconductor.

The main drawback of coumarins is, however, their limited stability inside the collector in the final working conditions: even though they are more stable than rhodamines, these chromophores in fact can undergo photodegradation (whose effects comprise the complete disappearance of the characteristic absorption spectra and the quenching of fluorescence emission) under continuous exposure to UV radiation. Among the possible degradation mechanism, the two clearly identified are: a) the disruption of the benzopyrone fragments of the structure by [2+2] photocycloaddition reaction, b) the reaction of the substituting groups (photosubstitution, photochemical oxidation). The photostability therefore becomes a central issue to be addressed, and in this regard, rylene imides become the most valuable fluorophores available for OLSCs.

Rylene dyes have been well known since 1913 as superior pigments and vat dyes thanks to their exceptional thermal, chemical, photochemical and photophysical stability. By virtue of their outstanding characteristics, along with high fluorescence quantum yields, high extinction molar coefficients and broad tunable absorption band covering the UV-VIS-NIR region of the spectrum, this class of molecules has been established as key chromophores for both traditional (coloration of textiles and consumer goods) and innovative high-level photonic applications [102]. When dealing with rylene imides a broad class of compounds is considered whose structures span from the simplest perylene monoimides (PMIs) and diimides (PDIs) to the IR absorbing hexarylene diimides (HDIs) and their dendritically expanded derivatives. The occurrence of a multitude of rylene imides is firstly determined by the intrinsic possibility of enlarging the central benzocondensed perylene system throughout annulations with further benzene/naphthalene units. In this regard, elongation of the longer molecular axis leads to the so-called *core-extension* of perylene imides towards its higher homologues, namely: terrylene, quaterrylene, pentarylene and hexarylene imides. An alternative consists in the lateral expansion of the perylene π-system along its short molecular axis (*core-expansion*) leading to benzoannulated compounds, also known as: coronene imides, dibenzocoronene imides. Higher homologues can indeed be obtained by combining the two enlargement mechanisms, designing a plethora of novel rylene dyes (Figure 13.14) [103].

The only relevant drawback of rylene imides is represented by their very low solubility, which requires specific further functionalizations of the rylene scaffold in order to obtain readily employable luminophores. Even 3,4,9,10-perylenetetracarboxylic dianhydride (PDA), the common precursor of all rylene dyes, is a substantial insoluble pigment. When the anhydride functionality is substituted with an imide group the simplest exponents of the rylene dye family, perylene monoimides (PMIs) and perylene diimides (PDIs), are obtained: unsubstituted PDI, formally originated from the condensation of the precursor anhydride with ammonia, is essentially insoluble in all common organic solvents. The reason for this behavior is the occurrence of a strong molecular aggregation due to the instauration of stable intermolecular interactions, specifically: hydrogen bonding between carboxylic oxygens and N-H imidic moieties, and extended π-π interactions between the aromatic benzoannulated core systems. A tight molecular face-to-face packing

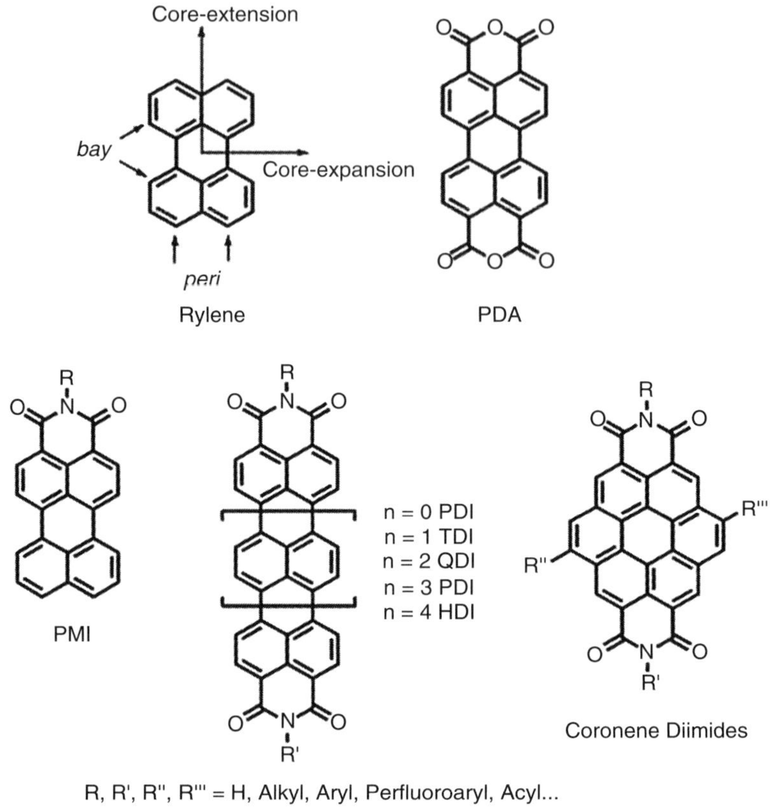

Figure 13.14 Structures of rylene dyes, illustration of the possible functionalization sites.

is obtained which is an obstacle to any interaction with the solvent molecules. Solubility can be increased through the introduction of long and branched alkylic chains at the amidic nitrogen atom. Even better results are obtained when bulky aromatic rings are introduced in this position [104]. The origin of this phenomenon at the molecular level lies in the almost perpendicular twisting of the aromatic groups with respect to the perylene core, due to unfavorable steric interactions. It is to be stressed how the functionalization at the imidic nitrogens, while fundamentally effecting the solubility and aggregation properties, has indeed almost no consequence on the relative stability of the compounds (even though specific functionalities such as perfluorinated rings can induce a slightly enhanced stability [105]) and no evident effects on the final optical properties. The reason for this behavior is connected to the fact that the nitrogen atoms of rylene imides are located on nodal planes of both molecular HOMO and LUMO, therefore the effects of imide substitution on the frontier orbital energies are only inductive in nature and no direct perturbation of their energy position is encountered, as demonstrated both experimentally and by theoretical modeling [106, 107]. The introduction of particular substituents at the imidic position is therefore the simplest and most widely diffuse functionalization design employed to confer other peculiar characteristics, such as: extended solubility, packing properties (self-assembly and self-organization tendency [108, 109]) and particular reactivity (e.g. polymerogenicity [110]). Taking into account the optical properties of PDIs (Figure 13.15), which are the preferred rylene dyes for LSCs, the absorption behavior displays characteristic vibronic structures in the transition from the singlet ground state to the first singlet excited state ($S_0 \rightarrow S_1$), indicating a substantial coupling of such transition with the vibration of the perylene skeleton [111]. This phenomenon is the consequence of the almost perfect planarity of the perylene core (practically flat when no substituents are

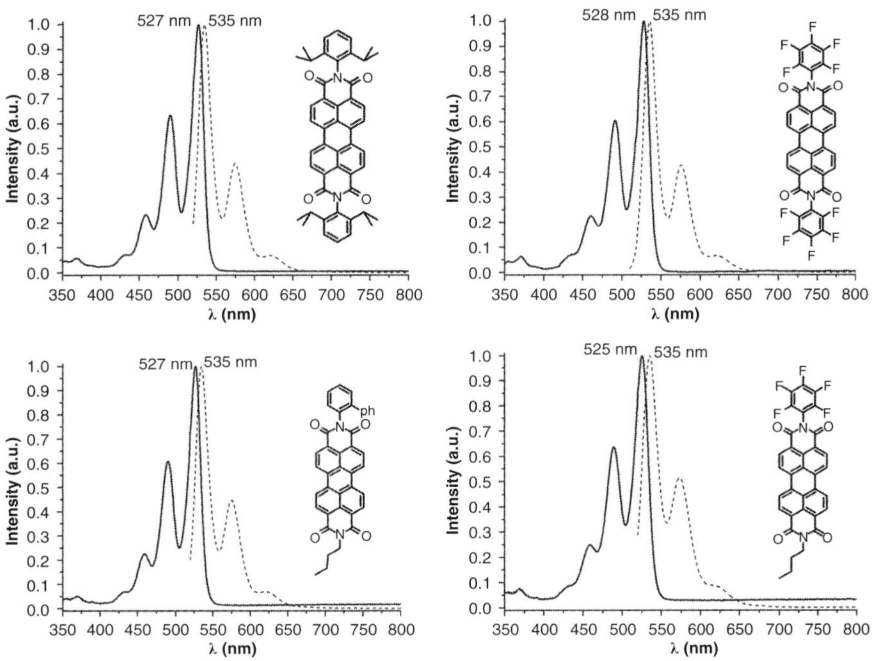

Figure 13.15 Absorption and emission spectra for variously N-substituted PDIs.

located on the benzocondensed rings), and is preserved in all the core-extended rylene diimide systems which therefore manifest the same structured pattern in the absorption spectra, bathochromically and hyperchromically shifted. The typical absorption spectrum for core-unsubstituted PDIs extends from 400 nm in the UV to around 550 nm in the VIS, with a maxima at 525 nm and a peak absorptivity ranging around 80000 L mol^{-1} cm^{-1}. The emission spectrum is a spectacular image of the absorption spectrum with a small Stokes shift (< 10 nm), a fluorescence quantum yields almost equal to unity and rather long singlet excited-state lifetimes (ca. 4 ns in common organic solvents) [112, 113]. These features are independent on the kind of substituent introduced on the nitrogen atoms.

Further modifications of the rylene imide scaffold require the functionalizations of the rylene core mainly at two different possible sites, namely *bay* and *peri* positions (Figure 13.15). In this case the nature and position of the substituents have a strong impact not only on the final molecular solubility (and aggregation) as for the N-functionalization but also on the optical and electronic properties, due to their direct influence on symmetry, planarity and energy levels of the central rylene core. The functionalization at the *bay* positions is considered the standard route pursued to gain far more soluble rylene diimides. Various examples of *bay* functionalized PDIs have been presented in the literature introducing various functionalities in these positions (e.g. phenoxy groups, halogen atoms, nitriles, alkynes) [113–115] depending on specific application, the most common functionalization for application as luminophores consisting in the introduction of phenoxy groups [116, 117]. Considering the simplest rylene diimides, *bay*-substituted derivatives manifest few major changes compared to unsubstituted PDIs, in both their absorption and emission properties: a) less defined vibronic structure, b) bathochromically shifted spectra, c) enhanced Stokes shift, e) broader bands, d) appearance of a $S_0 \rightarrow S_2$ transition in the UV region (symmetry forbidden for unsubstituted PDIs). This behavior is due to both electronic and steric effects, caused by the insertion of the phenoxy substituents: the electro-donating groups affect the relative position of the frontier orbitals and increase the charge-transfer character of the electronic transition. Moreover, their bulkiness induces a slight torsion and loss of planarity/symmetry of the the perylene core [118].

A commonly employed *bay*-substituted PDI for LSCs is the commercially available Lumogen® F Rot 305 developed and manufactured by BASF (Figure 13.16). According to G. Seybold of BASF

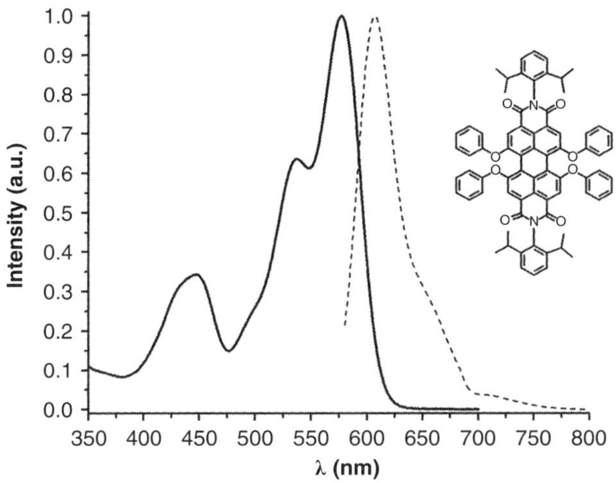

Figure 13.16 Structure and optical features of *bay* substituted perylene imides.

laboratories [78] the long-term stability of this class of compound is very good, as demonstrated by the very limited change of the original optical performances after 1.5 years of outdoor weathering of a PMMA plate doped with F Rot 305 dye: optical density drops by 7% and fluorescence drops by 10%. Thanks to 2,6-diisopropylphenyl moieties at the nitrogen atoms and the tetraphenoxy substitution at the *bay* positions, the solubility is indeed considerably enhanced, reaching an exceptional value of 100 g/l in ethyl acetate at 20°C. The absorption maximum in PMMA for this compound is 578 nm (absorption between 400–600 nm) and the yield of fluorescence is 96% with a fluorescence peak at 613 nm in the polymeric matrix. These data, while demonstrating the validity of such fluorescent dyes in term of stability, matrix compatibility, luminescence intensity and absorption efficiency, demonstrate the persistence of a still limited range of absorption and insufficient Stokes shift. Moreover, as already reported for other organic dyes previously seen, the emission wavelength poorly matches the optical bandgap of commonly employed PV cells. The latter issue can be easily overcomed introducing the considered stratagem of multi-dye LSCs or alternatively taking into account the possibility of employing core-extended exponent of the rylene imides family. When the benzocondensed rylene core is extended to give terrylene diimides (TDIs) and quaterrylene diimides (QDIs) what is observed is in fact a substantial red-shift of the absorption band with respect to PDIs, with a simultaneous enhancement of the molar absorptivity over the whole absorption range [119–121]. However, while TDIs still show an intense florescence emission (maxima between 670 nm and 710 nm depending on the *bay* substitution), almost unchanged with respect to PDIs, higher homologues lose their fluorescence dramatically, till complete quenching for NIR-absorbing hexarylene diimides. Nonetheless, rylen dyes remain the standard in LSC and most recent developments still focus on related structures.

Recently, the development of unsymmetrical structures has been considered a viable strategy to increase the Stokes shift, thus reducing reabsoprtion [122–124]. The strategy, recently proposed in the literature for the development of fluorophores for LSCs by Sanguineti *et al.* [125], aims to convey a dipolar character to the electronic structure of the perylene imides in order to induce a major difference between the electronic structure of the ground and first excited states. In this literature example, one of the two imides of a common PDI is replaced by a quinoxaline residue that, along with its positive effect on the Stokes Shift, also provides an additional high-energy absorption band in the UV region between 300–400 nm, thus improving the overall photon harvesting efficiency.

The resulting polarization is effectively reflected in an increase in the Stokes shift that reaches 70 meV, more than two times the typical value registered for symmetrical PDIs. The major drawback of this system is, however, the reduction of the fluorescence quantum efficiency (55%). In the same paper, a different route was pursued taking into consideration the functionalization at the *peri* positions of perylene monoimides. An easy control of the optical properties of such a system was underlined by Müllen *et al.* [126], demonstrating how it is possible to finely tune the absorption optical properties of PMIs by an accurate design of the substituents in *bay* and *peri* positions. The 9-substitited PMI introduced by our research group features the very peculiar donor dibenzazepine (DBA) whose central ring (possessing 8 π-electrons), makes it an antiaromatic system according to the Hükel definition. DBA therefore behaves like a very strong donor. DBA is also a rigid molecule which is particularly interesting for the suppression of the above reported detrimental decay pathways connected with vibrational and/or rotational transitions. This charge-transfer derivative shows, as expected, a broad absorption band (from 400 to 650 nm, with a maxima around 580 nm) with a weakly resolved vibronic structure. Even though the molar absorptivity of the compound is almost half that of a PDI but in

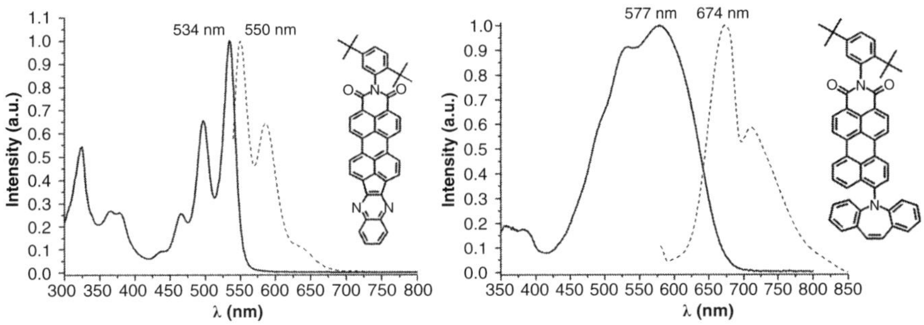

Figure 13.17 Asymmetric structured perylene imides as recently reported in ref. [125].

accordance with that of common *bay*-unsubstituted PMIs reported in the literature [126] (Figure 13.17), the Stokes shift exceeds 300 meV (not achievable by conventional PDIs approach), that renders it comparable with coumarins. Also in this case, the presence of the DBA ring is responsible for the formation of an additional high-energy absorption band at around 400 nm. Finally the fluorescence quantum yields are largely preserved, as demonstrated by the remarkably high value of 70% obtained from a PMMA slab embedded with the luminophore, the highest ever reported fluorescence quantum yield in the solid state for a push–pull perylene derivative. The preparation of such a prototype single-layer LSC through radical polymerization, the lack of degradation after 48 h at 120°C, and the photobleaching experiments, also demonstrate the high stability of the system in its working conditions.

13.4 Conclusion and Outlook

Luminescent solar concentrators represent a simple and low cost alternative to more common optical concentration solutions. The technology, originally proposed in the early 70s, in recent years has come back to the attention of the photovoltaic technologies community mainly due to the availability of innovative luminescent materials such as lanthanide chelates, quantum dots and high Stokes shift organic dyes. For simple systems working in single-layer configuration, the best solution for the matrix materials remains PMMA: it is in fact highly transparent in the visible and near-infrared region, it has a high refractive index (n = 1, 49) ensuring a limited escape cone. Moreover, the manufacturers guarantee PMMA stability for over 20 years.

The efficiency of the luminescent concentrator is directly influenced by the optical properties of phosphors, representing the actual engine of a LSC. The ideal luminophore should at the same time possess: a) panchromaticity; b) near unity luminescence efficiency; c) high stability; d) low cost and toxicity and most important of all, virtually no superimposition of absorption and emission spectra (high Stokes shift).

So far, none of the investigated luminophores have met all of the aforementioned requirements, and depending on the specific requirements, a trade off determines which one of the available luminophores is to be used. In detail, colloidal quantum dots can be extremely efficient in the absorption of light all over the visible spectrum but suffer from somewhat limited emission efficiency, low stability to environmental conditions, limited Stokes shift, and more importantly, very low compatibility with PMMA as well as other standard plastic matrixes. Lanthanide chelates on the other hand have huge Stokes shift, high stability and compatibility with PMMA, but suffer from very limited absorption capabilities (only the UV and the high energy portion of the Vis spectrum) and generally low emission efficiencies due to vibrational quenching.

Finally, performing organic dyes, with perylene dyes topping all of them, are efficient in absorbing and emitting light all over the Vis spectrum down to the NIR in cases of the most conjugated derivatives. They can be made compatible with PMMA, are generally of low toxicity and low cost (at least in prospective). Their main limitation remains a substantial self-absorption due to a still too limited, although recently significantly improved, Stokes shift. The future of this technology is still very much dependent on the development of better performing luminophores. So far the best trade off between the various requirements is represented by organic dyes, but new core-shell colloidal quantum dots as well as perfluforinated lanthanide chelates could outperform them in the near future.

Another important open issue deals with the stability of LSC under real working conditions. Derivatives of the Lumogen series and Lumogen R 305 in particular, seem to be able to comply with most stringent market requirements, whilst a final word on most recent performing luminophores cannot yet be given.

References

1. M.G. Debije, and P.P.C. Verbunt, *Advanced Energy Materials*, Vol. 2, 12–35, 2012.
2. T.C.B.A. Swartz, and A.H. Zewail, *Optics Letters*, Vol. 1, 73–75, 1977.
3. US Department of Energy, and NREL, http://www.nrel.gov/ncpv/images/efficiency_chart.jpg, 2012.
4. K.W.J.B. Wilfried G.J.H.M. van Sark, L.H. Slooff, A.J. Chatten, A. Büchtemann, A. Meyer, S.J. McCormack, R. Koole, D.J. Farrell, R. Bose, E.E. Bende, A.R. Burgers, T. Budel, J. Quilitz, M. Kennedy, T. Meyer, C. De Mello Donegá, A. Meijerink, and D. Vanmaekelbergh, *Optics Express*, Vol. 16, 21773–21792, 2008.
5. M. Grätzel, *Nature*, Vol. 414, 338–344, 2001.
6. S.K. Balasingam, M. Lee, M.G. Kang, and Y. Jun, *Chemical Communications*, Vol. 49, 1471–1487, 2013.
7. C.J. Brabec, S. Gowrisanker, J.J. Halls, D. Laird, S. Jia, and S.P. Williams, *Adv Mater*, Vol. 22, 3839–3856, 2010.
8. A. Mishra, and P. Bauerle, *Angewandte Chemie*, Vol. 51, 2020–2067, 2012.
9. W.G.J.H.M. van Sark, *Renewable Energy*, Vol. 49, 207–210, 2013.
10. D. Pile, *Nature Photonics*, Vol. 5, 710–710, 2011.
11. D.K.G. de Boer, *SPIE Newsroom*, Vol. 1–3, 2010.
12. A.B. Cristóbal López, A. Martí Vega, and A. Luque López, The sun tracker in concentrator photovoltaics, in: A.B. Cristóbal López, A. Martí Vega, and A. Luque López, eds., *Next Generation of Photovoltaics*, Springer-Verlag Berlin Heidelberg, pp. 61–93, 2012.
13. A. Goetzberger, *Applied Physics*, Vol. 16, 399–404, 1978.
14. A.M. Hermann, *Solar Energy*, Vol. 29, 323–329, 1982.
15. W.H. Weber, and J. Lambe, *Applied Optics*, Vol. 15, 2299, 1976.
16. Elsevier, http://www.scopus.com/home.url, 2013.
17. M.J. Currie, J.K. Mapel, T.D. Heidel, S. Goffri, and M.A. Baldo, *Science*, Vol. 321, 226–228, 2008.
18. R Reisfeld, D Shamrakov, and C. Jorgensen, *Solar Energy Materials and Solar Cells* Vol. 33, 417–427, 1994.
19. T. Dienel, C. Bauer, I. Dolamic, and D. Brühwiler, *Solar Energy*, Vol. 84, 1366–1369, 2010.
20. J.S. Batchelder, A.H. Zewail, and T. Cole, *Applied Optics*, Vol. 18, 3090–3110, 1979.
21. J.S. Batchelder, A.H. Zewail, and T. Cole, *Applied Optics*, Vol. 20, 3733–3754, 1981.
22. J.M. Drake, M.L. Lesiecki, J. Sansregret, and W.R. Thomas, *Applied Optics*, Vol. 21, 2945–2952, 1982.
23. Shin-ichi Hieda, Tadashi Kawabata, Shojiro Kuwabara, and M. Kurokawa, Methyl methacrylate syrup and production thereof, US Patent 6,448,354 B1, Mitsubishi Gas Chemical, September 10, 2002.
24. U. Rau, F. Einsele, and G.C. Glaeser, *Applied Physics Letters*, Vol. 87, 171101, 2005.
25. L.H. Slooff, E.E. Bende, A.R. Burgers, T. Budel, M. Pravettoni, R.P. Kenny, E.D. Dunlop, and A. Büchtemann, *physica status solidi (RRL) - Rapid Research Letters*, Vol. 2, 257–259, 2008.

26. A.A. Earp, G.B. Smith, P.D. Swift, and J. Franklin, *Solar Energy*, Vol. 76, 655–667, 2004.
27. F. Garcia-Santamaria, S. Brovelli, R. Viswanatha, J.A. Hollingsworth, H. Htoon, S.A. Crooker, and V.I. Klimov, *Nano Lett*, Vol. 11, 687–693, 2011.
28. F. Purcell-Milton, and Y.K. Gun'ko, *Journal of Materials Chemistry*, Vol. 22, 16687, 2012.
29. B.N. Pal, Y. Ghosh, S. Brovelli, R. Laocharoensuk, V.I. Klimov, J.A. Hollingsworth, and H. Htoon, *Nano Lett*, Vol. 12, 331–336, 2012.
30. K. Barnham, J.L. Marques, J. Hassard, and P. O'Brien, *Applied Physics Letters*, Vol. 76, 1197, 2000.
31. A.J. Chatten, K.W.J. Barnham, B.F. Buxton, N.J. Ekins-Daukes, and M.A. Malik, *Semiconductors*, Vol. 34, 909, 2004.
32. J. Bomm, A. Büchtemann, A.J. Chatten, R. Bose, D.J. Farrell, N.L.A. Chan, Y. Xiao, L.H. Slooff, T. Meyer, A. Meyer, W.G.J.H.M. van Sark, and R. Koole, *Solar Energy Materials and Solar Cells*, Vol. 95, 2087–2094, 2011.
33. C. de Mello Donega, *Chemical Society Reviews*, Vol. 40, 1512–1546, 2011.
34. Z. Krumer, S.J. Pera, R.J.A. van Dijk-Moes, Y. Zhao, A.F.P. de Brouwer, E. Groeneveld, W.G.J.H.M. van Sark, R.E.I. Schropp, and C. de Mello Donegá, *Solar Energy Materials and Solar Cells*, Vol. 111, 57–65, 2013.
35. P. Reiss, M. Protiere, and L. Li, *Small*, Vol. 5, 154–168, 2009.
36. S.J. Gallagher, B. Norton, and P.C. Eames, *Solar Energy*, Vol. 81, 813–821, 2007.
37. V. Sholin, J.D. Olson, and S.A. Carter, *Journal of Applied Physics*, Vol. 101, 123114, 2007.
38. M.G. Hyldahl, S.T. Bailey, and B.P. Wittmershaus, *Solar Energy*, Vol. 83, 566–573, 2009.
39. C. de Mello Donegá, *Physical Review B*, Vol. 81, 2010.
40. A. Thibon, and V.C. Pierre, *Anal Bioanal Chem*, Vol. 394, 107–120, 2009.
41. J. Yu, D. Parker, R. Pal, R.A. Poole, and M.J. Cann, *Journal of the American Chemical Society*, Vol. 128, 2294, 2006.
42. S.V. Eliseeva, and J.C. Bunzli, *Chemical Society Reviews*, Vol. 39, 189–227, 2010.
43. K. Binnemans, *Chem Rev*, Vol. 109, 4283–4374, 2009.
44. R. Reisfeld, and S. Neuman, *Nature*, Vol. 274, 144–145, 1978.
45. A. Kozlowska, M. Nakielska, J. Sarnecki, L. Lipinska, O. Jeremiasz, D. Podniesinski, and A. Malag, *Optica Applicata*, Vol. 41, 359–365, 2011.
46. G.A. Crosby, R.E. Whan, and R.M. Alire, *The Journal of Chemical Physics*, Vol. 34, 743, 1961.
47. J.C. Bunzli, and C. Piguet, *Chemical Society Reviews*, Vol. 34, 1048–1077, 2005.
48. L. Armelao, S. Quici, F. Barigelletti, G. Accorsi, G. Bottaro, M. Cavazzini, and E. Tondello, *Coordination Chemistry Reviews*, Vol. 254, 487–505, 2010.
49. C. de Mello Donegà, S. Alves Jr, and G.F.d. Sa, *Journal of Alloys and Compounds* Vol. 259, 422–426, 1997.
50. O. Moudam, B.C. Rowan, M. Alamiry, P. Richardson, B.S. Richards, A.C. Jones, and N. Robertson, *Chemical Communications*, Vol. 6649–6651, 2009.
51. O.L. Malta, H.F. Brito, J.F.S. Menezes, F.R. Goncalves e Silva, C. de Mello Donegà, S.A. Jr., and A. Böhm, *Chemical Physics Letters*, Vol. 282, 233–238, 1998.
52. J.C.G. Bunzli, *Chem. Rev.*, Vol. 110, 2729–2755, 2010.
53. K. Binnemans, Rare-earth beta-diketonates, in: J.K.A. Gschneidner, J.C.G. Bünzli, V.K. Pecharsky, eds., *Handbook on the Physics and Chemistry of Rare Earths*, Elsivier, Amsterdam, pp. 107–272, 2005.

54. Y. Hasegawa, T. Yasuda, K. Nakamura, and T. Kawai, *Japanese Journal of Applied Physics*, Vol. 47, 1192–1195, 2008.
55. J. Graffion, A.M. Cojocariu, X. Cattoën, R.A.S. Ferreira, V.R. Fernandes, P.S. André, L.D. Carlos, M. Wong Chi Man, and J.R. Bartlett, *Journal of Materials Chemistry*, Vol. 22, 13279, 2012.
56. T. Wang, J. Zhang, W. Ma, Y. Luo, L. Wang, Z. Hu, W. Wu, X. Wang, G. Zou, and Q. Zhang, *Solar Energy*, Vol. 85, 2571–2579, 2011.
57. X. Wang, T. Wang, X. Tian, L. Wang, W. Wu, Y. Luo, and Q. Zhang, *Solar Energy*, Vol. 85, 2179–2184, 2011.
58. A. Sanguineti, A. Monguzzi, G. Vaccaro, F. Meinardi, E. Ronchi, M. Moret, U. Cosentino, G. Moro, R. Simonutti, M. Mauri, R. Tubino, and L. Beverina, *Phys Chem Chem Phys*, Vol. 14, 6452–6455, 2012.
59. A. Monguzzi, A. Milani, L. Lodi, M.I. Trioni, R. Tubino, and C. Castiglioni, *New Journal of Chemistry*, Vol. 33, 1542, 2009.
60. A.K.Y. Jen, and L.R. Dalton, *Adv. Mater.*, Vol. 14, 1339–1365, 2002.
61. C. Kradjel, *Fresenius J Anal Chem*, Vol. 339, 65, 1991.
62. Y. Hasegawa, K. Murakoshi, Y. Wada, S. Yanagida, J. Kim, N. Nakashima, and T. Yamanaka, *Chemical Physics Letters*, Vol. 248, 8–12, 1996.
63. C. Bischof, J. Wahsner, J. Scholten, S. Trosien, and M. Seitz, *Journal of the American Chemical Society*, Vol. 132, 14334, 2010.
64. L. Beverina, M. Crippa, M. Sassi, A. Monguzzi, F. Meinardi, R. Tubino, and G.A. Pagani, *Chemical Communications*, Vol. 5103–5105, 2009.
65. G. Mancino, A.J. Ferguson, A. Beeby, N.J. Long, and T.S. Jones, *Journal of The American Chemical Society*, Vol. 127, 524–525, 2005.
66. M.A. El-Shahawy, and A.F. Mansour, *Journal of Materials Science: Materials in Electronics*, Vol. 7, 171–174, 1996.
67. M.I. Yanbastiev, *Zhurnal Prikladnoi Spektroskopii*, Vol. 17, 522–525, 1972.
68. R. Reisfeld, R. Zusman, Y. Cohen, and M. Eyal, *Chemical Physics Letters*, Vol. 147, 142–147, 1988.
69. R. Wehrspohn, J. Gutmann, J. Posdziech, M. Peters, L. Steidl, R. Zentel, H. Zappe, J.C. Goldschmidt, and A. Gombert, *Proceedings of SPIE*, Vol. 8438, 8438101 - 8438107, 2012.
70. N.A. Kuznetsova, and O.L. Kaliya, *Russian Chemical Journal*, Vol. 61, 1243–1267, 1992.
71. C.L. Mulder, P.D. Reusswig, A.P. Beyler, H. Kim, C. Rotschild, and M.A. Baldo, *Optic Express*, Vol. 18, A91-A99, 2010.
72. L.R. Wilson, and B.S. Richards, *APPLIED OPTICS*, Vol. 48, 212–220, 2009.
73. A.F. Mansour, M.G. El-Shaarawy, S.M. El-Bashir, M.K. El-Mansy, and M. Hammam, *Plymer International* Vol. 51, 393–397, 2002.
74. M.G. Debije, M.P. Van, P.P.C. Verbunt, M.J. Kastelijn, R.H.L. van der Blom, D.J. Broer, and C.W.M. Bastiaansen, *Applied Optics*, Vol. 49, 745–751, 2010.
75. A.F. Mansour, R.M. Ahmed, A.H. Bassyouni, and G.M. Nasr, *International Journal of Polymeric Materials*, Vol. 56, 651–662, 2007.
76. G. Griffini, L. Brambilla, M. Levi, M. Del Zoppo, and S. Turri, *Solar Energy Materials and Solar Cells*, Vol. 111, 41–48, 2013.
77. M.G. Debije, P.P.C. Verbunt, P.J. Nadkarni, S. Velate, K. Bhaumik, S. Nedumbamana, B.C. Rowan, B.S. Richards, and T.L. Hoeks, *Applied Optics*, Vol. 50, 163–169, 2011.
78. G. Seybold, and G. Wagenblast, *Dyes and Pigments*, Vol. 11, 303–317, 1989.

79. J.P. Kurtz, T. Grusenmeyer, L. Tong, G. Kosgei, R.H. Schmehl, J.T. Mague, and R.A. Pascal, *Tetrahedron*, Vol. 67, 7211–7216, 2011.
80. J. Sansregret, J.M. Drake, W.R.L. Thomas, and M.L. Lesiecki, *APPLIED OPTICS*, Vol. 22, 573–577, 1983.
81. O.A. Bozdemir, S. Erbas-Cakmak, O.O. Ekiz, A. Dana, and E.U. Akkaya, *Angewandte Chemie*, Vol. 50, 10907–10912, 2011.
82. R.F.S. Santarelli, G. Bianchi, Luminescent solar concentrators comprising disubstituted benzothiadiazole compounds, EP 2 557 606 A1, ENI S.p.A., February, 13 2013.
83. C. Botta, P. Betti, and M. Pasini, *Journal of Materials Chemistry A*, Vol. 1, 510, 2013.
84. S.M. Reda, *Electrochemical and Solid-State Letters*, Vol. 9, F73, 2006.
85. S.M. Reda, *Solar Energy*, Vol. 81, 755–760, 2007.
86. F. Vollmer, and W. Rettig, *Journal of Photochemistry an Photobiology A: Chemistry*, Vol. 95, 143–155, 1996.
87. M.J. Cook, A.P. Lewis, G.S.G. McAuliffe, V. Skarda, A.J. Thomson, J.L. Glasper, and D.J. Robbins, *Journal of the Chemical Society, Perkin Transactions 2*, Vol. 1293, 1984.
88. T. Saraidarov, V. Levchenko, A. Grabowska, P. Borowicz, and R. Reisfeld, *Chemical Physics Letters*, Vol. 492, 60–62, 2010.
89. A. Dubois, M.l Canva, A. Brun, F. Chaput, and J.P. Boilot, *Applied Optics*, Vol. 35, 3193–3199, 1996.
90. H. Zheng, X.Q. Zhan, Q.N. Bian, and X.J. Zhang, *Chemical communications*, Vol. 49, 429–447, 2013.
91. M. Beija, C.A. Afonso, and J.M. Martinho, *Chemical Society reviews*, Vol. 38, 2410–2433, 2009.
92. Z.R. Grabowski, and K. Rotkiewicz, *Chem. Rev.*, Vol. 103, 3899–4031, 2008.
93. J. Arden, G. Deltau, V. Huth, and U. Kringel, *Journal of Luminescence*, Vol. 48–49, 352–358, 1991.
94. R.F. Kubin, and A.N. Fletcher, *Journal of Luminescence*, Vol. 27, 455–462, 1982.
95. I. Baumberg, O. Berezin, A. Drabkin, B. Gorelik, L. Kogan, M. Voskobojnik, and M. Zaidman, *Polymer Degradation and Stability*, Vol. 73, 403–410, 2001.
96. A.A. Earp, G.B. Smith, J. Franklin, and P. Swift, *Solar Energy Materials and Solar Cells*, Vol. 84, 411–426, 2004.
97. J.C. Goldschmidt, M. Peters, A. Bösch, H. Helmers, F. Dimroth, S.W. Glunz, and G. Willeke, *Solar Energy Materials and Solar Cells*, Vol. 93, 176–182, 2009.
98. H. Hernandez-Noyola, D.H. Potterveld, R.J. Holt, and S.B. Darling, *Energy & Environmental Science*, Vol. 5, 5798, 2012.
99. S.T. Bailey, G.E. Lokey, M.S. Hanes, J.D.M. Shearer, J.B. McLafferty, G.T. Beaumont, T.T. Baseler, J.M. Layhue, D.R. Broussard, Y.-Z. Zhang, and B.P. Wittmershaus, *Solar Energy Materials and Solar Cells*, Vol. 91, 67–75, 2007.
100. D.J. Farrell, W.G.J.H.M.v. Sark, S.T. Velthuijsen, and R.E.I. Schropp, *Physica Status Solidi (C)*, Vol. NA-NA, 2010.
101. J.F. Geisz, and D.J. Friedman, *Semiconductor Science and Technology*, Vol. 17, 769–777, 2002.
102. T. Weil, T. Vosch, J. Hofkens, K. Peneva, and K. Mullen, *Angewandte Chemie*, Vol. 49, 9068–9093, 2010.
103. Y. Avlasevich, C. Li, and K. Müllen, *Journal of Materials Chemistry*, Vol. 20, 3814, 2010.

104. H. Langhals, R. Ismael, and O. Yuruk, *Tetrahedron*, Vol. 56, 5435–5441, 2000.
105. M.-M. Shi, H.-Z. Chen, J.-Z. Sun, J. Ye, and M. Wang, *Chemical Communications*, Vol. 1710, 2003.
106. C. Huang, S. Barlow, and S.R. Marder, *The Journal of Organic Chemistry*, Vol. 76, 2386–2407, 2011.
107. S.K. Lee, Y. Zu, A. Herrmann, Y. Geerts, K. Mullen, and A.J. Bard, *Journal of the American Chemical Society*, Vol. 121, 3513–3520, 1999.
108. A. Wicklein, A. Lang, M. Muth, and M. Thelakkat, *Journal of the American Chemical Society*, Vol. 131, 14442, 2009.
109. F. Würthner, C. Thalacker, and A. Sautter, *Advanced Materials*, Vol. 11, 754–758, 1999.
110. C. Ego, D. Marsitzky, S. Becker, J. Zhang, A. C. Grimsdale, K. Müllen, J. D. MacKenzie, C. Silva, and R.H. Friend, *Journal of the American Chemical Society*, Vol. 125, 437–443, 2003.
111. Z. Chen, U. Baumeister, C. Tschierske, and F. Wurthner, *Chemistry*, Vol. 13, 450–465, 2007.
112. A. Herrmann, and K. Mullen, *Chemistry Letters*, Vol. 35, 978–985, 2006.
113. F. Würthner, *Chem. Commun.*, Vol. 14, 1564–1579, 2004.
114. R. Schmidt, P. Osswald, M. Koenemann, and F. Wuerthner, *ChemInform*, Vol. 40, 735–746, 2009.
115. X. Zhan, A. Facchetti, S. Barlow, T.J. Marks, M.A. Ratner, M.R. Wasielewski, and S.R. Marder, *Adv Mater*, Vol. 23, 268–284, 2011.
116. H. Quante, P. Schlichting, U. Rohl, Y. Geerts, and K. Müllen, *Macromol. Chem. Phys.*, Vol. 197, 4029–4044, 1996.
117. C. Kohl, T. Weil, J. Qu, and K. Mullen, *Chemistry*, Vol. 10, 5297–5310, 2004.
118. J. Hofkens, T. Vosch, M. Maus, F. Köhn, M. Cotlet, T. Weil, A. Herrmann, K. Mullen, and F.C.D. Schryver, *Chemical Physics Letters*, Vol. 333, 255–263, 2001.
119. F. Holtrup, G. R. J. Muller, H. Quante, S. De Feyter, F. C. De Schryver, and K. Muellen, *Chemestry a European Journal*, Vol. 3, 219–225, 1997.
120. Y. Geerts, H. Quante, H. Platz, R. Mahrt, M. Hopmeier, A. Böhm, and K. Müllen, *Journal of Materials Chemistry* Vol. 8, 2357–2369, 1998.
121. F. Nolde, J. Qu, C. Kohl, N.G. Pschirer, E. Reuther, and K. Mullen, *Chemistry*, Vol. 11, 3959–3967, 2005.
122. H. Langhals, S. Sprenger, and M.T. Brandherm, *Liebigs Annalen* Vol. 481–486, 1995.
123. N. Pasaogullari, H. Icil, and M. Demuth, *Dyes and Pigments*, Vol. 69, 118–127, 2006.
124. Y. Zagranyarski, L. Chen, Y. Zhao, H. Wonneberger, C. Li, and K. Müllen, *Organic Letters*, Vol. 14, 5444–5447, 2012.
125. A. Sanguineti, M. Sassi, R. Turrisi, R. Ruffo, G. Vaccaro, F. Meinardi, and L. Beverina, *Chemical communications*, Vol. 49, 1618–1620, 2013.
126. C. Li, J. Schoneboom, Z. Liu, N.G. Pschirer, P. Erk, A. Herrmann, and K. Mullen, *Chemistry*, Vol. 15, 878–884, 2009.

14

PAn-Graphene-Nanoribbon Composite Materials for Organic Photovoltaics: A DFT Study of Their Electronic and Charge Transport Properties

Javed Mazher[1,*], Asefa A. Desta[1], and Shabina Khan[2]

[1]*Materials Science Program, Addis Ababa University, Addis Ababa, Ethiopia*
[2]*Govt. Motilal Vigyan Mahavidyalaya, Barkatullah University, Bhopal, India*

Abstract

Theoretical calculations of electronic and transport properties of graphene/conjugated polymer nanocomposites are presented by using density functional theory based ab-initio approach. The armchair and zigzag shaped graphene nanoribbons (GNRs) have been used for the formation of nanocomposites with chlorinated polyaniline strings in various geometric conformations. The ab-initio simulations have been carried out for the compositionally vivid samples of the polymer-GNR system with built-in two-probe device geometries. The calculated electronic bandgaps and density of states for the nanoribbons show a close match with the contemporary theoretical counterparts. Density of states and transmission spectra are examined for the charge transport behavior in the nanocomposite. Significant numbers of contributing states are found to be driven from contact electrodes into the nanocomposites. Observance of eigenstate conduction channels among the nanodevices, strong couplings, induction among different regions of the nanodevice and negative differential resistance are also rightly elucidated, which helps in understanding different modes of the charge transport among the nanocomposites. Finally, the work presented in this manuscript would be helpful in understanding the intrinsic properties of nanocomposite materials for the photovoltaics applications.

**Corresponding author*: mazher.javed@gmail.com

Atul Tiwari, Rabah Boukherroub, and Maheshwar Sharon (eds.) Solar Cell Nanotechnology, (357–408) 2014 © Scrivener Publishing LLC

Keywords: Graphene, organic photovoltaics, polyaniline, ab-initio DFT simulations, electronic and transport properties, nanodevices' simulations

14.1 Introduction

Graphene–aniline nanocomposites are well known charge transfer systems in which the graphene accepts the charge carriers from donor aniline [1]. Recent experiments have shown that the graphene nanostructructures can be easily dissolved in aniline without any prior chemical functionalization. Besides, such a graphene nanostructure-aniline solution can then be easily used, followed by simple electropolymerization step, to prepare thin films of graphene–polyaniline (Gr–PAn) nanocomposite on various conducting electrodes of preference like transparent conducting glasses, ITO (indium tin oxide), etc. [1, 2]. As prepared nanocomposite thin films can be directly used in a variety of applications from electrochemical sensing electrode material to the bulk heterojunction (BHJ) solar cells. The wonder material graphene, which is a relatively new allotrope among the carbon-based nanomaterials, has already been applied for fabricating composites with conducting polymers [3, 4]. As the graphene, carbon nanotubes and fullerenes (C-60) have a similar atomic structure with the ring symmetry, hence conjugated graphene sheets can be readily functionalized via non-covalent π stacking or covalent C–C coupling reactions with graphene nanostructures, through the formation of donor–acceptor complex, within the graphene plane [5]. Hence, in principle, the graphene should be the same good electron acceptor as carbon nanotubes or C-60, while aromatic amine like aniline is a fairly good electron donor and these donor type aromatic molecules can be π–π stacked on the graphene surface or edges. Thus graphene and polyaniline form a first-rate charge-transfer complex [1, 6]. Recently it has also been reported that the graphene voluntarily accepts electrons at very fast charge transfer rates, the photoinduced charge transfer rate in between organic aromatic donor to graphene is found to be 32 ps from quenched fluorescence decay lifetime experiments [7]. The authors therein also explored the possibility of multiple electron transfer to the graphene when attached with sulfonated zinc phthalocyanine. Nevertheless, the graphene-aniline charge transfer complex has some shortcomings also, which arise from the optical absorption band of aniline that is experimentally found to be in UV range at wavelengths of 234 nm, 255 nm and 287 nm [6, 7]. However, for the polymers of aniline, the absorption goes down up to 350 nm, which corresponds to blue absorption depending upon the monomer chain lengths [8]. Accordingly, the charge transfer material in its pure form is unable to absorb the low energy photonic range of a solar spectrum, this also marks the necessity of the proper light harvesting to increase the BHJ solar cell efficiencies. Nevertheless, the wide bandgap of polyaniline which varies from 1.8 eV to 3.5 eV is still a problem in achieving device efficiencies due to absorption of only shorter photonic wavelengths [8]. Apart from this, the high resistance of PAn composites is still a major bottleneck in realizing their efficient BHJ devices due to resistive energy losses among the as-harvested charge carriers [9]. Conversely, preparation of nanocomposites of PAn with long ribbons of graphene, in which the polymers are well-aligned with the ribbons by π–π stacking and are percolating throughout the composite material, can reasonably improvise the scenario of their electrical properties. Indeed, it is a high time to address these pertaining issues, especially in the context of nano-photovoltaic devices, where either due to the small size induced quantum confined transport effects or due to π–π stacking functionalization the electrical properties of these devices would change drastically. One of the aims of our current research is to propose a scheme of smart photovoltaic materials for their deployment as efficient light harvesters in the BHJ nanosolar cells and to simulate various electronic and transport properties of their two probe devices in which the as-modeled percolating two component nanosystems are sandwiched in between the conducting electrodes.

14.1.1 Organic Photovoltaic Technology

Photovoltaic (PV) technology offers a straightforward method for conversion of sunlight to electricity that is based on the photovoltaic effect—a pragmatic phenomenon of solar radiation interaction with matter—as observed by Henri Becquerel about two centuries ago [10]. The photovoltaic effect is a basic physical process through which a photovoltaic cell converts sunlight into electricity. The PV technology basically involves direct conversion of solar radiation into electricity without any mechanical moving parts, any noise, high temperatures or any pollution. Apart from these benefits a series of solar or PV cells also exhibit very long lifetime. Hence it works virtually by only using the enormous energy source, the SUN, which is absolutely free, ubiquitous, inexhaustible and a very flexible energy source. One can generate a huge amount of solar power ranging from microwatts to megawatts. PV cells are generally classified into three generations which indicate the order in which each becomes important. Currently, the popular solar cells are comprised of monojunction silicon wafer based photovoltaic devices—still capturing a majority of market share—including both single and multicrystalline layers, constitute the first generation of solar cells [11]. It is to be noted that this technology usually demands thick substrate layers for being more mechanically robust and due to presence of indirect energy transitions among the light harvesters. As a consequence, some major drawbacks manifest in this generation through increase in fabrication cost and reduction in solar conversion efficiency. Moreover, thin-film photovoltaic technologies, which are also referred to as second generation photovoltaics, are based on inorganic semiconductor materials that are more photon absorbing (solar radiation harvesting) than a crystalline silicon and show cost reduction due to easy processability at large area substrates [12]. Mostly, the thin-film PV devices deploy semiconductors that include II–VI compounds, amorphous $(Si)_n$ clusters, or chalcogenides such as ZnSe, CdS or CdTe, $CuInSe_2$ (CIS), $CuInGaSe_2$ (CIGS), etc. On the other hand, more recent organic-based approaches do not rely on conventional single p–n junctions and they are often referred to as third generation solar technology. Besides, the current thrust is on the development of the third generation of solar cells, which uses various novel phenomena of efficient light harvesting and high power conversion factors, such as, multijunction tandem layer solar cells, hot-carrier generation, intermediate energy gap devices, and of course, the bulk heterojunction (BHJ) organic photovoltaics [13]. Categorically, the third generation photovoltaic cells include: (i) dye-sensitized solar cells pioneered by Grätzel, which are electrochemical cells that require use of electrolytes; (ii) multijunction tendom cells fabricated from group IV and III– V semiconductor multilayers; (iii) an alternative hybrid approach, in which inorganic quantum dots are doped into a semiconducting polymer matrix or by combining nanostructured inorganic semiconductors such as TiO_2 with organic materials; (iv) lastly, all-organic solid-state approach, which is nowadays a thrust area of research because of the emphasis on green and environmentally friendly technologies [11–14]. Typical Organic photovoltaic

(OPV) cells are basic thin-film structures in which the organic photoactive materials are sandwiched in between two electrodes as active or photon harvesting layers. They have also attracted much attention in recent years due to their advantages in low-cost manufacturing, their light weight and good flexibility [15–17]. These advantages have created great interest and thus a great deal of research work is currently being devoted to increase their power conversion efficiencies and scale-up production processes [17, 18]. However, the power conversion efficiency of organic photovoltaic cells is still low compared to their inorganic counterparts. At present a significant number of research endeavors are dedicated to enhance the efficiency of organic photovoltaic cells by improving the work function of the electrode with different surface treatments or, alternatively, by incorporating other organic dopants as a hole or electron transporting layer along with inserting a buffer layer between the electrode (or hole) electron transporting layer [18].

Briefly, the OPVs can be described as thin-film devices derived from the use of one or more component layers based on π-conjugated (semiconducting) organic molecules, oligomers or polymers. Owing to the presence of composite nature of electron donating and accepting nanocomponents, these OPVs show varying excitonic PV features along with suitable redox energy levels, specifically, on sandwiching them in thin layers amongst the metallic contacts of the photovoltaic device [19]. The energy conversion process often has at least five fundamental steps [19, 20], which are as follows:

1. Photons are absorbed in the donor and excitons are created
2. excitons diffuse within the donor (acceptor) phase
3. excitons encounter an interface with the acceptor (donor) before decaying, and fast dissociation takes place, leading to charge generation and separation
4. as separated charge carriers are transported through the various percolated nanocomposite components to the electrodes via diffusion and/or drift processes
5. the charge carriers are collected at the electrodes

These steps ensure that a higher nonequilibrium conductivity of photovoltaic materials would yield higher values of a charge transfer and its collection at the electrodes, which obviously would facilitate more power generation from the solar cells. At present almost all efficient OPV cells are based on the donor-acceptor (DA) heterojunction (HJ) type structure, which is first demonstrated by Tang in 1986 [21]. Separately, two other OPV devices are single-layer and bilayer heterojunction devices. The former device consists of a semiconducting polymer layer which is sandwiched in between two metal electrodes of different work functions and its functioning can be described by the metal-insulator-metal (MIM) model. The second device consists of a donor and acceptor layer, where the polymer is applied with an electron acceptor layer on top, normally C-60. Such an arrangement easily facilitates lower recombination rate of the electron and hole in comparison to forward charge transport, which is a necessary requirement for PV power generation. Identically in the device one can also obtain an efficient charge separation because once the exciton gets dissociated, which by and large happens at the interface, the electrons travel towards the acceptor and the holes travel towards the donor material resulting in hefty photo-carrier generation. Besides, the typical diffusion length of excitons are in the range of 10 nm, thus, the maximum photoabsorption is achieved by keeping thickness of the light harvester layer more than 100 nm. This limits the efficiency of the abovementioned device. This difficulty has been overcome by using the new concept of bulk heterojunction [22].

14.1.2 Bulk Heterojunction Solar Cells

Organic photovoltaics materials—by the aid of bulk heterojunction concept—are yielding increasing efficient solar cells day by day [23, 24]. These bulk heterojunctions are achieved by blending donor and acceptor and they are increasingly based on conjugated polymers. Such blends also exhibit a large interface area and most excitons reach the donor/acceptor interface. Among the organic solar cells derived from blends of conjugated polymers (donor) and fullerenes (acceptor), former constituent absorbs the incident light. The absorption process in the polymer generates an exciton that can either relax back to the ground state or dissociate into an electron and a hole. Nevertheless, organic cells' excitonic diffusion lengths are very small (few nm) and the dissociation process only occurs at the donor/acceptor interface; consequent controlling of a structure of the active layer is very important for an efficient device fabrication. Figure 14.1(a) represents a comprehensive bulk heterojunction

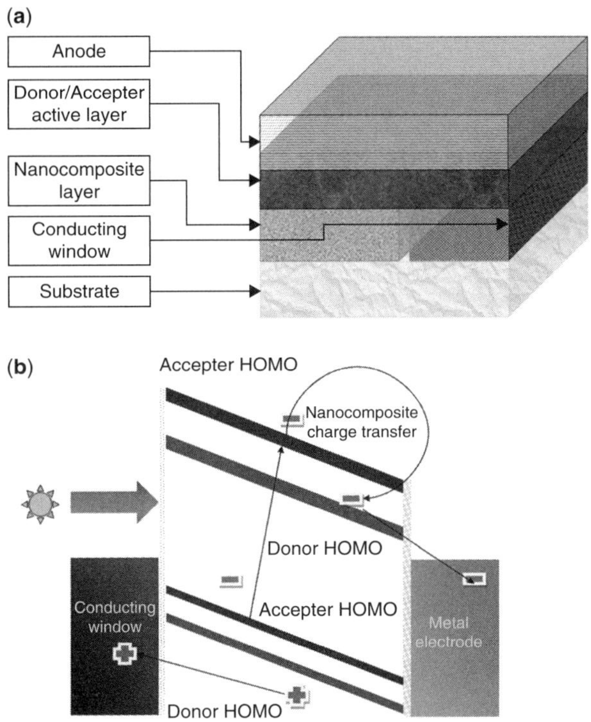

Figure 14.1 (a) Schematic of the typical bulk hetero-junction organic solar cell, in which the layered structures include solar cell electrodes and light harvester layers along with an active layer. The active layer comprises nanocomposite of conducting polymer and graphene nanoribbons as a donor/accepter blend, and (b) flatband diagram of the BHJ device depicting its working principle [25].

organic solar cell. In the given illustration, an active nanocomposite layer is sandwiched isolating the electron and hole acceptor bulk electrodes respectively, as depicted by an intimate blend of donor and acceptor organic molecules. The bulk heterojunction organic cell consists of at least four distinct layers on excluding the substrate, which may be glass or some flexible, transparent material as shown in Figure 14.1(a). Top surface of substrate covers cathode layer through which the light enters. Indium tin oxide (ITO) is a popular cathodic material due to its transparency. A layer of the conductive polymer mixture poly(3,4-ethylenedioxythiophene)/poly(styrenesulfonate) (PEDOT–PSS) may be applied between the cathode and the active layer. The PEDOT–PSS layer serves several functions. Not only does it serve as a layer to facilitate hole transportation in conjunction with blocking the excitonic movements, it also helps in reducing the surface roughness of ITO substrate by smoothening it. In addition to this, it also protects the active layer from oxygen-induced losses by preventing cathode material from diffusing into the active or organic nanocomposite layer that can lead to unwanted trap sites.

The PEDOT–PSS layer is deposited on top of the active layer. The active layer is responsible for light absorption, exciton generation/dissociation, and charge carrier diffusion. The active layer in the heterojunction device consists of two different types of materials, known as a donor and an acceptor; the corresponding materials can be conjugated polymers and carbon nanostructures [25, 26]. On top of the active layer is deposited an anode that is typically made of an aluminum. Alternatively, calcium, silver, or gold can also be used. Furthermore, a very thin layer of lithium fluoride (5–10 Å) is usually placed between the active layer and the aluminum anode. The layer of a lithium fluoride that does not seem to react chemically can serve as a protective layer between the metal and organic material. The electrode work functions' levels can be optimized for electron and hole extraction to selectively design a quality diode. An efficiency of the OPV-BHJ device is highly affected by an electronic bandgap of the constituents present in active layer apart from the energy level alignment. As depicted in an energy band diagram of the BHJ solar cell (see Figure 14.1b), photoinduced charge carriers are generated in the electron-donor material, which can be a conjugated polymer (e.g. polyaniline) and the electron-acceptor material can be a carbon nanostructure like a graphene nanoribbon. A difference in respective electron affinities in the two nanocomponents of the active organic layer is a reason for the photoinduced electron transfer. An energetically favorable transition occurs when the electron in the S1-excited state of the conjugated polymer transfers to the much more electronegative carbon nanostructure like GNR or C-60, thus resulting in an effective quenching of the excitonic photoluminescence of the polymer [27]. As a last step, charge carriers are extracted from the device through two selective contacts. A transparent indium tin oxide (ITO)-coated glass matching the HOMO level of the conjugated polymer (hole contact) is used on the illumination side, and evaporated thin lithium fluoride/aluminum metal contact layer matching the LUMO of GNR or PCBM (electron contact) is used on the other side. In the state-of-the-art devices, a thin (100 nm) hole injection layer and a highly doped PEDOT-PSS interfacial layer is used. Due to these stringent requirements of device components, at present the OPV device market is very small; their development is also partly hampered due to non-accomplishment of the present technological needs, namely, life-time, efficiency and costs [28].

The electron donor's (photon absorber material like PAn or P3HT) optical bandgap is defined by difference of its band levels of the highest occupied molecular orbital (HOMO) and lowest unoccupied molecular orbital (LUMO). If the photon is absorbed, an exciton is generated in the donor. An exciton is a quasi-neutral particle and contains a positive and negative charge. An exciton can diffuse at a maximum distance of 20 nm in its lifetime [2]. If the exciton does not reach the donor-acceptor interface in its lifetime, it would recombine, and the absorbed energy is converted into thermal energy or emitted as a photon and cannot be used for solar power generation. In proviso the exciton reaches the interface and gets dissociated via the electron transferring to the acceptor material. However, some energy is lost due to the difference in the LUMO levels of the donor and acceptor material. The as-generated charges are driven by drift and diffusion to the electrodes, where they are free to move through the external circuit. For efficient OPVs, however, the as-created charge carriers are required to be transported to the corresponding electrodes within their respective lifetimes, which mostly depend on the charge carrier's mobility in the given material. The graphene nanoribbon has a demonstrated width of 1–10 nm and thicknesse of 0.5 nm in addition to excellent electron acceptance and transport characteristics, which allows it to efficiently replace the C-60 in the active layer. It is also to be noted that carrier mobility in graphene is highest as of this date (>1 $cm^2V^{-1}s^{-1}$) and optical transmittance is as high as 99%. Moreover, an aspect ratio (length of ribbon over its width) of nanoribbon is very high, up to >30,000, and it is a far more promising material for demonstration of formation of acceptor molecule percolation network within the host PAn chain-like network cage. The development of such a smart nanocomposite has far reaching technological applications, not to mention again its use for improvised light harvesting inside OPVs.

14.1.3 Conjugated Polymers: Polyaniline (PAn)

A lot of attention in the field of polymer science has recently been paid to conjugated polymers, due to their characteristic optical, electrical, and magnetic properties. The most commonly studied classes of conducting polymers are polyacetylene, polythiophene, polypyrrole, polyaniline and their derivatives [29, 30]. Among

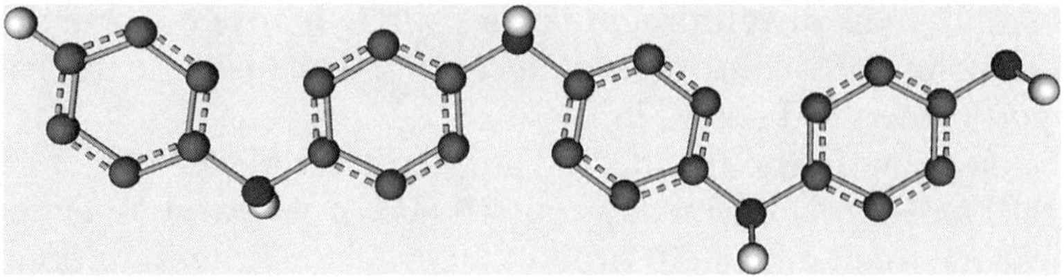

Figure 14.2 Chain of conducting polymer of polyaniline (PAn). The zigzag chain of the atomic conjugation backbone is obvious from the bonding pattern.

them, the polyaniline, as shown in Figure 14.2, is probably the most widely studied material due to its several unique properties [31, 32]. It exhibits several important technical features including ease of preparation, lightweight, low cost, better electronic-thermal-optical properties, high stability and solubility, good processibility, interesting redox properties associated with the chain heteroatoms, etc. The PAn is especially suitable for lightweight battery electrodes, hole-injecting electrodes in electroluminescent flexible light-emitting diodes, conductive adhesives electromagnetic shielding devices, electrocatalysts, anticorrosion coatings, microwave absorption, energy storage and transformation (erasable information storage, nonlinear optics) photonic indicators, membranes of precisely controllable morphology, organic actuators, supercapacitors, photovoltaics, transistors, transparent and plastic active and passive devices, and sensors [32–39].

The PAn is a semiconducting polymer with wide energy bandgap [38]. It belongs to a family of polymers with conjugated backbones [40, 41]. The PAn has three major synthesis forms, which differ from one another by a degree of oxidation 'y' that is a ratio of amine nitrogens over a total number of nitrogen atoms. Fully reduced leucoemeraldine base (LEB) form has the 'y' value of 1 and contains only a C–NH–C amine group in which the carbon 6-rings are phenyl rings with resonantly bonded benzene carbon. Half-oxidized emeraldine base (EB) has the 'y' value of 0.5 as a result of oxidation of half of the nitrogen in the imine form with nitrogen–carbon double bonds in which one-quarter of the carbon 6-rings change character and become quinoids containing carbon double bonds. Fully oxidized pernigraniline base (PNB) has the 'y' value of zero and all imine nitrogen with the carbon 6-rings consist of one-half phenyl and one-half quinoid rings. Each of these forms of PAn is a wide bandgap poor semiconductor with a bandgap of

3.6 eV for LEB and EB and about 1.4 eV for PNB [41–43]. However, the PAn-EB differs substantially from the LEB and the PNB in the sense that its conductivity can be tuned via protonic doping from 10^{-10} to 10^{2} S/cm [44].

Another key property of the conducting polymers is the presence of conjugated double bonds along their backbone, which simply is an alternate single and a double bond arrangement in between the carbon atoms of the chain. The conjugated unit of the chain contains a localized "sigma" (σ) orbital, leading to strong chemical bond and an internment electron, while the presence of double bond having a less localized "pi" (π) orbital gives weaker bond and a liberated electron which is free to hop around. However, intrinsically such conjugation in itself is not enough to induce the desired electrical properties in the polymer.

Previously, it has also been reported by several authors that any wide bandgap semiconductor can be made more useful photoabsorbent just by creating mid-bandgap states in it, so that its effective absorption lines can be increased for improving the light harvesting for the full color range of the solar spectrum. Such useful variants of the otherwise wide energy gap PAn can be prepared through various physical techniques of ion irradiation on a thin PAn film. Here one can irradiate the polymer by a beam of ions, and in principle it is possible to place any type of the dopant ions in proximity of aromatic rings of the polymeric backbone chains [45]. It should be noted that such a conjugated-backbone-ion interaction in the PAn brings some local distortion in the chain due to anion π effect and thus the mid-energy gap states could appear in the electronic structure of the polymer [46]. In a schematic given in Figure 14.3, we have tried to envisage consequences of an electronegative atom (say, chlorine) posed in proximity to the backbone on PAn's flatband energy diagram. As shown in Figure 14.3, the introduction of sub-band states would not only bring a strong light absorption band in PAn but would also increase the possibility of multiple electron transfer to the graphene nanoribbon, especially in light of recent reports that percolated nanographene can accept a number of electrons [1, 7].

14.1.4 Carbon Nanostructure: Graphene

Graphene is a two-dimensional (2D) allotrope of carbon that is considered as the mother of all graphitic allotropes because it is the building block for graphite (3D), fullerenes (0D), carbon nanotubes

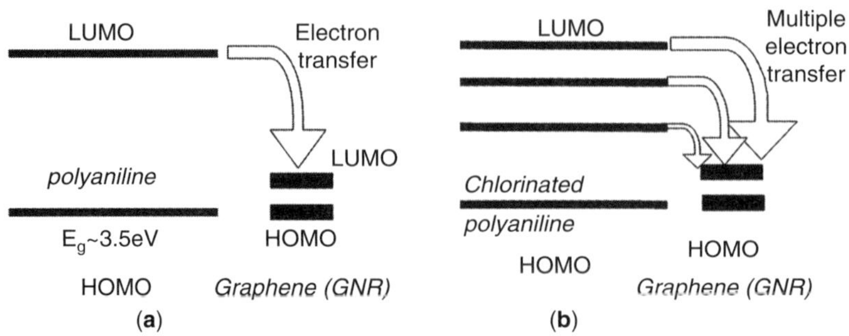

Figure 14.3 (a) Schematic depicting the possibility of single electron transfer from fundamental conduction band of the pure polyaniline to the graphene acceptor component of the nanocomposite. (b) In comparison, the possibility of multiple-charge transfer from the fundamental gap and various mid-gap states present in the chlorinated polyaniline to the graphene acceptor that is depicted for charge collection at the electrodes. Various sub-band states are expectably formed in the doped structure through local bond distortion due to proximity of the Cl with the conjugation backbone.

(1D), nanoribbons (1D), and anti dots (0D). Many interesting nanostructures can then just be formed either by stacking, wrapping, rolling-up, cutting of graphene sheet, or just by applying extreme pressure and temperature on graphene to transform the 2-dimensional sp^2 bonds into 3-dimensional sp^3 bonds [47]. Graphene, in which a monolayer of carbon atoms packed into a dense honeycomb crystal structure, is essentially a 2D layer of sp^2 bonded carbon. Carbon atoms of graphene are arranged in a hexagonal structure and have two atoms per unit cell. The carbon-carbon bond length in graphene is approximately 1.42 Å [48]. Graphene is a unique mixture of a semiconductor with zipped density of states and a gapless metal with unusual properties. It has extraordinary electronic excitations (low energy), which can be described in terms of Dirac fermions that move in a curved space. Unlike in the case of metals or common semiconductors, the electrons in graphene obey the Dirac equation rather than the Schrodinger equation [49]. New graphene physics leads to many exciting and unique properties such as high intrinsic mobility ($2 \times 10^5 cm^2\ V^{-1}S^{-1}$), and ballisticity up to a few millimeters, which give rise to a weak dependence of the mobility on a carrier concentration and temperature, high Young's modulus (~1.0 TPa), high thermal conductivity (~5000 $Wm^{-1}K^{-1}$), immense mechanical strength and flexibility, huge specific surface area (2630 m^2g^{-1}) and largest surface to volume ratio among all the materials on earth, very high optical transmittance of nearly 99%, rapid heterogeneous electron transfer; the graphene also readily exhibits both integer and fractional quantum Hall effect, zero effective mass of its charge carriers, and most importantly, graphene-nanodevices show spintronic and ballistic transport phenomenon even at room temperature [49, 50, 51]. Graphene is widely recognized as a promising material for future electronic devices. Much progress has already been accomplished in the fabrication and understanding of graphene's nanocomposites, spintronic devices and low on–off current ratio type of field-effect transistors (FETs) [52]. Graphene can be synthesized by various methods of a physical (mechanical exfoliation) and chemical (reduction of graphite oxide and epitaxial growth including chemical vapor deposition) nature. Recently Shukla and Mazher [2, 53] have reported two very useful methods of graphene synthesis by using an anodic bonding and an electrochemical intercalation techniques.

A structural flexibility of the graphene can be easily explained by using its atomic bonding. A sp^2 hybridization between one s orbital and two p orbitals lead to a trigonal planar structure in the graphene. Three out of four outer electrons for each carbon atom on its lattice are strongly bonded with respective C-neighboring atoms by σ orbitals [52, 54]. A $2p_z$ orbital of the fourth electron produces a weaker π bond with the neighboring carbon atom. The σ bonds form the covalent structure with honeycomb geometry. These strong bonds provide a desired mechanical strength, flexibility, and robustness to the graphene's lattice geometry. On the other hand, the π bonds, which describe the intrinsic electronic structure of graphene, are decoupled from the σ bands because of inversion symmetry, and they are closer to the Fermi energy in comparison to σ energy positions that separate the occupied and empty states. In a neutral graphene sheet, this is equal to zero energy since valence and conduction bands meet at a point known as the charge neutrality point, which results in conical valley-shaped bands that touch at two points denoted by their respective momentum vector K and K′ in the Brillouin zone, which converts the graphene into a gapless semiconductor. These points are also named as Dirac points. Thus, the graphene is regarded as a many body system in which the electrons are correlated from site to site ensuing in a rich collective behavior and strong correlations, which are manifested in the form of unique quantum effects and nanoelectronic properties [52, 54, 55].

As mentioned above the electrons in graphene obey a linear dispersion relation which can be expressed as, $E = hkv_F = pv_F$, where $p = \hbar k$ is momentum and V_F is Fermi velocity of electrons in the graphene ~$10^6 ms^{-1}$. Also, the energy-momentum relation implies that the effective rest mass is zero [since the energy of a particle having a rest mass m_0 and moving with a velocity u in a medium is $E = \sqrt{p^2u^2 + m_0^2u^4}$]. Thus, charge carriers in graphene have zero effective mass and simultaneously they move at a very high velocity equivalent to that of the speed of light. However, the all-purpose electrons are not actually massless, but the effective mass is a parameter that describes how an electron at particular wave vectors responds to applied forces. Since the velocity of electrons confined on the graphene remains constantly high, the parameter (effective mass) vanishes. The graphene could be a semiconductor or a metallic depending on the following factors: (a) Pristine graphene is intrinsically a zero-gap semiconductor because the density of states which can be derived as $D(E) = |E|/2\pi\hbar^2 V_F$ by using linear dispersion relation that seems to vanish at $E = 0$. But it is observed that the conductivity of the graphene is independent of the Fermi energy and the electron concentration, as long as variations in effective scattering strength are neglected [55]. Hence the graphene can also be regarded as a metal rather than a zero-gap semiconductor or semi-metal. (b) Functionalization process of the graphene rally rounds in creating a gap between the two principal bands (π and π*) of graphene. That is why graphene is sometimes considered as a hybrid lying in between a metal and a semiconductor [56]. Crystallographic feature of the graphene can be best described by a triangular lattice with a basis of two atoms per unit cell. Its lattice vectors can be written as Eq. 14.1,

$$a_1 = \frac{a}{2}\left(3, \sqrt{3}\right), a_2 = \frac{a}{2}\left(3, -\sqrt{3}\right) \tag{14.1}$$

where $a \approx$ **1.42Å** is the carbon-carbon distance. Its reciprocal-lattice vectors are given by Eq. 14.2.

$$b_1 = \frac{2\pi}{3a}\left(1, \sqrt{3}\right), b_2 = \frac{2\pi}{3a}\left(1, -\sqrt{3}\right) \tag{14.2}$$

The physics of graphene—as rightly referred by Dirac physics—is dominated by its crystal symmetry points of either K or K′, which

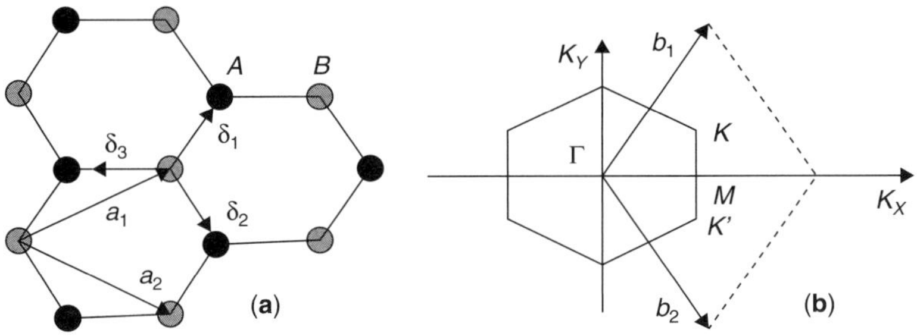

Figure 14.4 Honeycomb lattice and its Brillouin zone. (a) Lattice structure of the graphene showing two interpenetrating triangular lattices with a_1 and a_2 as lattice unit vectors, and δ_i (i=1, 2, 3) are nearest-neighbor vectors, (b) the corresponding Brillouin zone, in which Dirac cones of the linear dispersion curves are located at K and K′ points.

are at the corners of the graphene Brillouin zone (BZ) as shown in Figure 14.4, and that are often named as Dirac points. Their positions in momentum space are given by Eq. 14.3.

$$K = \left(\frac{2\pi}{3a}, \frac{2\pi}{3\sqrt{3}a} \right), K' = \left(\frac{2\pi}{3a}, -\frac{2\pi}{3\sqrt{3}a} \right) \tag{14.3}$$

The real space co-ordinates of the nearest-neighbor vectors can be given by Eq. 14.4,

$$\delta_1 = \frac{a}{2}\left(1, \sqrt{3}\right),\ \delta_2 = \frac{a}{2}\left(1, -\sqrt{3}\right),\ \delta_3 = -a\left(1, 0\right) \tag{14.4}$$

while the six second-nearest neighbors are located at $\delta_1' = \pm a_1, \delta_2' = \pm a_2, \delta_3' = \pm(a_2 - a_1)$

As mentioned above, the pristine graphene is a zero bandgap semiconductor with its electrons and holes behaving as massless fermions. But the crumpled or functionalized graphene is not, and in principle, the Fermi energy of functionalized graphene can be easily tuned from valence to conduction band via the external field effect which could be either electrical or chemical potential in nature. Incorporating a local gate can allow the induction of differential charge densities at different sample regions which could also induce a localized Fermi energy variation in which the E_F of one region lies in the valence band (p-type), while in the other region it lies in the conduction band (n-type) [57]. It would be particularly interesting to observe such effects of proximity of organic molecules with graphene, which is usually also the case in nanocomposites of organic polymers blended with graphene.

14.1.5 Graphene Nanoribbons (GNRs)

Graphene nanoribbons (GNRs) are narrow strips of the graphene sheet and are obtained by cutting graphene in the form of a quasi-1D wire along a given crystallographic direction [58, 59]. In the graphene nanoribbons, the presence of different types of boundary shapes, also called as edges, is used to modify the electronic structure of the material. The major effects are observed at the Fermi level, displaying unusual magnetic and transport features [60]. For many practical applications, such as in nanoelectronic devices, graphene needs to be patterned into the so-called graphene nanoribbons (GNRs). Typically, graphene is patterned by selectively removing the material by physical/chemical etching techniques [61, 62]. The electronic properties of GNRs are very sensitive to their width and edge geometry [62]. Dependence of edge geometry on the electronic structure of the GNRs has been mainly investigated using theoretical approaches, such as the tight-binding model and density functional theory (DFT) [63]. There are three basic shapes for the graphene edges: zigzag, armchair and hybrid. Nanoribbons with zigzag edges have less localized edge states at the Fermi energy in comparison to the armchair ribbons.

14.1.5.1 Types of Graphene Nanoribbons

a. Zigzag graphene nanoribbon (ZZGNR)
Zigzag nanoribbons have axial symmetry in a form of a single zigzag chain at both side edges along its length as shown in Figure 14.5. In the zigzag graphene nanoribbon, dangling σ-bonds at edges are assumed to be passivated by hydrogen atoms. A width of zigzag nanoribbon is approximately given by $W = (N/4)\ 3a$, where N is the number of atoms in the unit cell. A finite width of the ribbon produces a confinement of the electronic states near the Dirac points [64]. Dependence of electronic states on the width of the zigzag nanoribbon may be understood in terms of eigenstates of the Dirac Hamiltonian with appropriate boundary conditions. The crystallographic orientation of the ribbons also largely determines their electronic properties. The zigzag nanoribbons show a large degeneracy at the Fermi level. The band gap in ZZGNRs can also be

Figure 14.5 Zigzag graphene nanoribbon, abbreviated as ZZGNR, has the edge structures comprising single C-atom chains in zigzag pattern.

Figure 14.6 Armchair-graphene nanoribbon, abbreviated as ACGNR, has the edge structures comprising single C-atom chains in armchair pattern.

controlled by an external electric field. External fields due to gating or attachment of foreign entity can also result in the change of the contributing states in the ZZGNR [65].

b. Armchair graphene nanoribbon (ACGNR)
In an armchair-terminated nanoribbon, the edge consists of a line of dimmers (refer to Figure 14.6). The width is related to a number of atoms in the unit cell N through an expression $W = (N/4)\, a$. These ribbons do not have an axial symmetry [64]. The armchair-terminated graphene systems do not show zero-energy states (the Fermi energy is usually set to zero) [64, 65]. Electronic properties of the armchair nanoribbons depend strongly on the width; the electronic structure is less affected by ambient and edge-related interactions [66]. A crystallographic orientation of the ribbon largely determines their electronic properties. The electronic structure of the armchair ribbon that is not strongly modified by interactions can be exceptionally modified in the case of metallic ones where a small gap

opens [64–67], which is the case of wider ribbon. Similar to zigzag nanoribbons, electronic properties of the armchair nanoribbon may be understood in terms of eigenstates of the Dirac Hamiltonian with correct boundary conditions [67]. Depending on the value of the number of atoms 'm', the ACGNR can be classified in three categories, namely, m = 3p, 3p + 1 and 3p + 2 (;p is a positive integer) with widely varying electronic properties. The periodic ACGNR with m = 3p + 2 shows some metallic behavior, whereas the other two only show semiconducting behaviors. However, first-principle calculations within the local density approximation (LDA) show only the semiconducting behavior with a direct band gap at k = 0 for all of the three types of ACGNRs, and the gap Δ_a decreases with an increase in the width w_a, and follows a relation of w_a^{-1}. Nevertheless, in a LDA scheme (local density approximation) the ACGNR with $m = 3p + 2$ also shows a smaller gap in comparison to the other two families. The bandgap in ACGNRs is also found to be independent of the ribbon length [68]. However, in the case of *m = 3p and 3p + 2* ACGNRs, both HOMO and LUMO show delocalizations along the ribbon length spanning the central part. Therefore, the bandgap changes significantly with an increase in the ribbon length.

c. Hybrid graphene nanoribbon (HGNR)
Graphene nanoribbons produced by a chemical unzipping of carbon nanotubes are called as hybrid ribbons, which are characterized by a low-symmetry orientation of edges in contrast to a high-symmetry of the zigzag and armchair directions. An unusual presence of edge states are also found in their graphene edges. The values of the magnitude of their chiral vector (n, m) [68, 69] can be said to vary multiple times in a single nanoribbon. For the high-symmetry zigzag and armchair edges these vectors are (1, 0) and (1, 1), respectively. Thus, a hybrid graphene nanoribbon is a mixed nanoribbon which has either a head or tail made up of a one type of GNR, while the other end is composed of a different type of ribbon as shown in Figure 14.7. These structures mimic well-known edge-related disorders commonly found in synthesized GNR nanocomposites.

14.1.6 Nanocomposites and Their Percolation: GNR-Polyaniline Composites

To utilize the abovementioned novel properties of both nanocomponents, namely, GNR and PAn, it is necessary to prepare a

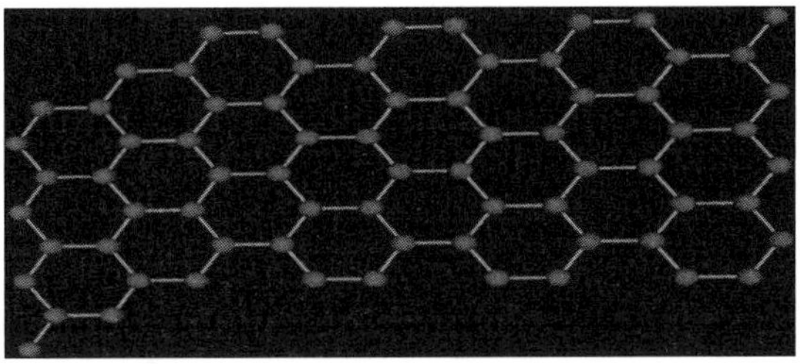

Figure 14.7 Hybrid graphene nanoribbon (HGNR) made up of two ends with different types of ribbons; right end is ACGNR and left is ZZGNR.

nanocomposite in which both the components conserve their individual character. We know that both the graphene and PAn have conjugated π-electrons; it would be interesting to observe any after effects of this π- π staking on electronic and transport properties of GNR-PAn nanocomposites from both the fundamental and the applications' point of view. Nevertheless, from the BHJ solar cells' applications viewpoint, small effect of π- π interactions in between GNR and PAn is better. Such effects are usually discernible in the density of states and the electron transmission spectrum of the nanocomposite's two-component system in comparison to that of a pure single component system. Additionally, if percolations happen, like the dendrite formation of the individual nanocomponents that takes place within the composite material, we can expect some superior transport properties of the bulk heterojunction solar cell. The BHJ solars cell's efficiency is expected to improve due to these newly formed percolative structures because some novel and efficient individual photoinduced charge transport channels would be created in the form of GNR ribbons and PAn percolating chains, which would scamper via percolation throughout the composite in between ITO and Al electrodes of BHJ cells. Thus, incorporation of graphene sheets into the PAn would definitely enhance the electrical conductivity of the nanocomposite without affecting photocharge generation rates and efficiencies in the light harvesting material, i.e., the PAn. Now it is important to theoretically or experimentally study the effects of π- π interactions either on the graphene's electronic transport property or on the electronic structure of the PAn-graphene nanocomposite. In fact, previous studies of UV–vis absorption on them already demonstrated that presence of graphene sheets augment and inflate the fundamental absorption band of the PAn, which is undoubtedly attributed to one of the π conjugation effects [70]. A further step to

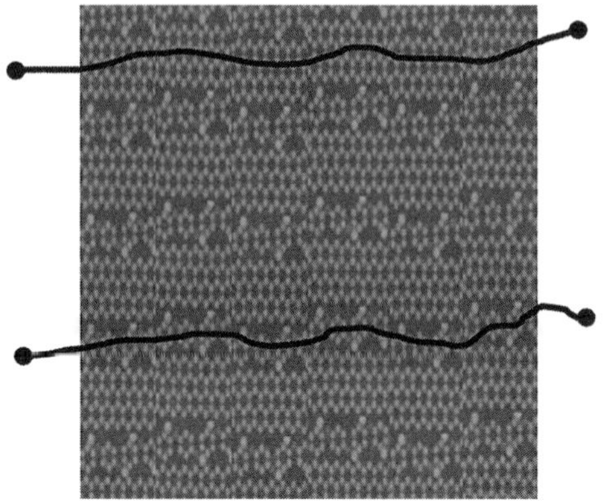

Figure 14.8 Graphene/polyaniline percolating nanocomposite. The top and bottom curved lines are to guide the eye, showing the percolation of graphene nanoribbons and PAn monomeric chains respectively. It is expected that the percolation facilitates in the photovoltaic charge generation and collection process.

multifunctional polymer nanocomposites is the development of well-aligned, percolating, two-component nanosystems instead of the dendrite-type percolation, as shown in Figure 14.8.

14.1.7 Origin of an Equilibrium Conductance in Nanodevices

It is a common observation among nanodevices that a zero bias conductivity appears at a Fermi energy position. The reason can be ascribed to discrete physical traits of nanodevices when sandwiched in between the charge injection electrodes with continuum physical characters. A small number of conduction states may appear at zero energy levels depending upon a competition that takes place amid two important competitive phenomena. The former effect is the Fano effect, which is a resonance effect that is thought to be an enhancement of a response of a system to an external excitation at a particular frequency. The Fano resonance (FR) is a universal physical phenomenon, which has historically been discovered in atomic physics as asymmetric line profiles in spectra of rare gases [71]. The Fano resonance is a consequence of interference amid a localized state and continuum that results in an antisymmetric line shape of the resonance created by the localized state [72]. Very recently, the Fano resonances have also been observed in condensed matter physics, nanomaterials sciences and nanoelectronics, both theoretically and experimentally [73]. The said phenomenon can also naturally arise in a coherent electrical transport through a nanostructured two-probe device, in which a central region of the device comprises the nanomaterial or a molecule having discrete energy states coupled with continuum states of electrodes. Such a situation creates the Fano resonances near the Fermi level that often strongly enhance an electron transmission through the nanodevice [74]. A necessary condition for the Fano resonance is a presence of at least two scattering channels in the form of a discrete level and a broad continuum band in proximity to each other, which is often the case of nanodevices' left and right electrodes and the central scattering region of the nanodevice [74]. The nature of two conduction channels in these mesoscopic systems is highly sensitive to geometries of the nanodevice under consideration.

The latter competitive effect is commonly called as strong LCR coupling. It takes place by an insertion of nanostructure, which is PAn-Graphene nanocomposite in the present study, in between the contact electrodes. This results in a sort of charge coupling that arises due to formation of strong bonds in between atomic contacts of the central scattering region (C) and the left (L) and the right (R) electrodes, which is covalent in character [75, 76]. Depending upon the amount of charge transfer that takes place in between two sides of the atomic contact, the strength of coupling increases. Because of the strong coupling some discrete physical traits usually found in the isolated nanostructure, like singular energy levels, become broadened in the strongly-coupled nanostructure. Occasionally, depending upon the amount of charge transfer from electrodes, if these broad energy levels are aligned with chemical potentials of attached electrodes, we usually get small conduction states at the Fermi level of the nanostructure. Hence, the band realignment of the nanostructure is responsible for low-energy charge carrier conduction in the two-probe devices [77].

Therefore, either a rectification in the nanodevice type of switching arises due to the Fano effects or the strong coupling induced singularities. One of the ways to explore their signatures is to study the density of states and nonequilibrium electron-transmission spectra of the nanodevices having a quantum confined central region. Usually, in nanoelectronic two-probe devices, the rectification does not happen by the Schottky barrier effect because the nature of both contact electrodes and the sample material in consideration are purely mesoscopic in nature. Moreover, various atomistic ab-initio modeling studies show that bulk contacts are absent in the nanodevice; instead only pure atomic scale contacts are formed [73, 74]. This implies that the I–V characteristics and the rectification behavior (if any) in the device mainly depends upon: (a) Electrode materials and their dimensionality, whether they are one dimensional or two dimensional, (b) singularities present in the density of electronic states of the nanostructure located at the central region (sample to be tested) of the device, (c) interfaces and a contact geometry of the electrode and the central region, and (d) the coupling strength between the two ends of the central region with electrodes [78]. Often the strong coupling results in sharp and intense DOS peaks near the Fermi level, and the weak coupling gives broader DOS resonant states. This effect has also been earlier reported in the density of states and the transmission energy spectra along with I-V measurements [73, 74], where a Ohmic contact is observed instead of a rectification at the nanometallic/nano-semiconductor atomic interface. Indeed, Fano line shapes are also observed in a differential electrical conductance (dI/dV) plot with bias voltage during a tunneling course using a scanning tunneling microscope and through an impurity atom on a metallic surface [71, 73, 74]. In contrast, the Fano antiresonance dips are reported in the density of states and the transmission spectrum of nanodevices at the Fermi level when an injection of states from electrodes to the central region is destructive in its interference with states present in the central region, which is manifested in the rectifying behavior of the nanodevices [74, 78].

14.1.8 Singularities Due to the Quantum Confinement of Nanostructures

Low dimensional systems with electron confinement such as 2-D quantum wells, 1-D quantum wires or 0-D quantum dots exhibit interesting new physical properties of discrete nature due to their unique electronic energy band structure, which often results in singularities of an electronic state density in respect to energy. This is reasonably well explained by various 1D to 3D free motion confinement models [76]. The reduced dimensionality of an electron gas has a profound influence on the nanowire or the atomic chain's density of states (DOS). A recent interest in quantum wires is arised partially from an inverse square-root relation of the DOS with its

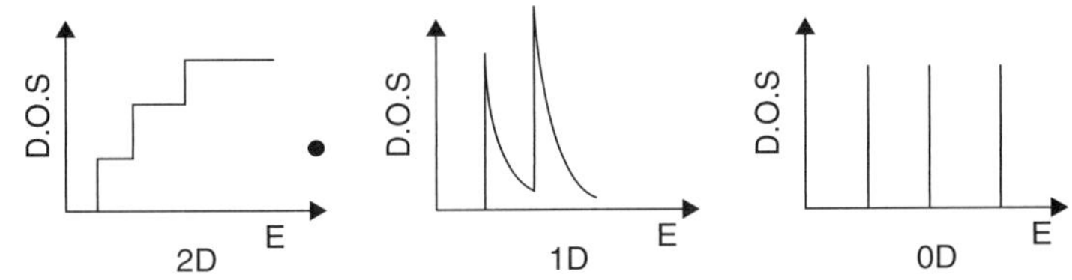

Figure 14.9 Density of states of two-dimensional (quantum-well), one-dimensional (quantum-wire) and zero-dimensional (quantum-dot) nanomaterials. The singularities in D.O.S. spectra are signatures of low-dimensionality.

electronic energy. It is interesting to note that the coulomb correlation effects in 1D systems further reduce interband transitions near the bandgap [79, 80], at least partially counteracting effects of singularities in the density of states. The discrete nature of electronic states in quantum structures and dynamics of relaxation of carriers between these discrete states are two of the areas of interest in majority of recent investigations in the field of nanoelectronics. Within the realm of nanoscience and nanotechnology, nanostructures-like nanowires, graphene nanoribbons, etc. play a special role because of their one dimensionality. When the nanomaterial's dimensions, such as GNR width or nanowire diameter, are made smaller, the singularity in the electronic density of states is developed at certain specific energy. The singularities exhibit size-dependent spectral positions in the DOS spectra, often referred to as van Hove singularities, where the electronic density of states become very large (see Figure 14.9), more closely resembling the case of molecules and atoms and appearing to be very different from the case of crystalline solids or even two-dimensional systems [81, 82]. In the limit of small diameter nanostructure with large aspect ratios such as nanowires or nanoribbons, different positions and shapes of these singular densities of states have dramatic effects on the nanoelectronic properties of the device.

14.2 Review of Computational Background

14.2.1 Modern Theoretical Methods: Ab-initio Nanocomposite Theory

Ab-initio techniques, or first principle approaches, are based on using a fundamental Hamiltonian of a given nanocomposite system for deriving their basic electronic and physio-chemical properties [83]. An exact solution of a resulting many-body (N-electron) Schrodinger equation of the nanocomposite is almost impossible and hence some approximations are deployed in such calculations. The approximations are purely technical in nature. A determinantal form of the nanocomposite wavefunction is used in a Hartree-Fock approximation and a representation of its single-particle ingredients is achieved from a finite basis set; still, the calculations are computationally costly for nanocomposites. Hence, a further technical simplification is required in the form of density functional theory to help in reducing the ab-initio calculation time. Nevertheless, the ab-initio simulation is characterized by the fact that its approximations do not introduce adjustable physical parameters [83, 84].

14.2.2 Density Functional Theory (DFT)

Use of electron density instead of wavefunction is first introduced by Thomas and Fermi [84, 85] as early as 1927 to effectively describe the electronic structure in atomic and molecular systems. However, this rudimentary model only takes into account the kinetic energies of involved constituents. In a way, the Thomas-Fermi and Hartree-Fock-Slater methods can be regarded as ancestors of modern DFT. The chief building blocks of these traditional methods are single-electron orbitals, and many-electron wavefunctions constructed from them. Thus, the chief elements of the modern DFT are the electron density $n(r)$ and fictitious single-particle orbitals. The density functional theory is phenomenally successful in finding solutions to the fundamental Schrodinger equation that describes the quantum behavior of atoms, molecules and nanomaterials. Gradually, the ab-initio DFT approach has spurted out from the clutches of a few physicists to the widespread scientific community, which is using the quantum mechanical theory as a tool in solving problems of chemistry, physics, materials science, chemical engineering, geology, and other disciplines. Conceptually, the DFT describes the ground-state energy of a system of interacting electrons in an external potential as a functional of the ground-state electronic density [84, 85], and the given material's electronic ground-state structure is given in terms of the electronic density distribution $n(r)$. In the Kohn-Sham DFT formulation [85] the ground-state density can be written in terms of the single-particle orbitals. The single particle orbital ϕ is obtained a priori by using atomic pseudopotentials [86].

Ideally, a pseudopotential is developed by considering an isolated atom of one element, but the resulting pseudopotential can then be used reliably for calculations that place this atom in the nanocomposite environment without further adjustment of the pseudopotential. This desirable property is referred to as the transferability of the pseudopotential. Current DFT codes typically provide a library of norm-conserving pseudopotentials that includes an entry for each (or at least most) elements in the periodic table.

The entire field of density functional theory rests on two fundamental mathematical theorems proved by Kohn and Hohenberg and a derivation of the set of equations by Kohn and Sham [87]. The former theorem of Hohenberg and Kohn proves that the ground-state energy from Schrodinger's equation is a unique functional of the electron density, and states that there exists a one-to-one mapping $E[n(r)]$ between the ground-state wavefunction and the ground-state electron density $n(r)$. The second theorem essentially describes a nature of this density functional along with its properties. Accordingly, the electron density that minimizes the energy of the overall functional is the true electron density corresponding to the full solution of the nanocomposite Hamiltonian. A useful way to write down the functional of the nanosystem can be described by the later Hohenberg - Kohn theorem in terms of a single-electron wave function $\psi_i(r)$. These functions collectively define the electron density $n(r) = \sum f_i |\psi_i(r)|$ among the nanocomposites. The energy functional can be written as Eq. 14.5,

$$E[\{\psi_i\}] = E_{known}[\{\psi_i\}] + E_{XC}[\{\psi_i\}] \tag{14.5}$$

where the functional split into two parts, $E_{known}[\{\psi_i\}]$ and everything else, E_{XC} . The "known" terms include four contributions (Eq. 14.6):

$$E_{known}[\{\psi_i\}] = \frac{h^2}{m} \sum_i \int \psi_i^* \nabla^2 \psi_i d^3r + \int V(r) n(r) d^3r + \frac{e^2}{2} \int\int \frac{n(r)n(r')}{|r-r'|} d^3r d^3r' + E_{ion} \tag{14.6}$$

The terms on the right are, in order of, the electron kinetic energies, the Coulomb interactions between the electrons and the nuclei, the Coulomb interactions between pairs of electrons, and the Coulomb interactions between pairs of nuclei respectively. The other term, $E_{XC}[\{\psi_i\}]$, that is the exchange-correlation functional is defined to include all the quantum mechanical effects present in the nanocomposites, which are not included in the known term. This undefined exchange-correlation energy functional can be expressed by Kohn and Sham equations, which have the form of Eq. 14.7.

$$\left[\frac{\hbar^2}{2m}\nabla^2 + V(r) + V_H(r) + V_{XC}(r)\right]\psi_i(r) = \varepsilon_i\psi_i(r) \tag{14.7}$$

These equations are superficially similar to a more complete description of the Schrodinger equation for the nanocomposite, which is Eq. 14.8.

$$\left[\frac{\hbar^2}{2m}\sum_{i=1}^{N}\nabla_i^2 + \sum_{i=1}^{N}V(r_i) + \sum_{i=1}^{N}\sum_{j<i}U(r_i, r_j)\right]\psi = E\psi \tag{14.8}$$

The only main difference in the Kohn-Sham equations are missing summations that appear inside the full Schrodinger equation (Eq. 14.8). This is because the solutions of the Kohn-Sham equations are single-electron wavefunctions. On the left-hand side of the Kohn-Sham equations there are three potentials, V, V_H, and V_{XC}. The first of these also appeared in the full Schrodinger equation (Eq. 14.8) and in the "known" part of the total energy functional given in Eq. 14.6. This potential defines the interaction between an electron and the collection of atomic nuclei arranged in the given nanocomposite geometric model. The second is called the Hartree potential, defined by Eq. 14.9.

$$V_H(r) = e^2\int\frac{n(r')}{|r - r'|}d^3r' \tag{14.9}$$

This potential describes the Coulomb repulsion in between the electron of nanocomposite, that is considered in one of the above Kohn-Sham equations, and the total electron density defined by all electrons of the nanocomposite system. The Hartree potential includes a so called self-interaction contribution because the electron that we are describing in the Kohn-Sham equation is also a part of the total electron density, so a part of V_H involves a Coulomb interaction between the electron and the density itself. The self-interaction is unphysical, and the correction for it is one of several effects that are lumped together into the final potential in the Kohn-Sham equations, V_{XC}, which defines exchange and correlation contributions to the single electron equations. V_{XC} can formally be defined as a "functional derivative" of the exchange-correlation energy (Eq. 14.10).

$$V_{XC}(r) + \frac{\delta E_{XC}(r)}{\delta n(r)} \tag{14.10}$$

The strict mathematical definition of a functional derivative is slightly more subtle than the more familiar definition of a function's derivative, but conceptually we can think of this just as a regular derivative. The functional derivative is written using δ to emphasize that it is not quite identical to a normal derivative. The exchange correlation approximation uses only the local density to define the approximate exchang-correlation functional, so it is called the local density approximation (LDA). LDA is the oldest and simplest way of solving the variational equations for molecular structures [88]. To solve the Kohn-Sham equations, we further need to define the Hartree potential, and we need to know the electron density of the nanocomposite under consideration. But to find the electron density, we must know the single-electron wavefunctions, and to know these wavefunctions we must solve the Kohn-Sham equations. To break this circle, the problem of nanocomposite system is usually treated in an iterative way as outlined in the following algorithm, also called as the Hohenberg-Kohn-Sham scheme.

1. Define an initial, trial electron density, $n(r)$.
2. Solve the Kohn-Sham equations defined using the trial electron density to find the single-particle wavefunctions for the nanocomposite, $\psi_i(r)$.
3. Calculate the electron density defined by the Kohn-Sham single-particle wavefunctions from step 2, $n_{KS}(r) = 2\sum_i \psi_i^*(r)\psi_i(r)$.
4. Compare the calculated electron density in the nanocomposite, $n_{KS}(r)$, with the electron density used in solving the Kohn-Sham equations of the nanocomposite, $n(r)$. If the two densities are the same, then this is the ground-state electron density, and it can be used to

compute the total energy of the nanocomposite. If the two densities are different, then the trial electron density must be updated in some way. Once this is done, the process begins again from step 2.

In summary, the ab-initio DFT methods have the advantage that they have to converge, and not necessarily in monotonic fashion, to the exact answer in the given basis set and the correlation limit. Hence, the ab-initio DFT method can essentially be summarized in the following five points: (1) all calculations are done in a linear combination of atomic orbitals type basis set resembling Gaussian functions, (2) a rigorous, orbital-dependent, energy functional for the exchange and correlation in the nanocomposite is taken from wavefunction theory, (3) there is a convergence to the right answer within the basis set and correlation limit for the nanocomposite's properties, (4) the Kohn-Sham determinant as calculated in the method provides a correct and correlated density consistent with corresponding functional for the nanocomposite and, lastly, (5) the potentials are multiplicative and localized as required by the DFT, but the functionals are nonlocal [88].

14.2.3 Nonequilibrium Green's Function (NEGF)

The NEGF method for calculation of quantum transport among the open nanosystems is a complicated mathematical/numerical routine. Nevertheless, due to versatility and numerical stability, it is becoming a rapidly popular method in comparison to the other transfer-matrix-based approaches. The NEGF method provides an ease of density evolution of the semi-infinite electrodes with small definite nanocomposite system under characterization in the nanodevice [89]. Furthermore, the Green's function approach can be generalized to many-body quantum theory, allowing the inclusion of electron–lattice as well as electron–electron interactions within a unified and systematic formalism [90]. Nonequilibrium Green's function methods are usually employed as a basic theory to build quantitative models for studying the transport properties in nanoscale conductors under voltage difference and for quantum device simulation [91]. It is used for calculating current, effective potential and charge densities in nanoscale (both molecular and semiconductor) conductors under bias. In order for any program to simulate a quantum device, a self-consistent solution of a transport equation must be performed. The calculations of the electron density and the effective potential in the nanocomposite are iterated till they converge to a self-consistent value. The I–V characteristics have been calculated with the Landauer formula (Eq. 14.11) [92],

$$I = \frac{2e^2}{h} \int_{-\infty}^{\infty} T(E)\left[f_l(E) - f_r(E)\right] dE \tag{14.11}$$

where e is the electron charge, h is Plank's constant, and $f_l(E)$, $f_r(E)$ represent the Fermi distribution functions at the left and right electrodes, respectively. The term $T(E)$ stands for the transmission coefficient as a function of the electron energy E.

14.3 Atomistic Computational Simulations: Modeling and Methodology

14.3.1 Atomistix Toolkit (ATK): Ab-initio DFT Software Package for Nanosystems

The ab-initio DFT implementation of ATK software is based on the SIESTA code (Spanish Initiative for Electronic Simulations with Thousands of Atoms). The ATK facilitates the SIESTA computer program implementation for performing ab-initio electronic energy structure calculations and molecular-dynamics-based structure relaxation simulations for the nanocomposite materials. In brief, the workflow in the ATK calculator can be described in the following steps:

- One can deploy either local density approximation (LDA) or general gradient approximation (GGA) for nanocomposite calculations in cluster and pseudo-periodic configurations for the calculation of exchange-correlation terms in the ATK-based Kohn-Sham self-consistent density functional method.
- ATK uses norm-conserving pseudopotentials in their fully nonlocal form.
- Linear combination of atomic orbital forms the basis set in ATK numerical calculations that allow the wide choice of single-zeta, double-zeta, and angular momenta polarization among the orbitals.

- It projects the electron wavefunctions and density onto a real-space grid of the nanocomposite in order to calculate the Hartree and exchange-correlation potentials.
- Fast and feasible simulations of nanocomposites theoretical models containing up to several hundred atoms are possible through ATK with modest workstations and parallel computing clusters.

It routinely provides the total and partial energy, atomic forces, stress tensor, electric dipole moment, band structure, electron density and atomic, orbital and bond populations (Mulliken). In addition to this, the ATK provides the ability to model the open boundary nanosystems where ballistic electron transport is taking place. Using the ATK program one can compute electronic transport properties, such as the zero bias conductances and the I–V characteristic of a nanocomposite in contact with two electrodes at different electrochemical potentials. The present software release includes the possibility of performing calculations of nanoelectronic transport properties [93], which deploys a finite material structure sandwiched in between two semi-infinite metallic leads and gives a solution of the electronic structure of such an open system device of the nanocomposite. A finite bias can be applied between both leads to drive a finite current. In brief, the calculations using ATK involve first the device's electronic density elucidation from the DFT Hamiltonian using Green's functions techniques followed by I–V calculations. The atomistix toolkit also provides various other interfaces that are found to be useful for performing electronic transport calculations of nanocomposites. Besides, a library of nanocomposite modeling techniques can be used to calculate a wide range of properties of organic and polymeric nanoscale systems [94].

14.3.2 Nanodevice Characteristics Simulation: PAn and GNR-PAn Composites

We have modeled a set of four different two-probe devices (A, B, C & D), which are electrically connected through metallic Lithium atomic-chain electrodes of high electrical transmitivity even at higher voltages. The central region of the as-prepared nanocomposites' two-probe device is packed with various samples; pristine chlorinated PAn (A), nanocomposite of zigzag graphene ribbon and

PAn (B), nanocomposite of arm-chair graphene ribbon and PAn (C), nanocomposite of hybrid graphene ribbon and PAn (D), in various conformations. Our nanocomposite conformations are mainly comprised of parallel plane and in-plane conformations. The parallel plane is the situation when both the PAn chain and the nanoribbon are in different planes but in proximity with each other in the direction of transmission of device while the later conformation lies in the same plane.

The device B has a nanocomposite placed in its central region, in which PAn is in parallel conformation with the graphene nanoribbons. Similarly, devices C & D have the nanocomposites with in-plane and parallel conformations of graphene, respectively. Figure 14.10 shows the first device with a central region of the polyaniline in relax-pose atomic positions (in Å) at tolerance of less than 0.001 eV. The PAn is modeled in the chlorinated form to increase energy mid-bandgap states for achieving improved light harvesting and photoinduced charge transfers as previously discussed in the introductory section. The light harvesting in the PAn chain is expected to increase with its proximity to highly electronegative chlorine atom. The schematics of modeled nanocomposite devices (B–D) are shown in Figure 14.11.

In order to carry out the study of the electronic structure and transport properties of the nanocomposite devices – A to D, a virtual environment in the nano-regime simulation performs the calculations of the various physical properties. Using the methodology described below, we have performed a mixed set of simulational experiments comprising density of states (DOS), transmission spectrum (TS), transmission coefficient (TC), total energy (TE), eigenvalue (EV), eigenstates (ESs), and I–V characteristics. The basic parameters used to simulate the different two-probe device samples

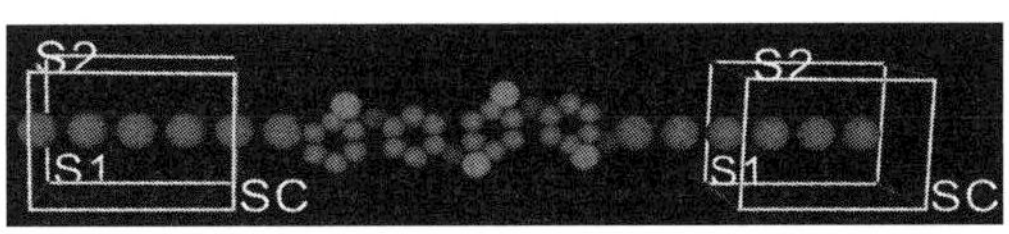

Figure 14.10 Two-probe device A in which a central region constitutes of a modeled structure of the polyaniline (PAn) chain and the proximate chlorine (Cl) atoms. In the central region, C-C, N-C and N=C bond lengths are set to be 1.42 Å, 1.47 Å and 1.29 Å, respectively, while the N:Cl^- distance is 1.75 Å. The Li-Li interatomic distance of source-drain electrodes is 2.9 Å. The left and right electrodes are very high conductivity Li-atomic chains that are metallic in character and provided with up to three electron transmission channels.

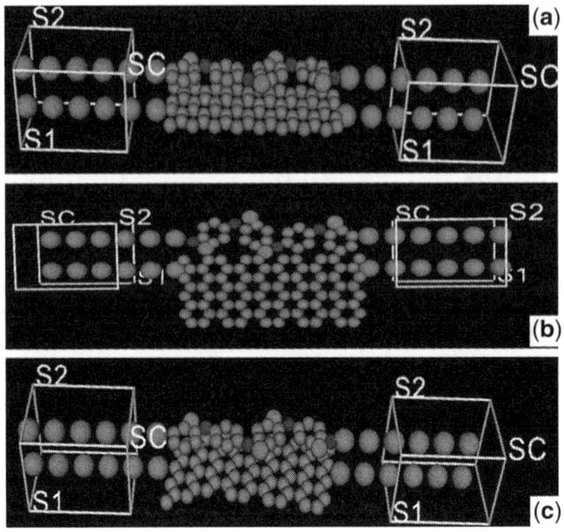

Figure 14.11 (a) Two-probe device B in which the central region constitutes of a modeled structure of ZZ-GNR/PAn nanocomposite, in which the nanoribbon is in parallel conformation with PAn monomeric chain. (b) Two-probe device C with the AC-GNR/PAn nanocomposite, in which the GNR has in-plane conformation with PAn monomeric chain. (c) Two-probe device (D) of H-GNR/PAn nanocomposite in parallel conformation. The hybrid nanoribbons are aligned exactly in parallel planes with the PAn monomeric chain.

in the virtual environment are as follows: the two-probe system is consistent for the entire set of samples and is comprised of linear combination of single zeta-type Gaussian orbitals; norm-conserving pseudopotentials are used for all the chemical species involved; the k-point grid for Brillioun zone integration is taken sufficiently large along the axis of the device. The optimized k-grid values are found to be (1, 1, 500) for the k_A, k_B, and k_C, respectively; eigenstate occupation values of both the electrodes of the two-probe devices is set at the same temperature value of 300 K; equilibrium conductance is calculated at the zero electrode voltage bias; the optimized electron density mesh cut-off of the nanocomposite is found to be 2040 eV; the Perdew-Zunger type of exchange-correlation function is taken for the nanocomposite calculations within localalized density approximation; the self-consistency field calculations are achieved by total energy criterion of the nanocomposite with tolerance values of the order of 10^{-5} eV and 500 maximum iterative steps; the diagonal iteration mixing is performed by using Pulay algorithm for the nanocomposite's Hamiltonian, and finally; a numerical calculation file is created under the self-consistent label, which stores results of each of the iteration steps. For the I–V measurement, we have taken the device structures, numerical calculation files of nanocomposites with the help of numpy functions of python scripts [95–97]. We have introduced the bias of the value in range of [-1, 3] V across the devices and we got the values of respective currents for all the nanocomposite samples, with the help of nonequilibrium Green's function (NEGF) method.

14.4 Results and Discussions

Now we intend to present and discuss the results that we have obtained using ab-initio DFT LDA NEGF simulations on modeled nanocomposite samples. Various calculations for the density of states (DOS), transmission spectrum (TS), energy eigenstates (ES) and I-V measurements are performed. A choice of Li linear atomic chains for the electrode material that is used to connect the bias in all of our two-probe devices is deliberate and consistent. Because firstly we wish to reduce the computation time and the use of chain with a few atoms as electrical contact is a good option. Secondly these atomic strings are perfectly metallic nanowires [98–100] which are even also expected to show ballistic conductance up to $3G_0$ values (G_0 being a conductance quanta) owing to their unique strong bonding character. Around a Fermi level, these atomic wires show a band structure that confirms a doubly degenerate π band and σ^*band crossing the Fermi level [101]. It is to be noted that the graphene nanoribbon/organic semiconductor nanocomposites synthesized by various physical or chemical routes often have random distributions. This is because in such blended nanomaterials the nanoribbon is usually aligned spontaneously with a polyaniline in different possible conformations. Two exemplary conformations are one in which the axis of the PAn is aligned parallel to the nanoribbon length either in the same plane or in the two different planes but parallel to each other. In fact the blended GNR-PAn nanocomposites are frustrated systems in which the rate of composite formation far exceeds the natural process of energy minimization. So even in such blended systems, the self-assembly allows a possibility of a presence of the percolating structures owing to the axial alignment of the two components of the composite [102]. The percolation effect induces a number of nanoribbon-PAn composite units to join each other end-to-end in the form of chain-like orientations. The percolatively-oriented nanocomposites are also expected to show high conductivity and mobility [103]. Room temperature mobility of the percolatively-oriented conducting polymer transistors has already been reported to be as high as 1cm^2/Vs [104].

Mimicking the perfect axial alignment criteria of convergence for quasi-Newtonian forces used in the geometrical optimization of modeled nanocomposite structures is not followed in the present work. The only convergence criterion we have used is the total energy for GNR-PAn axially-aligned nanostructures. This is a necessary requirement to observe the electronic and transport properties of the aligned nanostructures in the percolation regime. The reason for this is that the geometrical optimization and nanostructural alignment are often competitive processes in the percolating systems.

Table 4.1 summarizes the results of transmission coefficients for the hole and electron conduction channels and total energies that are the sum of all the band structure energies for the device – A to D – as shown in the Figures 14.10 and 14.11. Now we intend to present our results of density of states, transmission spectrum, eigenstates (#0) corresponding to that of holes in valence band,and eigenstates (#1) corresponding to that of electrons in the conduction band, and I–V characteristics are calculated for all four devices.

14.4.1 Device Characteristics of Chlorinated PAn

A plot of the density of states vs energy for the polyaniline is shown in Figure 4.12(a). The energy range is located from -1 eV to 3 eV.

Table 14.1 Transmission coefficients for conduction and valence channels and total energies of various nanocomposite devices.

Device No	Transmission coefficient		Total energy of the device
	0(hole channel)	1(electron channel)	
A	0.0266	0.6233	–6642.0637 eV
B	0.0279	0.1793	–16749.3760 eV
C	0.0014	0.2762	–16516.6774 eV
D	0.0612	0.1771	–17443.9975 eV

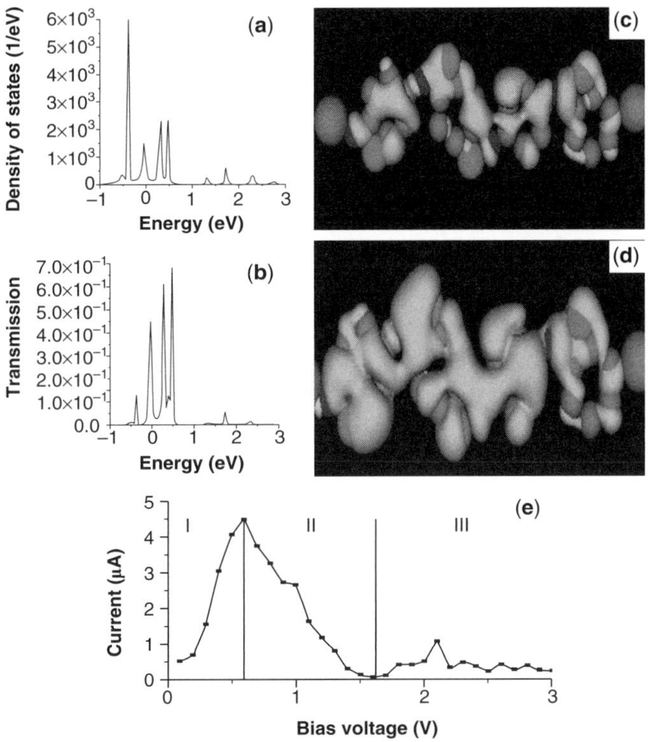

Figure 14.12 Electronic and transport properties of PAn monomeric chain in device A. (a) Density of states, (b) charge transmission spectrum, (c) hole eigenstate transmission channel, (d) electron eigenstate transmission channel, and (e) I–V characteristics.

There is an occurrence of sharp peaks in the DOS across -0.39 eV and 0.34 eV and around the Fermi energy level (0 eV). A peak at ~-0.4 eV corresponds to contributing states arise from a π valence band of the organic semiconductor. The other peaks at the upper end of the Fermi level placed in between an energy range of ~0.34 eV to ~2.6 eV and can be attributed to a π conduction band state and mid-gap states induced by the dopant (Cl) in the conjugated conducting chain of PAn. Hence as expected with the addition of chlorine proximity to the PAn conjugation backbone, the ab-initio DFT-LDA calculations have yielded a DOS spectrum with rich singularity features without any significant contribution to the equilibrium conductivity as shown in the Figure 4.12(a). Hence the materials are in a semiconducting state with a small peak at the Fermi level although decorated with a large number of inter-gap states, which would definitely improve the central region's photon absorption for incident solar light.

A resonant peak exactly at the zero point of energy spectrum with a significant DOS value of ~1500 is observed. Such resonant features have also been reported earlier to be typically found in several two-probe devices and are more commonly known as DOS resonance peaks or simply Fano peaks. Very recently, Furst *et al.* [105] have proposed that matching of localized states present in the central region, which is the polyaniline in this modeled device sample, and the contributing states in the electrodes, which are left and right Li-linear atomic chains, is the origin of this Fano symmetry that gives rise to resonant peaks in the vicinity of the Fermi level.

We have also observed a similar transport channel at the Fermi level in our transmission spectra, which is attributed to a presence of the Fano symmetry in this device. However, we are sure that this is not a chlorine dopant-induced Fano effect, because the dopants usually interfere their localized states (atomic orbitals of Cl atom) with that of the central region (contributing states in polyaniline conjugated backbone) yielding an anti-Fano type of effect or a dip in DOS spectrum near to the E_F [72, 105]. If we look carefully at the DOS spectrum the presence of a lower resonant peak at -0.39 eV closer to Fermi level arises due to hole delocalization in polyaniline which is stronger than the upper resonant peak at 0.34 eV. This effect can only be ascribed to the hole transport character of the as-modeled chlorinated polyaniline where π band related hole eigenstate transmission channel is more effective due to sp^2p_z configuration of C-orbitals which overlap along the length of conjugation [106]. But due to the effect of chlorination in PAn, we can also see several other resonant peaks at the upper levels. Also, an observance of broader and stronger upper resonant peaks nearer to the Fermi level in the transmission spectrum are signatures of presence of several electron transport channels in this modeled PAn structure [107]. Our observance is also further corroborated from our studies of individually resolved electron and hole eigenstate conduction channels [108] depicted in Figures 4.12(c) and (d). It is clear from the images of the two conduction channels 0 and 1 that both of them are very strong channels, but the equivalent conductance of channel (1) is approximately twenty three times more than that of channel (0). The respective values of transmission coefficients at zero bias are given in Table 14.1. Hence the modeled structure can effectively behave as an efficient conduit for both the hole and electron transport.

We have further studied a nonequilibrium conductance of the PAn nanodevice by applying a bias of -1 to 3 V in between source (left electrode) and drain (right electrode). I–V characteristics of the device are extracted from the values of nonequilibrium conductance as described by using the Landauer's relation as given in Eq. 14.11. As obtained, I–V measurement of the polyaniline is shown in Figure 4.12(e). Clearly, the I–V curve follows some non-Ohmic conduction mechanisms, which is typical to all nanostructural materials. This is because of a presence of van Hove singularities in the density of states of one-dimensional nanowires owing to the quantum size confinement effect [109]. The non-Ohmicity induced phenomenon of negative differential resistance (NDR) occurs spontaneously in our result after an Ohmic region of 0 to ~ 0.5 V, which is further followed by a plateau of saturated current from ~ 1.5 V onwards. Hence, clearly we have three distinct I–V features: (I) nearly Ohmic region but with differential conductance, (II) non-Ohmic region with NDR, and (III) current saturation as shown in the figure. The sharp ascent of the current in region (I) is obviously arising from the presence of contributing states due to a number of mid-gap states in the chlorinated polyaniline. However, the PAn is also strictly a nanostructure, hence most of the upper resonant states as discussed in DOS and transmission spectrums still have only a few numbers of singularities and the conduction channels decay very rapidly after 0.5 V. Hence we observe a true NDR regime of the PAn two-probe nanodevice from V > 0.5 V until higher energy electron channels comes into this picture at V > 1.8 V. Nevertheless there is also a small T(E) channel induced by chlorine proximity-related transmission oscillations which gives rise to a small I–V peak at ~2 V and a complete saturation of the current thereafter with a very small aggregate value of charge transport.

14.4.2 Device Characteristics of ZZGNR-PAn Nanocomposite

The electronic structure of the nanodevice (B) is clear from Figure 4.13(a) which is a plot of DOS vs energy for a system of the zigzag graphene nanoribbon blended with the chlorinated polyaniline when they are in parallel conformation. The energy range is set in between -1 to 3 eV. In fact a Fano-type peak is also observed in ZZ nanocomposite sample which is centered at Fermi level in both DOS and T(E) spectrum. An exceptional resonant state is observed at

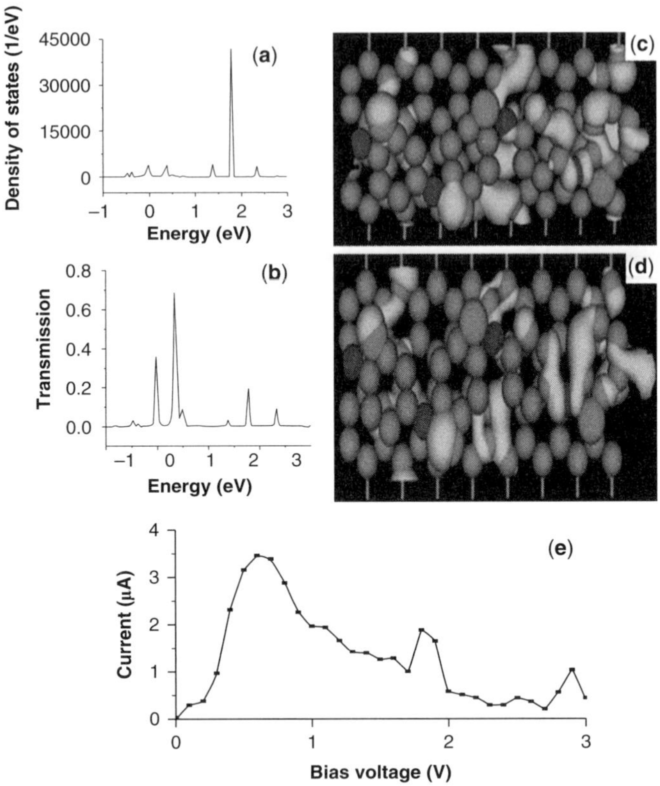

Figure 14.13 Electronic and transport properties of nanocomposite of zigzag graphene nanoribbon and PAn in the device B.

~2.0 eV along with several other singularities amid the contributing states. This would obviously give rise to several transmission channels. The infinitely long and sufficiently wide ZZGNRs are supposed to be conductive due to their edge states which provide a through channel for conduction [110]. Usually these ZZGNR edge states are found near the Fermi level (at $E_F = 0$, in the energy corrected spectrum) where the mid-gap of the material is expected. Hence, the edge states are a good source of equilibrium conductance in ZZGNR. Park *et al.* have reported at least three 100% eigen channel conduction in ZZGNR for a clear window of ± 0.2 eV around E_F. But the ZZGNR with finite length and small width behaves differently in a two-probe device configuration [111]. Nonetheless, the prominent near zero point transmission is from channel 0 for holes and channel 1 for electrons, respectively. The transmission channel at ~0.35 eV is strongest due to presence of number of resonant states at the upper part of the

Fermi level. At the same time, this nanocomposite device also exhibits some degree of semi-metallicity because the transport channel is present at zero point. Hence, the eigenstate transmission decomposition for the electron is more continuous as compared to that for the hole delocalized eigenstates in Figures 4.13(c) and (d). I–V curve for this sample shows a good steep rise of current commencing from zero finite bias and reaches its maximum at ~0.65 V. This steepest ascent can be attributed to both the equilibrium contribution states and reemergence of the metallicity in the ZZGNR in the nanocomposite material, as well as to resonant modes that exist in the central region owing to the PAn's contributing states injection, further evident from overlapping eigenstate decomposition. We ascribe the peculiar behavior of this nanocomposite to the significant π-π interactions amongst the two nanocomponents. Also, in low dimensional systems like graphene nanoribbons the charge transport mostly depends upon the quantum transport properties, in which the lowest unoccupied molecular orbital of the central region, i.e., π^* band of ZZGNR, should be perfectly aligned with the conduction band of electrodes for better charge transport [112]. However, an average NDR regime starts showing up in the I–V after ~0.65 V which continues up to 2 V, with some oscillatory kinks arising from the higher voltage resonant states. Since these discrete resonant DOS states are very high, up to ~45,000, there are several NDR regimes in our I–V. Nonetheless, the device performance of ZZGNR-PAn nanocomposites is found to be better than that of the pristine ZZGNR. This implies that there is a strong injection of states among the different components of nanocomposite device at high voltages.

14.4.3 Device Characteristics of ACGNR-PAn Nanocomposite

The DOS vs energy plot of a armchair-graphene nanoribbon with a polyaniline when configured in plane is shown in Figure 4.14(a); the energy is set from -1 eV to 3 eV. The fact that our pristine ACGNR sample is best coupled with the lithium electrodes among all other samples is also manifested in the form of a number of strong resonant peaks that are found in both DOS and T(E) spectra, which provide a good number of electronic conduction channels in the device. However, there is an asymmetric distribution of contributing states around the Fermi level, which is also an important feature of our ACGNR device (C) and this minimizes the NDR effect to

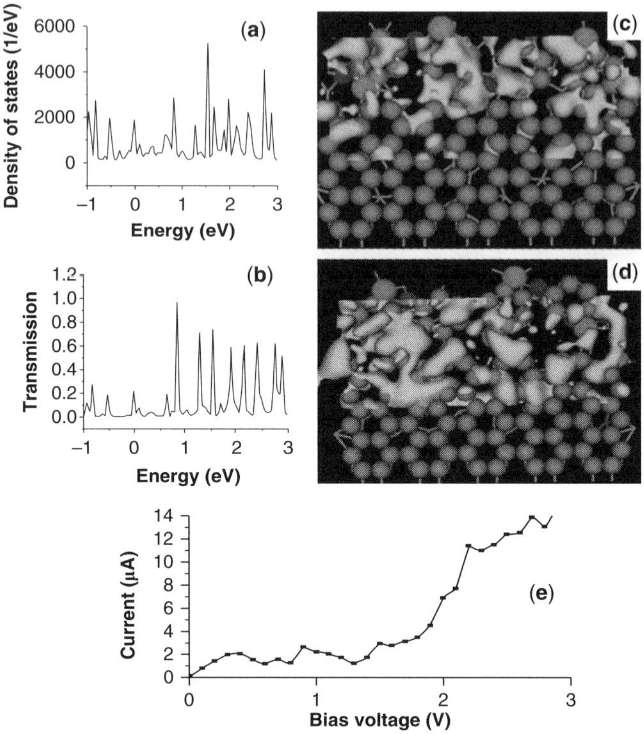

Figure 14.14 Electronic and transport properties of nanocomposite of armchair and PAn in device C.

some extent. Amongst ACGNR nanostructures, the edge-state kind of conduction channels is absent due to a reduced π-π coupling at the edge and large effective C-neighbor distances in the armchair edge. Thus ACGNRs are inherently semiconducting with direct bandgaps [113]. However the bandgap of ACGNR changes with respect to width and length of ACGNR. Typically, for the ACGNR of finite length of 6 unit cells, the calculated bandgaps are 0.49, 1.034, 0.709, 0.195, 1.433 and 0.29 eVs for corresponding widths of 17, 10, 9, 8, 7 and 5 carbon atoms [114].

From the I–V curve, we can see a continuous rise in the average current with no significant NDR; only the current plateau is found between the voltage values of 0.5 to 1.7 V, after which the current increases steadily. This is an exceptional result which mimics a bulk-like or pseudo-Ohmic character of the nanodevice. The current increases exponentially after the value of 1.9 V, but at the same time there is a significant value of current at lower voltages.

Hence the rectifying behavior cannot be ascribed to the effects of Schottky barrier where the leakage current is usually very small. In our case the rectification can be understood from the presence of a large number of peaks at higher energies in the DOS and the transmission energy spectra. Also we can see a significant current transport of 14 μA at 2.7 V in these strictly axially-aligned nanosystems. By the way, this is the largest value of transport obtained in the present set of studies.

14.4.4 Device Characteristics HGNR-PAn Nanocomposite

In DOS vs energy plot as shown in a Figure 4.15(a) the energy is set from -1 eV to 3 eV. There is an occurance of tailing edge effect on both sides of the equilibrium peak in the DOS across -0.32 eV and

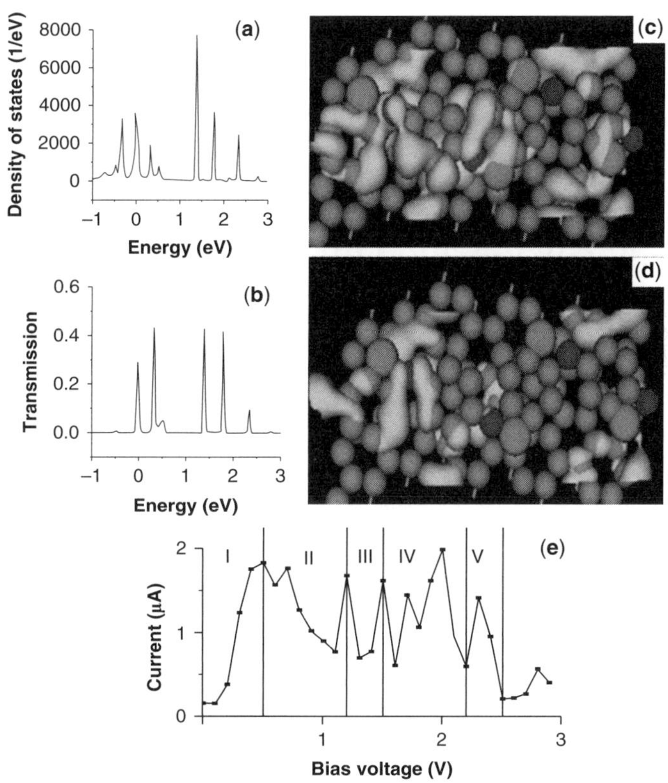

Figure 14.15 Electronic and transport properties of hybrid nanoribbon and polyaniline nanocomposite in device D.

0.30 eV around the Fermi level (0 eV), which indirectly signifies the extent of coupling in this hybrid nanocomposite. A total energy gap that appears across the Fermi energy level is 0.62 eV, along with an equilibrium conduction peak. Nevertheless, this device shows the reduced injection of resonant states from the electrodes. Still the coupling is manifested by strong resonance state at the zero point.

Actually, this device is also unique because most of the resonant states are tens of thousands in number, which ensures well-defined, strong, singular transport channels, yielding several accurate NDR switching regimes. Hence it is easy to predict that these are the ideal characteristics for a high-performance nanoswitch/nanotransistor that can easily control ultrafast dynamics of modern nanoelectronics, owing to high mobilities of respective nanocomponents, rather than simply deploying in nano solar cell technology. However, the device yields a maximum current of ~2 μA among different NDR regimes as depicted in the I–V curve, and thus is also useful for solar cell fabrication along with the D5 and D6 devices. This observation is also evident from the continuous eigenstate channels in Figure 4.15(c) and (d), and the continuous current rise from zero bias up to ~0.5 V. On further increasing the current, secondary NDR regimes originate as shown in I–V measurements in Figure 4.15(e).

14.5 Conclusion and Outlook

We have studied the electronic structure and transport properties of nanodevices of chlorinated polyaniline (PAn), graphene nanoribbons (GNR) in various edge conformations, and GNR-PAn nanocomposite in-plane and in-parallel alignment device configurations for the very purpose of their deployment as nanocomposite active layers in a bulk heterojunction (BHJ) solar cell for smart light harvesting. From the decomposed transport eigenstate channels, chlorinated PAn is found to be an efficient transport channel with noticeable values of equilibrium conductance. From the DOS and the transmission energy spectra we have found the existence of a number of energy inter-gap states, along with fundamental transition states, that vary in the energy regime of 0.4 to 3.2 eV, which corresponds well to a complete scale of UV-Vis-IR energies. We deem that such a rich electronic band structure of the chlorinated PAn would indeed be helpful for the efficient photon harvesting in the full solar energy spectrum. We also found that all the transport channels corresponding to these energy states are active, along with sufficient electronic transmission near the Fermi energy. Such behavior of the modeled donor material is ascribed to the proximity of a chlorine atom with PAn conjugated backbone. The equilibrium conduction states are also found in graphene nanoribbon two-probe devices with edge conformations of zigzag, armchair and hybrid forms. Studies of DOS(E) and T(E) spectra of nanodevices indicate a higher number of contributing mid-gap states and site-selective transport channels, which also reveal the rationale of deployment of these GL-PAn nanocomposite in the BHJ type of nano solar cells. All of the three nanocomposite samples (B, C & D) are found to exhibit excellent charge transport characteristics that are found to be particularly suitable for the photovoltaic applications. They are comprised of: (a) the PAn formulates nanocomposite with zigzag GNR when configured in parallel, (b) the PAn formulates nanocomposite with arm-chair GNR when configured in the same plane, and (c) the PAn formulates nanocomposite with hybrid GNR when configured in parallel. All of the devices show high charge transfer values in their respective transport channels along with significant density of states in the energy range of photovoltaic interest (0 eV to 3 eV). These results also corroborate the obvious higher solar power yield excitations from the GNR-PAn solar cell device.

Another interesting aspect of the present studies is that we have found blueprints of singularities in our ab-initio simulations for most of the nanostructures. A phenomenon of negative differential resistance (NDR) is quite often observed during our simulation process, which we have verified rigorously. It is interesting to note that the novel quantum features in our samples could be attributed to the presence of singularities in the density of states and transmission spectrum. However, small peaks in the DOS and the T(E) can be easily ascribed either to the presence of Fano-like modes of resonances near the Fermi level, or to the enhanced coupling in between the electrodes and the central region of the device in our samples. The NDR is also effortlessly observed in most of our I–V measurements. However, we do not observe any significant rectification in our devices either due to the strong coupling/Fano induced conduction modes, or due to the pointed nature of DOS features leading to a rapid decay of the current. The observance of significant electronic transmission values near the Fermi level is still ambiguous in nature. We have also found an effect of the nanocomposite samples showing severe divergence from the respective physical properties of the individual components; we have attributed such an adherence to the significant π-π interaction amongst the C-ring constituents of the nanocomposite. Nevertheless, in the device (C), a different type of rectification is found in the I–V characteristics, which is attributed to the presence of higher energy electronic states in transmission. Needless to say, the work performed in this piece of research is of high importance for the optimization of nano OPV solar cell fabrication and its functioning.

Finally, the authors would like to mention some suggestions for improvement of the photovoltaic device performance of these PAn/graphene nanocomposites as a future extension of the current piece of research. Firstly, the effects of different cationic/anion dopants on the equilibrium charge transfer through the as-modeled PAn two-probe devices. Subsequently, the study of the nanoswitch transistor characteristics in the device (D) should be done for eigenstate channel decomposition in various NDR regimes for a better understanding of active transport channels. And, to conclude, the simulations of these novel nanocomposites, the as-observed absence of the Fano effect or the strong LCR coupling in the pristine nanoribbon systems need further investigations.

References

1. Z. Zheng, Y. Du, Q. Feng, Z. Wang, and C. Wang, Facile method to prepare pd/graphene–polyaniline nanocomposite and used as new electrode material for electrochemical sensing, *Journal of Molecular Catalysis A: Chemical*, Vol. 353, pp. 80–86, 2012.
2. J.M. Khan, R. Kurchania, and V.K. Sethi, Synthesis and characterization of two dimensional graphene lamellae based PAn nanocomposites, *Thin Solid Films*, Vol. 519, No. 3, pp. 1059–1065, 2010.
3. J. Yan, T. Wei, B. Shao, Z. Fan, W. Qian, M. Zhang, and F. Wei, Preparation of a graphene nanosheet/polyaniline composite with high specific capacitance, *Carbon*, Vol. 48, No. 2, pp. 487–493, 2010.
4. Q. Wu, Y. Xu, Z. Yao, A. Liu, and G. Shi, Supercapacitors based on flexible graphene/polyaniline nanofiber composite films, *Acs Nano*, Vol. 4, No. 4, pp. 1963–1970, 2010.
5. Y. Xu, L. Zhao, H. Bai, W. Hong, C. Li, and G. Shi, Chemically converted graphene induced molecular flattening of 5, 10, 15, 20-tetrakis (1-methyl-4-pyridinio) porphyrin and its application for optical detection of cadmium (ii) ions, *Journal of the American Chemical Society*, Vol. 131, No. 37, pp. 13490–13497, 2009.

6. D.F. Perepichka, F. Wudl, S.R. Wilson, Y. Sun, and D.I. Schuster, The dissolution of carbon nanotubes in aniline, revisited, *Journal of Materials Chemistry*, Vol. 14, No. 18, pp. 2749–2752, 2004.
7. X.-F. Zhang and Q. Xi, A graphene sheet as an efficient electron acceptor and conductor for photoinduced charge separation, *Carbon*, Vol. 49, No. 12, pp. 3842–3850, 2011.
8. A. Bhattacharya and A. De, Conducting composites of polypyrrole and polyaniline a review, *Progress in Solid State Chemistry*, Vol. 24, No. 3, pp. 141–181, 1996.
9. J. Kalowekamo and E. Baker, Estimating the manufacturing cost of purely organic solar cells, *Solar Energy*, Vol. 83, No. 8, pp. 1224–1231, 2009.
10. R.H. Bube, *Photovoltaic Materials*, Vol. 1. Imperial College Press London, 1998.
11. P.-L. Ong and I.A. Levitsky, Organic/iv, iii-v semiconductor hybrid solar cells, *Energies*, Vol. 3, No. 3, pp. 313–334, 2010.
12. G. Gomard, E. Drouard, X. Letartre, X. Meng, A. Kaminski, A. Fave, M. Lemiti, E. Garcia-Caurel, and C. Seassal, Two-dimensional photonic crystal for absorption enhancement in hydrogenated amorphous silicon thin film solar cells, *Journal of Applied Physics*, Vol. 108, No. 12, pp. 123102–123102, 2010.
13. B. Kippelen and J.-L. Brédas, Organic photovoltaics, *Energy & Environmental Science*, Vol. 2, No. 3, pp. 251–261, 2009.
14. M.A. Green, Third generation photovoltaics: Ultra-high conversion efficiency at low cost, *Progress in Photovoltaics: Research and Applications*, Vol. 9, No. 2, pp. 123–135, 2001.
15. J. Li, J. Liu, C. Gao, and G. Chen, Nanocomposite hole-extraction layers for organic solar cells, *International Journal of Photoenergy*, Vol. 2011, 2010.
16. M.-G. Kang, H.J. Park, S.H. Ahn, T. Xu, and L. Guo, Toward low-cost, high-efficiency, and scalable organic solar cells with transparent metal electrode and improved domain morphology, *Journal of Selected Topics in Quantum Electronics, IEEE*, Vol. 16, No. 6, pp. 1807–1820, 2010.
17. S.H. Kim, M. Jang, H. Yang, J.E. Anthony, and C.E. Park, Organic field-effect transistors: Physicochemically stable polymer-coupled oxide dielectrics for multipurpose organic electronic applications, *Advanced Functional Materials*, Vol. 21, No. 12, pp. 2197–2197, 2011.
18. N. Senthilkumar, H.-S. Kang, D.-W. Park, and Y. Choe, Improving efficiency of organic photovoltaic cells using pedot: Pss and mwcnt nanocomposites as a hole conducting layer, *Journal of Macromolecular Science, Part A: Pure and Applied Chemistry*, Vol. 47, No. 5, pp. 484–490, 2010.
19. P.G. Nicholson and F.A. Castro, Organic photovoltaics: principles and techniques for nanometre scale characterization, *Nanotechnology*, Vol. 21, No. 49, p. 492001, 2010.
20. G.F. Burkhard, E.T. Hoke, S.R. Scully, and M.D. McGehee, Incomplete exciton harvesting from fullerenes in bulk heterojunction solar cells, *Nano letters*, Vol. 9, No. 12, pp. 4037–4041, 2009.
21. J. Xue, Perspectives on organic photovoltaics, *Polymer Reviews*, Vol. 50, No. 4, pp. 411–419, 2010.
22. J. Bernede, Organic photovoltaic cells: history, principle and techniques, *Journal of the Chilean Chemical Society*, Vol. 53, No. 3, pp. 1549–1564, 2008.

23. A. Cheknane, T. Aemouts, and M. Merad Boudia, Modelling and simulation of organic bulk heterojunction solar cells, *Revue des* Énergies Renouvelables, pp. 83–90, 2007.
24. V. Shrotriya, G. Li, Y. Yao, T. Moriarty, K. Emery, and Y. Yang, Accurate measurement and characterization of organic solar cells, *Advanced Functional Materials*, Vol. 16, No. 15, pp. 2016–2023, 2006.
25. T.L. Benanti and D. Venkataraman, Organic solar cells: An overview focusing on active layer morphology, *Photosynthesis Research*, Vol. 87, No. 1, pp. 73–81, 2006.
26. S. Morita, A.A. Zakhidov, and K. Yoshino, Doping effect of buckminsterfullerene in conducting polymer: Change of absorption spectrum and quenching of luminescene, *Solid State Communications*, Vol. 82, No. 4, pp. 249–252, 1992.
27. N. Sariciftci, L. Smilowitz, A. Heeger, F. Wudl, et al., Photoinduced electron transfer from a conducting polymer to buckminsterfullerene, *Science*, Vol. 258, No. 5087, pp. 1474–1476, 1992.
28. K.M. Coakley and M.D. McGehee, Conjugated polymer photovoltaic cells, *Chemistry of Materials*, Vol. 16, No. 23, pp. 4533–4542, 2004.
29. C. Chiang, C. Fincher Jr, Y. Park, A. Heeger, H. Shirakawa, E. Louis, S. Gau, and A.G. MacDiarmid, Electrical conductivity in doped polyacetylene, *Physical Review Letters*, Vol. 39, No. 17, pp. 1098–1101, 1977.
30. G. Rimbu, I. Stamatin, C. Jackson, and K. Scott, The morphology control of polyaniline as conducting polymer in fuel cell technology, *Journal of Optoelectronics and Advanced Materials*, Vol. 8, No. 2, p. 670, 2006.
31. A.J. Epstein and J. Yue, Polyaniline compositions, processes for their preparation and uses thereof, Aug. 11 1992. US Patent 5,137,991.
32. S.D. Patil, S. Raghavendra, M. Revansiddappa, P. Narsimha, and M.A. Prasad, Synthesis, transport and dielectric properties of polyaniline/co3o4 composites, *Bulletin of Materials Science*, Vol. 30, No. 2, pp. 89–92, 2007.
33. D.N. Huyen, N.T. Tung, N.D. Thien, L.H. Thanh, et al., Effect of tio2 on the gas sensing features of tio2/pani nanocomposites, *Sensors*, Vol. 11, No. 2, pp. 1924–1931, 2011.
34. U. Sajeev, C.J. Mathai, S. Saravanan, R.R. Ashokan, S. Venkatachalam, and M. Anantharaman, On the optical and electrical properties of rf and ac plasma polymerized aniline thin films, *Bulletin of Materials Science*, Vol. 29, No. 2, pp. 159–163, 2006.
35. A.B. Afzal, M. Akhtar, M. Nadeem, M. Ahmad, M. Hassan, T. Yasin, and M. Mehmood, Structural and electrical properties of polyaniline/silver nanocomposites, *Journal of Physics D: Applied Physics*, Vol. 42, No. 1, p. 015411, 2009.
36. A. Varela-Álvarez, J.A. Sordo, and G.E. Scuseria, Doping of polyaniline by acid-base chemistry: Density functional calculations with periodic boundary conditions, *Journal of the American Chemical Society*, Vol. 127, No. 32, pp. 11318–11327, 2005.
37. N. Gospodinova and L. Terlemezyan, Conducting polymers prepared by oxidative polymerization: Polyaniline, *Progress in Polymer Science*, Vol. 23, No. 8, pp. 1443–1484, 1998.
38. A.A. Ansari, M. Khan, M.N. Khan, S.A. Alrokayan, M. Alhoshan, and M. Alsalhi, Optical and electrical properties of electrochemically deposited

polyaniline/ceo2 hybrid nanocomposite film, *Journal of Semiconductors*, Vol. 32, No. 4, p. 043001, 2011.
39. J. Stejskal, R. Gilbert, et al., Polyaniline: Preparation of a conducting polymer (iupac technical report), *Pure and Applied Chemistry*, Vol. 74, No. 5, pp. 857–867, 2002.
40. F. Yakuphanoglu, E. Basaran, B. Senkal, and E. Sezer, Electrical and optical properties of an organic semiconductor based on polyaniline prepared by emulsion polymerization and fabrication of ag/polyaniline/n-si schottky diode, *The Journal of Physical Chemistry B*, Vol. 110, No. 34, pp. 16908–16913, 2006.
41. A.K. Mishra and P. Tandon, Dft study and heat capacity of polyaniline pernigraniline base, *The Journal of Physical Chemistry B*, Vol. 113, No. 29, pp. 9702–9707, 2009.
42. A. Parsa, S. Hosseini, and M. Asefoddoleh, Anion exchange properties and kinetic behavior of polyaniline-coated silica gel for anion phosphates, *Eur J Sci Res*, Vol. 26, pp. 369–375, 2009.
43. M.H. Lee, G. Speyer, and O.F. Sankey, Theory of electron transport through single molecules of polyaniline, *Journal of Physics: Condensed Matter*, Vol. 19, No. 21, p. 215204, 2007.
44. A. Elsayed, M.M. Eldin, A. Elsyed, A.A. Elazm, E. Younes, and H. Motaweh, Synthesis and properties of polyaniline/ferrites nanocomposites, *Int. J. Electrochem. Sci*, Vol. 6, pp. 206–221, 2011.
45. E. Kurmaev, M. Katsnelson, A. Moewes, M. Magnuson, J. Guo, S. Butorin, J. Nordgren, D. Ederer, and M. Iwami, Spectroscopic observation of polaron-lattice band structure in the conducting polymer polyaniline, *Journal of Physics: Condensed Matter*, Vol. 13, No. 17, p. 3907, 2001.
46. M. Chipara, G. Aldicaa, D. Huib, M. Dimoniec, K. Laud, L. Georgescue, I. Munteanue, and H. Marascoiuc, Electron spin resonance and dc electrical investigations on chlorine doped polyanilines, *Journal of Optoelectronics and Advanced Materials*, Vol. 6, No. 1, pp. 297–305, 2004.
47. A.C. Neto, F. Guinea, N. Peres, K. Novoselov, and A. Geim, The electronic properties of graphene, *Reviews of Modern Physics*, Vol. 81, No. 1, p. 109, 2009.
48. R. Verdejo, M.M. Bernal, L.J. Romasanta, and M.A. Lopez-Manchado, Graphene filled polymer nanocomposites, *Journal of Materials Chemistry*, Vol. 21, No. 10, pp. 3301–3310, 2011.
49. Y.H. Hu, H. Wang, and B. Hu, Thinnest two-dimensional nanomaterial graphene for solar energy, *ChemSusChem*, Vol. 3, No. 7, pp. 782–796, 2010.
50. Y. Sun, Q. Wu, and G. Shi, Graphene based new energy materials, *Energy & Environmental Science*, Vol. 4, No. 4, pp. 1113–1132, 2011.
51. Y. Zhu, S. Murali, W. Cai, X. Li, J. W. Suk, J.R. Potts, and R.S. Ruoff, Graphene and graphene oxide: Synthesis, properties, and applications, *Advanced Materials*, Vol. 22, No. 35, pp. 3906–3924, 2010.
52. X.-Y. Fan, R. Nouchi, L.-C. Yin, and K. Tanigaki, Effects of electron-transfer chemical modification on the electrical characteristics of graphene, *Nanotechnology*, Vol. 21, No. 47, p. 475208, 2010.
53. A. Shukla and J. Mazher, Method of fastening lamellae of a lamellar material on a suitable substrate, Jan. 3 2013. US Patent 20,130,000,833.
54. M. Pumera, Graphene-based nanomaterials and their electrochemistry, *Chemical Society Reviews*, Vol. 39, No. 11, pp. 4146–4157, 2010.

55. S. Sahoo, Quantum hall effect in graphene: Status and prospects, *Indian Journal of Pure and Applied Physics*, Vol. 49, No. 6, p. 367, 2011.
56. Y.-M. Lin, V. Perebeinos, Z. Chen, and P. Avouris, Electrical observation of subband formation in graphene nanoribbons, *Physical Review B Condensed Matter and Materials Physics*, Vol. 78, No. 16, p. 161409R, 2008.
57. B. Özyilmaz, P. Jarillo-Herrero, D. Efetov, D.A. Abanin, L.S. Levitov, and P. Kim, Electronic transport and quantum hall effect in bipolar graphene pnp junctions, *Physical Review Letters*, Vol. 99, No. 16, p. 166804, 2007.
58. M. Zwierzycki and S. Krompiewski, Ab initio study of the edge states of graphene nanoribbons in the presence of electrodes, *Acta Physica Polonica-Series A General Physics*, Vol. 118, No. 5, p. 856, 2010.
59. J.E. Padilha, R.B. Pontes, A.J. Da Silva, and A. Fazzio, Ixv curves of boron and nitrogen doping zigzag graphene nanoribbons, *International Journal of Quantum Chemistry*, Vol. 111, No. 7-8, pp. 1379–1386, 2011.
60. R. Faccio, P.A. Denis, H. Pardo, C. Goyenola, and Á.W. Mombrú, Mechanical properties of graphene nanoribbons, *Journal of Physics: Condensed Matter*, Vol. 21, No. 28, p. 285304, 2009.
61. L. Jiao, X. Wang, G. Diankov, H. Wang, and H. Dai, Facile synthesis of high-quality graphene nanoribbons, *Nature nanotechnology*, Vol. 5, No. 5, pp. 321–325, 2010.
62. N. Kharche, Y. Zhou, K.P. O?Brien, S. Kar, and S.K. Nayak, Effect of layer stacking on the electronic structure of graphene nanoribbons, *ACS Nano*, Vol. 5, No. 8, pp. 6096–6101, 2011.
63. K. Nakada, M. Fujita, G. Dresselhaus, and M.S. Dresselhaus, Edge state in graphene ribbons: Nanometer size effect and edge shape dependence, *Physical Review B*, Vol. 54, No. 24, p. 17954, 1996.
64. K.D. Sattler, *Handbook of Nanophysics: Clusters and Fullerenes, Vol.* 2. CRC press, 2010.
65. G. Abadir, K. Walus, and D. Pulfrey, Bias-dependent amino-acid-induced conductance changes in short semi-metallic carbon nanotubes, *Nanotechnology*, Vol. 21, No. 1, p. 015202, 2010.
66. T. Li, S. Chang, and M.-F. Lin, Electron transport in nanotube-ribbon hybrids, *The European Physical Journal B*, Vol. 70, No. 4, pp. 497–505, 2009.
67. S. Dutta and S.K. Pati, Novel properties of graphene nanoribbons: A review, *Journal of Materials Chemistry*, Vol. 20, No. 38, pp. 8207–8223, 2010.
68. O.V. Yazyev and S.G. Louie, Topological defects in graphene: Dislocations and grain boundaries, *Physical Review B*, Vol. 81, No. 19, p. 195420, 2010.
69. C. Tao, L. Jiao, O.V. Yazyev, Y.-C. Chen, J. Feng, X. Zhang, R.B. Capaz, J.M. Tour, A. Zettl, S.G. Louie, et al., Spatially resolving edge states of chiral graphene nanoribbons, *Nature Physics*, Vol. 7, No. 8, pp. 616–620, 2011.
70. X. Zhou and Z. Liu, A scalable, solution-phase processing route to graphene oxide and graphene ultralarge sheets, *Chemical Communications*, Vol. 46, No. 15, pp. 2611–2613, 2010.
71. B. Babic and C. Schönenberger, Observation of fano resonances in single-wall carbon nanotubes, *Physical Review B*, Vol. 70, No. 19, p. 195408, 2004.
72. U. Fano, Effects of configuration interaction on intensities and phase shifts, *Physical Review*, Vol. 124, No. 6, p. 1866, 1961.

73. H. Zheng, Z. Wang, T. Luo, Q. Shi, and J. Chen, Analytical study of electronic structure in armchair graphene nanoribbons, *Physical Review B*, Vol. 75, No. 16, p. 165414, 2007.
74. A.E. Miroshnichenko, S. Flach, and Y.S. Kivshar, Fano resonances in nanoscale structures, *Reviews of Modern Physics*, Vol. 82, No. 3, p. 2257, 2010.
75. A. Nitzan and M.A. Ratner, Electron transport in molecular wire junctions, *Science*, Vol. 300, No. 5624, pp. 1384–1389, 2003.
76. Z. Dai, X. Shi, X. Zheng, X. Wang, and Z. Zeng, Ab initio investigations of the transport properties of a ge_ {7} cluster, *Physical Review B*, Vol. 75, No. 15, p. 155402, 2007.
77. Y. Lee, M. Brandbyge, M. Puska, J. Taylor, K. Stokbro, and R. Nieminen, Electron transport through monovalent atomic wires, *Physical Review B*, Vol. 69, No. 12, p. 125409, 2004.
78. Y. Li, J. Yao, C. Liu, and C. Yang, Theoretical investigation on electron transport properties of a single molecular diode, *Journal of Molecular Structure: THEOCHEM*, Vol. 867, No. 1, pp. 59–63, 2008.
79. J. Le Rouzo, I. Ribet-Mohamed, N. Guérineau, R. Hadar, M. Tauvy, E. Rosencher, and S. Chuang, Van hove singularities in intersubband transitions in multiquantum well photodetectors, Äpplied *Physics Letters*, Vol. 88, No. 9, pp. 091117–091117, 2006.
80. Y.J. Sun, S. Souma, W.J. Li, T. Sato, X. G. Zhu, G. Wang, X. Chen, X. C. Ma, Q.K. Xue, J.F. Jia, et al., Van hove singularities as a result of quantum confinement: The origin of intriguing physical properties in pb thin films, *Nano Research*, Vol. 3, No. 11, pp. 800–806, 2010.
81. M. Kemerink, P. Koenraad, P. Christianen, A. Geim, J. Maan, J. Wolter, and M. Henini, Enhancement of spin-dependent hole delocalization in degenerate asymmetric double quantum wells, *Physical Review B*, Vol. 53, No. 15, p. 10000, 1996.
82. R. Waser, *Nanoelectronics and Information Technology: Advanced Electronic Materials and Novel Devices*. John Wiley & Sons Ltd, Chichester, 2005.
83. R.M. Dreizler and E. Engel, *Density Functional Theory: An Advanced Course, Theoretical and Mathematical Physics*. Springer-Verlag Berlin Heidelberg, 2011.
84. C.J. Cramer, *Essentials of Computational Chemistry: Theories and Models*. Wiley, 2005.
85. B. Tong and L. Sham, Application of a self-consistent scheme including exchange and correlation effects to atoms, *Physical Review*, Vol. 144, No. 1, p. 1, 1966.
86. W. Kohn, A.D. Becke, and R.G. Parr, Density functional theory of electronic structure, *The Journal of Physical Chemistry*, Vol. 100, No. 31, pp. 12974–12980, 1996.
87. D. Sholl and J.A. Steckel, *Density Functional Theory: A Practical Introduction*. Wiley-Interscience, 2011.
88. Z. Ao and Q. Jiang, Density functional theory calculations on interface structures and adsorption properties of graphenes: A review, *Open Nanoscience Journal*, Vol. 3, pp. 34–55, 2009.
89. R.J. Bartlett, V.F. Lotrich, and I.V. Schweigert, Ab initio density functional theory: The best of both worlds?, *The Journal of Chemical Physics*, Vol. 123, p. 062205, 2005.

90. A. Pecchia, G. Penazzi, L. Salvucci, and A. Di Carlo, Non-equilibrium green's functions in density functional tight binding: Method and applications, *New Journal of Physics*, Vol. 10, No. 6, p. 065022, 2008.
91. P.G. Sankar, Mechanical and electrical properties of single walled carbon nanotubes: A computational study, *European Journal of Scientific Research*, Vol. 60, No. 3, pp. 342–358, 2011.
92. Z. Yang, B. Wen, R. Melnik, S. Yao, and T. Li, Geometry dependent current-voltage characteristics of zno nanostructures: A combined nonequilibrium green's function and density functional theory study, *Applied Physics Letters*, Vol. 95, No. 19, pp. 192101–192101, 2009.
93. J.M. Soler, E. Artacho, J.D. Gale, A. Garcá, J. Junquera, P. Ordejón, and D. Sánchez-Portal, The siesta method for ab initio order-n materials simulation, *Journal of Physics: Condensed Matter*, Vol. 14, No. 11, p. 2745, 2002.
94. M. Brandbyge, J. Mozos, P. Ordejon, J. Taylor, and K. Stokbro, Density-functional method for nonequilibrium electron transport, *Physical Review B: Condensed Matter and Materials Physics*, Vol. 65, No. 11, p. 165401, 2002.
95. M. Summerfield, *Programming in Python 3: A Complete Introduction to the Python Language*. Addison-Wesley Professional, 2009.
96. M.F. Sanner et al., Python: A programming language for software integration and development, *J Mol Graph Model*, Vol. 17, No. 1, pp. 57–61, 1999.
97. G. Van Rossum and F.L. Drake, *The Python Language Reference Manual*. Network Theory Ltd., 2011.
98. Q. Han, B. Cao, L. Zhou, G. Zhang, and Z. Liu, Electrical transport study of single-walled zno nanotubes: A first-principles study of the length dependence, *The Journal of Physical Chemistry C*, Vol. 115, No. 8, pp. 3447–3452, 2011.
99. B. Larade, J. Taylor, H. Mehrez, and H. Guo, Conductance, iv curves, and negative differential resistance of carbon atomic wires, *Physical Review B*, Vol. 64, No. 7, p. 075420, 2001.
100. J.-L. Mozos, C. Wan, G. Taraschi, J. Wang, and H. Guo, Quantized conductance of si atomic wires, *Physical Review B*, Vol. 56, No. 8, pp. R4351–R4354, 1997.
101. N. Lang and P. Avouris, Carbon-atom wires: charge-transfer doping, voltage drop, and the effect of distortions, *Physical Review Letters*, Vol. 84, No. 2, pp. 358–361, 2000.
102. D. Stauffer and A. Aharony, Introduction to percolation theory, 1992.
103. I. Osaka, R. Zhang, G. Sauvé, D.-M. Smilgies, T. Kowalewski, and R.D. McCullough, High-lamellar ordering and amorphous-like π-network in short-chain thiazolothiazole- thiophene copolymers lead to high mobilities, *Journal of the American Chemical Society*, Vol. 131, No. 7, pp. 2521–2529, 2009.
104. B. Hamadani, D. Gundlach, I. McCulloch, and M. Heeney, Undoped polythiophene field-effect transistors with mobility of 1 $cm^2V^{-1}s^{-1}$, *Applied Physics Letters*, Vol. 91, No. 24, pp. 243512–243512, 2007.
105. J.A. Fürst, M. Brandbyge, A.-P. Jauho, and K. Stokbro, Ab initio study of spin-dependent transport in carbon nanotubes with iron and vanadium adatoms, *Physical Review B*, Vol. 78, No. 19, p. 195405, 2008.
106. L. Pauling, The nature of the chemical bond, *Journal of the American Chemical Society*, Vol. 53, No. 4, pp. 1367–1400, 1931.

107. P.J. Brown, D.S. Thomas, A. Köhler, J.S. Wilson, J.-S. Kim, C.M. Ramsdale, H. Sirringhaus, and R.H. Friend, Effect of interchain interactions on the absorption and emission of poly (3-hexylthiophene), *Physical Review B*, Vol. 67, No. 6, p. 064203, 2003.
108. M. Brandbyge, M.R. Sørensen, and K.W. Jacobsen, Conductance eigenchannels in nanocontacts, *Physical Review B*, Vol. 56, No. 23, p. 14956, 1997.
109. B. Raquet, M. Goiran, N. Negre, J. Leotin, B. Aronzon, V. Rylkov, and E. Meilikhov, Quantum size effect transition in percolating nanocomposite films, *Physical Review B*, Vol. 62, No. 24, p. 17144, 2000.
110. J. Park, H. Yang, K.-S. Park, and E.-K. Lee, Effects of nonmagnetic impurities on the spin transport property of a graphene nanoribbon device," *The Journal of Chemical Physics*, Vol. 130, p. 214103, 2009.
111. H. Cheraghchi and K. Esfarjani, Negative differential resistance in molecular junctions: Application to graphene ribbon junctions, *Physical Review B*, Vol. 78, No. 8, p. 085123, 2008.
112. J. Taylor, H. Guo, and J. Wang, Ab initio modeling of open systems: Charge transfer, electron conduction, and molecular switching of a C^{60} device, *Physical Review B*, Vol. 63, No. 12, p. 121104, 2001.
113. Y.-W. Son, M.L. Cohen, and S.G. Louie, Energy gaps in graphene nanoribbons, *Physical Review Letters*, Vol. 97, No. 21, p. 216803, 2006.
114. D.V. Kosynkin, A.L. Higginbotham, A. Sinitskii, J.R. Lomeda, A. Dimiev, B.K. Price, and J.M. Tour, Longitudinal unzipping of carbon nanotubes to form graphene nanoribbons, *Nature*, Vol. 458, No. 7240, pp. 872–876, 2009.

15

Analytical Modeling of Thin-Film Solar Cells – Fundamentals and Applications

Kurt Taretto

The National University of Comahue – National Council of Science and Technology (CONICET), Neuquen, Argentina

Abstract

Analytical modeling is a key tool in solar cell design and characterization. This chapter introduces fundamental concepts, equations and strategies applicable to thin-film solar cell modeling. Assuming knowledge of basic semiconductor and junction theory, models for key solar cell structures are developed, with a special focus on the formulation of closed-form equations of current-voltage characteristics. Models for the horizontal *pn* junction, vertical *pn* multijunction, nanorod multijunction and *pin* junction are derived on the basis of a universal set of fundamental equations, showing their salient features and establishing comparisons between the different designs. Relevant considerations for the interpretation of experimental solar cell data are also drawn from the different models.

Keywords: Solar cell modeling, *pn* junction, vertical junction, multijunction, nanorod solar cell, nanowire solar cell, pin solar cell, p-i-n solar cell, efficiency, current-voltage characteristics, semiconductor equations, tunneling recombination

15.1 Introduction

The optimization of solar cells requires characterization and modeling techniques that allow the detection of key power loss

**Corresponding author*: kurt_taretto@yahoo.com

Atul Tiwari, Rabah Boukherroub, and Maheshwar Sharon (eds.) Solar Cell Nanotechnology, (409–446) 2014 © Scrivener Publishing LLC

mechanisms. Depending on the required description of the physics governing a specific device, one may resort to numerical modeling for a high detail level, or to analytical modeling, with a lower degree of physical depth. Both strategies have strengths and weaknesses: numerical solutions may give very complete, closely approximated physical pictures, but at the cost of a rather large number of simulation parameters, which are often not known. Analytical models require fewer parameters, and often yield closed-form equations that explain the behavior of the solar cell in a manageable way, sometimes at the cost of a more confined range of validity (e.g. temperature, solar cell dimensions, material parameters, etc.). However, while numerical simulations are necessarily confined to particular solutions, analytical models offer a generic mathematical description of the device without specifying its material, temperature, etc., and are therefore the starting point for understanding and predicting the response and primal limitations of solar cells.

This contribution attempts a walkthrough on the fundamentals of analytical solar cell modeling, developing the equations for thin-film homo- and heterojunction *pn* cells, vertical multijunction (leading to vertical nanorod cells), and *pin*-type solar cells. Although the basic theory applies to solar cells in general, particular physics that should not be neglected at the nanoscale are highlighted along the text. The reader should be familiarized with elemental semiconductor material and *pn*-junction theory, including Boltzmann statistics.

The next section introduces typical solar cell configurations, describes the electrical output parameters and the equivalent circuit. In Section 15.3 the basic semiconductor equations are given, establishing fundamental solar cell design criteria, while Section 15.4 develops analytical models for the mentioned solar cell designs.

15.2 Basics

The first task upon physical solar cell analysis is to sketch the energy band diagram (EBD) of the device. Since an electrical current is established when electrons and holes move oppositely, solar cells resort to an asymmetric EBD configuration that induces charge separation. Figure 15.1 shows three EBDs found in solar cells ranging from bulk silicon to thin-film organic semiconductor, and nanorod solar cells. Part (a) shows the band diagram of a *pn*-homojunction,

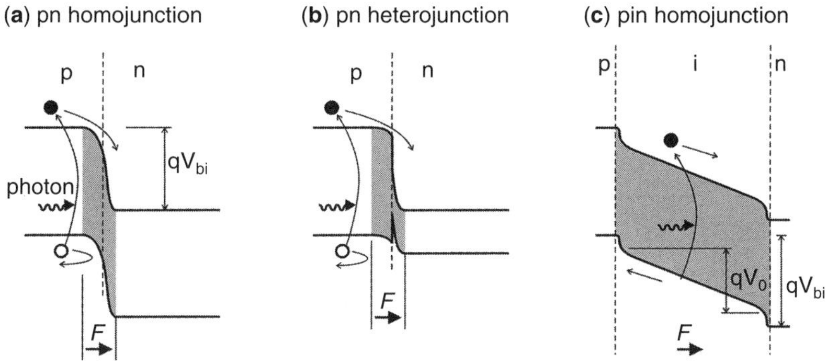

Figure 15.1 Energy band diagrams for the (a) *pn* homojunction solar cell, (b) *pn* heterojunction solar cell, and (c) *pin* solar cell, showing a photogenerated electron-hole pair that is separated by virtue of the asymmetry of the bands, followed by movements of electron and hole in opposite directions towards the contacts (not shown) at the limits of the cell.

where *p*- and *n*-type doped semiconductors are brought into contact, resulting in a step-like EBD with built-in voltage V_{bi}. The asymmetry that induces separation of photogenerated charge carriers is precisely the energy step: electrons photogenerated at the *p*-side perceive the nearby *n*-side as a sink, while the opposite is valid for holes. Thus, electrons photogenerated in the *p*-type region diffuse to the *n*-region (to the right in Figure 15.1) and finally reach the electron contact (not shown in the figure), where they exit the cell towards the load. Electrons generated in the *n*-type region can only move towards the electron contact, readily exiting the cell. Holes diffuse in the opposite direction (to the left in Figure 15.1) and flow out through the hole contact.

The homojunction uses only one semiconductor material with differently doped regions. Figure 15.1b shows a *pn* heterojunction, where the junction is formed by two materials with different bandgap, again profiting from asymmetry for charge separation.

Since *pn* junctions imply the formation of a space-charge region (SCR), an electrical field *F* appears around the SCR, shown in gray in Figure 15.1. Typical widths of the SCR range from 100 nm to 500 nm. Away from the junction, where the energy bands are horizontal, *F* vanishes. Although not fundamental for charge separation, the electric force within the SCR also contributes to charge separation.

Charge separation by the electrical field is strongly exploited in the *pin* solar cell shown in Figure 15.1c, where an intrinsic (*i*) layer is placed between an *n*- and a *p*-type region, or, alternatively, between two conductors with different work functions. Since intrinsic material lacks on sufficient free carriers to develop a junction with a doped layer, the electric field runs from the *n*-layer to the *p*-layer (or alternate contact layers), traversing the intrinsic layer. Normally, the *i*-layer is made much thicker than the doped layers so that most electron-hole pairs are photogenerated there, the field becoming the main source for charge separation.

The realization of any of the junctions explained above is possible under a variety of device geometries. Figure 15.2 shows cross sections of a horizontal *pn* solar cell (a), a horizontal *pin* (b) and a vertical multijunction *pn* solar cell (c), including contact layers. In all three examples, the top contact is the electron extracting contact, while the bottom contact extracts holes. The top contact is a transparent conductive material, while the back contact is a metallic, light-reflecting material. The grey areas represent the SCRs, where the electric field is found. The cell shown in Figure 15.2c consists of periodically repeating *n* and *p* layers, effectively connected in parallel, making up a vertical multijunction solar cell.

The absorbed photons produce electron-hole pairs that contribute to photocurrent, if they are not lost by recombination before reaching the contacts. The main power loss processes are sketched in Figure 15.3a. The input light power suffers from optical losses (insufficient light absorption, reflection, etc.), leaving some available

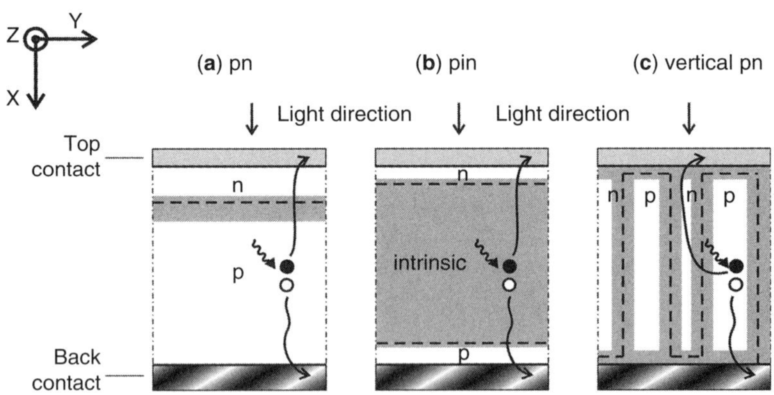

Figure 15.2 Cross sections of (a) a horizontal *pn* junction, (b) a *pin* junction and, (c) a vertical multijunction solar cell. Light comes from above and absorbed photons produce electron-hole pairs that may be separated and contribute to photocurrent.

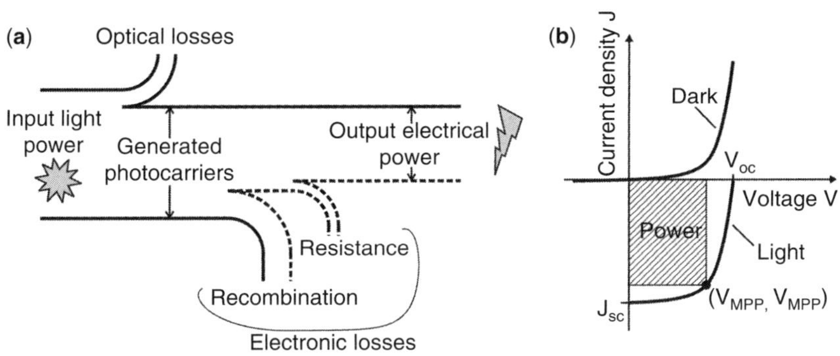

Figure 15.3 Power flow diagram of a solar cell (a), showing the incoming power partly lost by optical losses, recombination, and resistance effects. The remaining power is obtained in form of an electrical current flowing through some external load. The corresponding current density (*J*)/voltage(*V*) characteristics is shown in part (b) under dark and light conditions.

flux of photogenerated carriers, which are partly lost to recombination, or dissipate power through resistive effects. The remaining charge carriers do output the solar cell and deliver electrical power to an external load. The dashed lines represent the fact that electronic losses depend on the voltage *V* at the terminals, and should therefore be looked at as "mobile boundaries" in the diagram. The electronic losses increase with the voltage between the terminals of the solar cell (higher load resistance).

Following Figure 15.3b, in the dark, the *J(V)* curve obeys an exponential law (termed "diode law") and corresponds to the recombination current of the externally injected electrons and holes. Upon light incidence, the photocurrent "shifts" the dark *J(V)* curve to the fourth quadrant, denoting power delivery. In some cases, the photocurrent depends on voltage, and therefore the application of light does not merely imply a homogeneous shift of the dark *J(V)* curve. This is known as voltage-dependent photocarrier collection, a phenomenon specially noticed in thin-film devices.

At the maximum power point (MPP) of the *J(V)* curve, the corresponding current density J_{MPP} and voltage V_{MPP} yield the maximum power density $P_{MPP} = J_{MPP} \times V_{MPP}$, defining the fill factor *FF*

$$FF = \frac{V_{MPP} J_{MPP}}{V_{OC} J_{SC}}, \tag{15.1}$$

where V_{OC} is the open-circuit voltage and J_{SC} the short-circuit current density. With the incident light power density P_{ph}, the power conversion efficiency η reads

$$\eta = \frac{V_{OC}\left|J_{SC}\right|FF}{P_{ph}}. \tag{15.2}$$

The analytical description of any given solar cell starts off with its circuital model, describing the *J(V)* curve, which is developed below.

15.2.1 Equivalent Solar Cell Circuit and Current-Voltage Equation

The *J(V)* curve can be described by a circuit of lumped electronic components, as shown in the schematic of Figure 15.4. The current source represents the photocurrent density J_{ph} delivered by the illuminated cell, the 1,2…*N* diodes correspond to the current loss by recombination of electron-hole pairs, now decomposed into *N* possible physical mechanisms. We also include resistance losses through the specific shunt (parallel) and series resistances R_p and R_S, respectively, here assumed as lumped elements.

The *J(V)* equation obeying the circuit in Figure 15.4 is expressed by

$$J = \sum_{j=1}^{N} J_{0,j}(V)\left(\exp\left(\frac{V - JR_S}{n_{id,j}V_t}\right) - 1\right) + \frac{V - JR_S}{R_P} - J_{ph}(V). \tag{15.3}$$

The second term represents the shunt current through R_p, while the sum is the total, dark current through the diodes, where $J_{0,j}$ is

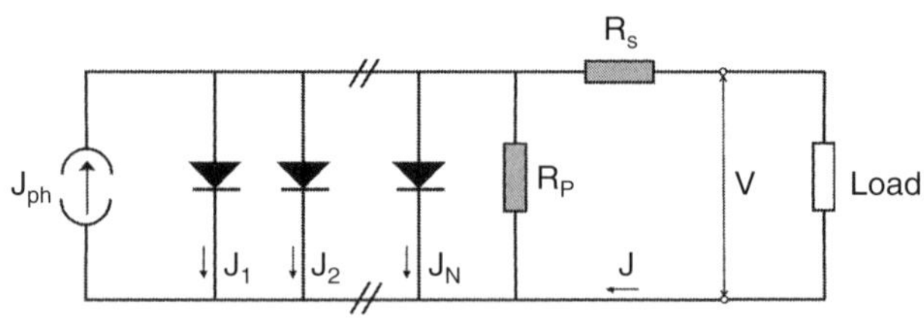

Figure 15.4 Equivalent circuit model for a generic solar cell, described as a combination of a current source, several diodes connected in parallel, specific shunt and series resistances R_p and R_S, respectively, and the load.

the reverse saturation current density of the j-th diode, $n_{id,j}$ is the j-th diode ideality factor (sometimes denoted by an 'A'), V_t the thermal voltage given by $V_t = kT/q$, where k is Boltzmann's constant, T the cell temperature, and q the elemental electric charge. Although Eq. 15.3 shows an explicit voltage dependence in the exponential term only, in general, $J_{0,j}$ and J_{ph} depend on voltage.

Experimentally, it is common practice to deduce an effective ideality n_{id}^{eff} at forward bias (e.g. for $V > 3V_t$) by plotting

$$n_{id}^{eff}(V) = \frac{1}{V_t}\frac{dV}{d\ln(J)}, \tag{15.4}$$

which implicitly rationalizes the sum of the individual diode currents in Eq. 15.3 as a single, effective dark current with voltage-dependent n_{id}^{eff}. When comparing the obtained n_{id}^{eff} to theoretical n_{id} values, it is often possible to infer the dominating recombination mechanism. For example, in standard silicon solar cells, a transition from $n_{id}^{eff} \cong 2$ at low voltages to $n_{id}^{eff} \cong 1$ at higher voltages around V_{OC} is observed. The physical models (developed below) predict that this transition corresponds to a dominating recombination in the SCR at low voltage to recombination in the neutral p and n regions at higher voltages.

The number N of diodes necessary to describe $J(V)$ curves of actual solar cells depends on the specific material, device design, operating conditions (such as temperature or illumination), or preparation conditions. But it often turns out that one or two exponential terms from Eq. 15.3 are sufficient to reproduce $J(V)$ curves accurately. This leads to the ideal single diode model ($N = 1$), idealized with no resistance losses and voltage independent n_{id}, J_0 and J_{ph}. It constitutes a first-order tool to quantify solar cell loss mechanisms. The $J(V)$ equation of the ideal single diode solar cell reads

$$J(V) = J_0\left[\exp\left(\frac{V}{n_{id}V_t}\right) - 1\right] - J_{ph}. \tag{15.5}$$

The electrical output parameters that go into the efficiency η are readily extracted from this expression. At $J = 0$, V_{OC} is given by

$$V_{OC} = n_{id}V_t \ln\left(1 + J_{ph}/J_0\right) \cong n_{id}V_t \ln\left(J_{ph}/J_0\right), \tag{15.6}$$

while at $V = 0$ we have the short circuit current density

$$J_{SC} = -J_{ph} \,. \tag{15.7}$$

Since in practice, under solar illumination conditions, it holds that $J_0 << J_{ph}$ and the ideality is of the order of $n_{id} \cong 1..3$, reducing J_0 by as many orders of magnitude as possible is crucial to solar cell efficiency. The incidence of n_{id} is further discussed below.

The power density $P = V \times J$ predicted by the ideal single diode equation delivers the maximum delivered power point (MPP) at the voltage V_{MPP} according to [1]

$$V_{MPP} \cong V_{OC} - n_{id} V_t \ln\left(1 + V_{OC} / n_{id} V_t\right), \tag{15.8}$$

and the current density at the *MPP* according to

$$J_{MPP} \cong -\frac{J_{ph}}{1 + n_{id} V_t / V_{OC}}, \tag{15.9}$$

which contains the ideal maximum $J_{MPP} = J_{SC}$. These equations show that n_{id} reduces both, J_{MPP} and V_{MPP}, and hence also the efficiency η, while higher V_{OC} leads to higher η through both J_{MPP} and V_{MPP}. With Eqs. 15.8 and 15.9 the fill factor is given by the approximate expression

$$FF \cong \frac{V_{OC} / n_{id} V_t - \ln\left(1 + V_{OC} / n_{id} V_t\right)}{1 + V_{OC} / n_{id} V_t}, \tag{15.10}$$

which delivers *FF* with a relative error of 12% to 4% in the range of $V_{OC} / n_{id} V_t$ from 10 to 25, respectively. We see that *FF* depends solely on V_{OC} and n_{id}, both representative of recombination losses.

So far we analyzed the case of the ideal, single junction solar cell with no resistance losses. When dealing with non-ideal, specially with experimental, non-optimized devices, the one-diode model must include the resistance losses in order to explain experimental *J(V)* curves. Despite its simplicity, the resulting "real" one diode model succeeded in fitting *J(V)* curves of solar cells ranging from nanoscale bulk heterojunction organic, to fine-grained, thin-film $CuInGaSe_2$ solar cells or even high-bandgap GaAs junctions. Further, explicit equations for the *MPP* parameters are also found including resistance losses [1], allowing for handy predictions on efficiency improvements attainable through individual or combined improvements in R_S, R_P, J_0, n_{id}, and J_{ph}.

A more profound understanding of the incidence of material quality and cell design on efficiency is only possible through physical models, which are derived from the fundamental semiconductor equations, given below.

15.3 Fundamental Semiconductor Equations

15.3.1 Electric Field and Free Carrier Currents

We now deploy the fundamental semiconductor equations in one spatial dimension, since the most practical analytical models are restricted to this situation. The first equation to be considered is the stationary Maxwell equation relating the electric field F to the local charge concentration r, known as Poisson's equation [2]

$$\frac{dF}{dx} = \frac{\rho}{\varepsilon_0 \varepsilon_S}, \tag{15.11}$$

where

$$\rho = q\left(p - n + N_{dopants} + N_{defects}\right), \tag{15.12}$$

being p the concentration of holes in the valence band of the semiconductor, and n the concentration of electrons in the conduction band. Taking into account atomic doping with impurities, $N_{dopants}$ represents the net (positively and negatively) ionized impurity concentration. Atomic defects or impurities capable of trapping charges are included by the net defect concentration $N_{defects}$.

The electric field can be expressed is given by $F = -d\psi / dx$, where ψ is the electrostatic potential, which is related to the energy bands through $\psi = -E / q$, where E is the potential energy of electrons read on an EBD.

The free carrier concentrations n and p in Eq. 15.12 can be written as a function of densities of band states and quasi-Fermi energies. According to Boltzmann statistics,

$$n = N_C \exp\left(\frac{E_{Fn} - E_C}{kT}\right), \tag{15.13}$$

and

$$p = N_V \exp\left(\frac{E_V - E_{Fp}}{kT}\right), \tag{15.14}$$

where $N_C(N_V)$ is the effective conduction(valence) band density of states, $E_{Fn}(E_{Fp})$ the electron(hole) quasi-Fermi level, and $E_C(E_V)$ the energy at the edge of the conduction(valence) band, being the electronic bandgap defined by $E_g = E_C - E_V$.

Under thermodynamic equilibrium, e.g., a semiconductor in the dark with no charge injection from external sources, the quasi-Fermi levels join into a single Fermi level E_F, i.e. $E_F = E_{Fn} = E_{Fp}$. With the equilibrium electron and hole concentrations termed n_0 and p_0, respectively,

$$n_0 p_0 = N_C N_V \exp(-E_g / kT) = n_i^2, \tag{15.15}$$

being n_i defined as the intrinsic carrier concentration, a constant for a given material at a given temperature. Out of equilibrium, Eqs. 15.13 and 15.14 yield the *np* product

$$np = n_i^2 \exp\left(\frac{E_{Fn} - E_{Fp}}{kT}\right). \tag{15.16}$$

Importantly, in semiconductor junctions, the energy difference $E_{Fn} - E_{Fp}$, termed "quasi-Fermi level splitting," is related to the voltage V between the terminals of the junction. The quasi-Fermi level splitting of **majority carriers** at each contact obeys [3]

$$qV = E_{Fn} - E_{Fp}. \tag{15.17}$$

In particular, the open circuit voltage V_{OC} corresponds to a majority carrier quasi-Fermi level splitting of qV_{OC} between contacts.

Low-injection conditions are satisfied in devices where, under operation, the increases in carrier concentrations upon illumination or polarization (bias application to a junction) verify $\Delta n, \Delta p << n_0$ for *n*-type regions and $\Delta n, \Delta p << p_0$ in *p*-type regions.

The electrical current flowing in a semiconductor is rationalized as a flow of conduction band electrons and valence band holes, with the corresponding local electron and hole current densities J_n and J_p, respectively. In solar cells, the forces acting on carriers are due to the electrical field and concentration gradients. Hence, the one-dimensional current densities are expressed by [2]:

$$
\begin{aligned}
J_n &= q\mu_n nF + qD_n \frac{dn}{dx}, \\
J_p &= q\mu_p pF - qD_p \frac{dp}{dx},
\end{aligned}
\tag{15.18}
$$

where μ_n (μ_p) is the electron(hole) drift mobility, and D_n (D_p) is the electron(hole) diffusion constant, which are related by Einstein's relation according to $D_{n/p} = V_t \mu_{n/p}$.

The total current density is given by

$$
J = J_n + J_p, \tag{15.19}
$$

which for one-dimensional devices under steady state is constant for all x. Replacing by n and p given by Eqs. 15.13 and 15.14 and using Einstein's relation, J_n and J_p can be expressed as:

$$
\begin{aligned}
J_n &= \mu_n n \frac{dE_{Fn}}{dx}, \\
J_p &= \mu_p p \frac{dE_{Fp}}{dx}.
\end{aligned}
\tag{15.20}
$$

15.3.2 Steady-State Continuity Equations

The continuity principle for particle flux allows formalizing the steady-state balance between generation and recombination of electron and holes, which under steady-state is expressed by:

$$
\begin{aligned}
\frac{dJ_n}{dx} &= -q\left(G_{ph} - R_{total}\right), \\
\frac{dJ_p}{dx} &= q\left(G_{ph} - R_{total}\right).
\end{aligned}
\tag{15.21}
$$

Here, we introduce the optical photogeneration rate G_{ph}, which accounts only for the carriers generated by light absorption, while the net recombination rate R_{total} accounts for the difference between total carrier recombination and thermal generation.

Taking Poisson's equation, differentiating Eqs. 15.18 and replacing the derivatives in Eqs. 15.21, we obtain a system of three differential equations and three unknowns, namely $n(x)$, $p(x)$ and $F(x)$, which describe entirely the behavior of a given device (see also Eq. 15.25). Alternatively, if J_n and J_p are expressed in terms of the quasi-Fermi levels (cf. Eqs. 15.20), the unknowns become E_{Fn}, E_{Fp} and ψ.

Modeling the generation rate G_{ph} depends on the excitation radiation (mono- or polychromatic, non-polarized or polarized, etc.) and on the usually complex optical design of the solar cell. This leads to sophisticated modeling schemes [4] that are beyond the scope of this chapter. Here, we bear to the essence of photogeneration through basic approaches, adopting two photogeneration models: either a position-independent (i.e. G_{ph} = const.) or an exponentially decaying G_{ph} given by Beer's law [2]

$$G_{ph}(x) = \phi_0 a \exp(-ax), \tag{15.22}$$

which holds for a monochromatic, unpolarized plane-wave radiation reaching the cell surface at a normal angle, with a photon flux ϕ_0. Here, a is the absorption coefficient of the semiconductor, which depends on photon energy $h\nu$, and increases from its threshold energy called the optical gap E_g^{opt} towards higher $h\nu$. Despite its simplicity, Eq. 15.22 bears a fundamental design criterion: since up to the depth $x = a^{-1}$ about 2/3 of the photon flux is absorbed, it is convenient to define the **absorption length** by $L_a = a^{-1}$. Thus, efficient solar cells require an absorber thickness d of at least $d = L_a$ to absorb 2/3 of the radiation (at a given photon energy $h\nu$). Since the solar spectrum shows a distribution over $h\nu$, absorbers would basically require $d \geq L_a$, with L_a corresponding to the wavelength at the maximum photon flux in the spectrum.

The case of a position-independent G_{ph} occurs for weakly absorbed light situations, that is, $d << L_a$. In this case, light bounces several times between the front surface and the back reflector before being completely absorbed. Despite its simplicity, assuming G_{ph} = const. enables simple yet insightful solar cell models, in many cases valid even under solar operation conditions.

Turning to carrier recombination processes, we notice that several recombination mechanisms can occur concomitantly: radiative recombination, defect-level recombination, and Auger recombination. Under low injection conditions, each recombination mechanism adds independently to the total net recombination rate. Since recombination occurs at least with the presence of an electron and a hole, recombination rates are proportional to the product $n \times p$. This constitutes the main source of non-linearity in semiconductor device equations, and analytical solutions always require linearization strategies. Fortunately, under low injection conditions (defined above), electrons as minority carriers ($n << p$) recombine at a linear rate

$$R_{total} = \frac{n - n_0}{\tau_n}, \tag{15.23}$$

where τ_n is the electron lifetime. Analogously, holes as minorities ($p << n$) recombine at a rate

$$R_{total} = \frac{p - p_0}{\tau_p}, \tag{15.24}$$

being τ_p the hole lifetime. The recombination lifetimes $\tau_{n/p}$ are understood as effective lifetimes that contain all recombination mechanisms. If recombination takes place at a surface or interface, the recombination lifetimes are replaced by surface recombination velocities $S_{n/p}$, giving an areal (instead of volumetric) recombination rate.

Having linearized R_{total}, it is now possible to formulate linear differential equations for minority carriers under low injection conditions. Replacing J_n and J_p from Eqs. 15.18 in the continuity equations from Eqs. 15.21, we obtain [2]:

$$\begin{aligned} G_{ph} - \frac{n - n_0}{\tau_n} + n\mu_n \frac{dF}{dx} + \mu_n F \frac{dn}{dx} + D_n \frac{d^2 n}{dx^2} &= 0, \\ G_{ph} - \frac{p - p_0}{\tau_p} - p\mu_p \frac{dF}{dx} - \mu_p F \frac{dp}{dx} + D_p \frac{d^2 p}{dx^2} &= 0. \end{aligned} \tag{15.25}$$

Together with Poisson's equation, these expressions constitute the foundation for analytical solar cell modeling.

15.3.3 Considerations for Excitonic Solar Cells

Let us briefly mention that in excitonic solar cells, photogeneration produces excitons—free carriers are only obtained after exciton dissociation. The optical generation rate (producing excitons) must now be distinguished from the free-carrier generation rate. Evidently, excitonic solar cells require not only an asymmetry in the EBD for charge separation, but also some exciton-dissociating artifact. Normally, this is achieved through band discontinuities occurring at interfaces between different semiconductors: if the discontinuity energy is higher than the exciton binding energy, then efficient exciton dissociation is possible. In conclusion, the fundamental equations require an additional diffusion and continuity equation for excitons [5, 6].

15.3.4 Characteristic Lengths

The equations above immediately deliver two major concepts in solar cell design. First, in a field-free material, carriers move only by diffusion. If excess minority carriers are steadily generated at a given location, they diffuse away from the point of generation, recombining as they move. The resulting carrier concentration profile is an exponential decay law (under low injection conditions). The decay constant is the **diffusion length** of minority carriers, given by

$$L_n = \sqrt{D_n \tau_n}, \tag{15.26}$$

for electrons, and by

$$L_p = \sqrt{D_p \tau_p}, \tag{15.27}$$

for holes. Thus, at a distance much farther away than $L_{n/p}$ from the point of photogeneration, no excess minority carriers are found. Alternatively, if carriers are homogeneously photogenerated in a solar cell based on diffusion, they can reach the junction and contribute significantly to photocurrent if the distance from the point of photogeneration to the junction is not larger than $L_{n/p}$.

A second fundamental case arises when considering transport of carriers under sufficiently strong electrical fields that diffusion turns negligible. Similarly to the diffusion case, this again yields an exponential decay of carrier concentration away from the point of photogeneration, and the decay constant is the **drift length** of minority carriers, given by [3]

$$\ell_n = \mu_n \tau_n |F|, \tag{15.28}$$

for electrons, and

$$\ell_p = \mu_p \tau_p |F| \tag{15.29}$$

for holes. Thus, solar cells based on drift require values of $\ell_{n/p}$ comparable to the layer thickness in order to contribute significantly to photocurrent.

So far, we defined three fundamental lengths: absorption length L_a, diffusion lengths $L_{n/p}$ and drift lengths $\ell_{n/p}$. These lengths enable us to express two fundamental criterions for efficient solar cell design:

1. with an absorber thickness d, efficient light absorption requires $d \geq L_a$,
2. with a distance to the junction d_j, efficient photocarrier collection requires $L_{n/p} \geq d_j$ for a diffusion-based cell, or $\ell_{n/p} \geq d_j$ for a drift-based cell.

15.3.5 Tunneling Recombination Occurring at the Nanometer Scale

Tunneling recombination is a special case of defect recombination which can severely limit device performance, and must be specially considered in nanoscaled devices [7]. At strong electric fields appearing in highly doped *pn* or sufficiently thin *pin* junctions, carriers may tunnel into a defect energy level in the gap and recombine, as shown in Figure 15.5a. Such tunneling-enhanced recombination leads to an increased recombination rate that depends on $|F|$, and hence on the applied voltage V. This effect is introduced into the electron and hole lifetimes as seen by [8]

$$\tau_{n,p} = \frac{\tau_{0n,p}}{1+\Gamma_{n,p}}, \tag{15.30}$$

where $\tau_{0n,p}$ are the zero field band-to-defect transition times for electrons and holes, respectively, and the function gamma always verifies $\Gamma_{n,p} > 0$, accounting for the reduction of lifetime due to tunneling. Following ref. [8], for electrons or holes gamma is given by

$$\Gamma_{n,p} = 2\sqrt{3\pi}\frac{|F|}{F_\Gamma}\exp\left(\frac{|F|^2}{F_\Gamma^2}\right) \tag{15.31}$$

where $F_\Gamma = (24m^*)^{1/2}(kT)^{3/2}/q\hbar$, being m^* the effective tunneling mass for either electrons or holes. For defects in the SCR of a *pn* junction, *F* may be high enough to yield $\Gamma_{n,p} >> 1$, strongly reducing lifetime. Figure 15.5b shows the ratio $\tau_{n,p}/\tau_{0n,p}$ as a function of $|F|$, with the ratio between m^* and the free electron mass m_0 as a parameter. The shaded area corresponds to typical values of $|F|$ at zero bias occurring in SCRs of moderately-doped *pn* junctions, or *pin* junctions with similarly doped *n* and *p*-layers and *i*-layer thickness below 300 nm. Evidently, the combination of defects levels with electrical fields can strongly reduce carrier lifetime. In the

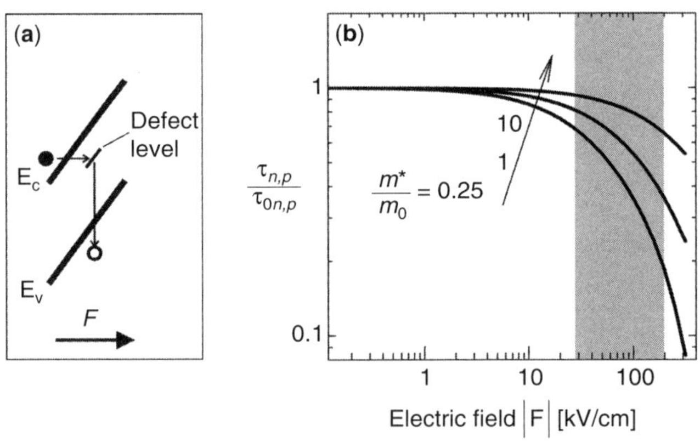

Figure 15.5 Elemental tunneling recombination process (a), where an electron tunnels into a defect level and then recombines with a hole. The tunneling-assisted recombination yields smaller carrier lifetimes $\tau_{n,p}$, depending on the intensity of the electrical field $|F|$, (b). The grey area extends over the range of electrical field values found in semiconductor junctions.

following models, it is implicitly assumed that carrier lifetime can be limited by tunneling recombination.

15.4 Analytical Models for Selected Solar Cells

15.4.1 Horizontal PN Junction

Figure 15.6a shows the cross section of a *pn* solar cell with the junction parallel to the cell surface, i.e., a horizontal single junction (HSJ). Assuming that the cell extends infinitely in *z* and *y*, a one-dimensional model using the above equations is sufficient. In Figure 15.6a we also define the total cell thickness d, the SCR width W, the width of the neutral $p(n)$-region $d_p(d_N)$, and the front(back) surface recombination velocity of holes(electrons) $S_p(S_n)$. The layer facing sunlight (*n*-type in this case) has a smaller width than the back *p*-layer, placing the junction near the upper surface, where most carriers are photogenerated (according to Beer's law). Figure 15.6b shows the EBDs of the *pn* cell under thermodynamic equilibrium,

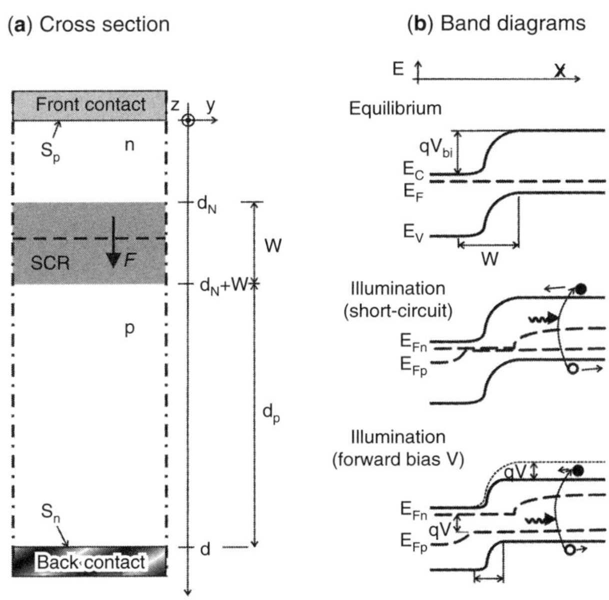

Figure 15.6 Cross sectional view of a slab of horizontal junction *pn* solar cell with the *n*-side oriented to the front of the cell (a) and, (b) energy band diagrams along the *x*-axis at equilibrium, illumination at short-circuit and with forward bias.

illumination at short-circuit and at some bias polarization, showing the Fermi levels in each case.

The electrical contacts are supposed to be designed in such a way that majority carriers experience only an ohmic resistance effect when flowing through them. This resistance is neglected in the physical model since it can be introduced later through R_S in the circuital model.

In order to formulate the continuity equations, the cell is rationalized in three parts, governed by different physics: the neutral part of the n-type region, the SCR, and the neutral part of the p-type region. The solution procedure that enables analytical solutions is to formulate the total current $J = J_n + J_p$ at one of the edges of the SCR, where the mathematical expressions for the solution turn out to be particularly simple. Taking for example the SCR edge from the n-type side, at $x = d_N$, the total current is given by

$$J = J_n(d_N) + J_p(d_N), \tag{15.32}$$

where J_n is majority carrier current and J_p minority carrier current at this location. In the neutral regions, it is possible to show that the minority carriers move by diffusion, while majority carriers move through a small drift field that implies a voltage drop much smaller than V. Since there is no analytical solution to this nearly negligible field in the neutral regions, the evaluation of the majority carrier currents is not straightforward, but must instead be circumvented using an approximation. This is achieved by expressing the continuity principle for electrons at $x = d_N$ in integral form, resulting

$$J_n(d_N) = J_n(d_N + W) + \int_{d_N}^{d_N+W} \frac{dJ_n}{dx} dx = J_n(d_N + W) - \int_{d_N}^{d_N+W} q(G_{ph} - R_{tot}) dx, \tag{15.33}$$

which is simply Kirchoff's current law applied to the electron current at $x = d_N$, where the integral represents the current lost by recombination in the SCR. We are now left with two minority carrier diffusion currents evaluated at each side of the SCR, namely $J_p(d_N)$ and $J_n(d_N + W)$, and the integral, which will be treated later. The diffusion currents are solved formulating the continuity Eqs. 15.25 with $F = 0$, yielding two independent second order ordinary differential equations. The following set of boundary conditions (BC) is assumed:

$$J_n(d) = qD_n \left.\frac{dn_p}{dx}\right|_d = -qS_n(n_p(d) - n_{p0}),$$
$$p_n(d_N) \cong p_{n0}\exp\left(\frac{V}{V_t}\right),$$
$$n_p(d_N + W) \cong n_{p0}\exp\left(\frac{V}{V_t}\right), \tag{15.34}$$
$$J_p(0) = -qD_p \left.\frac{dp_n}{dx}\right|_0 = -qS_p(p_n(0) - p_{n0}),$$

where the subscripts 'n' and 'p' in the carrier concentrations mean *n*- or *p*-type region, respectively. The first and fourth BC indicates that at the contacts, a flux of minority carriers is lost by recombination at a rate given by the recombination velocities S_n and S_p. The second and third BC are the strongest assumptions of the model, adopted from *pn*-diode theory in the dark. These indicate that at the edges of the SCR, n_p and p_n differ from the equilibrium values p_{n0} and n_{p0} by the factor $\exp(V/V_t)$. This corresponds to the shift by qV in E_C and E_V that must be introduced in Eqs. 15.13–15.14, assuming that E_{Fn} and E_{Fp} are nearly flat in the SCR (see Figure 15.6b). Although this is well understood in dark conditions, one might expect larger concentrations upon photogeneration. But if mobility is sufficiently high in the material, photogenerated carriers arriving at the edge of the SCR are pulled by the field instantaneously to the opposite side, avoiding accumulation [9]. The solutions *n(x)* and *p(x)* are of the form $y = f(x) + A\cosh(Bx) + C\sinh(Bx)$, and the boundary conditions enable to obtain the integration constants and finally evaluate $J_p(d_N)$ and $J_n(d_N + W)$ as diffusion currents.

The integral in Eq. 15.33 is obtained analytically by modeling R_{tot} assuming defect recombination using Shockley-Read-Hall's model with the defect energy near the center of the bandgap and having identical electron/hole capture times [2]. The integration of G_{ph} contributes $qG_{ph}W$ to the total photocurrent, while the integration of R_{tot} goes into the total dark current, as shown below.

Finally, the total current density in the *pn* solar cell is given by [3]

$$J = \underbrace{(J_{0,N} + J_{0,P})(e^{V/V_t} - 1) + J_{0,SCR}\sinh(V/2V_t)}_{J_{dark}(V)} - J_{photo}, \tag{15.35}$$

where we separated dark from light current components. The saturation current density from recombination in the SCR is given by

$$J_{0,SCR} = \frac{\pi}{2} \frac{qn_i W}{\tau_{SCR}} \frac{V_t}{V_{bi} - V} \tag{15.36}$$

where τ_{SCR} is the defect-recombination lifetime of carriers in the SCR, the built in voltage V_{bi} is given by Eq. 15.56 below, and W depends on V according to [3]

$$W = \sqrt{\frac{2\varepsilon_0\varepsilon_S}{q} \frac{N_D + N_A}{N_D N_A} \left(V_{bi} - V\right)}, \tag{15.37}$$

both equations holding for $V < V_{bi}$. Furthermore, the dependence of W with V can often be neglected in the evaluation of the $J(V)$ curve.

In Eq. 15.35, the saturation current densities from the quasi-neutral regions are given by

$$J_{0,N} = \frac{qD_p n_i^2}{N_D L_{p,inj}}, \tag{15.38}$$

and

$$J_{0,P} = \frac{qD_n n_i^2}{N_A L_{n,inj}}, \tag{15.39}$$

where we find the electron and hole injection diffusion lengths

$$L_{n,inj} = L_n \left(\frac{\cosh(d_P / L_n) + (S_n L_n / D_n)\sinh(d_P / L_n)}{\sinh(d_P / L_n) + (S_n L_n / D_n)\cosh(d_P / L_n)} \right), \tag{15.40}$$

and

$$L_{p,inj} = L_p \left(\frac{\cosh(d_N / L_p) + (S_p L_p / D_p)\sinh(d_N / L_p)}{\sinh(d_N / L_p) + (S_p L_p / D_p)\cosh(d_N / L_p)} \right) \tag{15.41}$$

The injection diffusion lengths can be shorter or longer than the bulk diffusion length, or than the layer thicknesses d_N, d_P (see next section). As a consequence, the saturation current densities $J_{0,N}$ and $J_{0,P}$, and thus the open-circuit voltage V_{OC}, are limited by surface recombination when $L_n > d_P$ and $L_p > d_N$.

Given that the SCR current shows an ideality $n_{id} = 2$ (cf. Eq. 15.35), and the currents from the neutral region exhibit $n_{id} = 1$, under forward bias ($V > 0$) the exponential term corresponding to the neutral regions is much larger than the SCR term. However, in, e.g. silicon *pn* cells, it holds that $(J_{0,N} + J_{0,P}) << J_{0,SCR}$, and then the first part of the total dark current (as a rule of thumb, below $V < V_{OC}/2$) is determined by the current in the SCR, while only at higher voltages the currents from the neutral regions dominate.

Now we proceed to the photocurrent in Eq. 15.35. Assuming a constant photogeneration rate, the photocurrent in Eq. 15.35 is given by

$$J_{photo} = qG_{ph}(L_{p,coll} + L_{n,coll} + W), \tag{15.42}$$

where we find the collection diffusion lengths

$$L_{p,coll} = L_p\left(\frac{\sinh(d_N / L_p) + (S_p L_p / D_p)\left(\cosh(d_N / L_p) - 1\right)}{\cosh(d_N / L_p) + (S_p L_p / D_p)\sinh(d_N / L_p)}\right), \tag{15.43}$$

and

$$L_{n,coll} = L_n\left(\frac{\sinh(d_P / L_n) + (S_n L_n / D_n)\left(\cosh(d_P / L_n) - 1\right)}{\cosh(d_P / L_n) + (S_n L_n / D_n)\sinh(d_P / L_n)}\right). \tag{15.44}$$

With increasing bulk diffusion lengths and decreasing surface recombination velocities, $L_{n,coll}$ and $L_{p,coll}$ saturate at the thicknesses d_P and d_N, respectively, yielding the maximum photocurrent

$$J_{photo}^{max} = qG_{ph}d, \tag{15.45}$$

being d the total cell thickness.

15.4.1.1 PN Junction under Moderately Absorbed Light

So far, we dealt with homogeneous generation, which is a reasonable assumption for weakly absorbed light, or $d << L_a$. In the regime of moderately absorbed light it holds that $d > L_a$, and the generation rate at normal incidence follows Eq. 15.22 for the case of monochromatic illumination.

With the boundary conditions again given by Eqs. 15.34, the same dark current as Eq. 15.35 is obtained in this case, while the photocurrent $J_{photo} = J_{photo,N} + J_{photo,P} + J_{photo,SCR}$ under monochromatic illumination is given by the components [3]

$$J_{photo,SCR}(V) = q\phi_0 \exp\left(-ad_N\right)\left(\exp\left(-aW\right)-1\right), \qquad (15.46)$$

for the SCR,

$$J_{photo,N} = q\phi_0 \frac{aL_p}{(aL_p)^2-1} \times \left(\frac{(S_pL_p/D_p + aL_p) - \left((S_pL_p/D_p)\cosh(d_N/L_p) + \sinh(d_N/L_p)\right)e^{-ad_N}}{(S_pL_p/D_p)\sinh(d_N/L_p) + \cosh(d_N/L_p)} - aL_pe^{-ad_N} \right), \qquad (15.47)$$

for the n-type region, and

$$J_{photo,P} = q\phi_0 \frac{aL_n}{(aL_n)^2-1} \exp\left(-a(d_N+W)\right) \times \left(aL_n - \frac{(S_nL_n/D_n)\left(\cosh(d_P/L_n) - e^{-ad_P}\right) + \sinh(d_P/L_n) + aL_ne^{-ad_P}}{(S_nL_n/D_n)\sinh(d_P/L_n) + \cosh(d_P/L_n)} \right), \qquad 15.48)$$

for the p-type region. In some cases, it may prove useful to write these in terms of the absorption length $L_a = a^{-1}$. The maximum photocurrent predicted within this model is obtained for vanishing $S_{n,p}$ and $d_{N,P}/L_{p,n} << 1$, $ad >> 1$, yielding the intuitive value $J_{photo}^{max} = q\phi_0$.

A particularly useful application of the above monochromatic, moderately absorbed light model is the model of the internal quantum efficiency *IQE*, which is defined as the ratio between the extracted charge carrier flux to the absorbed photon flux. Therefore, the spectral *IQE* is given by

$$IQE(\lambda) = \frac{J_{photo}(\lambda)/q}{\phi_0(\lambda)}, \tag{15.49}$$

where $\phi_0(\lambda) = A(\lambda)\phi_{in}(\lambda)$, and $\phi_{in}(\lambda)$ is the incident photon flux, and $A(\lambda)$ is the optical absorbance defined by $A(\lambda) = 1 - R(\lambda) - T(\lambda)$, with $R(\lambda)$ and $T(\lambda)$ the global optical reflectance and transmittance, respectively (typically, solar cells are opaque, i.e. $T(\lambda)$ =0). Analyzing measured *IQE* spectra using Eqs. 15.46 through 15.49 allows to extract diffusion lengths and even recombination velocities [10].

Spectrally dependent illumination, e.g., solar illumination, requires to express the wavelength dependence of the photon flux and the optical properties of the cell, i.e., $\phi_0 \to \phi_0(\lambda)$ (in photons per unit area and now also per unit wavelength), $\alpha \to \alpha(\lambda)$, and $A \to A(\lambda)$. The total photocurrent $\Im_{photo}$ obtained upon spectral illumination is found by integrating $J_{photo}(\lambda)$ in λ across the given spectrum, i.e.,

$$\Im_{photo} = \int J_{photo}(\lambda) d\lambda. \tag{15.50}$$

15.4.1.2 Open-Circuit Voltage Limitations

With the equations developed so far, it is possible to obtain several useful relations that help to establish the physical picture of the HSJ *pn*-cell. For simplicity, let us assume that $d_N << d_P$ and $W << d_P$, i.e., a *p*-region much wider than the SCR and the top *n*-type layer. Additionally, we assume $N_D >> N_A$, implying that practically the whole SCR develops in the *p*-type layer. If SCR recombination turns negligible, the total saturation current density of the *pn* cell is given by $J_0 = J_{0,P}$ and since for neutral region recombination n_{id} = 1, the open-circuit voltage (cf. Eq. 15.6) is given by

$$V_{OC} = V_t \ln\left(J_{photo} / J_{0,P}\right). \tag{15.51}$$

Assuming a homogeneous G_{ph}, the above assumptions lead to $J_{photo} = qG_{ph}L_{n,coll}$, yielding

$$V_{OC} = V_t \ln\left(\frac{qG_{ph}N_A L_{n,inj} L_{n,coll}}{D_n n_i^2}\right). \tag{15.52}$$

With ideal surfaces ($S_n = S_p = 0$), we obtain $L_{n,coll} \times L_{n,inj} = L_n^2 = D_n \tau_n$, hence

$$V_{OC} = V_t \ln\left(\frac{G_{ph}N_A \tau_n}{n_i^2}\right). \tag{15.53}$$

This ideal V_{OC} equation reflects some open-circuit basics: increasing lifetime and generation rates yield higher V_{OC}. The incidence of doping must be treated with more caution, because if radiative recombination dominates, then $\tau_n \propto N_A^{-1}$. Eq. 15.53 further shows that V_{OC} is ultimately related to lifetime rather than diffusion length since carrier transport vanishes at open circuit in a cell without surface recombination. The link between V_{OC} and the electronic bandgap E_g is also found when replacing n_i^2 with Eq. 15.15, yielding

$$V_{OC} = \frac{E_g}{q} + V_t \ln\left(\frac{G_{ph}N_A \tau_n}{N_C N_V}\right). \tag{15.54}$$

Since, under low injection conditions, the quotient in the logarithm is smaller than one, we obtain a fundamental limitation: $V_{OC} < E_g / q$.

For the case of moderately absorbed light and assuming $L_a >> W$, with $W, d_N << d_P$ the total photocurrent obeys $J_{photo} \cong J_{photo,P}$. Writing $J_{photo,P} = q\phi_0 L_n L_a^{-1} f(L_n, d_P, S_n, L_a)$, where f is deduced from Eq. 15.48 and $f < 1$, we obtain

$$V_{OC} = V_t \ln\left(\frac{\phi_0 N_A \tau_n}{n_i^2 L_a} f(L_n, d_P, S_n, L_a)\right), \tag{15.55}$$

which shows the spectral dependence of V_{OC} through $L_a = 1/a$. This has been exploited as a characterization technique that works complementary to the well known quantum efficiency characterization. Additionally, this relationship predicts some dependence of V_{OC} on transport parameters (contained in D_n through $L_n = \sqrt{D_n \tau_n}$), specially for high S_n, showing that in cells with high D_n, more electrons tend to reach the surface where they can recombine.

From the equations above, we can also establish a relationship between V_{OC} and the built-in voltage V_{bi} [3]

$$V_{bi} = V_t \ln\left(\frac{N_A N_D}{n_i^2}\right), \tag{15.56}$$

yielding $n_i^2 = N_A N_D \exp(-V_{bi} / V_t)$, resulting

$$V_{OC} = V_{bi} + V_t \ln\left(\frac{J_{photo} L_{n,inj}}{qD_n N_D}\right). \tag{15.57}$$

It is possible to show analytically that under low-injection conditions, the second term in this equation is always negative. For example, with $G_{ph} = const.$ and $J_{photo} \cong J_{photo,P}$, we obtain

$$V_{OC} = V_{bi} + V_t \ln\left(\frac{G_{ph} L_{n,coll} L_{n,inj}}{D_n N_D}\right). \tag{15.58}$$

In the ideal case of $S_n = 0$, $L_{n,coll} \times L_{n,inj} = D_n \tau_n$, and hence

$$V_{OC} = V_{bi} + V_t \ln\left(\frac{G_{ph} \tau_n}{N_D}\right). \tag{15.59}$$

Since the amount $G_{ph}\tau_n$ equals the excess photogenerated carrier concentration Δn far from the SCR, the low-injection condition ($\Delta n < N_D$) directly implies $G\tau_n < N_D$, which finally yields $V_{OC} < V_{bi}$. Thus, under low injection conditions, V_{OC} is not limited by the gap but by the built-in voltage of the junction.

15.4.2 PN Heterojunction

In horizontal, heterojunction *pn* solar cells, the *J(V)* curve can be modeled similarly to the homojunction case. An important addendum is that, the heterointerface, which occurs within the SCR, may be a defect rich interface that serves as an additional recombination path [9]. Under low injection, this adds an interface recombination current with $n_{id} = 1$, and saturation current density $J_{0,int}$. The *J(V)* characteristics is thus given by

$$J = \underbrace{(J_{0,N} + J_{0,P} + J_{0,int})(e^{V/V_t} - 1) + J_{0,SCR}\sinh(V/2V_t)}_{J_{dark}(V)} - J_{photo}. \quad (15.60)$$

The saturation current density $J_{0,N}$ must be evaluated by Eq. 15.38 replacing $n_i \rightarrow n_{i,N}$, while $J_{0,P}$ is evaluated by Eq. 15.39 replacing $n_i \rightarrow n_{i,P}$. In the case where electrons are majorities at the heterointerface, $J_{0,int}$ can be expressed by [11]

$$J_{0,int} = qS_{int}N_{V,P}\exp\left(-\frac{\Phi_{b,P}}{kT}\right), \quad (15.61)$$

where S_{int} is the interface recombination velocity, $N_{V,P}$ is the effective density of electron states of the valence band of the *p*-type material, and $\Phi_{b,P}$ is the energy barrier for holes seen from the *p*-type material towards the heterointerface [11].

15.4.3 Vertical PN Multijunction

The vertical-multijunction (VMJ) solar cell is rationalized as a periodic arrangement of individual cells connected in parallel (see Figure 15.2c). Figure 15.7 shows a slab of a VMJ, which extends along the *z*-coordinate, containing one "unit cell" with length $l = l_N + l_P + W$, where l_N and l_P are the half-lengths of the neutral *n* - and *p*-type layers, respectively, and *W* is given by Eq. 15.37. The periodicity allows to model only the *J(V)* characteristics of a unit cell, shown between dotted lines in Figure 15.7, and then obtain the total current of the cell considering the parallel connection.

For simplicity, let us assume that the carrier flow in the *x* direction is independent from the flow in the *y* direction, enabling a *quasi* one-dimensional analysis [12]. This requires the thin *n(p)* layers at the top(bottom) to be much thinner than the absorber thickness *d*, and more importantly, the carrier concentration gradients along *x* being negligible compared to the gradients along the *y*-direction. Further, we are implicitly assuming that majority carrier transport along the center of each cell (dotted lines in Figure 15.7) does not affect electrostatics, which is compatible to low injection conditions.

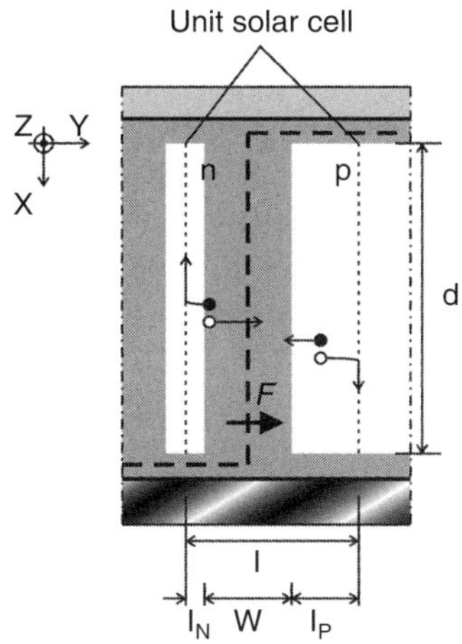

Figure 15.7 Cross section of a vertical multijunction solar cell showing only a unit cell (between dotted lines). The whole solar cell is supposed to extend infinitely into the *z* and *y* directions.

The above assumptions allow formulating the one-dimensional semiconductor equations only in the *y*-direction, following the strategy previously seen for the HSJ: obtain the minority carrier diffusion currents at each edge of the SCR, add the contribution to the current from the SCR, and obtain the current density J_x^{cell} *per* cell *per* differential element *dx*. The boundary conditions are: at the edge of the SCR, the minority carrier concentrations are multiplied by the factor $\exp(V/V_t)$. Additionally, we require the solutions for *n* and *p* to be periodic in *y* with periodicity *l*, or, equivalently, *dn*/*dy* and *dp*/*dy* be zero at the center of each neutral region (i.e. along the dotted lines in Figure 15.7). Once the current J_x^{cell} along a differential element *dx* is obtained, integration from $x = 0$ to $x = d$ yields the current density *per* cell, and taking into account that there are l^{-1} cells *per* unit length along *y*, the total VMJ current density *J* is given by

$$J = \frac{1}{l}\int_0^d J_x^{cell}\,dx = \frac{1}{l}\int_0^d \left(J_{x,dark}^{cell} - J_{x,photo}^{cell}\right)dx\,. \tag{15.62}$$

Since with the above assumptions the dark current does not depend on *x*, while the photocurrent may depend on *x* or not depending on the optical model assumed, the total current density can be written as

$$J = \frac{d}{l} J_{dark}^{cell} - \frac{1}{l} \int_0^d J_{photo}^{cell} dx. \quad (15.63)$$

Analogously to the HSJ, we can expand this expression with the different components of the dark current, obtaining

$$J = \frac{d}{l} \left[(J_{0,N} + J_{0,P})(e^{V/V_t} - 1) + J_{0,SCR} \sinh(V/2V_t) \right] - J_{photo}, \quad (15.64)$$

where J_{photo} is the total photocurrent (negative term in Eq. 15.63), and $J_{0,N}$, $J_{0,P}$ are evaluated from Eqs. 15.38 and 15.39 from the HSJ, but using the injection diffusion lengths

$$L_{p,inj} = L_p \coth(l_N / L_p), \quad (15.65)$$

and

$$L_{n,inj} = L_n \coth(l_P / L_n). \quad (15.66)$$

Notice that these expressions take the form of the HSJ when $S_n = S_p = 0$ in Eqs. 15.40 and 15.39. Further, $J_{0,SCR}$ in Eq. 15.64 is identical to the corresponding HSJ expression (Eq. 15.36).

Assuming $G_{ph} = const.$, the photocurrent is given by

$$J_{ph} = qG_{ph} d \frac{L_{p,coll} + L_{n,coll} + W}{l}, \quad (15.67)$$

where the collection diffusion lengths are given by

$$L_{p,coll} = L_p \tanh(l_N / L_p), \quad (15.68)$$

and

$$L_{n,coll} = L_n \tanh(l_P / L_n). \quad (15.69)$$

The quotient in Eq. 15.67 saturates at '1' for small $l_{N,P} / L_{p,n}$, and represents the collection efficiency. The maximum photocurrent density is therefore $J^{max}_{photo} = qG_{ph}d$, identically to the HSJ. However, here is where the advantage of the vertical junction is directly seen: it is sufficient to choose $l_{N,P} < L_{p,n}$ to obtain a high photocarrier collection, with no fundamental restrictions posed on d. Thus, the VMJ allows to choose $d \cong L_a$ and disregard d from $L_{p,n}$, opening the path to efficient cells with low diffusion length materials.

We arrive at similar conclusions for the case of a non-uniform G_{ph}. With a generation rate given by Eq. 15.22, the photocurrent density is given by [12]

$$J_{photo}(V) = q\phi_0\left[1-\exp(-ad)\right]\frac{L_{p,coll}+L_{n,coll}+W}{l}. \tag{15.70}$$

Again, we see that the choice of a large d does not affect the choice of l_N and l_P contained in $L_{p,coll}$ and $L_{n,coll}$. The maximum photocurrent is in this case $J^{max}_{photo} = q\phi_0$, identical to the HSJ cell when G_{ph} decays exponentially.

Let us analyze V_{OC}, assuming, for simplicity, $l_P >> l_N$ and $N_D >> N_A$, and neglecting any contribution from the SCR. with $G_{ph} = const.$, we obtain $V_{OC} = V_t \ln(G_{ph}N_A\tau_n / n_i^2)$, equal to V_{OC} in the HSJ cell under comparable assumptions (cf. Eq. 15.53). Physically different results are obtained with an exponentially decaying generation rate (J_{photo} given by Eq. 15.70), where we obtain [12]

$$V_{OC} = V_t \ln\left(\frac{\phi_0[1-\exp(-d/L_a)]N_A\tau_n}{dn_i^2}\right). \tag{15.71}$$

This carries a major difference to the HSJ case: if $d = L_a$, V_{OC} decreases linearly with $ln(d)$. This poses a fundamental limitation to maximizing J_{ph} through increasing d.

Figure 15.8 shows a comparison between the electrical output parameters of HSJ vs. VMJ solar cells as a function of d/L_a, assuming a photon flux ϕ_0 =3×10[17] cm[-2]s[-1], L_a = 100 μm, N_A = 10^{17} cm^{-3}, $L_n < L_a$, and standard silicon parameters. For simplicity, the thickness of the n-type and SCR layers in both cells has been neglected. The VMJ cell was assumed to be properly designed by assuming that $l_P = L_n$. Figure 15.8c shows that the VMJ cell runs at higher efficiency than the HSJ when $d/L_a > 1$, a consequence of choosing $L_n < L_a$, where the VMJ cell outperforms the HSJ through a much higher photocurrent, even with markedly lower V_{OC}.

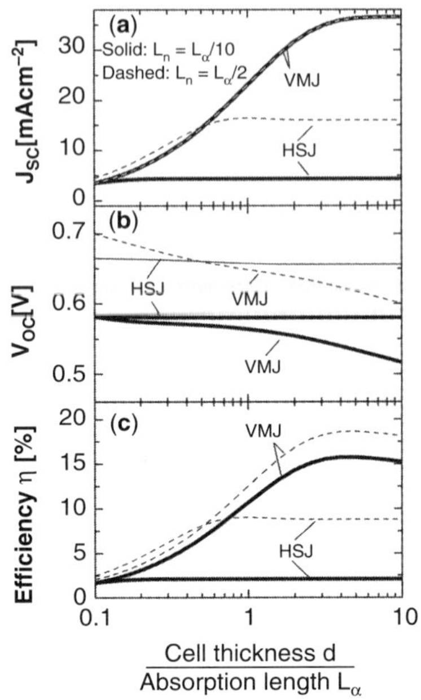

Figure 15.8 Comparison of horizontal single-junction (HSJ) *vs.* vertical multi-junction (VMJ) short-circuit current density J_{SC} (a), open-circuit voltage V_{OC} (b), and efficiency *h* (c), as a function of the ratio between the cell thickness *d* and the absorption length L_a, assuming standard silicon material parameters. Since it was assumed that $L_n < L_a$, the VMJ cell shows higher efficiency than the HSJ when $d/L_a > 1$, despite V_{OC} decreasing with thickness in the VMJ cell.

15.4.4 Considerations for Nanorod Solar Cells

Vertical nanorod (VNR) solar cells are formed by semiconductor pillars (or wires) standing as a two-dimensional array on a substrate, with the junction occurring along either the axial or radial direction [13]. Here, we discuss the radial junction layout, since it offers more advantages than the axial arrangement. Similarly to the VMJ cell, VNR cells enable a decoupling between absorption length and carrier collection lengths. This geometry can also be modeled analytically [7], and in the case of exponentially decaying photogeneration rate along rods of height *d*, it is found that $J_{photo} \propto 1 - \exp(-\alpha d)$ and $J_0 \propto d$ [14], which again leads to decreasing V_{OC} when $d > L_\alpha$, but also to high photocurrent improvements compared to HSJ cells when $L_{n,p} << d$ (provided the rod radii are of the order of $L_{n,p}$). What is new to the nanorod design, is that VNR cells inherently have much lower reflectance and higher absorbance [15] than conventional planar

(VMJ or HSJ) solar cells. However, if the packing of the nanorods on the substrate is not sufficiently high, photovoltaically dead areas appear. This has been introduced into the modeling by using an effective rod length, that is, making the replacement $d \to f_{area} \times d$, where $f_{area} < 1$ is the fraction of nanorod base area to substrate area [7]. Nevertheless, even with area factors as low as $f_{area} = 0.05$, nanorod solar cells can outperform HSJ cells when $L_{n,p} << d$ [7].

But as opposed to the VMJ solar cell, in the VNR cell each individual cell does not share its doped layers with the next cell; their outer doped layer is necessarily exposed to the medium surrounding each rod. If the recombination velocity at the (cylindrical) surface of the rod is high, then nanorods with radii smaller than $L_{n,p}$ may suffer from surface recombination, requiring surface passivation treatments.

15.4.5 PIN Junction

Upon modeling *pin* solar cells, the main routes to analytical solutions are decoupling the electric field from the current equations, and linearizing the recombination rate. At low voltages, it can be assumed that carriers move only by drift, which allows modeling the photocurrent. Higher voltages lower the built-in field, and around open circuit, diffusion currents cannot be neglected.

15.4.5.1 Enhanced Uniform Field Approximation

If the *i*-layer in a *pin* cell is surrounded by highly conductive regions and the charged defect density in the *i*-layer is much smaller than the doping concentrations, then the electrical field is controlled by the doped layers, and can be assumed uniform, i.e., $F = const.$ This condition applies to most *pin* solar cells, with the exception of defect-rich amorphous silicon cells. A first-order estimate for the electrical field in the *i*-layer is given by $F = (V_{bi} - V)/d$, which corresponds to the spatial mean value of *F*, an approximation widely used from inorganic planar [16, 17] to organic *pin* heterojunction solar cells [18]. A more precise calculation of *F* is readily accessible when formulating Poisson's equation in the *i*-layer [19], yielding the potential drops at the *p-i* and *i-n* interfaces, and the potential drop V_0 in the *i*-layer (See Figure 15.1c). Thus, *F* is calculated by

$$F = \frac{V_0 - V}{d}, \tag{15.72}$$

where V_0 is given by [19]

$$V_0 = 4V_t \left[\ln\left(\sqrt{2} \frac{V_{bi}}{V_t} \frac{L_i}{d} \right) - 1 \right], \tag{15.73}$$

and $L_i = \sqrt{\varepsilon_s V_t / (2qn_i)}$ is the intrinsic Debye-Hückel length in the *i*-layer. Eq. 15.73 is valid under the assumption $d < L_i$, which is readily satisfied, e.g., in silicon *pin* cells, where at room temperature $L_i \approx 24$ μm.

15.4.5.2 Drift-Diffusion Model

A closed-form *J(V)* curve is possible assuming a "symmetrical *pin* solar cell," implying: identical doping in the *n* and *p* layers, $N_A = N_D$, giving $V_{bi} = V_t \ln(N_A / n_i)$, identical front/back effective surface recombination velocities S_{eff}, $G_{ph} = const.$, *F* given by Eq. 15.72, and the use of a single, effective lifetime τ and mobility μ (and diffusion coefficient *D*) for electrons and holes [19]. Furthermore, the contributions from the much thinner *n* and *p* layers are neglected, leaving us with the problem of solving the carrier profile within the *i*-layer only. Within the symmetrical *pin* cell, only one minority carrier concentration profile must be known in half of the *i*-layer to obtain the total *J*. Since at the center of the *i*-layer, placed at $x = 0$, it holds that $J_n = J_p$, the total current is directly given by $J = 2J_n(0)$. The boundary conditions are

$$\begin{aligned} n(0) &= n_i \exp(V / 2V_t), \\ J_n(-d/2) &= qS_{eff}\left(n(-d/2) - n_0(-d/2)\right), \end{aligned} \tag{15.74}$$

expressing that the quasi-Fermi levels in the center at $x = 0$ are split by the applied voltage *V* (first equation), and that at the *p-i* interface at $x = -d/2$ (second equation), excess electrons recombine with an effective velocity S_{eff}. When formulating the equation for electrons in Eq. 15.25 taking $F = const.$, the solution for *n(x)* is of the form $y = f(x) + A + B\exp(-\beta_1 x / d) + C\exp(\beta_2 x / d)$. The *J(V)* curve is finally given by

$$J = J_{dark}(V) - J_{photo}(V) \quad (15.75)$$

with the photocurrent given by [19]

$$J_{photo}(V) = qG_{ph}\frac{2L^2}{d}\times \left\{\left[\beta_2 + \frac{S_{eff}d}{D} + \left(\frac{V-V_0}{V_t} - \frac{S_{eff}d}{D}\right)e^{-\beta_1/2}\right]\times A_1 - \beta_1\right\}, \quad (15.76)$$

where we find the dimensionless parameters

$$\beta_{1,2} = \sqrt{\left(\frac{L}{d}\right)^{-2} + \left(\frac{V-V_0}{2V_t}\right)^2} \pm \left(\frac{V-V_0}{2V_t}\right), \quad (15.77)$$

and

$$A_1 = \frac{\beta_1 + \beta_2}{\beta_2 + \frac{S_{eff}d}{D} + \left(\beta_1 - \frac{S_{eff}d}{D}\right)e^{-\frac{\beta_1+\beta_2}{2}}}, \quad (15.78)$$

being $0 < A_1 < 1$ for $V < V_0$. It is possible to show that Eq. 15.76 correctly leads to the maximum photocurrent $J_{photo}^{max} = qG_{ph}d$. The first salient feature is that the photocurrent depends markedly on voltage. The dark current density is given by [19]

$$J_{dark}(V) = \frac{2qn_iD}{d}\times\left[\left(\beta_1 + \frac{\frac{S_{eff}d}{D} - \beta_1}{e^{\frac{\beta_1+\beta_2}{2}}}A_1\right)\times\left(e^{\frac{V}{2V_t}} - \left(1 - \frac{V_0VL^2}{V_t^2d^2}\right)^{-1}\right) + \frac{V}{V_t}\times\left(1 - \frac{V_0VL^2}{V_t^2d^2}\right)^{-1}\right]. \quad (15.79)$$

By inspection, we identify an ideality $n_{id} = 2$ seen in the exponent of $\exp(V/2V_t)$, but the remaining terms containing V modify the shape of J_{dark} in such a way that a logarithmic plot of J_{dark} vs. V may show an effective $n_{id} = 1.2$ to 1.8.

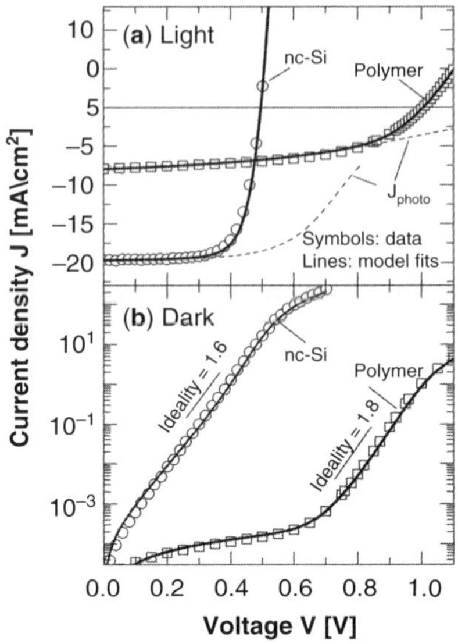

Figure 15.9 Current-voltage characteristics of inorganic, nc-Si *pin* solar cell (circles) and polymer solar cell (squares) fitted by the drift-diffusion model (solid lines) under light (a) and dark (b) conditions. The dashed lines show the corresponding voltage-dependent photocurrent J_{photo}.

Figure 15.9a shows fits to $J(V)$ curves of illuminated nanocrystalline silicon (nc-Si) [20] and polymer bulk-heterojunction [21] solar cells, while Figure 15.9b shows the corresponding dark $J(V)$ curves. The fits were performed using Eqs. 15.72, 15.73 and 15.73 through 15.77 with identical parameters for light and dark curves, resulting μ = 23(2×10^{-4}) cm^2/Vs, τ = 67(3200) ns, S_{eff} = 9.6×10^3(10^6) cm/s, and R_S = 0.63(13) Ωcm^2 for the nc-Si(polymer) cell, which are in agreement with literature values. The dashed lines in part (a) show the voltage-dependent J_{photo} from Eq. 15.76 resulting in each case. From the dark experimental curves we can extract an effective ideality of n_{id} = 1.6 for the nc-Si and n_{id} = 1.8 for the organic solar cell, which is naturally reproduced by the model.

15.5 The Importance of the Temperature Dependence of V_{OC}

Since V_{OC} is linked to J_0 through Eq. 15.6 and the temperature dependence $V_{OC}(T)$ is measured with relative ease, the $V_{OC}(T)$

curve is a popular tool for identifying the dominating recombination mechanism of solar cells [22]. It helps to investigate recombination bottlenecks and thus suggest a path for improving cell perfomance. As can be deduced from the models studied so far, the dark saturation current density J_0 always contains a Boltzmann factor of the form $\exp(-E_A/kT)$, where E_A is defined as the activation energy of the dark current. In *pn* solar cells for example, the saturation current density from the neutral regions shows a temperature dependence $J_{0,N/P} \propto \exp(-E_g/kT)$, while for the SCR we have $J_{0,SCR} \propto \exp(-E_g/2kT)$, and for interface recombination $J_{0,int} \propto \exp(-\Phi_{b,P}/kT)$. Since for neutral region and interface recombination $n_{id} = 1$ and for the SCR $n_{id} = 2$, we can propose a generic expression for the dark current under forward bias ($V > 3V_t$)

$$J_{dark} = J_{00} \exp\left(\frac{qV - E_A}{n_{id}kT}\right), \tag{15.80}$$

with J_{00} grouping factors that depend weakly on temperature. If we assume that only one recombination path dominates the $J(V)$ curve around V_{OC}, we can write

$$V_{OC} = \frac{E_A}{q} - n_{id}V_t \ln\left(\frac{J_{00}}{J_{photo}}\right). \tag{15.81}$$

Plotting measured $V_{OC}(T)$ data, we can obtain E_A from the V_{OC} axis intercept at $T = 0$. For example, if the ideality (measured, e.g., by a J_{SC} vs. V_{OC} curve [22]) is around $n_{id} = 2$ and E_A from $V_{OC}(T)$ matches $E_A \cong E_g/2$, then we can assert that the limiting recombination mechanism is SCR recombination.

More generally, each recombination mechanism gives a particular combination of n_{id} (sometimes temperature-dependent) and activation energies E_A, depending on the dominating recombination path (e.g. bulk, heterojunction interface [11, 23], or surface), and physical mechanisms, ranging from radiative recombination, tunneling-enhanced recombination at defect levels [24], Auger, multi-defect level, to defect band-tail recombination [25]. Additionally, chemically inhomogeneous samples also leave characteristic traces on $V_{OC}(T)$ curves [26].

15.6 Conclusions and Outlook

Despite the variety and complexity of solar cells, simple yet physically meaningful models are required as a starting point for the analysis and prediction of their photovoltaic performance. This chapter outlines models for thin-film solar cells that are capable of describing their behavior in terms of analytical mathematical expressions. The presented models describe the current-voltage characteristics of horizontal *pn*, *pin*, and vertical *pn* multijunction solar cells. These are developed on the basis of fundamental semiconductor device equations. A number of issues to be considered when modeling nanometric solar cells such as tunneling recombination through defect energy states, or excitonic effects in bulk-heterojunction cells are also introduced.

The emphasis is made on identifying fingerprint physical properties of each type of solar cell type: the relationship between solar cell output and characteristic minority carrier's drift/diffusion lengths, characteristic light absorption thickness, and physical design distances (thickness, junction width, etc.), as well as the physical meaning of the ideality factor, the temperature dependence of the open circuit voltage, and the bias voltage dependence of the photocurrent. Further, the logical paths towards equations revealing fundamental limitations of each solar cell design are discussed.

Acknowledgements

The author acknowledges M. Soldera for carefully reading the manuscript and A. Koffman-Frischknecht for polymer cell data.

References

1. K. Taretto, M. Soldera, and M. Troviano, *Progress in Photovoltaics: Research and Applications*, Vol. DOI: 10.1002/pip.2235, 2012.
2. S.M. Sze, and K.K. Ng, *Physics of Semiconductor Devices, 3 Ed.*, Wiley-Interscience, Hoboken, 2006.
3. J. Nelson, *The Physics of Solar Cells*, Imperial College Press, London, 159, 2003.
4. R. Brendel, and A. Goetzberger, *Thin-Film Crystalline Silicon Solar Cells*, John Wiley & Sons, Weinheim, 2003.
5. T. Kirchartz, B.E. Pieters, K. Taretto, and U. Rau, *Journal of Applied Physics*, Vol. 104, 094513–094519, 2008.

6. T. Kirchartz, J. Mattheis, and U. Rau, *Physical Review: Condensed Matter and Materials Physics*, Vol. 78, 235320–235313, 2008.
7. K.R. Catchpole, S. Mokkapati, and F.J. Beck, *Journal of Applied Physics*, Vol. 109, 084519–084519-084515, 2011.
8. G.A.M. Hurkx, D.B.M. Klaassen, and M.P.G. Knuvers, *IEEE Transactions on Electron Devices*, Vol. 39, 331–338, 1992.
9. A.L. Fahrenbruch, and J. Aranovich, Heterojunction phenomena and interfacial defects in photovoltaic converters, in: B.O. Seraphin, eds., *Solar Energy Conversion*, Springer, Heidelberg, pp. 257–326, 1979.
10. R. Brendel, and U. Rau, *Journal of Applied Physics*, Vol. 85, 3634–3637, 1999.
11. N. Jensen, U. Rau, R.M. Hausner, S. Uppal, L. Oberbeck, R.B. Bergmann, and J.J.H. Werner, *Journal of Applied Physics*, Vol. 87, 2639–2645, 2000.
12. A. Gover, and P. Stella, *IEEE Transactions on Electron Devices*, Vol. 21, 351–356, 1974.
13. E.C. Garnett, M.L. Brongersma, Y. Cui, and M.D. McGehee, *Annual Review of Materials Research*, Vol. 41, 269–295, 2011.
14. B.M. Kayes, H.A. Atwater, and N.S. Lewis, *Journal of Applied Physics*, Vol. 97, 114302-114311, 2005.
15. E. Garnett, and P. Yang, *Nano Letters*, Vol. 10, 1082–1087, 2010.
16. R.S. Crandall, *Journal of Applied Physics*, Vol. 54, 7176–7186, 1983.
17. S. Hegedus, D. Desai, and C. Thompson, *Progress in Photovoltaics: Research and Applications*, Vol. 15, 587–602, 2007.
18. J. Xue, B.P. Rand, S. Uchida, and S.R. Forrest, *Journal of Applied Physics*, Vol. 98, 124903-124909, 2005.
19. K. Taretto, *Progress in Photovoltaics: Research and Applications*, Vol. DOI 10.1002/pip.2325, 2012.
20. B.E. Pieters, H. Stiebig, M. Zeman, and R.A.C.M.M. van Swaaij, *Journal of Applied Physics*, Vol. 105, 044502-044510, 2009.
21. D. Veldman, O.Ipek, S.C.J. Meskers, J. Sweelssen, M.M. Koetse, S.C. Veenstra, J.M. Kroon, S.S. van Bavel, J. Loos, and R.A.J. Janssen, *Journal of the American Chemical Society*, Vol. 130, 7721–7735, 2008.
22. U. Rau, D. Abou-Ras, and T. Kirchartz, *Advanced Characterization Techniques for Thin Film Solar Cells*, John Wiley & Sons, Weinheim, 2011.
23. H. Wilhelm, H.-W. Schock, and R. Scheer, *Journal of Applied Physics*, Vol. 109, 084514-084516, 2011.
24. U. Rau, *Applied Physics Letters*, Vol. 74, 111–113, 1999.
25. E.A. Schiff, *Solar Energy Materials and Solar Cells*, Vol. 78, 567–595, 2003.
26. S. Siebentritt, *Solar Energy Materials and Solar Cells*, Vol. 95, 1471–1476, 2011.

16

Efficient Organic Photovoltaic Cells: Current Global Scenario

Sandeep Rai[1,*] **and Atul Tiwari**[2]

[1]*Department of Chemical Technology, Shroff S.R. Rotary Institute of Chemical Technology, At-Vataria, Ankleshwar, India*
[2]*Department of Mechanical Engineering, University of Hawaii at Manoa, Honolulu, Hawaii, USA*

Abstract

Organic photovoltaic, comprising conjugated polymers blended with a fullerene derivative, have recently attracted intense interest. These photovoltaic (PV) energy technologies can contribute to the development of environmentally friendly, renewable energy production, and the significant reduction of the carbon dioxide emission associated with fossil fuels and biomass. Such a combination of conjugated polymers and fullerene has emerged as a promising alternative for the production of inexpensive and flexible photovoltaic cells. Due to the solubility of polymers, these can be economically printed for larger scale production when compared to inorganic solar cells. The development of new materials with low optical gaps such as the combination of polymer-fullerene has incrementally resulted in a gradual but significant increase in solar cell efficiency. In this chapter, we briefly review some of the developments in photovoltaics, especially in polymers and fullerenes with a focus on the economics of solar energy across the globe.

Keywords: Solar cells, organic photovoltaic, solar energy, fullerene, polymer, economics of solar energy

**Corresponding author*: sadeep1964@yahoo.co.in

Atul Tiwari, Rabah Boukherroub, and Maheshwar Sharon (eds.) Solar Cell Nanotechnology, (447–474) 2014

16.1 Introduction

Projected global energy demand will double within the next 50 years. Fossil fuels are running out at an extremely fast rate and are the main source for the increased concentration of carbon dioxide in the earth's atmosphere. Solar cell manufacturing based on the technology of crystalline silicon devices has grown by approximately 40% per year, and this growth rate is further accelerating. Inorganic materials such as gallium-arsenide (GaAs), cadmium-telluride (CdTe), silicon (Si), and cadmium-indium-selenide (CIS) are the active materials that are used for the current commercial fabrication of solar cells which have power conversion efficiency varying from 8 to 29% (Table 16.1) [1]. There are two types of technologies used for these solar cells. The silicon solar cells or crystalline solar cells are made of either mono- or polycrystalline silicon or GaAs. About 85% of the current PV market is shared by these crystalline solar cells. Improvised thin-film technologies were based on amorphous silicon, CdTe, and CI(G)S.

However, the prevailing status of PV based on inorganic materials is not in good shape as it hardly contributes to the energy market, being too expensive. The large production costs of the silicon solar cells are one of the major hurdles for commercialization.

Table 16 .1 Status of the power conversion efficiencies as reached for inorganic solar cells and the technology used to prepare these solar cells. Source: Photovoltaic Network for the Development of a Roadmap for PV (PV-NET).

Semiconductor material	Power conversion efficiency [%]	Technology
Mono-crystalline silicon	20-24	Crystalline
Poly-crystalline silicon	13-18	Thick and thin-film
Gallium-arsenide	20-29	Crystalline
Amorphous silicon	8-13	Thin-film
Cadmium telluride	10-17	Thin-film
Cadmium indium selenide	10-19	Thin-film

Even with the reduced production costs, large-scale production of the current silicon solar cells would be limited by the inadequacy of some elements required, e.g., solar-grade silicon. Although the price of silicon solar cells has dropped lately, the costs are still too high to economically cope with other sources of power generation. Therefore, efforts are under way in the field of inorganic thin films to develop technologies based on cheaper production processes.

Conjugated polymers and novel molecules have been part of the new approach for solar cell development. Conjugated materials are fundamentally the organic molecules consisting of alternate double bonds. The development based on conjugated materials-based electronics was triggered in 1977 when Heeger, MacDiarmid, and Shirakawa [2–4] discovered that the conductivity of the conjugated polymer polyacetylene (PA) (Figure 16.1) could be increased by several orders of magnitude upon oxidation with iodine [5]. The PA was followed by the discovery of electroluminescence in a poly(p-phenylene vinylene) (PPV) (Figure 16.1) by Burroughes *et al.* in 1990 [6]. In the year 2002, Phillips Corporation released their first light-emitting product based on the electroluminescence in conjugated polymers. The light-emitting product made of conjugated molecules was later introduced by a joint venture of the Kodak and Sanyo companies in Japan.

The properties of the conjugated macromolecules could be tuned to meet the requirement of a low bandgap material. Moreover, conjugated polymers (Figure 16.1) display electronic properties similar to those of traditional semiconductors and conductors with facile processing and mechanical flexibility available in plastics. This new class of materials has, therefore, attracted considerable attention because of the huge potential of providing environmentally safe, flexible, lightweight and inexpensive electronics.

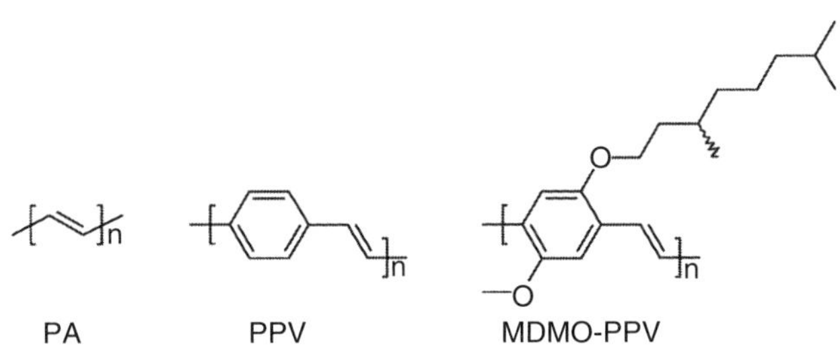

Figure 16.1 Molecular structures of the conjugated polymers trans-polyacetylene (PA), poly(p-phenylene vinylene) (PPV), and a substituted PPV (MDMO-PPV).

The cost reduction mainly resulted in facile solution-based processing that needs solubilized polymers. PPV is hardly soluble, but the side groups attached to the backbone increase the solubility of the polymer to a significant amount. Additionally, the nanoscale morphology that affects the optoelectronic properties of these polymer films can be controlled by selecting suitable side groups.

The recent developments in inkjet printing, micro-contact printing, and other soft lithography techniques has further improved the opportunity of conjugated polymers for low-cost fabrication of large-area integrated devices on both rigid and flexible substrates. Further, architectures to overcome expected electronic scale-up problems related to thin-film organics are being developed.

Unlike conjugated polymers, conjugated molecules can be thermally evaporated under high vacuum. This deposition technique is quite expensive compared to solution processing and, therefore, less attractive for commercialization. The improvements and developments in photovoltaic technologies and the role of plastic electronics have drawn considerable attention towards the adoption of plastics in solar cell manufacturing. It is expected that plastic solar cells will create an entirely new and large market in the field of cheap electronics.

Organic semiconductors are generally carbon-based molecular materials, including both low molecular weight monomers and high molecular weight polymers pivoted around sp^2 hybridized carbon atoms. The electrons interacting within p_z orbitals on neighboring carbon atoms constitute a π electronic system. It is worth noting that the energy gap between the highest of the filled π orbitals (also known as the highest occupied molecular orbital, or HOMO) and the lowest unoccupied molecular orbital, or LUMO, ranges between 1–3 eV. This gap is shorter compared to the gap between filled and unfilled levels of π conjugated carbon molecule that can be easily accessed by the visible light. The weakly bound delocalized π orbitals are free to hop between the molecules and generate charges either electrically or optically. Moreover, the molecular packing disorder suggests that the energies of electronic states on different molecules could be different. This results in the distribution of the HOMO and LUMO energies. The LUMO and HOMO energy states could act as charge traps. In such case, due to weak electronic interaction between the charge traps and conjugated segments, the resulting charge mobilities could be much lower than those in crystalline inorganic semiconductors.

In photovoltaic applications, the excited states are localized, and this validates the prominent phenomenon that differentiates organic and inorganic semiconductors. An organic semiconductor on photon by absorption generates exciton (a neutral excited state) which is within a volume of a few nm^3 and is localized on a single molecule or a single conjugated segment. Since the Coulomb energy binding such localized charges is high, it is difficult for the exciton to dissociate into a separate electron and hole; this is because of the spatial confinement and the low dielectric permittivity ($\varepsilon \approx 3 - 4$) of organic semiconductors. The exciton in a crystal of silicon has a diameter of tens of nm and a binding energy of around 0.1 eV, which allows charge pairs to be generated spontaneously at room temperature. As a result of this, the exciton in an organic semiconductor tends to decay to the ground state within a few nanoseconds, sometimes with the release of light, and does not dissociate into charges. To overcome this problem in organic solar cells and photodiodes, a second molecular component (the electron acceptor) is introduced which has a higher affinity for electrons, e.g., fullerene. So, when an exciton is photogenerated within the first donor phase close to this second type of molecule, the electron is attracted to the higher affinity molecule due to the free energy difference. The difference in free energies ΔG between the exciton and the charge-separated state provides the lost exciton binding energy (Figure 16.2a,b) [7].

Once the charges are located on different types of molecules, the electron on an acceptor molecule and the hole remaining on a donor molecule, the charges initially form a geminate pair in which they are held by Coulombic forces. Once this pair is separated, the two types of charge may travel through the respective phase and may be collected by a low-resistance contact to the external circuit. This requires that the two phases percolate through the medium.

Photocurrent and photovoltage generation in an organic donor-acceptor device requires the following:

- Visible light must be absorbed in the donor or acceptor material or both.
- The width of domains of pure donor or acceptor material must be shorter than the distance an exciton diffuses before decaying ($< \sim 10$ nm) in order for most photo generated excitons to dissociate, i.e., the two components should be sufficiently well mixed.

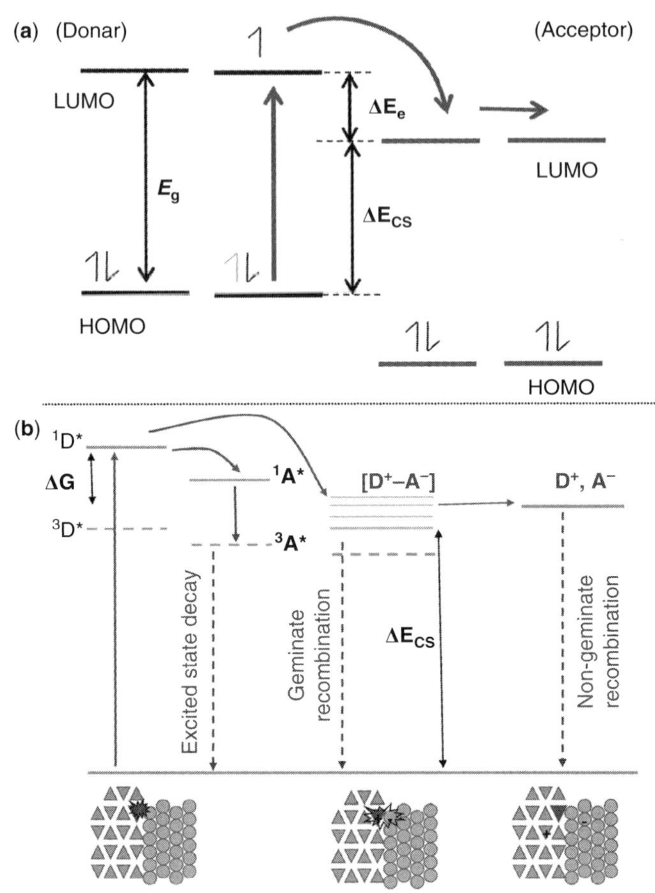

Figure 16.2 a) Schematic diagram showing an organic solar cell based on heterojunction between the electron donor and electron acceptor. b) State diagram of the total energy of the states involved in charge pair generation.

- Both phases should form continuous percolating networks that connect the bulk of the film to the electrodes. The electrodes should be electronically different, such that electrons are preferentially collected at one and holes at the other, in order to provide a direction for the photocurrent. Such selectivity can be achieved using one high and one low work function electrode.

Becquerel [8] discovered the photovoltaic effect, i.e., light-induced voltage in 1839 when he shed light onto an AgCl electrode in an electrolyte solution. The first solid-state solar cell was demonstrated

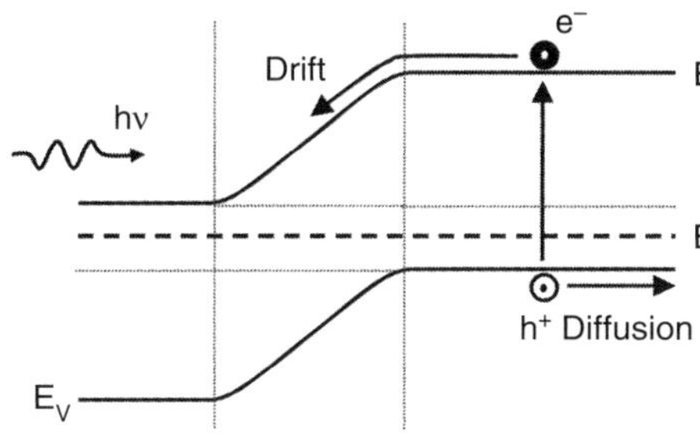

Figure 16.3 Wafer -Si solar cells: Principle.

by Fritts in 1883 by depositing a thin layer of Au on Se semiconductor, which served as the light absorber to convert photons into electron-hole pairs. The photo-excited charge carriers separated the internal electric field in the Au/Se Schottky junction. Inorganic solar cells even today have light absorption and charge separation as two fundamental processes. Ohl [9] patented modern solar cells in 1946 which were subsequently demonstrated by Chapin, Fuller, and Pearson in 1954 at Bell Laboratories. Figure 16.3 [10] illustrates the principle of their cell which employed a single-crystal Si (sc-Si) wafer for light absorption and a p-n junction for charge separation and had an efficiency of ~5%. Absorption of photon is throughout the wafer and extends outside the depletion region. So, charge separation makes diffusion of photo-excited charge carried toward the depletion region. As a result of this, the quality of the entire wafer becomes critical in terms of minority carrier lifetime and carrier mobility for high efficiency. Contradictory to this, in most of the CMOS devices it is the quality of the surface region that matters most since the operation takes place within a thin layer (<100 nm) of the wafer surface. In spite of the steady improvement in the efficiency of solar cells since 1954, the majority of today's commercial solar cells are very much similar to the Bell Laboratories' cell, utilizing a Si wafer in either single-crystal or polycrystalline form for light absorption and a p-n junction for charge separation.

Figure 16.4 [10] shows a cross section along with fabrication process of a common structure for sc-Si cells. It starts with a p-type Si(100) wafer with resistivity of ~1 Ω-cm. The wafer is etched, in either NaOH or KOH, to form microscale pyramids with facets on the surface. The textured surface leads to reduction of sunlight reflection since reflected light from a pyramid often strikes another pyramid, thus creating a second chance for incidence. A

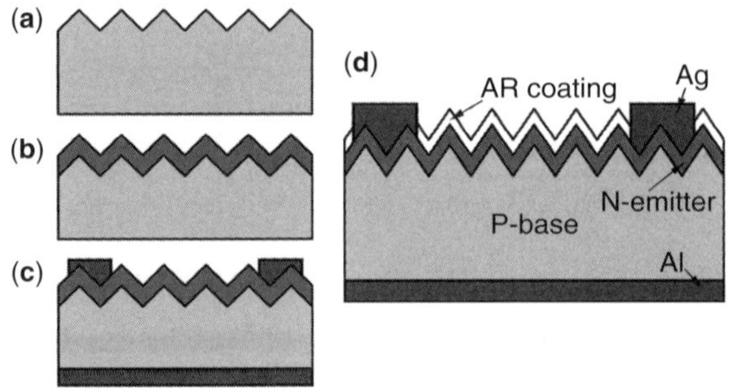

Figure 16.4 Cross-sectional view and fabrication process of single-crystal Si solar cells: (a) Charge separation. Figure shows different processes, i.e., (b) n-emitter diffusion, (c) metallization, and (d) antireflection coating.

p-n junction is formed by diffusing n-type dopants, often P, into the front side of the wafer for a typical depth of ~0.5 μm. The back-side electrical contact is formed by reacting a screen-printed Al film and the wafer to form an AlSi compound.

In this cell, the front finger contact is also screen printed with an Ag paste. The typical width of the contact fingers is ~200 μm, and the separation between two fingers is ~3 mm, to maintain a balance between minimum resistance and minimum light shadowing. The final step involves deposition of a transparent material, often SiNx, on the wafer as an antireflection coating by plasma-enhanced chemical vapor deposition (PECVD).

Figure 16.5 [10] shows the continual improvement in the efficiency for various solar cell technologies over the last three decades. As of now, the reported efficiency of sc-Si cells has reached 24.7%, while that of polycrystalline Si (poly-Si) cells is at 20.3%. The maximum recorded efficiency for all solar cell technologies is held at 40.8%, obtained with a so-called tandem cell. This cell involves stacking of three p-n junctions made of GaxIn1-xAs or GayIn1-yP with different compositions. Each junction has a different bandgap and absorbs light in a certain portion of the solar spectrum. The efficiency of this cell was measured with a concentrator under 326 Suns. Under one Sun, the reported efficiency of this triple junction cell is 33.8%. In addition to wafer Si, several thin-film solar cell technologies have also been commercialized, including Si in either amorphous (a-Si) or microcrystalline (μc-Si) form and metal chalcogenides (CdTe and CuInxGa1- xSe2(CIGS)). Their record efficiencies were obtained in the range between 12% and 20%. Though the progress in efficiencies of these laboratory-size solar cells has been remarkable, the commercial-size cells and modules

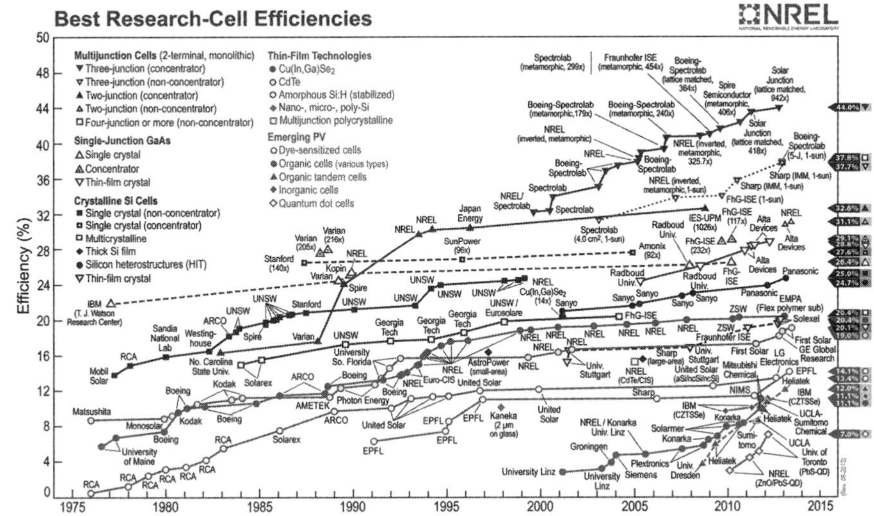

Figure 16.5 Current best research-cell efficiencies versus time for various technologies. (*Source: National Renewable Energy Laboratory, 2012, www.nrel.gov*)

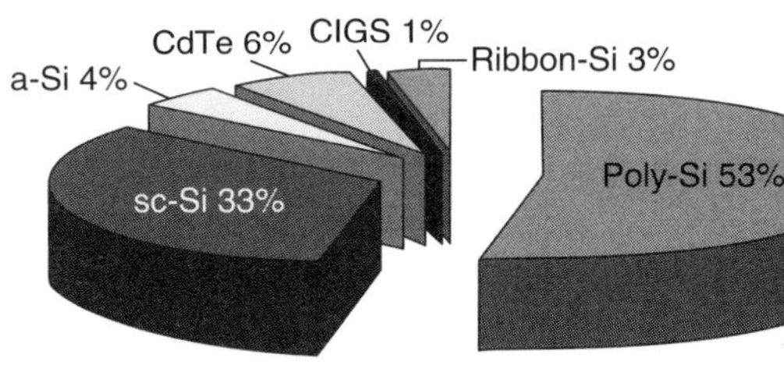

Figure 16.6 Solar Cell Technologies – Market Scenario in 2007.

could typically achieve only one-half to two-thirds of these recorded efficiencies. Eliminating the efficiency gap is one of the subjects of ongoing research, especially by the solar cell industry.

Figure 16.6 [10] shows the market share of different solar cell technologies in 2007. Poly-Si with 53% has the largest, followed by sc-Si at 33% and all the thin-film technologies at 11% of the market.

16.2 Current Developments in OPVs

The significant advances achieved for Organic Photo Voltaic (OPV) technology are the results of exhaustive development work on four major fronts: (i) clarity in understanding the photon-to-electron conversion mechanism; (ii) development of new materials having tailored energy levels and solubility as required for the end application; (iii) optimal microstructures in the active layer achieved by new processing techniques; and (iv) new improved devices with novel interfacial layers. In this section, we briefly review the recent developments pertaining to materials, specially polymers, fullerenes and nanomaterials. The main objectives in the field of OPV are achieving high efficiency, long-term stability and low cost. Low cost can be achieved by combining a low bill of raw materials with high productivity roll-to-roll (R2R) manufacturing techniques. It is essential that compared to Si-based solar cells, organic solar cells should be less expensive and easier to manufacture due to lower energy consumption, free from vacuum process, and the possibility of direct patterning during coating. Applications based on organic semiconductors, such as light-emitting diodes and photovoltaic devices, have strict and precise requirements of the layer properties, such as thickness, uniformity and over-layer accuracy. Various printing and coating techniques are applied and developed for the deposition of thin layers of organic semiconductors. Galagan *et al.* [11] reported a technology for the roll-to-roll production of OPV and also proposed further improvements in this technology.

Bulk heterojunction solar cells based on a series of variable alkyl chain length substituted donor–acceptor (D–A) conjugated polymers with thiophene ring and benzenetriazole moiety as the donor and the acceptor, respectively, have been investigated [12]. Evaluation of the optical and electrochemical properties showed that the absorption onsets and the energy levels of the copolymers were not dependent of alkyl substitution. D–A copolymers PTBT-DA6, PTBT-DA10 and PTBTDA12 have been successfully investigated in terms of their photovoltaic performances. Among all polymers PTBT-DA10 showed the best performance with a Power Conversion Efficiency (PCE) of 1.23%. Further optimizations such as using additives and different alkyl-alkoxy chains on the polymer backbone can lead to higher PCEs and readily synthesizable PTBT-DA polymers which are exceptional candidates for organic photovoltaics.

A paper presents [13] results for photovoltaic performance obtained from the application of different bulk heterojunction blends onto flexible polypropylene (PP) substrates for textile applications. A new device which employed a transparent metal cathode (ca. 10 nm LiF/Al) and highly conductive PEDOT:PSS- and Ag-based anode was achieved in this study. These results show that polymer-based solar cell materials and manufacturing techniques are suitable and applicable for flexible and non-transparent textiles, especially tapes and fibers, with transparent outer electrodes.

Blankenburg and coworkers [14] reported for the first time extremely promising results in up-scaling coating processes for thin flexible polymer solar cell (PSC) application. Two functional layers for PSC devices, the conducting poly(3,4-ethylenedioxythiophene):polystyrene sulfonate and the photoactive one (poly-3-hexylthiophene/ [6,6]-phenyl-C61-butyric acid methyl ester), could be successfully prepared by continuous R2R wet coating of low-viscosity solutions and dispersions on ITO-coated PET substrate. It was concluded that BHJ-PSC can be produced with adjustable coating thicknesses on plastic foil using a R2R slot die coating process for the preparation of transparent conducting and photoactive layers.

Two main research and technological areas of development in both organic and inorganic solar cells are the inclusion of plasmonic metal nanoparticles to boost the efficiency of thin-film solar cells by increasing light absorption, and the incorporation of graphene as transparent conductive electrode to replace indium tin oxide (ITO) films. It was demonstrated that both the plasmon resonance of gold nanoparticles and graphene are valuable approaches to improve performance of both silicon and organic solar cells.

The incorporation of photovoltaics (PVs) into clothing opens up the possibility of having a truly mobile and handy source of energy that can be used for charging batteries for communications equipment, monitoring and lighting.

Krebs *et al.* [15] presented two strategies for incorporation of polymer PVs into clothing. The first strategy involved the simple incorporation of a polymer PV on a polyethyleneterphthalate (PET) substrate directly into the clothing as a structural element. The second strategy was more elaborate and involved the lamination of a thin layer of polyethylene (PE) onto a textile material followed by plasma treatment and application of a PEDOT electrode onto the textile material. Subsequent screen printing of the active material and evaporation of the final electrode completed the device as a polymer PV that is fully integrated into the textile material. These strategies of the incorporation of polymer PVs into garments and textiles may find the application in "intelligent clothing."

A roll-to-roll process enabling fabrication of polymer solar cells comprising five layers on flexible substrates is presented by Krebs [16]. While the process efficiently met all the objectives set out and solves many of the problems currently encountered in attempts to upscale and transfer the polymer solar cell technology to industry, there is a sacrifice in performance due to poor light transmission of the PEDOT:PSS-silver grid composite electrode.

Commercially available semiconductor quantum dots have been encapsulated [17] within a poly(methyl, methacrylate) polymer matrix to form a luminescent downshifting layer. It was shown that core/shell semiconductor quantum dots are a strong contender for use in luminescent downshifting on CdTe photovoltaic devices.

Hoth *et al.* [18] showed that with spray coating technology it is possible to deposit all solution processable layers of organic solar cells. With this technology, the hole transporting layer poly(3,4-ethylene dioxythiophene) doped with polystyrene sulphonic acid (PEDOT:PSS) can also be deposited. It was demonstrated that a fully-sprayed coated device with 2.7% PCE can be developed.

A series of ester functionalized [19] polythieno[3,4-b] thiophene derivatives (P1–P3) were prepared by Stille cross-coupling polymerization. From P1 to P3, the bandgap becomes wider and the HOMO energy level becomes lower with increasing the content of the thiophene-3,4-dicarboxylate moiety in the polymer backbone. Though the final PCE of the prepared OPV devices based on the three copolymers is not high (the best is 1.02%), a high Voc up to 0.71 V can be achieved, suggesting that incorporation of thiophene-3,4-dicarboxylate moiety into the polythieno[3,4 b] thiophene backbone is a feasible way to gain high voltage OPVs.

Kroon *et al.* [20] reported a new class of quinoxaline- and pyridopyrazine-based polymers for solution-processable photovoltaics. Three new TQ1 (this polymer, known as poly[2,3-bis(3-(octyloxy)phenyl) qui- noxaline-alt-thiophene]) derivatives were developed in which the acceptor structure has been modified. It was demonstrated that small structural alterations on the original TQ1 structure results in significantly different electronic properties. The synthesis route is robust towards changes in side-chain chemistry, other heteroatoms, as well as partial fluorination. Thus, in principle, a vast library of polymers such as TQ1 can readily be generated that will all have different properties. The polymers synthesized resulted in moderate to substantial photovoltaic performance. In this study, devices based on TQ-8A displayed an improved fill factor when compared to TQ1-based devices.

Two semiconducting copolymers [21] based on porphyrin compounds, poly[9,9-dioctylfluorene-co-5,15-bis(hexoxybenzyl)-10,20-bis(benzo-4-yl)porphyrin] (PFPor) and poly[9-(heptadecan-9-yl)carbazole-co-5,15-bis(hexoxybenzyl)-10,20-bis(benzo-4-yl) porphyrin] (PCPor), were used as organic photovoltaic materials. The two polymers did not exhibit exceptionally high PCEs if absorption properties of these materials were reinforced by enlarging a conjugation length. However, by improving charge transferred property through adopting efficient spacer materials in polymer backbones, it is expected that new high-performance organic photovoltaic materials could be developed. Materials based on phthalimide were successfully synthesized through the Suzuki coupling reaction for the OPVs. Although, the two polymers did not exhibit unusually high PCEs if phthalimide derivatives with reinforced electron affinity due to introducing electron withdrawing functional groups, are adopted in polymer backbones, and expected that new high performance organic photovoltaic materials could be developed.

A donor-acceptor type low bandgap [22] conjugated polymer that was composed of carbazole and 4,7-dithiazolylbenzothiadiazole (DTzBT), i.e., Poly [N-9′-heptadecanyl-2,7-carbazole-alt-5,5-(4′,7′-di-2-thiazol-2′,1′,3′-benzothiadiazole)](PCDTzBT)] (PCDTzBT), was synthesized through the Suzuki coupling polymerization for organic photovoltaics. These copolymers exhibited low bandgaps of 1.85 eV and well-balanced HOMO and LUMO energy levels of −5.54 and −3.65 eV, respectively. Although the low molecular weight prohibited PCDTzBT from having as high a PCE value as PCDTBT, it is expected that PCDTzBT will improve the photovoltaic properties of devices by optimizing polymerization conditions and with the introduction of alkyl chains.

A solution-processable [23], low-cost, and environmentally-friendly bulk heterojunction photovoltaic device based on ZnO nanoparticles/P3HT hybrid material was developed by Li and coworkers. The performance of the solar cells can be significantly improved through interfacial modification with the semiconducting liquid crystal ligands.

In their work, Li and coworkers [24] intramolecularly incorporated the mesogenic cyanobiphenyl to photo-cross-linkable polythiophenes for the purpose of highly stable and ordered liquid-crystalline conjugated materials. A clear advantage of the photo-crosslinking strategy is that crosslinking can be decoupled from thermal annealing. This valuable feature for BHJ solar cells require some annealing to achieve optimum nanomorphology before crosslinking, and freeze the optimum morphology to preserve long-term performance.

A series of soluble poly(perylene diimide)s [25] were synthesized and evaluated in organic photovoltaic cells. These polymers were synthesized through the polycondensation of perylene-3,4,9,10-tetracarboxylic dianhydride (PTCDA) with a number of poly(ethylene glycol) (PEG)- or poly(propylene glycol) (PPG)-based diamine comonomers. These new polymers exhibit good performance in solution processable applications as stable electron-conductive layers with large light absorbtivities.

Synthesis of conjugated polymers [26] bearing indacenodithiophene and cyclometalated platinum (II) units and their application in organic photovoltaics was reported by Liao *et al.* The absorption bands of the metal-based polymers were significantly red-shifted relative to those of the metal-free polymers. Notably, the higher performance of Pt-containing polymers device is not attributed to the triplet contribution, but the significant red–shifted (80–90 nm) absorption bands of the Pt-containing polymers, resulting in superior overlap with the solar spectrum which was consistent with the current. When designing such conjugated polymers, one should take into consideration the negative effects of large dihedral angles, which disrupt intermolecular interactions and charge mobility.

A novel copolymer [27] system was developed with high crystallinity and photocrosslinkable building blocks for p–p intermolecular interactions that has, an alternating copolymer of liquid-crystalline nature and heat/solvent resistance characteristics. By copolymerization of 2,5-bis(3- bromododecylthiophen-2-yl)thieno[3,2-b]thiophene (BbTTT) monomer with thiophene and thieno[3,2-b]thiophene via Stille reaction, two novel copolymers, PBbTTT-T and PBbTTT-TT, were synthesized.

You and coworkers [28] reported the development of a high-performance low bandgap polymer (bandgap of 1.4 eV), poly[2,7-(5,5-bis-(3,7-dimethyloctyl)-5H-dithieno[3,2-b:20,30-d]pyran)-alt-4,7-(5,6-difluoro-2,1,3-benzothia diazole)] with a bandgap of 1.38 eV, high mobility and deep HOMO. As a result, a single-junction device shows high external quantum efficiency of more than 60% and spectral response that extends to 900 nm, with power conversion efficiency of 7.9%. The polymer can be used to process tandem solar cell from solution with certified 10.6% power conversion efficiency under standard reporting conditions. This is the first certified polymer solar cell efficiency reported beyond 10%.

16.2.1 Development of Low Optical Gap Materials

Number of strategies have been pursued to extend the absorption range of donor polymers for OPV. It includes extension of the conjugation length of π-conjugated segments, i.e. replacing phenyl rings with thiophene rings which facilitate to planarize the polymer backbone through a reduced steric effect or by using a bridging atom to planarize neighboring monomers [29]. It reduces optical gap further by enabling the formation of π-stacked aggregates [30]. Another approach deals with varying heteroatoms in the conjugated backbone, for e.g., replacing the sulphur atom in polythiophene with selenium, which leads to a lowered LUMO energy but unchanged HOMO, due to the influence of the heteroatom on the LUMO in this molecule [31]. The recently used route involves combining electron rich (e.g., thiophene,pyridine,pyrrole containing) and electron poor (e.g., benzothiadiazole [BT]) units in the polymer backbone in a 'push-pull' structure. The HOMO of this copolymer is dominated by the HOMO of the electron deficient unit and the LUMO by that of the electron rich unit. This strategy is widely adopted because of large range of 'push' and 'pull' units that can be used and the ability to control copolymer HOMO and LUMO energies independently. Remarkable advances in OPV efficiency were made using push-pull copolymers based on dithiophene - BT structures [32], carbazole – BT structures [33], dithienobenzene - modified BT structures [34] diketopyrrolopyrrole - thiophene structures [35] (Fig. 16.7) [7].

In the case of fullerenes there is weak optical absorption in the visible area on account of the high symmetry of the C_{60} cage and resulting symmetry forbidden optical transitions. This defect in can be enhanced by replacing the symmetric C_{60} cage with one of lower symmetry such as C_{70}which leads to strong blue-green absorption by C_{70} fullerenes and for this reason they are now widely used in combination of low gap polymers to complement the polymer absorption and span the visible spectrum.

16.2.2 Designing of Higher IP Polymers and Lower EA Acceptors

The push-pull copolymer approach is useful in designing polymers with lowered HOMO energy (*Voc*) since it allows control of the polymer HOMO and LUMO separately. It is also reported that

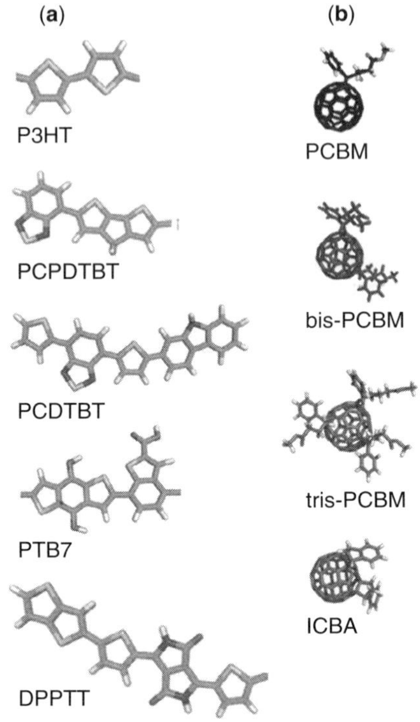

Figure 16.7 Chemical structures of the polymers, P3HT, PCPDTBT, PCDTBT, PTB7 and a DPP-co-thienothiophene polymer (without side chains for clarity). (b) Chemical structures of the fullerenes, adducts of PCBM and indene fullerene (ICBA).

push-pull polymers may allow charge pair separation at a lower value of ΔG than other classes of polymers [36].

Other approaches to push down the polymer HOMO include substitution of bridging carbon atoms by the more electronegative silicon [34] and building torsion into the polymer backbone. However, the latter approach has the negative effect of increasing the optical gap.

Another successful route to higher *V*oc is achieved by reducing the fullerene electron affinity with the addition of more side chains to derivatives such as PCBM [35], the indene-fullerene [36], and also to metal hydride occupied endohedral fullerenes [37]. This approach has produced excellent results with the crystalline polymer P3HT but has not yet succeeded with any amorphous polymer. The possible reasons are thought to be the inferior microstructure, and consequently transport, formed by the amorphous fullerene adduct and a possible effect on charge generation due to the reduced ΔG [38].

16.2.3 Control of Blend Microstructure

Efficient photocurrent generation for the ideal polymer:fullerene can be achieved by sufficiently fine mixing of the phases so that exciton dissociation is achieved. A high degree of molecular ordering and connectivity within the donor and acceptor phases helps in charge transport. In addition to this, a certain degree of phase segregation assists both the competition between charge collection and recombination [42] and the competition between charge separation and geminate recombination. Ideally, the size of domains should be comparable to the length over which excitons diffuse, which is typically between some few nm (P3HT), 10 nm for polymers [43], and for crystalline molecular semiconductors [44] in several tens of nm. Numerous studies [45] have shown that charge separation yield increases when at least one of the phases is crystalline/well ordered, possibly due to the greater delocalization of electronic states and lower resulting Coulombic binding energy.

In practice, the structure of a blend film cast from a single solution is sensitive to many factors (e.g., solvent, casting method, temperature, self organizing properties of materials, interaction between donor and acceptor, any additives, etc.) and therefore difficult to control. Many strategies are used to control microstructure include building in a structural motif to encourage self organization of polymer chains, as with regioregular P3HT [46], and with the addition of processing additives [32] or thermal annealing regimes [32, 46] that control the solidification dynamics. The degree of segregation or intermixing is strongly depends on the polymer structure, in particular, upon the possibility of fullerene intercalation between polymer chains either because of their amorphous chain packing [47] or may be because sufficiently wide side chain spacing that permits fullerene intercalation [48]. The study of phase behavior can help to define the conditions in which a preferred blend microstructure will be realized [49]. Generally, all blend approaches lead to thermodynamically unstable structures with the possible problem of disintegration of phases over time or on exposure to heat [50]. Structures with better defined phase behavior can be developed using multi block copolymers where the length of the blocks in the polymer backbone determine the domain sizes in the blend film [51].

Organic/inorganic hybrid structures seem to be a most efficient route to the control of microstructure, since a rigid and stable inorganic framework can be built [52]. For the control of three-dimensional phase structure, the vertical segregation of phases within the blend film should be desirably controlled since this may assist in electrode selectivity and in directing the photocurrent. Other approaches include the choice of electrodes combined with post deposition processing [53] and methods to apply different successive layers from solution [54].

A huge amount of research has been conducted into microscopic and combined microscopic/spectroscopic structural probes including AFM, TEM, XRD, SE, Raman, and PET due to the complexity of blend film microstructure and the difficulty of determining structure at nm length scales.

The new photoactive materials are the need of the hour for PV devices with longer service lifespan. The main objectives to be achieved are the design and development of organic semiconductors with improved photochemical stability and blend formulations with greater robustness under heating.

16.3 Economics of Solar Energy

Figure 16.8 [55] depicts PV production from 1990–2010, showing the distribution by geographical area (based on reports from PV News,

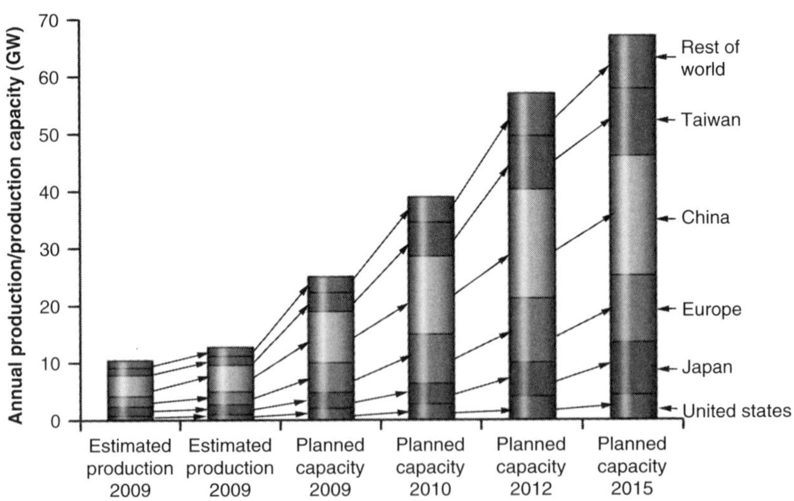

Figure 16.8 Current and planned production and capacity of photovoltaic cells.

Figure 16.9 One of the houses of Europe showing PVs installed on its roof.

1980–2011, GTM Research, San Francisco). This significant and exceptional growth of the technology despite the sluggish world economy bolsters the fact that PV is truly a "real" business now, proving itself as a competitor in the electricity sector, and is expected to exhibit substantial annual growth in capacity, production, and installations in coming years.

Interest in rooftop application (Figure 16.9) [56] started in the USA in the late 1970s. Japan took over the running in the mid-1980s with the construction of a test bed for over 200 residential systems on Rokko Island in 1986. After several years of evaluating technical issues related to their grid connection, a subsidised installation program was launched by the Japanese government in 1993. By the end of 1999, over 20,000 rooftop photovoltaic systems had been installed in urban areas of Japan, with a massive 1.5 million installed during 2010.

It is a well known fact that the ultimate photovoltaic technology needs to be highly efficient, low cost, and environmentally friendly. It is also expected that depending upon applications, technology should meet additional requirements such as flexibility, portability, etc. It can now be concluded that nanomaterials-based photovoltaics have the potency to address these needs, and with the enormous efforts being made across the globe, significant progress is expected in the future. Lee in his review on the current status and projected advancement of concentrated photovoltaics (CPV) concluded that utilizing larger cells at higher concentrations would provide greater opportunities for reducing the net cost of ownership while increasing the production of electricity. The thermal management in CPV lowers the cost of ownership per watt of electricity generated, mainly through improving two parameters: 1) efficiency and 2) reliability. As per the detailed analysis of the future of photovoltaics using the learning curve, electricity from PV can become competitive between 2015 and 2040. It can also be highlighted that not investing in PV is much riskier than promoting the diffusion of PV. In Europe, solar energy is the fastest growing energy technology and PV farms are integrated in housing. As a matter of fact, the production of solar cells and PV systems have doubled every two years since 2002, yet solar electricity covered only 0.26% of EU electricity consumption in 2008. Southern EU countries have enormous potential for solar energy due to increased sunshine. Solar energy may potentially cover up to 12% of EU's energy demand by 2020.

The turnover of PV energy was estimated at EUR 10 billion in 2007. Assuming an average PV system cost of 2.5 €/W, the global annual revenue can be estimated at €44 billion in 2010. The European Photovoltaic Industry Association projected this revenue to increase to €200 billion for an annual market size of around 135 GWs, provided that some cost reduction targets are achieved.

The cost per kW is still considered high (around 2.5 €/W), and higher than the conventional energy production in the EU. However the downward trend in cost observed in the last few years is mainly due to increasing efficiency. As per the forecast of the European Photovoltaic Energy Platform, PV energy may become competitive in a limited and selected group of markets by 2013 and across the entire EU by 2020.

16.3.1 Scenario in US

As per the report of Solar Energy Industries Association, USA, the US now has over 6,400 MWs of installed solar electric capacity, enough to power more than 1 million average American households. The average cost of a completed PV system has dropped by 33 percent in the third quarter of 2012 compared to the beginning of 2011. The average price of a solar panel has declined by 58 percent since the first quarter of 2011. As the solar industry grows, its impact on the economy in the US will be apparent. The increase in value of solar installations has positively impacted the US economy.

In 2011, solar installations were valued at $8.4 billion, compared to $6 billion in 2010.

16.3.2 Solar Potential in India

India is blessed with rich solar energy resource. The average intensity of solar radiation received on India is 20 MW/km square (megawatt per kilometer square). With a geographical area of 3.287 million km square, this is estimated to 657.4 million MW. However, theoretically, a mere 12.5% of the total land area amounting to 0.413 million km square can be used for solar energy installations. As per the estimation, 10% use of this area can result in the available solar energy to 8 million MW per year. However, in India the efficiency of conversion of solar energy to useful energy is low. As per the order of the Central Electricity Regulatory Commission, the life of a solar PV plant has been accounted for 25 years, and accordingly the tariff has been calculated for 1MW solar PV power projects over a period of 25 years. The cost of one unit of solar electricity has been calculated based on this assumption. Accordingly, the tariff leveled on solar energy has been decided as INR 15.39 by the regulatory commission [57]. Similarly, the cost of electricity over a period of 25 years generated by solar thermal power projects has been calculated. The tariff leveled on solar energy by the Solar Thermal Power Project, therefore, has been decided as INR 15.04 by the regulatory commission based on that assumption [57]. As of now, photovoltaic efficiency and manufacturing costs have not reached the point where photovoltaic power generation can compete with conventional coal-, gas-, and nuclear-powered facilities. As per the current estimates, the cost of solar power is approximately four times that of conventionally (coal-based thermal power station) produced power. There are various factors that are affecting the cost of production and distribution of conventional electrical power. Due to the wide range of these variables, some applications of photovoltaic power are becoming economically superior to conventional systems.

16.3.3 Global Solar Cell Demand

Figure 16.10 [58] presents data about when certain countries have reached and will reach grid parity. It shows, for example, that countries with higher electricity prices, such as Germany, Denmark, Italy,

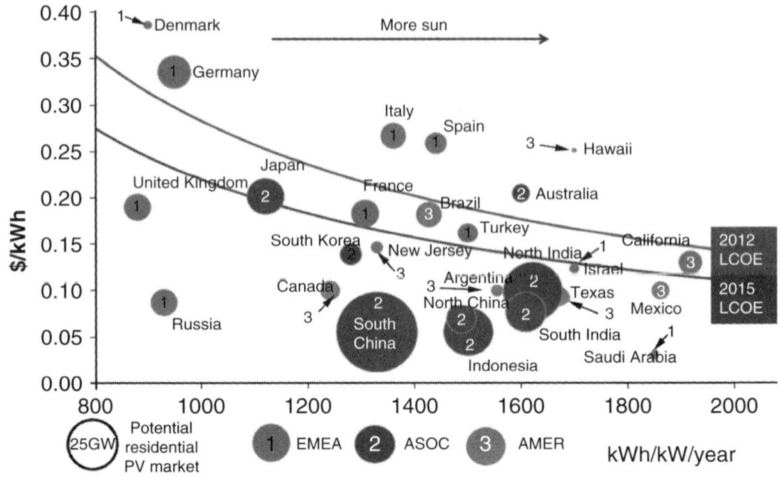

Figure 16.10 Depiction of residential PV price parity-size; bubbles indicate market size.

Spain and parts of Australia have already reached socket parity, (defined here as the point where a household can make 5% or more return on investment in a PV system just by using the energy generated to replace household energy consumption) while countries like Japan, France, Brazil and Turkey are expected to reach it by 2015.

16.4 Conclusions and Future Trends in Photovoltaic

Conjugated polymers offers unique opportunity to combine the optoelectronic properties of semiconductors along with the mechanical properties and have processing advantages of polymers. As discussed throughout this chapter in detail that the optical and electronic characteristics of conjugated polymers are primarily dependent on the nature of the molecular conjugation which may be easily designed or tailored via organic synthesis. Currently, the main efforts in the field are focused on the development of conjugated polymers with lower optical bandgap that are solvent-processable and are also reasonably stable at ambient conditions.

The new materials under development are used in conjunction with electron acceptors, such as fullerenes and CNTs, for preparation of functional hybrids which are active layer in BHJ POPV. The performance of the active layer depends not only on the structure of the different components but also on their spatial organization at the nano- to mesoscales. The morphology determined by interfacial and intermolecular interactions between the components may or may not promote the desired performance of the active layer. While morphology performance relations have been thoroughly investigated in fullerene-based hybrids, the effect of CNTs on the polymer-fullerene system is relatively less studied. The inherent structural and morphological heterogeneity of as-produced CNT powders, and the sensitivity of the polymer- CNT system to processing conditions, contribute to the wide variations in the reported observations and sometimes lead to misleading and controversial conclusions. Furthermore, the fundamental physics of charge transfer and charge transport between conjugated polymers and CNT as well as the self-assembly and phase morphology of conjugated polymers-CNT hybrids is still not understood thoroughly. More precisely, it is not yet clear which type of CNT, under which conditions, can be used as an interface under what conditions for exciton dissociation, charge transporters (holes or electrons), or contribute indirectly to the self-organization of the conjugated polymer layer. The progress and research that has been accomplished over the last decade in regulating the structure and properties of CNT will become the basic guidelines for researchers to further design and carry out well-focused experiments towards significant progress in understanding the fundamentals of optoelectronic behavior in polymer-CNT hybrids. The development of particular processing pathways each adequate for a different end product, coupled with progress in characterization techniques of the resulting hybrids, is expected to significantly improve the properties and the consequential performance of CNT-based POPV. The future of conjugated polymers CNT hybrids as active materials in POPV seems to be exceptionally bright but will certainly depend on the pace of progress in parallel areas of fundamental research and design of technologically feasible solutions for structural engineering at the nano- to the mesoscale of CNT-conjugated polymers hybrids.

PV systems are expensive and still need market support to compete with low-cost conventional electricity supply. The main research challenge and focus are therefore associated with significantly reducing costs, for the modules themselves but also for balance-of-system (BOS) costs, including in particular the inverters used for grid connection. Increase in conversion efficiency of these modules is another priority. Array mounting costs reflects the total module area of the system and thus BOS costs can be relatively reduced if efficiencies are increased. For novel but potentially extremely low-cost dye and plastic cells, there is a need to increase efficiency and develop designs with long-term stability. The biggest challenges to both dye and polymer cells are there tendency to degrade on exposure to ultraviolet radiation, atmospheric humidity and oxygen. Thin-film cell designs as discussed above are well developed; now the primary research challenge for these designs is to develop better and lower cost production technologies. The large area uniformity of deposition in this type of cell is particularly noteworthy because nonuniformity results in losses due to cell mismatch which can significantly degrade their performance.

There is an increasing need for accurate yield prediction tools with the growth of PV market as potential purchasers of such systems would be extremely interested in understanding more precisely the economics of technology. In the early stages of the growth of grid-connected PV, purchasers might not be concerned with the financial aspects, but later on as the market matures this trend will certainly change. In fact, accurate and easy-to-use yield estimation tools are required, and research is underway in many laboratories to address this issue. It is predicted that PV will make a significant contribution to the electricity supply in the long term. In a country like the UK which has a modest solar resource, estimations show that widescale deployment of PV on existing roofs could generate a substantial proportion of electricity needs.

Researchers have shown that very high performance, space-type cells could be combined with solar concentrators to produce cost-effective electricity. There are also designs available in which fluorescence is being used to capture and concentrate the light falling on a window to its edge, where it can be converted efficiently, in part because of its narrowband spectrum.

Internal energy storage may also be envisioned in future solar cell materials, e.g., analogous to a photoelectrochemical solar cell (PEC) with internal storage. Various battery technologies, e.g., metal hydrides, and nanotechnologies could represent some of many possible ways of increasing the energy storage density. Future solar cell materials may also be envisioned as thin laminate or paint layers suitable for application by paint brush or spray. However, there is still a need for "intelligent design." To attain dream of low cost, highly efficient and robust PVs technology, dedicated efforts are required to provide the technical expertise, resources, creativity and innovation. Current trends indicate that photovoltaic-based electricity will definitely prosper and lead to the Age of Neoteric Energy and beyond.

References

1. A. Jäger-Waldau, *PV Status Report 2003*, European Commission, EUR 20850 EN, 2003.
2. A.J. Heeger, *Angewandte Chemie International Edition*, Vol. 40, 2591–2611, 2001.
3. A.G. MacDiarmid, *Angewandte Chemie International Edition*, Vol. 40, 2581–2590, 2001.
4. H. Shirakawa, *Angewandte Chemie International Edition*, Vol. 40, 2574–2580, 2001.
5. C.K. Chiang, C.R. Fincher, Jr., Y.W. Park, A.J. Heeger, H. Shirakawa, E.J. Louis, S.C. Gau, and A.G. MacDiarmid, *Physical Review Letters*, Vol. 39, 1098–1101, 1977.
6. J.H. Burroughes, D.D.C. Bradley, A.R. Brown, R.N. Marks, K. Mackay, R.H. Friend, P.L. Burns, and A.B. Holmes, *Nature*, Vol. 347, 539–541, 1990.
7. J. Nelson, Materials Today, Vol. 14, 10, 462-470, 2011.
8. E. Becquerel, *Comptes Rendus*, Vol. 9, 561–567, 1839.
9. R.S. Ohl, and L. Silver, Light-Senstive Electric Device, USPTO, 2,402,662.
10. M. Tao, The Electrochemical Society Interface, Vol. 17, 4, 30-35, 2008.
11. Y. Galagan, I.G. de Vries, A.P. Langen, R. Andriessen, W.J.H. Verhees, S.C. Veenstra, and J.M. Kroon, *Chemical Engineering and Processing: Process Intensification*, Vol. 50, 454–461, 2011.
12. D. Baran, A. Balan, T. Stubhan, T. Ameri, L. Toppare, and C.J. Brabec, *Synthetic Metals*, Vol. 162, 2047–2051, 2012.
13. A. Bedeloglu, A. Demir, Y. Bozkurt, and N.S. Sariciftci, *Synthetic Metals*, Vol. 159, 2043–2048, 2009.
14. L. Blankenburg, K. Schultheis, H. Schache, S. Sensfuss, and M. Schrödner, *Solar Energy Materials and Solar Cells*, Vol. 93, 476–483, 2009.
15. F.C. Krebs, M. Biancardo, B. Winther-Jensen, H. Spanggard, and J. Alstrup, *Solar Energy Materials and Solar Cells*, Vol. 90, 1058–1067, 2006.
16. F.C. Krebs, *Organic Electronics*, Vol. 10, 761–768, 2009.
17. S.D. Hodgson, W.S.M. Brooks, A.J. Clayton, G. Kartopu, V. Barrioz, and S.J.C. Irvine, *Nano Energy*, Vol. 2, 21–27, 2013.
18. C.N. Hoth, R. Steim, P. Schilinsky, S.A. Choulis, S.F. Tedde, O. Hayden, and C.J. Brabec, *Organic Electronics*, Vol. 10, 587–593, 2009.
19. X.-L. Hu, L.-J. Zuo, Y.-X. Nan, M. Helgesen, O. Hagemann, E. Bundgaard, M.-M. Shi, F.C. Krebs, and H.-Z. Chen, *Synthetic Metals*, Vol. 162, 2005–2009, 2012.
20. R. Kroon, R. Gehlhaar, T.T. Steckler, P. Henriksson, C. Müller, J. Bergqvist, A. Hadipour, P. Heremans, and M.R. Andersson, *Solar Energy Materials and Solar Cells*, Vol. 105, 280–286, 2012.

21. J.Y. Lee, H.J. Song, S.M. Lee, J.H. Lee, and D.K. Moon, *European Polymer Journal*, Vol. 47, 1686–1693, 2011.
22. J.Y. Lee, K.W. Song, H.J. Song, and D.K. Moon, *Synthetic Metals*, Vol. 161, 2434–2440, 2011.
23. F. Li, W. Chen, K. Yuan, and Y. Chen, *Organic Electronics*, Vol. 13, 2757–2762, 2012.
24. X. Li, L. Chen, Y. Chen, F. Li, and K. Yao, *Organic Electronics*, Vol. 13, 104–113, 2012.
25. Z. Liang, R.A. Cormier, A.M. Nardes, and B.A. Gregg, *Synthetic Metals*, Vol. 161, 1014–1021, 2011.
26. C.-Y. Liao, C.-P. Chen, C.-C. Chang, G.-W. Hwang, H.-H. Chou, and C.-H. Cheng, *Solar Energy Materials and Solar Cells*, Vol. 109, 111–119, 2013.
27. S.Y. Nam, E.Y. Park, T.-D. Kim, S. Cho, J.-G. Park, and K.-S. Lee, *Current Applied Physics*, Vol. 11, e44-e48, 2011.
28. J. You, L. Dou, K. Yoshimura, T. Kato, K. Ohya, T. Moriarty, K. Emery, C.-C. Chen, J. Gao, G. Li, and Y. Yang, *Nature Communication*, Vol. 4, 1446, 2013.
29. D. Mühlbacher, M. Scharber, M. Morana, Z. Zhu, D. Waller, R. Gaudiana, and C. Brabec, *Advanced Materials*, Vol. 18, 2884–2889, 2006.
30. C.J. Brabec, M. Heeney, I. McCulloch, and J. Nelson, *Chemical Society Reviews*, Vol. 40, 1185–1199, 2011.
31. M. Heeney, W. Zhang, D.J. Crouch, M.L. Chabinyc, S. Gordeyev, R. Hamilton, S.J. Higgins, I. McCulloch, P.J. Skabara, D. Sparrowe, and S. Tierney, *Chemical Communications*, Vol. 0, 5061–5063, 2007.
32. J. Peet, J.Y. Kim, N.E. Coates, W.L. Ma, D. Moses, A.J. Heeger, and G.C. Bazan, *Nat Mater*, Vol. 6, 497–500, 2007.
33. S.H. Park, A. Roy, S. Beaupre, S. Cho, N. Coates, J.S. Moon, D. Moses, M. Leclerc, K. Lee, and A.J. Heeger, *Nat Photon*, Vol. 3, 297–302, 2009.
34. Y. Liang, Z. Xu, J. Xia, S.-T. Tsai, Y. Wu, G. Li, C. Ray, and L. Yu, *Advanced Materials*, Vol. 22, E135-E138, 2010.
35. H. Bronstein, Z. Chen, R.S. Ashraf, W. Zhang, J. Du, J.R. Durrant, P. Shakya Tuladhar, K. Song, S.E. Watkins, Y. Geerts, M.M. Wienk, R.A.J. Janssen, T. Anthopoulos, H. Sirringhaus, M. Heeney, and I. McCulloch, *Journal of the American Chemical Society*, Vol. 133, 3272–3275, 2011.
36. T. Clarke, A. Ballantyne, F. Jamieson, C. Brabec, J. Nelson, and J. Durrant, *Chemical Communications*, Vol. 0, 89–91, 2009.
37. E. Wang, L. Wang, L. Lan, C. Luo, W. Zhuang, J. Peng, and Y. Cao, *Applied Physics Letters*, Vol. 92, 033307–033303, 2008.
38. M. Lenes, M. Morana, C.J. Brabec, and P.W.M. Blom, *Advanced Functional Materials*, Vol. 19, 1106–1111, 2009.
39. Y. He, H.-Y. Chen, J. Hou, and Y. Li, *Journal of the American Chemical Society*, Vol. 132, 5532–5532, 2010.
40. R.B. Ross, C.M. Cardona, D.M. Guldi, S.G. Sankaranarayanan, M.O. Reese, N. Kopidakis, J. Peet, B. Walker, G.C. Bazan, E. Van Keuren, B.C. Holloway, and M. Drees, *Nat Mater*, Vol. 8, 208–212, 2009.
41. M.A. Faist, P.E. Keivanidis, S. Foster, P.H. Wöbkenberg, T.D. Anthopoulos, D.D.C. Bradley, J.R. Durrant, and J. Nelson, *Journal of Polymer Science Part B: Polymer Physics*, Vol. 49, 45–51, 2011.

42. J.M. Frost, F. Cheynis, S.M. Tuladhar, and J. Nelson, *Nano Letters*, Vol. 6, 1674–1681, 2006.
43. J.J.M. Halls, K. Pichler, R.H. Friend, S.C. Moratti, and A.B. Holmes, *Applied Physics Letters*, Vol. 68, 3120–3122, 1996.
44. R.R. Lunt, J.B. Benziger, and S.R. Forrest, *Advanced Materials*, Vol. 22, 1233–1236, 2010.
45. D. Veldman, O.z. Îpek, S.C.J. Meskers, J.r. Sweelssen, M.M. Koetse, S.C. Veenstra, J.M. Kroon, S.S.v. Bavel, J. Loos, and R.A.J. Janssen, *Journal of the American Chemical Society*, Vol. 130, 7721–7735, 2008.
46. Y. Kim, S. Cook, S.M. Tuladhar, S.A. Choulis, J. Nelson, J.R. Durrant, D.D.C. Bradley, M. Giles, I. McCulloch, C.-S. Ha, and M. Ree, *Nat Mater*, Vol. 5, 197–203, 2006.
47. N.C. Cates, R. Gysel, J.E.P. Dahl, A. Sellinger, and M.D. McGehee, *Chemistry of Materials*, Vol. 22, 3543–3548, 2010.
48. N.C. Cates, R. Gysel, Z. Beiley, C.E. Miller, M.F. Toney, M. Heeney, I. McCulloch, and M.D. McGehee, *Nano Letters*, Vol. 9, 4153–4157, 2009.
49. C. Müller, T.A.M. Ferenczi, M. Campoy-Quiles, J.M. Frost, D.D.C. Bradley, P. Smith, N. Stingelin-Stutzmann, and J. Nelson, *Advanced Materials*, Vol. 20, 3510–3515, 2008.
50. S. Bertho, G. Janssen, T.J. Cleij, B. Conings, W. Moons, A. Gadisa, J. D'Haen, E. Goovaerts, L. Lutsen, J. Manca, and D. Vanderzande, *Solar Energy Materials and Solar Cells*, Vol. 92, 753–760, 2008.
51. R.C. Hiorns, E. Cloutet, E. Ibarboure, A. Khoukh, H. Bejbouji, L. Vignau, and H. Cramail, *Macromolecules*, Vol. 43, 6033–6044, 2010.
52. J. Boucle, P. Ravirajan, and J. Nelson, *Journal of Materials Chemistry*, Vol. 17, 3141–3153, 2007.
53. M. Campoy-Quiles, T. Ferenczi, T. Agostinelli, P.G. Etchegoin, Y. Kim, T.D. Anthopoulos, P.N. Stavrinou, D.D.C. Bradley, and J. Nelson, *Nat Mater*, Vol. 7, 158–164, 2008.
54. T.A.M. Ferenczi, N. Jenny, B. Colin, M.B. Amy, C.-Q. Mariano, M.B. Felix, and D.C.B. Donal, *Journal of Physics: Condensed Matter*, Vol. 20, 475203, 2008.
55. L.L. Kazmerski, Solar Photovoltaics Technology: No Longer an Outliner, in A. Sayigh Ed., Comprehensive Renewal Energy, Vol. 1, 13-30, 2008.
56. Solar cell- Supplier and Manufacturers A-Z, www.speedace.info/solar_cells.htm (accessed on March, 01, 2013).
57. Renewable Energy Tariff for Financial Year 2011–12, Order-256–2010, Central Electricity Regulatory Commission, India, 2012.
58. A. Jäger-Waldau, "Status and perspectives of thin film photovoltaics," in A. Bosio and A. Romeo, eds., *Thin Film Solar Cells: Current Status and Future Trends*, Nova Publisher, New York, ch. 1, pp. 1–24, 2010.

17

Real and Reactive Power Control of Voltage Source Converter-Based Photovoltaic Generating Systems

S. Mishra* **and P. C. Sekhar**

Department of Electrical Engineering, Indian Institute of Technology Delhi, New Delhi, India.

Abstract

The PV generator power output is highly influenced by the solar insolation and hence it has to be controlled properly for maximum possible power extraction. The increase in penetration of distributed generators which are on the load side of the power system, like photovoltaic sources, poses a serious threat to the system and its associated devices by raising the voltage levels of the network due to the reverse power flow in the lines. The conventional voltage regulation schemes which are formulated for the unidirectional power flow, from the centralized generator to the load, are not effective for the voltage regulation with reverse power flow from the distributed generators. Hence, a novel control scheme has been formulated in this study for the mitigation of voltage rise with a voltage source converter-based PV generator. Further, the PV generator has been proposed to be controlled as a reactive power supplying ancillary.

Keywords: Decoupled control, distributed generation, maximum power point tracking, PV generation, real and reactive power control, voltage regulation, voltage source converter

Corresponding author: sukumar@ee.iitd.ac.in

Atul Tiwari, Rabah Boukherroub, and Maheshwar Sharon (eds.) Solar Cell Nanotechnology, (475–504) 2014 © Scrivener Publishing LLC

17.1 Introduction

The arena of connection of dispersed generators (DGs) to the grid which transforms the distribution system into an active network from its existing passive state, is gaining prominence. The active distribution networks involve bidirectional power flow, from the source as well as load sides. It permits the installation of relatively small scale generators including the energy storages at the medium and low voltage distribution levels. Most of these DGs are renewable energy sources such as photovoltaic (PV), wind, fuel cells, etc., with a small number of conventional diesel generators.

Among the available DGs, photovoltaic (PV) systems have drawn prominent attention both from industry and academic communities in the last three decades. In spite of high capital cost and poor conversion efficiency, the PV systems can be treated as a viable alternate for energy generation due to their good list of merits and potential long-term benefits. The PV systems have already become an integral part of power generation systems, especially in DGs, and there is an anticipation of more penetration worldwide. All the renewable energy generators including the PV generators are controlled to operate at their maximum possible power outputs due to their eco-friendly nature and high capital cost involvement. Considering the PV generator, being a photosensitive device, its output is highly influenced by the incident insolation on the PV module which is the power generating device of the PV generator. The output of the PV generator is also affected by its operating temperature to some extent, however, not as much as it is influenced by the incident solar insolation. Another key issue that is to be noted here is that the power output of the PV generator (P_{PV}) will vary according to the terminal voltage (V_{dc}) where it is operating, and the current (I_{PV}) what it is delivering to the external circuit, hence exhibits a nonlinear relationship between the P_{PV} and the operating voltage (V_{dc}) or the current (I_{PV}) as shown in Figure 17.1. The power as well as the current output of any PV generator increases with increase in insolation with all other parameters unchanged.

The PV generator output current and its operating voltages are interdependent on each other. It is important to note here that, even though the P_{PV} varies with V_{dc}, there is a unique operating voltage, known as maximum power point (MPP) voltage (V_{MPP}), where the P_{PV} will be at its maximum possible value for a specific insolation and temperature. This concludes that the PV generator has to be operated at that particular MPP to extract the maximum possible

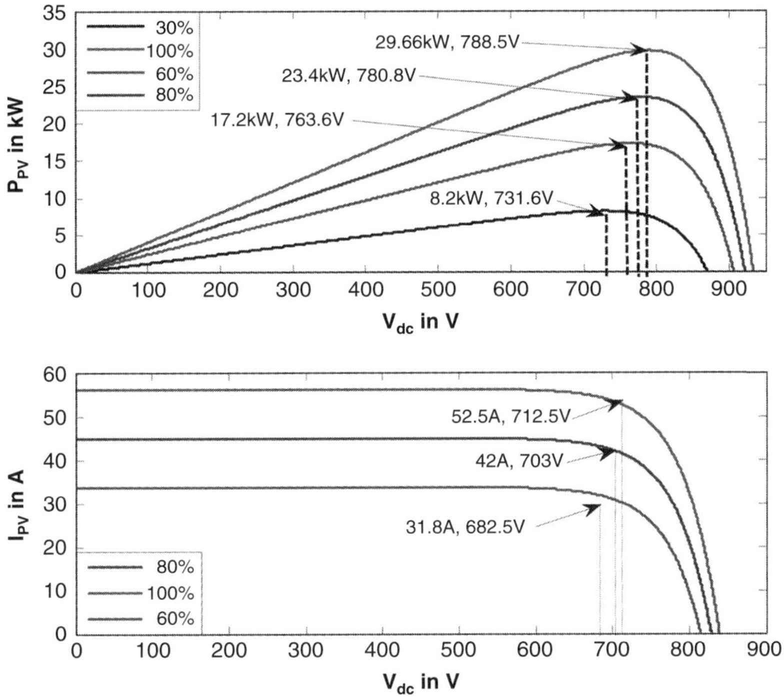

Figure 17.1 Variation of the PV generator power and its MPP current with operating voltage at various incident insolations.

power from the PV generator. However, this MPP is not a constant operating point and will vary depending on the incident insolation and the operating temperature. Hence, a mechanism to detect the unique operating point (MPP) for varying operating conditions to meet the objective of maximum power extraction from the PV generator, known as *maximum power point tracking* (MPPT) is necessary. Due to the nature of intermittency and fluctuating power outputs, the PV generators are to be controlled properly so that there is complete evacuation of available maximum power without affecting the performance of the system and the different devices connected to it.

The P-V characteristics of any PV array are nonlinear as shown in Figure 17.1 in which 100% of irradiation corresponds to 1000W/m^2. The objective of the MPPT algorithm is to provide the extreme point on these nonlinear characteristic, which is nothing but maximum power point (MPP).

For stable and satisfactory operation of any power system, it is essential to have a regulated frequency along with the voltage at their nominal values. High penetration of DGs is advisable to meet the increasing demands of strict greenhouse gas emission regulations and cannot be avoided; however, it poses a severe voltage regulation problem due to the reverse power flow from DG which can result in the rise of distribution line voltage level. The conventional voltage regulation methods used by utility side regulators are designed for unidirectional power flow from the centralized power plants, and may no longer be valid for the voltage control. In addition, the evolution of smart grid/microgrid demands automated quick active and reactive power control with minimum human intervention.

17.2 State of Art

The rise in the voltage can be mitigated by storing the surplus power, which is responsible for the voltage rise, using suitable storage devices. However, it incurs high initial and periodic maintenance costs. As suggested in [1] and [2], the DG-injected power can be curtailed to regulate the voltage rise; however, it diminishes the energy efficiency of the PV system straightforwardly. Determination of control settings for the devices such as tap changing transformers, shunt capacitors and line regulators which are used for voltage regulation in distribution systems at the substations is a challenging task in a system with DGs [3, 4]. The problem will be more severe in the case of multiple DGs with PV and wind as power sources, as they result in frequent voltage regulations which will damage the voltage regulating devices due to their fluctuating nature [5, 6]. To control the under load tap changer, a multiple line drop compensation method was proposed in [7]. Reduction of voltage regulation by loss minimization was addressed in [8, 9]; reconfiguration of network reconfiguration was considered in [6, 10]. The voltage control with line droop compensator (LDC) still poses a serious problem whose compensation is based on the estimation of drop in the voltage at the remote point considering the measurements at the voltage regulator. Since these regulators are designed without considering intermediate power injections, the line with intermediate DG may experience excessive low or high voltages towards the feeder ends, based on location and capacity of the DG and its net power injection to the grid. However some literature proposes optimal tap setting [11] which requires the information of all the nodes, whereas others propose methods to estimate the output of remote DG [12] which is not clear about the case with multiple DGs. Hence, in a system with high DG penetration, especially with fluctuating power sources such as PV and wind, the voltage regulation at the distribution level is a critical task which is to be addressed carefully. As the MPP changes with changing weather conditions, tracking MPP and evacuating the maximum available power without affecting other parameters of the system is highly essential. Keeping this in view, the transient and dynamic performance analysis of PV systems and their controllers' investigations for varying weather conditions, its interactions with the grid/network and changing loads can be treated as of timely importance.

17.3 Proposed Solution

The MPPT mechanism is essential to detect the MPP and provide the reference set points for the controller. A good number of MPPT algorithms have been proposed in the literature which will differ in their performance, simplicity in hardware realization, application area and cost of implementation, etc. Different MPPT algorithms evolved over time and detailed reviews are presented in [13, 14] starting from 1968 to 2012. Among the different MPPT techniques, the perturb and observe (P&O) method is a widely used algorithm which is considered in this study owing to its simplicity, low cost and easy implementation.

The primary function of a PV-based DG is to deliver the real power to the utility grid at constant power factor. In the constant power factor based operation, which is considered to be appropriate for DGs, the reactive power output will vary according to the real power output. Since the drop in the line will vary according to the total current; the voltage at the PCC will vary accordingly. Even if the DG is operated at unity power factor, it causes lower line losses but a rise in the voltage. Hence, in the UPF operation, the PV output is curtailed to mitigate the voltage rise, and results in not only loss of energy but limits the rate of penetration as well. However, according to [15], the constant voltage operation is a better strategy to increase the penetration rate. Even though a wide variety of DGs are available and integrated with the grid, all of them are not capable of controlling the voltage with a minimal effect on real power feeding. However, the capability of DGs, which are integrated with the distribution grid through voltage source converters (VSCs), to control the terminal voltage due to the evolution of decoupled and control theories without affecting the real power injection has not been seriously considered yet. This can be achieved by controlling the reactive power feeding of the VSC which incurs minimal extra hardware and cost, without any complexity and just by modifying the control philosophy. Even though the operation policies do not support the control of voltage by DGs as on today, the anticipated high PV penetration which necessitates the control/supply of the reactive power from the PV-based DGs as an additional ancillary service, may drive the change in policies in the near future. Hence in this study, a PV-based grid integrated DG supplying the local load is considered, which uses decoupled controller for VSC to control the dc as well as ac side dynamics. The P&O-based MPPT is used to extract the maximum possible real power. The capability of supplying a negative as well as positive reactive power of the VSC-based PV generator is utilized to supply the controlled reactive power according to the specified power factor as an ancillary. To suppress the voltage rise in the system due to the reverse power flow from the PV generator, a novel drop/rise estimation scheme is proposed to control the reactive power flow with due attention to parameter uncertainties and disturbances. The robustness of the developed control scheme is demonstrated for the specified objectives of extracting maximum possible real power from the PV generator, achieving zero voltage regulation at PCC and operating the PV generator as an ancillary service to supply the reactive power at specified power factors under different operating conditions.

17.4 Modeling of the PV Generator

The system considered is of a VSC interfaced rooftop PV generator supplying the grid, controlled by decoupled controller whose schematic block diagram is shown in Figure 17.2. The basic element of any PV generator is the PV cell whose equivalent circuit model with current source representing the photo current (I_{ph}), a diode, a parallel resistance to consider the leakage current and a series resistance to account for internal resistance to the current flow as shown

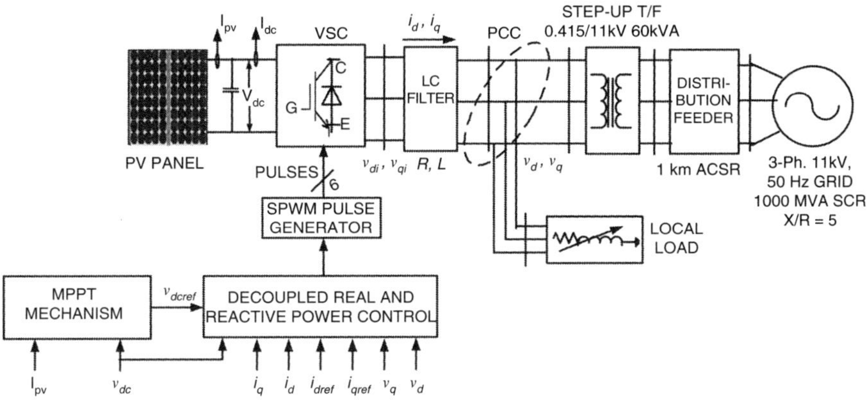

Figure17.2 Schematic block diagram of rooftop PV generator connected to grid; interfaced through VSC.

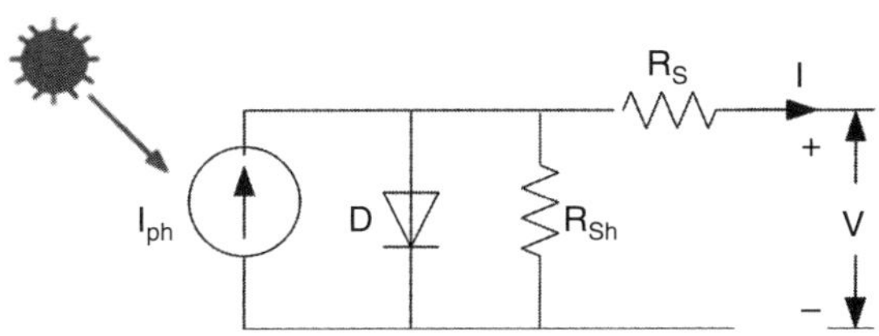

Figure 17.3 Equivalent circuit model of a PV cell.

in Figure 17.3, is the widely accepted general model. The output current of the PV cell can be given by (17.1),

$$I = I_{ph} - I_{rs}\left(\exp\left(\frac{q}{kT_cA}(V+IR_s)\right)-1\right)-\frac{(V+IR_s)}{R_{sh}} \quad (17.1)$$

where, I_{ph} is photocurrent, I_{rs} is the cell reverse saturation current (of *p-n* junction), q is charge of an electron (1.602×10^{-19}C), k is a Boltzmann's constant (1.38×10^{-23}J/K), T_c is working temperature of the cell, A is the ideality factor, R_{sh} is shunt resistance and is inversely proportional to the leakage current, and R_s is series resistance. The photo current or light-generated current varies with the cell's working temperature T_c, and irradiation S, is given by (17.2),

$$I_{ph} = \left[I_{sc}+K_I\left(T_c-T_{ref}\right)\right]S \quad (17.2)$$

where, I_{SC} is the cell's short-circuit current at reference temperature T_c, and irradiation S (W/m^2), K_I is short-circuit current temperature coefficient of the cell, and T_{ref} is the cell's reference temperature. For a PV cell with good manufacturing standards in terms of materials used and the process of manufacturing the leakage and series resistances can be neglected, since these have a negligible effect on the generator's efficiency. Hence,

$$I = I_{ph} - I_{rs}\left(\exp\left(\frac{qV}{kT_cA}\right) - 1\right) \tag{17.3}$$

For a PV module with N_p number of parallel cells and N_s number of series cells the output current can be given by (17.4),

$$I_{pv} = N_pI_{ph} - N_pI_{rs}\left(\exp\left(\frac{q}{kT_cA}\frac{V_{dc}}{N_s}\right) - 1\right) \tag{17.4}$$

where V_{dc} is output voltage and I_{PV} is output current of PV module. Hence, the power output of the PV generator is:

$$P_{pv} = f(V_{dc}, S, T_c) = V_{dc} * I_{PV} \tag{17.5}$$

All the other parameters are given in Table 17.1.

Table 17.1 PV Module Parameters.

Parameter	**Value**
Number of parallel and series cells, N_p and N_s	5, 1050
Cell's short circuit current, I_{SC}	8.03 A
Cell reverse saturation current, I_{rs}	1.2e-7 A
Charge of an electron, q	1.602×10^{-19}C
Boltzmann's constant, k	1.38×10^{-23}J/K
Ideality factor, A, Cell's reference temp, T_{ref}	1.92, 300^{0}K
Short circuit current temperature coefficient of cell, K_I	0.0017

The other system parameters are:

Transformer : 415 V/11 kV, 50 kVAr
Distribution line : ACSR conductor of 1km length, X/R=3.5
Gird : 11 kV, 50 Hz, SCMVA-1000, X/R=5
PI controller parameters : K_{pdc} = - 0.0102, K_{idc} = - 0.102, $K_{pd} = K_{pq}$ = 0.80677, $K_{id} = K_{iq}$ = 1; L = 0.80677 mH, R= 1mΩ.

17.5 Control of the PV Generator

The decoupled controller of VSC is designed to deliver the maximum possible real power under varying insolations. The backbone for the real power control is MPPT mechanism which governs the decoupled controller of VSC by providing V_{dcref}. The decoupled controller controls the dc link voltage V_{dc} according to the V_{dcref}, thereby ensuring the maximum possible power extraction from the PV generator at all operating conditions. For the reactive power control, two distinct control philosophies are considered such as: (i) constant power factor control and (ii) zero voltage regulation control. For reactive power control, in zero voltage regulation mode, a novel drop/rise estimation scheme is proposed with due attention to parameter uncertainties and disturbances. The reference currents (i_{dref} and i_{qref}) are derived using the *vector-based approach* which are further used to derive modulation indices (m_d and m_q) for the control VSC of PV generator.

The main control objectives addressed in this study are: 1) control of the dc link voltage for quick and accurate tracking of MPP in changing insolations and operating at MPP in steady insolations, and 2) control of reactive power either to maintain the power factor as an ancillary service or to regulate the voltage. All these control objectives are achieved through co-operative nested control architecture with interleaved inner current control and outer voltage control loops. The main advantage of this approach is that it is autonomous control and can control the local bus voltage with locally measurable signals thereby addressing the load fluctuations effectively when compared with other voltage regulating devices.

17.5.1 Maximum Power Point Tracking and the Perturb and Observe Algorithm

The MPPT ensures the evacuation of maximum possible power from the PV generator all the time. Among the different MPPT mechanisms, the P&O is the widely adopted one due to its merits, as aforementioned. In P&O the voltage is varied with a small value, known as perturbation ΔV. This ΔV could be a positive value or a negative value. Once the system gets settled, the tracking mechanism will generate its next perturb. It decides the polarity of next perturb only after observing the effect of earlier perturb, as its name

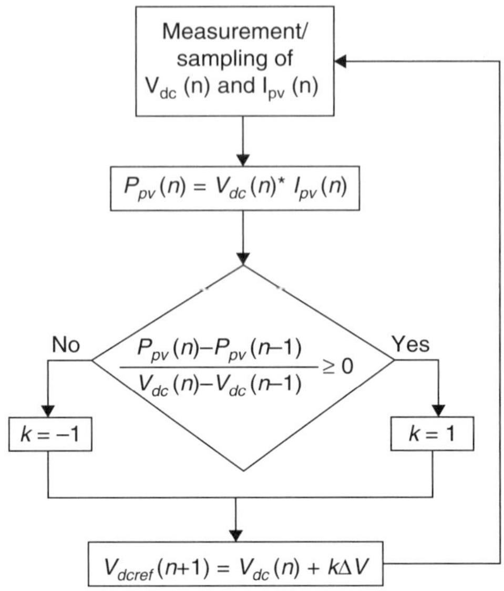

Figure 17.4 Perturb and Observe MPP tracking algorithm.

suggests (P&O). Lastly, the system will settle at a voltage that corresponds to the maximum power. The P&O algorithm changes the voltage at constant sampling rate to track the maximum power irrespective of change in operating conditions [16]. The flow chart of the P&O algorithm is shown in Figure 17.4.

17.5.2 Control of Reactive Power Output from the PV System

As aforementioned, since the PV generator is interfaced with the grid through a VSC, one can control the reactive power output from the PV generator without affecting the real power transfer. This reactive power control could be based on constant power factor (*p.f.*) mode or voltage regulation mode.

Constant power factor mode of operation:

In constant *p.f.* mode, the ratio of real power transfer (W or kW) to the total apparent power (VA or kVA) is constant. Since the PV power output is subjected to variations with changing environment, the reactive power output also needs to be controlled accordingly to maintain constant *p.f.* In constant *p.f.* mode of operation, the reactive power reference Q_{ref} can be calculated using Eq. 17.6.

$$Q_{ref} = P_{pv} * \tan(\cos^{-1}(pf)) \tag{17.6}$$

If the VSC is operating at unity power factor, $Q_{ref} = 0$.

Zero voltage regulation mode of operation:

Depending on the direction and amount of power flow (either of P or Q or both) it incurs some drop/rise in the voltage, which results in nonzero voltage regulation. The phasor representation of the system for voltage regulation is shown in Figure 17.5.

If it is assumed that when there is no power flow, the voltage at the PCC will be at $V_{nominal} = 415\text{V}$ (*say*), then the drop in the system voltage (change/deviation of voltage at PCC from its nominal value) is given by,

$$\Delta v = \sqrt{\Delta v d^2 + \Delta v q^2} \tag{17.7}$$

where $\Delta v_d = i_d R + (-i_q X)$ and, $\Delta v_q = -(-i_q R) + i_d X$, R and X are resistance and reactance of power flow path respectively, up to the PCC where the voltage is to be regulated (from Figure 17.5). Consider $\Delta v_d = i_d R + (-i_d X)$

$$\Delta v_d = \left(\frac{P}{v_d}\right) R + \left(\frac{Q}{v_d}\right) X, \tag{17.8}$$

$$\Rightarrow \Delta v_d = \left(\frac{PR + QX}{v_d}\right) \tag{17.9}$$

where P and Q are real and reactive power feedings which are flowing in the line.

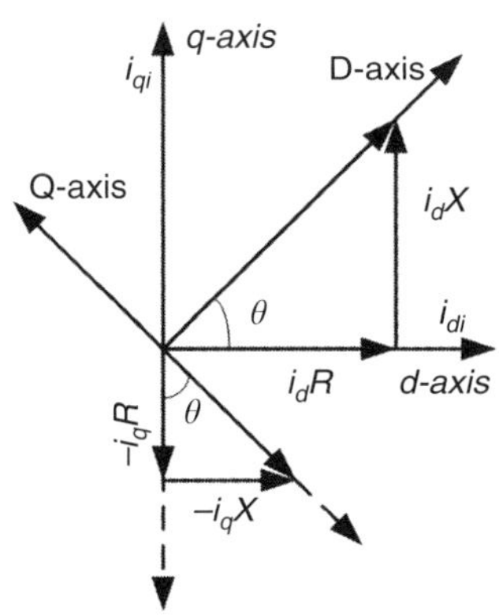

Figure 17.5 Phasor representation of the system for voltage regulation.

Similarly,

$$\Delta v_q = -(-i_q R) + i_d X \tag{17.10}$$

$$\Delta v_q = \left(\frac{-Q}{v_d}\right)R + \left(\frac{P}{v_d}\right)X \tag{17.11}$$

$$\Rightarrow \Delta v_q = \left(\frac{-QR+PX}{v_d}\right) \tag{17.12}$$

Net voltage regulation Δv will be the resultant of Δv_d and Δv_q. Hence, in unity *p.f.* operation, the voltage regulation is purely governed by the real power delivered as the reactive power is equal to zero. If the PV system is operating at *p.f.* other than unity, the net Δv will be:

$$\Delta v = \sqrt{\Delta v_d^2 + \Delta v_q^2} = \sqrt{\left(\frac{PR+QX}{v_d}\right)^2 + \left(\frac{-QR+PX}{v_d}\right)^2} \tag{17.13}$$

In order to attain zero voltage regulation, both the Δv_d and Δv_q needs to be made zero.
Consider,

$$\Delta v_d = \frac{PR+QX}{v_d} = 0 \tag{17.14}$$

$$\Rightarrow PR = -QX \tag{17.15}$$

Hence, Q_{ref} that is to be injected to make $\Delta v_d = 0$, can be decided as:

$$Q_{ref} = -\frac{PR}{X} \tag{17.16}$$

Similarly, to make $\Delta v_q = 0$,

$$\Delta v_q = \left(\frac{-QR+PX}{v_d}\right) = 0 \tag{17.17}$$

$$\Rightarrow P(R^2 + X^2) = 0 \tag{17.18}$$

$\Delta v_q = 0$ only when, both R and X are zero. In other way one can get $\Delta v_q = 0$ only if $P = 0$, which is nothing but zero real power injection by the PV system. Hence, to compensate for this drop and thereby

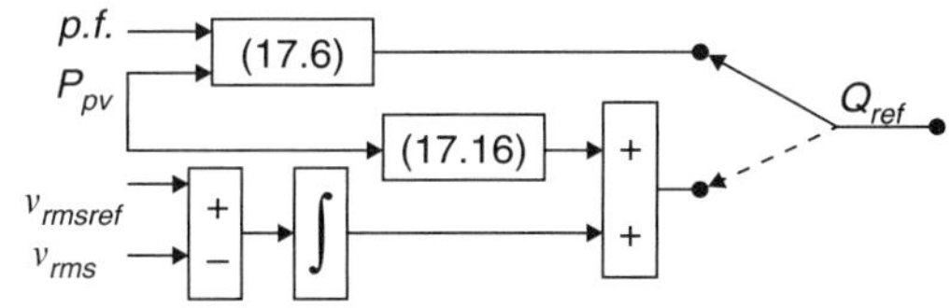

Figure 17.6 Reference decision reactive power control under different operating modes.

to attain zero voltage regulation at PCC, an extra amount of reactive power needs to be injected by the PV system, even with non-zero real power injection. This will easily and effectively be carried out by a simple integrator, which also will take care of parameter uncertainties and their estimation errors.

Altogether, the reactive power control and its reference decisions according to different operating modes are as shown in Figure. 17.6

17.5.3 DC Link Voltage Control for Maximum Power Extraction

The V_{dcref}, where the PV generator is controlled to be operated will be decided by the MPPT mechanism as explained in the preceding section. A PI controller is used to derive the Δi_{dcref} after processing the error between the actual dc voltage, V_{dc} and V_{dcref}, as in [17]. In this proposed scheme, the dc link voltage is controlled to ensure the stable operation of the PV system along with maximum power extraction from it. Once the controller is provided with I_{dcref}, the system will extract specified dc current subjected to the delay involved with the controller and the VSC which can be approximated as first order transfer function [18, 19]. The control loop for DC link voltage regulation is as shown in Figure 17.7. Applying KCL at the dc link capacitor (Figure 17.2):

$$I_{dc} = I_{pv} - I_c \tag{17.19}$$

$$I_{dc} = I_{pv} - C\frac{dV_{dc}}{dt} \tag{17.20}$$

The Bode plot approach is followed in this study to tune the parameters of the PI controller for the DC link voltage control. The Bode plot of the DC voltage control loop for the system considered

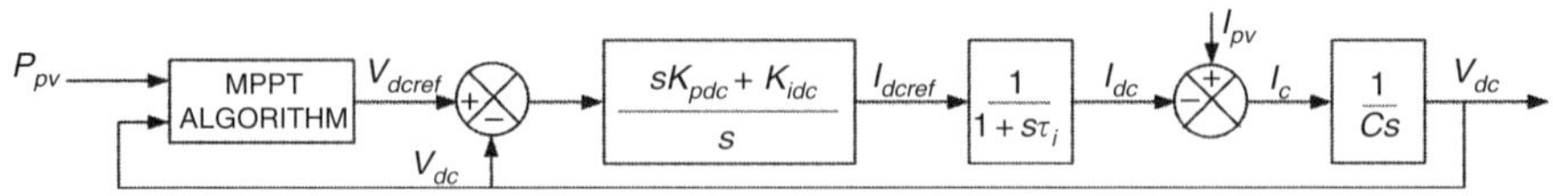

Figure 17.7 Block diagram for the DC voltage control loop.

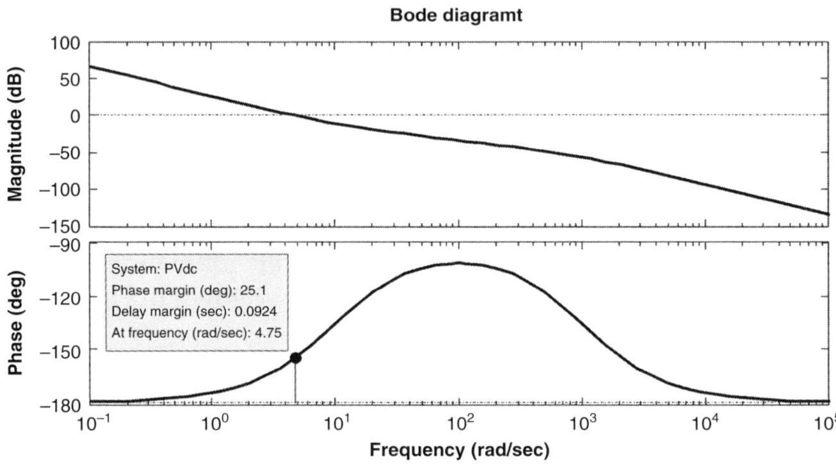

Figure 17.8 Bode plot for the DC link voltage control loop of the system.

is shown in Figure. 17.8. The PI controller parameters are chosen such that the phase margin is 25^0. All the parameters are given in Table 17.1.

17.5.4 Reference Current Generation for Voltage Source Converter

The phasor representation of the considered PV-generating system for real and reactive power control is shown in Figure 17.9. The controller should act such that the entire power on the dc side needs to be evacuated to maintain the V_{dc} at its reference value, if the loss in the inverter and the filter is neglected. Hence, the D-axis current (ac side of the VSC) can be calculated as $V_{dc}I_{dcref}/v$. The Q-axis current will be equal to Q_{ref}/v, Q_{ref} is the amount of reactive power that the VSC is directed to supply, which will be decided according to the mode of reactive power control, as explained in the preceding sections. Hence, the d and q axis reference currents i_{dref} and i_{qref} can be calculated as Eq. 17.21.

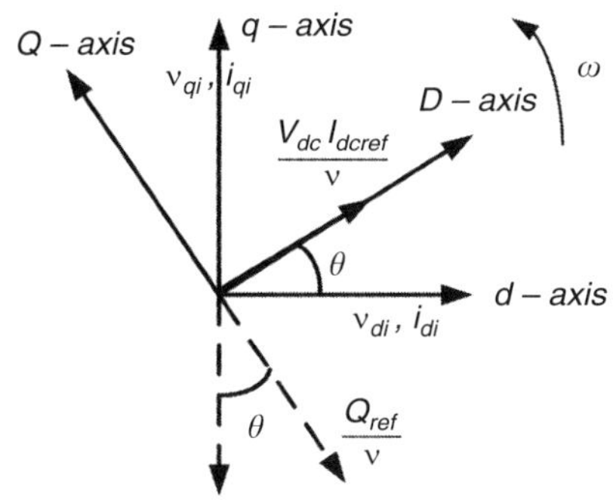

Figure 17.9 Phasor representation of system for referencing current calculations.

$$\begin{bmatrix} i_{dref} \\ i_{qref} \end{bmatrix} = \begin{bmatrix} \cos\theta & \sin\theta \\ \sin\theta & -\cos\theta \end{bmatrix} \begin{bmatrix} \frac{V_{dc} I_{dcref}}{v} \\ \frac{Q_{ref}}{v} \end{bmatrix} \tag{17.21}$$

where $v = \sqrt{v_d^2 + v_q^2}$ and $\theta = \tan^{-1}\left(\frac{v_q}{v_d}\right)$

17.5.5 Decoupled Control of VSC-based PV Generating System

The basic block diagram for the proposed control scheme is shown in Figure 17.10. From the inverter end to the PCC, the system considered can be represented mathematically as:

$$v_i = \dot{m}\frac{V_{dc}}{2} = R\dot{i} + L\frac{d\dot{i}}{dt} + \dot{v} \tag{17.22}$$

All the three-phase ac signals are transformed into proper dq frame counterparts with the help of phased-locked loop (PLL). Hence, the controllers process dc equivalents only, not the original sinusoidally varying signals. The above Eq. 17.22 can be rewritten in dq domain as:

$$m_d \frac{V_{dc}}{2} = Ri_d + L\frac{di_d}{dt} + v_d - \omega L i_q \tag{17.23}$$

$$m_q \frac{V_{dc}}{2} = Ri_q + L\frac{di_q}{dt} + v_q + \omega L i_d \tag{17.24}$$

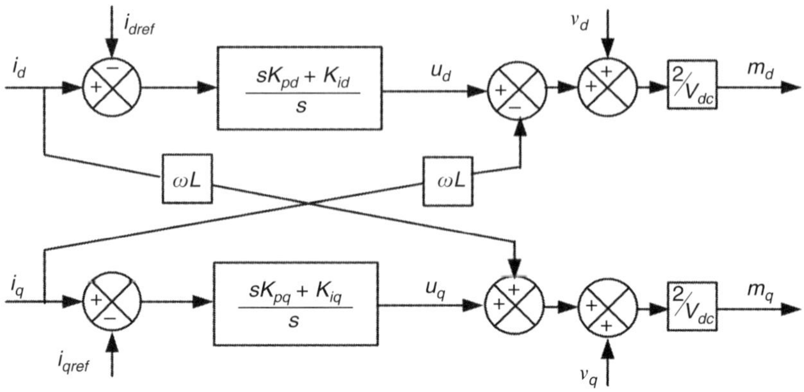

Figure 17.10 Control block diagram of the VSC for decoupled control.

$$\Rightarrow L\frac{di_d}{dt} = -Ri_d + \left(-v_d + \omega Li_q + m_d\frac{V_{dc}}{2}\right) \tag{17.25}$$

$$\Rightarrow L\frac{di_q}{dt} = -Ri_q + \left(-v_q - \omega Li_d + m_q\frac{V_{dc}}{2}\right) \tag{17.26}$$

From Eqs. 17.25 and 17.26, it can be observed that the dynamics of the currents i_d and i_q are nonlinear and coupled through a factor ωL. In order to avoid this coupling and to linearize the dynamics m_d and m_q can be determined by the following control laws [19] whose control structure is as shown in Figure 17.10.

$$m_d = \frac{2}{V_{dc}}\left(u_d + v_d - \omega Li_q\right) \tag{17.27}$$

$$m_q = \frac{2}{V_{dc}}\left(u_q + v_q + \omega Li_d\right) \tag{17.28}$$

Hence Eqs. 17.25 and 17.26 can be rewritten as:

$$\Rightarrow L\frac{di_d}{dt} = -Ri_d + u_d \tag{17.29}$$

$$\Rightarrow L\frac{di_q}{dt} = -Ri_q + u_q \tag{17.30}$$

From the above decoupled Eqs. 17.29 and 17.30, it can be concluded that the currents i_d and i_q can be controlled by controlling u_d and u_q. Since the control plants for these currents are the same:

$$\frac{sK_{pd}+K_{id}}{s}=\frac{sK_{pq}+K_{iq}}{s} \quad (17.31)$$

The proportional and integral gains can be calculated as,

$$K_{pd}=K_{pq}=K_iL \quad (17.32)$$

$$K_{id}=K_{iq}=K_iR \quad (17.33)$$

$$K_i=\frac{1}{\tau_i} \quad (17.34)$$

where, τ_i denotes the response time of the VSC which is considered to be 1ms.

17.6 Validation of the Proposed Control Architecture

In this section the capability of the PV generator and the developed control philosophies for the aforementioned three objectives are validated under different operating conditions. The incident insolation is assumed to be at 50% (1000 W/m^2 corresponds to 100%) for the first 2s and is changing at every 2s. The insolation is changing from 50% to 100%, from 100% to 80%, from 80% to 90%, from 90% to 60% and from 60% to 100% at 2s, at 4s, at 6s, at 8s and at 10s, respectively, as shown in Figure 17.11.

17.6.1 Control of Real Power Feeding of PV Generator and PCC Voltage under Reverse Power Flow in a Distribution System

In this case study the developed control scheme is examined to control the real power feeding by the PV generator interfaced through VSC such that the maximum possible power will be extracted from the PV generator under varying insolations. The capability of the VSC-interfaced PV generator along with the developed control strategy to regulate the PCC voltage to its nominal value is examined under varying reverse power (real) flows in a distribution system are examined in this case study. Figure 17.12 shows the extracted real power from the PV generator against its ideal values, neglecting the losses in the converters. It can be seen from Figure 17.12 that

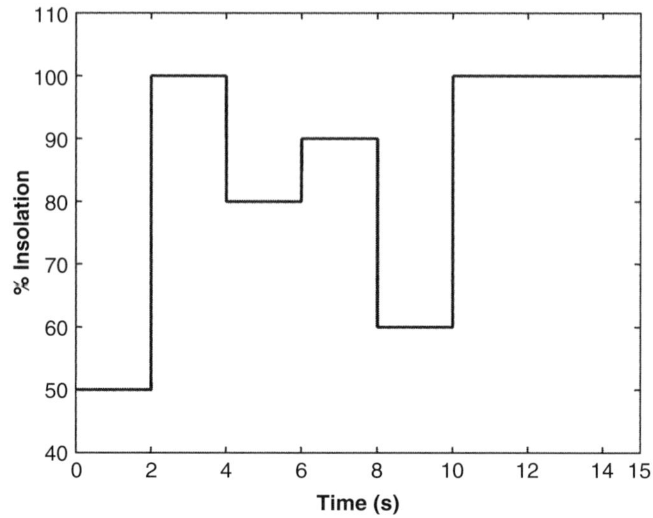

Figure 17.11 Variation of incident insolation on the PV module.

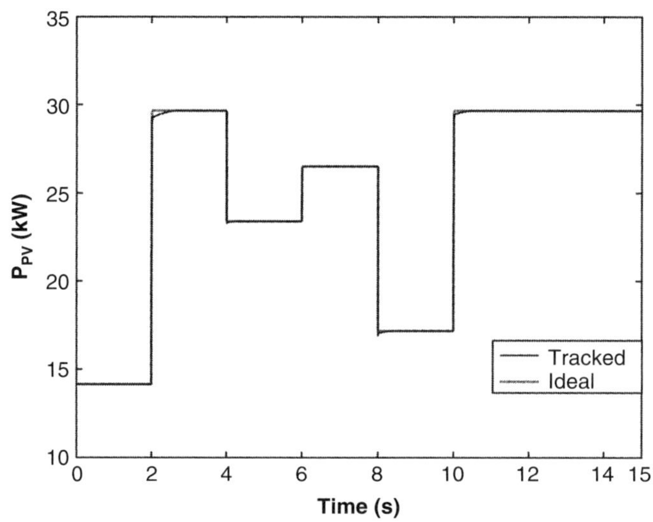

Figure 17.12 Power evacuated from the PV generator under varying incident insolations.

the developed controller is able to extract the maximum possible power from the PV generator at all the incident insolations, irrespective of insolation and its change. The developed controller is able to track the new MPP very quickly and is maintaining that level of power as output from the PV generator till the incident insolation changes, proving the excellent tracking and stability characteristics

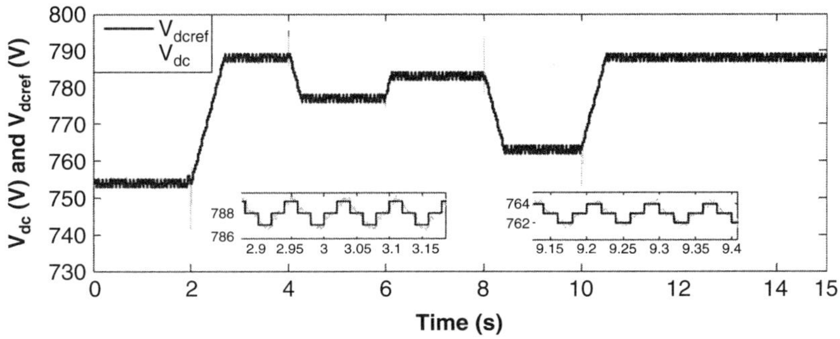

Figure 17.13 Reference dc voltage provided by MPPT mechanism and the actual operating voltage of the PV generator.

of the developed controller. The reference dc voltage and the actual operating dc voltages are shown in Figure 17.13. The MPP voltage will rise with an increase in insolation and vice-versa, which can be confirmed from Figure 17.13. An important conclusion that can be drawn from Figure 17.13 is that the MPPT mechanism is perfectly tracking the reference voltage and is maintaining three level operations at each and every insolation level, once the system gets settled. The quick tracking of MPP and the three level operations ensures the proper execution of P&O algorithm and the capability of the controller to follow the reference provided by the MPPT mechanism.

The real and reactive power outputs from the VSC are shown in Figure 17.14. It can be observed from Figure 17.14 that the real power delivered by the VSC varies according to the power that is extracted from the PV generator, which in turn varies with the insolation. In this study case, the reactive power from the PV generator is controlled to maintain the PCC voltage at its nominal value under all operating conditions. The aggregate of local loads of real power 10 kW and 5 kVAr (0.9 *p.f.* approximately) whose rated voltage is 415 V is assumed to be connected at PCC. It is changing its consumption to 12.5 kW and 6 kVAr at 5s and is coming back to its earlier value at 6s as shown in Figure 17.15. Figure 17.16 shows the PCC voltage for the period of study under different real power feedings from the PV generator. The real and reactive power supplied by the PV generator to the grid after the local load is shown in Figure 17.17. Till the 2s, the PV generator is supplying a reactive power 4 kVAr and remaining 1 kVAr out of 5 kVAr load supplied by the grid as shown in Figures 17.14 and 17.17. The amount of reactive power supplied by the PV generator is decided by the

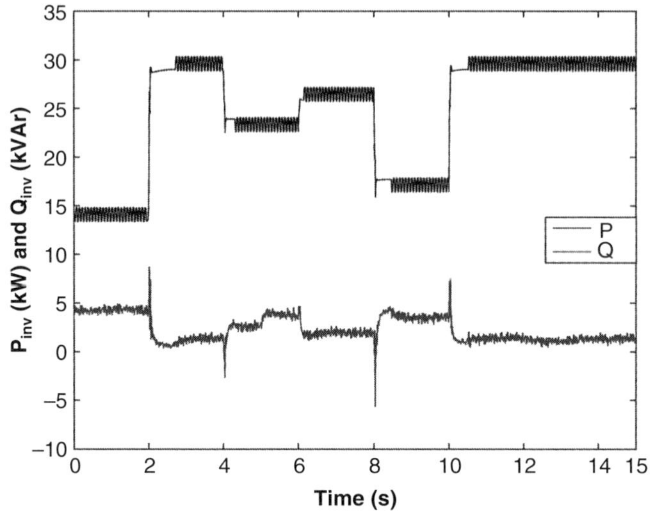

Figure 17.14 Real and reactive powers supplied by the PV generator interfaced through VSC.

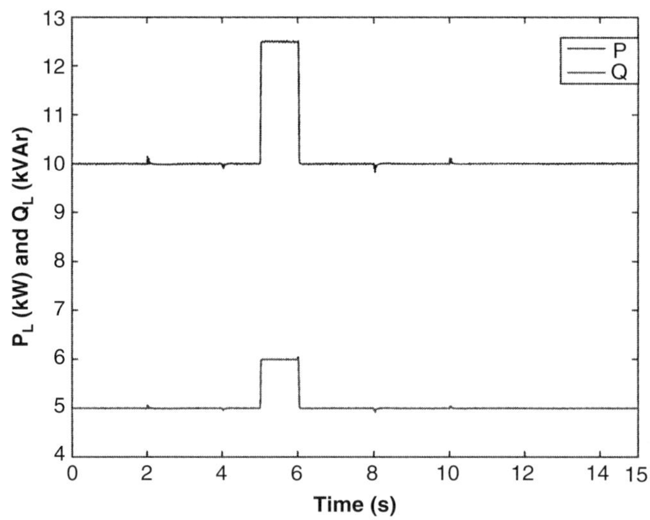

Figure 17.15 The variation of aggregate of local loads connected at PCC.

controller to maintain the PCC voltage at 415 V. From 2s to 4s there is an increased real power supply from the PV generator due to increased insolation as shown in Figures 17.11, 17.12, 17.14 and 17.17. As it results in increased real power flow in the reverse direction in a distribution system from the DG (PV generator), the PCC

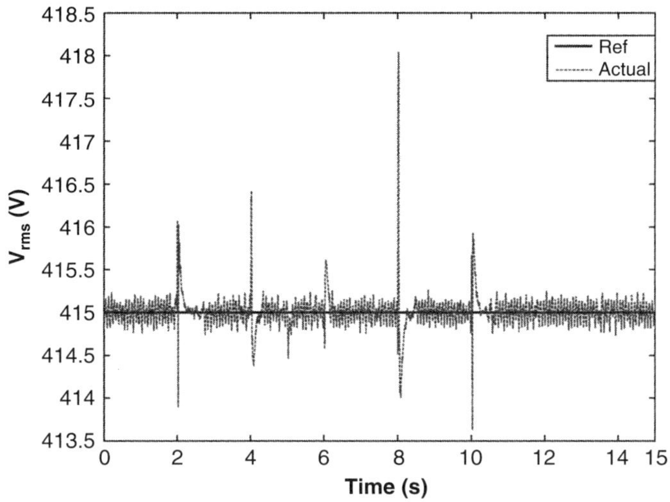

Figure.17.16 The PCC voltage under varying reverse power flows from the DG (PV generator).

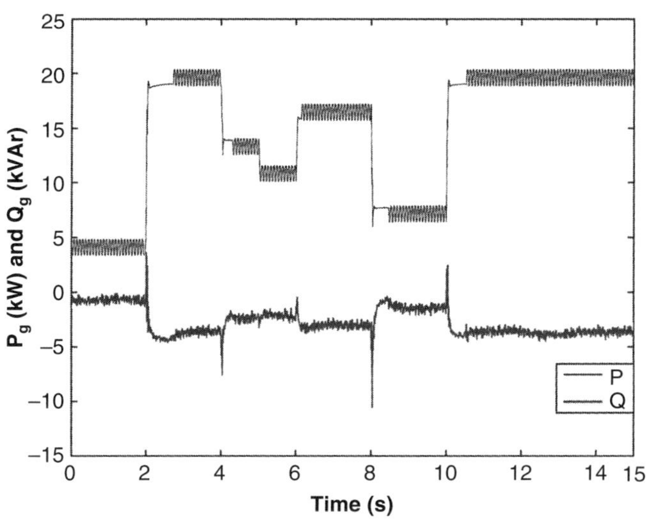

Figure 17.17 The real and reactive powers supplied by the PV generator to the grid after the local load.

voltage tends to rise above its nominal value of 415 V as shown in Figure 17.16. However, the developed controller immediately detects this rising tendency in PCC voltage and takes the necessary control action to regulate reactive power output from the PV generator as shown in Figures 17.14 and 17.16.

A critical phenomenon that is to be observed here is that, with the increase in real power supply form the PV generator, the reactive power delivered by it is decreasing. It can be well explained as, with the increase in the real power feeding from the DG, the PCC voltage will rise.

In this study, the PV generator is supplying the reactive power (part which will go to local load) to regulate the PCC voltage before the increase of the real power. With the increase of real power supply, the PCC voltage tends to rise, hence the net reactive power required to maintain zero voltage regulation at PCC will be less when compared with low real power supply, which is very much clear from Figures 17.12, 17.14 and 17.16.

An important issue that is to be noted here is that the PV generator which is acting as DG together with the grid will serve the local load. For example, the PV generator reactive power is regulated to maintain the voltage at PCC, due to which it may not deliver the total required reactive power by the load; however, the deficiency or excess will be supplied or absorbed by the grid from the other end of the PCC, as shown in Figures 17.14, 17.15 and 17.17. This is applicable to both real as well as reactive powers and hence the networks with DGs are known to be active distribution networks, as mentioned in the introduction. When there is a change in local load consumption, the grid feeding by the DG (here the PV generator) will be adjusted accordingly. For example, the local consumption is increasing by an amount of 2.5 kW at 5 s as shown in Figure 17.15. Since the PV generator output is constant at 23.38 kW (approximately) before and after the disturbance, the grid feeding will get diminished by the same amount of 2.5 kW as shown in Figure 17.17.

The PCC voltage shown in Figure 17.16 depicts the strictly regulated PCC voltage at 415 V under different incident insolations and varying local loads. Despite varying insolations which affect the delivered real power, the rise of PCC voltage levels and varying loads which affect the voltage drop in various sections of the network; the developed control strategy along with the designed controllers are excellent in maintaining the PCC voltage at its nominal value of 415 V as shown in Figure 17.16. Modulation index of the VSC, as shown in Figure 17.18, is always less than the maximum permissible value of 1.1 and depicts the capability of the developed controller in restricting the VSC to linear operating region.

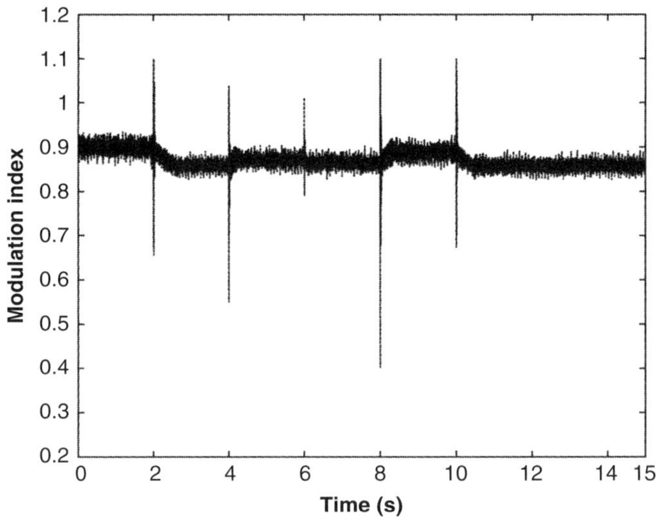

Figure 17.18 The modulation index of the VSC for varying real and reactive power extractions from the PV generator.

17.6.2 Control of Real and Reactive Power Feedings of PV Generator

In this case study the real power feeding of the PV generator is controlled at its maximum possible value where the reactive power is controlled according to the specified *p.f.*, thereby controlling the PV generator as a reactive power supporting ancillary. In this case study two different scenarios, operating the PV generator at 0.9 leading *p.f.* and 0.9 lagging *p.f.*, are considered to validate the capability of VSC-based PV generator and the developed control strategy to supply or absorb the reactive power, thereby operating PV generator as an ancillary of continuously variable reactive power support. Figure 17.19 shows the real power supplied by the PV generator which is controlled at its maximum possible value for the varying insolation as that of Figure 17.11. The real power is the same as that of Figure 17.14. Since the PV generator is controlled at 0.9 lagging *p.f.*, the PV generator absorbs the reactive power. For every variation in the real power (due to insolation change), a similar magnitude variation, but in an opposite direction, will be there in the reactive power as the real and reactive power flows are in opposite direction to each other. For instance,

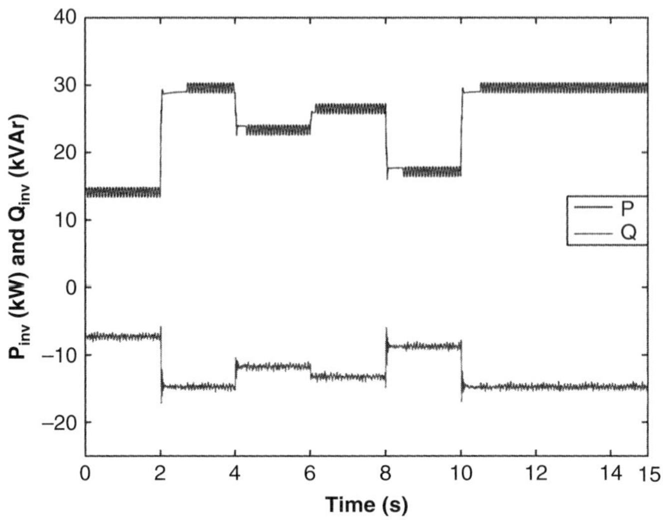

Figure 17.19 Real and reactive powers supplied by the PV generator interfaced through VSC operating at 0.9 lagging *p.f.*

the real power at 2s is increasing due to an increase in insolation for which the reactive power will increase, but in a negative direction, to maintain the constant ratio of real power to the apparent power (*p.f.*).

The PCC voltage for 0.9 lagging *p.f.* operation is shown in Figure 17.20. Since the reactive power is absorbed by the PV generator, the PCC voltage is always less than its nominal voltage of 415 V. The deviation of PCC voltage from its nominal value depends on the amount of negative reactive power supplied or the amount of reactive power absorbed. In other words, the larger the magnitude of absorbed reactive power, the wider the deviation of PCC voltage from its nominal value. Figure 17.21 shows the real and reactive power supplied by the PV generator under 0.9 leading *p.f.* operation. Since the PV generator is supplying the reactive power, its polarity is positive, the same as that of the real power, and varies with the variation in the real power to maintain constant *p.f.* The PCC voltage under this operating mode is shown in Figure 17.22. Since the reactive power is supplied by the PV generator in this mode, the PCC voltage is higher than the nominal voltage and depends on the amount of the reactive power injection into the system.

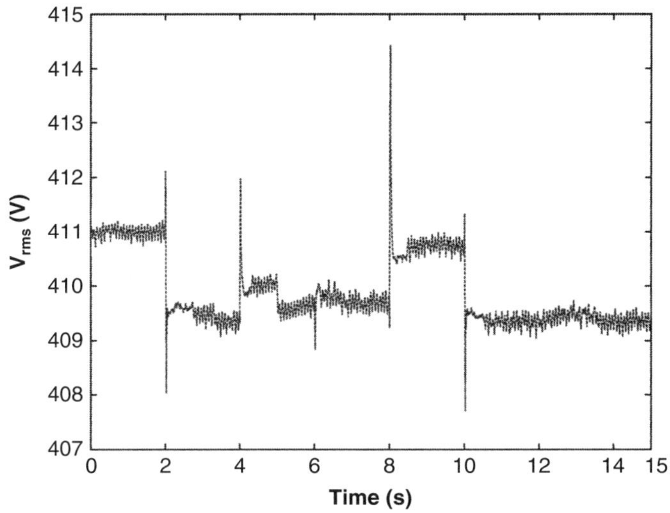

Figure 17.20 Variation of voltage at PCC due to the variation in reactive power feeding for 0.9 lagging *p.f.*

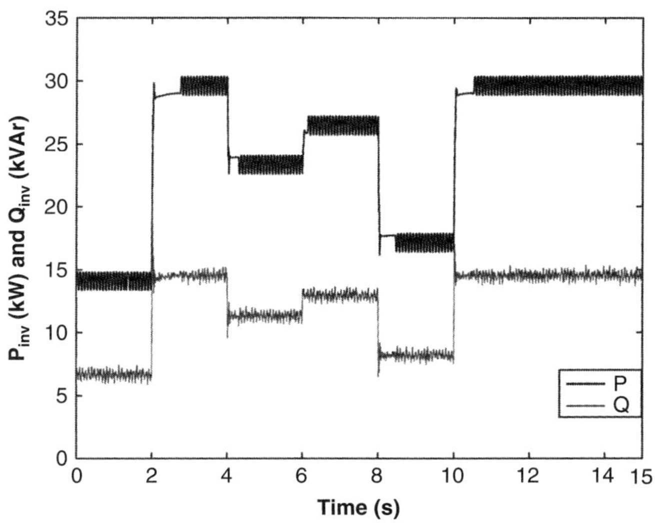

Figure 17.21 Real and reactive powers supplied by the PV generator interfaced through VSC operating at 0.9 leading *p.f.*

The reactive power supplies and the corresponding PCC voltages for different operating power factors with varying incident insolations are shown in Figures 17.23 and 17.24, respectively.

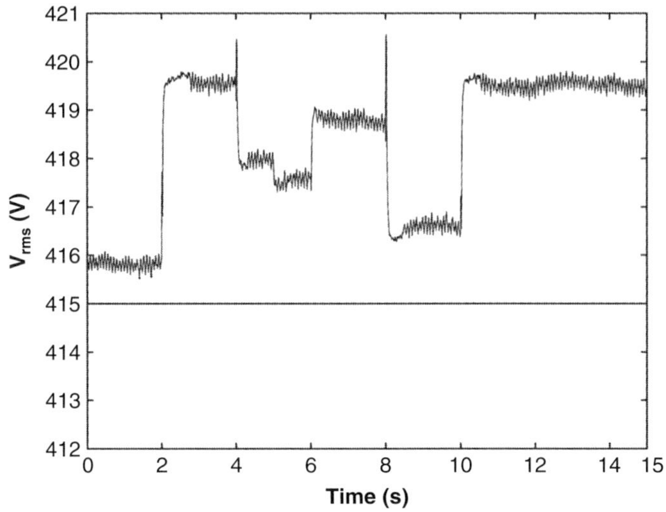

Figure 17.22 Variation of voltage at PCC due to the variation in reactive power feeding for 0.9 leading *p.f.*

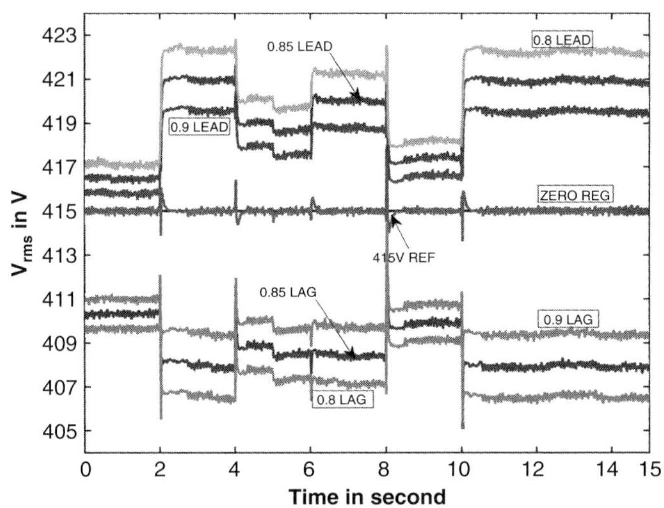

Figure 17.23 Variation of voltage at PCC for varying real power feedings at different operating *p.f.*s

It can be concluded from Figures 17.23 and 17.24 that the poorer the power factor the higher the reactive power supplies and wider the deviation of PCC voltage from its nominal value. These

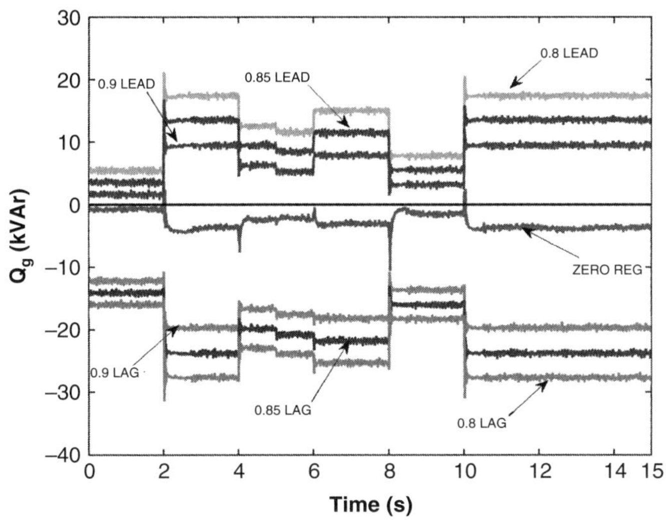

Figure 17.24 Variation of reactive power feedings at different operating *p.f.*s for varying real powers due to insolation changes.

responses confirm the capability of the VSC-based PV generator in supplying positive as well as negative reactive power.

17.7 Conclusion and Outlook

The penetration of DGs in the system is increasing at a high rate and more penetration is anticipated. It poses a severe challenge to voltage regulation as the reverse power flow from the DGs in the distribution network raise the voltage levels of the distribution network. Since the PV generators are interfaced thorough VSC to the grid, it has the capability to serve as reactive power supporting ancillary. In this chapter a control philosophy is developed to regulate the PCC voltage at its nominal value without affecting the real power feedings to the system, which are controlled at maximum possible power outputs at varying insolations. The PV generator is controlled to supply the reactive power in such a way as to nullify the voltage rise, thereby achieving the zero voltage regulation. The developed control strategy has also been examined for supply of lagging as well as leading reactive powers at different operating power factors. The developed control strategy exhibits excellent performance by meeting all the control objectives.

References

1. R. Tonkoski, L.A.C. Lopes, and T.H.M. El-Fouly, Coordinated active power curtailment of grid connected PV inverters for overvoltage prevention, *IEEE Trans. on Sust. Energy*, Vol. 2, No. 2, pp. 139–147, Apr. 2011.
2. C.L. Masters, Voltage rise: The big issue when connecting embedded generation to long 11 kV overhead lines, *Power Eng. J.*, Vol. 16, pp. 5–12, 2002.
3. H.E. Farag, E.F. El-Saadany, and R. Seethapathy, A two ways communication-based distributed control for voltage regulation in smart distribution feeders, *IEEE Trans. on Smart Grid*, Vol. 3, No. 1, pp. 271–281, Mar. 2012.
4. R.A. Walling, R. Saint, R.C. Dugan, J. Burke, and L.A. Kojovic, Summary of distributed resources impact on power delivery systems, *IEEE Trans. Power Del.*, Vol. 23, No. 3, pp. 1636–1644, Jul. 2008.
5. W. Jewell and R. Ramakumar, The effects of moving clouds on electric utilities with dispersed photovoltaic generation, *IEEE Trans. Energy Conv.*, Vol. EC-2, No. 4, pp. 570–576, 1987.
6. D.L. Garrett and S.M. Jeter, A photovoltaic voltage regulation impact investigation technique: Part I - model development, *IEEE Trans. Energy Conv.*, Vol. 4, No. 1, pp. 47–53, 1989.
7. J.H. Choi and J.C. Kim, Advanced voltage regulation method of power distribution systems interconnected with dispersed storage and generation systems, *IEEE Trans. Power Sys.*, Vol. 16, No. 2, pp. 329–334, 2001.
8. T. Hoff and G.S. Shugar, The value of grid-support photovoltaics in reducing distribution system losses, *IEEE Trans. Energy Conv.*, Vol. 10, No. 3, pp. 569–576, 1995.
9. V.H. Méndez Quezada, J.R. Abbad, and T.G. San Román, Assessment of energy distribution losses for increasing penetration of distributed generation, *IEEE Trans. Power Sys.*, Vol. 21, No. 2, pp. 533–540, 2006.
10. A. Augugliaro, L. Dusonchet, S. Favuzza, and E.R. Sanseverio, Voltage regulation and power losses minimization in automated distribution networks by an evolutionary multi objective approach, *IEEE Trans. Power Sys.*, Vol. 19, No. 3, pp. 1516–1527, 2004.
11. S.N. Liew and G. Strbac, Maximising penetration of wind generation in existing distribution networks, *IEE Proc.-Gener. Transm. Distrib.*, Vol. 149, pp. 256–262, 2002.
12. M. Fila, G.A. Taylor, J. Hiscock, M.R. Irving, and P. Lang, Flexible voltage control to support distributed generation in distribution networks, in: *Proc. 43rd Int. Universities Power Eng. Conf. (UPEC 2008)*, pp. 1–5.
13. T.Esram and P.L.Chapman, Comparison of photovoltaic array maximum power point tracking techniques, *IEEE Trans. Energy Conv.*, Vol. 22, No. 2, pp. 439–449, Jun. 2007.
14. B.Subudhi and R.Pradhan, A comparative study on maximum power point tracking techniques for photovoltaic power systems, *IEEE Trans. on Sust. Energy*, Vol. 4, No. 1, 2012.
15. P.N. Vovos, A.E. Kiprakis, A.R. Wallace, and G.P. Harrison, Centralized and distributed voltage control: Impact on distributed generation penetration, *IEEE Trans. Power Sys.*, Vol. 22, No. 1, pp. 476–483, 2007

16. S. Mishra and P.C. Sekhar, TS fuzzy based adaptive perturb algorithm for MPPT of a grid connected single stage three phase VSC interfaced PV generating system, *IEEE PES General Meeting, 2012.*
17. S. Mishra and P.C. Sekhar, Sliding mode based feedback linearizing controller for a PV system to improve the performance under grid frequency variation, *International Conference on Energy, Automation and Signal, ICEAS - 2011,* pp. 106–112.
18. H. Mahmood, J. Jiang, Modeling and control system design of a grid connected VSC considering the effect of the interface transformer type, *IEEE Trans. on Smart Grid,* Vol. 3, No. 1, pp. 122–134, Mar.2012.
19. A.Yazdani and P.P.Dash, A control methodology and characterization of dynamics for a photovoltaic (PV) system interfaced with a distribution network, *IEEE Trans. on Power Delivery*, Vol. 24, No. 3, pp. 1538–1551, Jul. 2009.

Index

Also of Interest

Check out these published and forthcoming related titles from Scrivener Publishing

Advanced Carbon Materials and Technology
Edited by Ashutosh Tiwari and S.K. Shukla
Forthcoming January 2014. ISBN 978-1-118-68623-2

Responsive Materials and Methods
State-of-the-Art Stimuli-Responsive Materials and Their Applications
Edited by Ashutosh Tiwari and Hisatoshi Kobayashi
Published 2013. ISBN 978-1-118-68622-5

Nanomaterials in Drug Delivery, Imaging, and Tissue Engineering
Edited by Ashutosh Tiwari and Atul Tiwari
Published 2013. ISBN 978-1-118-29032-3

Biomimetics
Advancing Nanobiomaterials and Tissue Engineering
Edited by Murugan Ramalingam, Xiumei Wang, Guoping Chen, Peter Ma, and Fu-Zhai Cui
Published 2013. ISBN 978-1-118-46962-0

Atmospheric Pressure Plasma Treatment of Polymers
Edited by Michael Thomas and K.L. Mittal
Published 2013. ISBN 978-1-118-59621-0

Polymers for Energy Storage and Conversion
Edited by Vikas Mittal
Published 2013. ISBN 978-1-118-34454-5

Encapsulation Nanotechnologies
Edited by Vikas Mittal
Published 2013. ISBN 978-1-118-34455-2

Biomedical Materials and Diagnostic Devices Devices
Edited by Ashutosh Tiwari, Murugan Ramalingam, Hisatoshi Kobayashi and Anthony P.F. Turner
Published 2012. ISBN 978-1-118-03014-1

Intelligent Nanomaterials
Processes, Properties, and Applications
Edited by Ashutosh Tiwari Ajay K. Mishra, Hisatoshi Kobayashi and Anthony P.F. Turner
Published 2012. ISBN 978-0-470-93879-9

Integrated Biomaterials for Biomedical Technology
Edited by Murugan Ramalingam, Ashutosh Tiwari, Seeram Ramakrishna and Hisatoshi Kobayashi
Published 2012. ISBN 978-1-118-42385-1

Integrated Biomaterials in Tissue Engineering
Edited by Murugan Ramalingam, Ziyad Haidar, Seeram Ramakrishna, Hisatoshi Kobayashi, and Youssef Haikel
Published 2012. ISBN 978-1-118-31198-1

The Physics of Micropdroplets
Jean Berthier and Kenneth Brakke
Published 2012. ISBN 978-0-470-93880-0

Antioxidant Polymers
Synthesis, Properties and Applications
Edited by Giuseppe Cirillo and Francesca Iemma
Published 2012. ISBN 978-1-118-20854-0

Introduction to Surface Engineering and Functionally Engineered Materials
Peter Martin
Published 2011. ISBN 978-0-470-63927-6

Handbook of Bioplastics and Biocomposites Engineering Applications.
Edited by Srikanth Pilla
Published 2011. ISBN 978-0-470-62607-8

Biopolymers: Biomedical and Environmental Applications
Edited by Susheel Kalia and Luc Avérous
Published 2011. ISBN 978-0-470-63923-8

Renewable Polymers
Synthesis, Processing, and Technology
Edited by Vikas Mittal
Published 2011. ISBN 978-0-470-93877-5

Miniemulsion Polymerization Technology
Edited by Vikas Mittal
Published 2010. ISBN 978-0-470-62596-5

Polymer Nanotube Nanocomposites
Synthesis, Properties, and Applications
Edited by Vikas Mittal.
Published 2010. ISBN 978-0-470-62592-7